Methods in Molecular Biology

Series Editor
John M. Walker
School of Life and Medical Sciences
University of Hertfordshire
Hatfield, Hertfordshire, AL10 9AB, UK

For further volumes:
http://www.springer.com/series/7651

ATM Kinase

Methods and Protocols

Edited by

Sergei V. Kozlov

University of Queensland Centre for Clinical Research (UQCCR),
University of Queensland, Herston, QLD, Australia

Humana Press

Editor
Sergei V. Kozlov
University of Queensland Centre
for Clinical Research (UQCCR)
University of Queensland
Herston, QLD, Australia

ISSN 1064-3745 ISSN 1940-6029 (electronic)
Methods in Molecular Biology
ISBN 978-1-4939-6953-1 ISBN 978-1-4939-6955-5 (eBook)
DOI 10.1007/978-1-4939-6955-5

Library of Congress Control Number: 2017936986

Printed on acid-free paper

This Humana Press imprint is published by Springer Nature
The registered company is Springer Science+Business Media LLC
The registered company address is: 233 Spring Street, New York, NY 10013, U.S.A.

Preface

Ataxia-telangiectasia (A-T) is a rare and severe genetic disorder affecting children. Since the discovery and cloning of the ATM (ataxia-telangiectasia, mutated) gene causing the disease in 1995, A-T has become an ever expanding field of research that has been enriched by contributions from the broad community of scientists attracted to the field by the fundamental role played by the ATM kinase in DNA damage signaling and diverse cellular processes. This has led to the development of a great number of protocols related to studies of the ATM gene and protein, which have been scattered across many published papers and books, including the *Methods in Molecular Biology* series. It is a timely undertaking by the *Methods in Molecular Biology* program to collate the essential protocols in A-T research and present them in a single volume. The positive response received from the majority of scientists we have contacted to share their protocols for the book is a testament to the collaborative spirit of the A-T research community. Many colleagues made the poignant observation that, in 2015, we celebrated 20 years since the discovery of the ATM gene and that it is now important to reflect on discoveries, which have been made possible by the invention and application of many new techniques and approaches in A-T research.

We have attempted to present a reasonably comprehensive collection of protocols by our contributors within the space limitations of a single volume. These limitations precluded us from giving sufficient attention to A-T animal models, which, without a doubt, deserve a separate volume due to their importance and complexity of techniques used. We do apologize for any unintentional omissions.

We hope this book will be a handy desktop reference for both seasoned A-T researchers and postgraduate students, as it demonstrates the breadth of recent developments in A-T studies. We also hope to ignite and attract the interest of colleagues from diverse fields to A-T research in an effort to bring their expertise and fresh ideas to resolve many A-T puzzles still waiting to be pieced together.

For all scientists working in the A-T field, the ultimate goal is to alleviate the suffering of A-T children and their families. We hope that our humble effort to collate technological and methodological advances in ATM and DNA damage research will facilitate this goal by helping scientists to utilize these techniques in their labs.

This book would not have been possible without generous contributions of many scientists, who shared their knowledge, for which I am very grateful. I am also sincerely grateful to the series editor, Professor John Walker, for his help, advice, and patient guidance in preparing this volume. I am indebted to Professor Martin Lavin for his relentless effort to advance ATM research and his continuing support over the years.

Herston, QLD, Australia *Sergei V. Kozlov*

Contents

Preface . . . . *v*
Contributors . . . . *xi*

1 Assaying Radiosensitivity of Ataxia-Telangiectasia 1
Hailiang Hu, Shareef Nahas, and Richard A. Gatti

2 Assaying for Radioresistant DNA Synthesis, the Hallmark Feature of the Intra-S-Phase Checkpoint Using a DNA Fiber Technique 13
Amanda W. Kijas and Martin F. Lavin

3 ATM Gene Mutation Detection Techniques and Functional Analysis 25
Guillaume Rieunier, Catherine Dubois D'Enghien, Alice Fievet, Dorine Bellanger, Dominique Stoppa-Lyonnet, and Marc-Henri Stern

4 An HTRF® Assay for the Protein Kinase ATM 43
Phillip Adams, Jonathan Clark, Simon Hawdon, Jennifer Hill, and Andrew Plater

5 ATM Kinase Inhibitors: HTS Cellular Imaging Assay Using Cellomics™ ArrayScan VTI Platform 57
Catherine Bardelle and Joanna Boros

6 Image-Based High Content Screening: Automating the Quantification Process for DNA Damage-Induced Foci 71
Yi Chieh Lim

7 Analyzing ATM Function by Electroporation of Endonucleases and Immunofluorescence Microscopy 85
Keiji Suzuki

8 Quantitative and Dynamic Imaging of ATM Kinase Activity by Bioluminescence Imaging 97
Shyam Nyati, Grant Young, Brian Dale Ross, and Alnawaz Rehemtulla

9 Zn(II)–Phos-Tag SDS-PAGE for Separation and Detection of a DNA Damage-Related Signaling Large Phosphoprotein 113
Eiji Kinoshita, Emiko Kinoshita-Kikuta, and Tohru Koike

10 Identification of ATM Protein Kinase Phosphorylation Sites by Mass Spectrometry 127
Mark E. Graham, Martin F. Lavin, and Sergei V. Kozlov

11 Studies of ATM Kinase Activity Using Engineered ATM Sensitive to ATP Analogues (ATM-AS) 145
Masato Enari, Yuko Matsushima-Hibiya, Makoto Miyazaki, and Ryo Otomo

12 Functional Characterization of ATM Kinase Using Acetylation-Specific Antibodies 157
Yingli Sun and Fengxia Du

13 Identification of ATM-Interacting Proteins by Co-immunoprecipitation and Glutathione-S-Transferase (GST) Pull-Down Assays 163
Amanda L. Bain, Janelle L. Harris, and Kum Kum Khanna

14 ATM Activation and H2AX Phosphorylation Induced by Genotoxic Agents Assessed by Flow- and Laser Scanning Cytometry 183
Hong Zhao, H. Dorota Halicka, Jorge Garcia, Jiangwei Li, and Zbigniew Darzynkiewicz

15 Peptide Immunoaffinity Enrichment with Targeted Mass Spectrometry: Application to Quantification of ATM Kinase Phospho-Signaling 197
Jeffrey R. Whiteaker, Lei Zhao, Regine M. Schoenherr, Jacob J. Kennedy, Richard G. Ivey, and Amanda G. Paulovich

16 Mass Spectrometry-Based Proteomics for Quantifying DNA Damage-Induced Phosphorylation. 215
Marina E. Borisova, Sebastian A. Wagner, and Petra Beli

17 Statistical Analysis of ATM-Dependent Signaling in Quantitative Mass Spectrometry Phosphoproteomics 229
Ashley J. Waardenberg

18 ChIP Technique to Study Protein Dynamics at Defined DNA Double Strand Breaks 245
Jie Wen and Patrick Concannon

19 Studies of the DNA Damage Response by Using the Lac Operator/Repressor (LacO/LacR) Tethering System 263
Rossana Piccinno, Marta Cipinska, and Vassilis Roukos

20 Imaging of Fluorescently Tagged ATM Kinase at the Sites of DNA Double Strand Breaks 277
Anthony J. Davis, Shih-Ya Wang, David J. Chen, and Benjamin P.C. Chen

21 Live Cell Imaging to Study Real-Time ATM-Mediated Recruitment of DNA Repair Complexes to Sites of Ionizing Radiation-Induced DNA Damage. 287
Burkhard Jakob and Gisela Taucher-Scholz

22 Analyzing Heterochromatic DNA Double Strand Break (DSB) Repair in Response to Ionizing Radiation. 303
Karolin Klement and Aaron A. Goodarzi

23 Phenotypic Analysis of ATM Protein Kinase in DNA Double-Strand Break Formation and Repair 317
Elisabeth Mian and Lisa Wiesmüller

24 Monitoring DNA Repair Consequences of ATM Signaling Using Simultaneous Fluorescent Readouts 335
Adrian Wiegmans

25 Noncanonical ATM Activation and Signaling in Response to Transcription-Blocking DNA Damage 347
Jurgen A. Marteijn, Wim Vermeulen, and Maria Tresini

26 Study of ATM Phosphorylation by Cdk5 in Neuronal Cells. 363
Hua She and Zixu Mao

27 DNA Damage Response in Human Stem Cells and Neural Descendants 375
Jason M. Beckta, Bret R. Adams, and Kristoffer Valerie

28 A Patient-Specific Stem Cell Model to Investigate the Neurological Phenotype Observed in Ataxia-Telangiectasia 391
Romal Stewart, Gautam Wali, Chris Perry, Martin F. Lavin, Francois Féron, Alan Mackay-Sim, and Ratneswary Sutharsan

29 Lentiviral Reprogramming of A-T Patient Fibroblasts to Induced Pluripotent Stem Cells 401
Sam Nayler, Sergei V. Kozlov, Martin F. Lavin, and Ernst Wolvetang

30 Monitoring the ATM-Mediated DNA Damage Response in the Cerebellum Using Organotypic Cultures 419
Efrat Tal and Yosef Shiloh

Index *431*

Contributors

BRET R. ADAMS • *Department of Radiation Oncology, Miller School of Medicine, University of Miami, Miami, FL, USA*

PHILLIP ADAMS • *Eurofins Pharma Discovery Services UK Limited, Gemini Crescent, Dundee Technology Park, Dundee, UK*

AMANDA L. BAIN • *QIMR Berghofer Medical Research Institute, Brisbane, QLD, Australia*

CATHERINE BARDELLE • *Discovery Sciences iMed, AstraZeneca, Global HTS Centre, Macclesfield, Cheshire, UK*

JASON M. BECKTA • *Department of Radiation Oncology and the Massey Cancer Center, Virginia Commonwealth University, Richmond, VA, USA; Yale University School of Medicine, New Haven, CT, USA*

PETRA BELI • *Institute of Molecular Biology (IMB), Mainz, Germany*

DORINE BELLANGER • *Inserm U830, Institut Curie - Section de Recherche, Paris, France*

MARINA E. BORISOVA • *Institute of Molecular Biology (IMB), Mainz, Germany*

JOANNA BOROS • *Lead Discovery Center GmbH, Dortmund, Germany*

BENJAMIN P.C. CHEN • *Division of Molecular Radiation Biology, Department of Radiation Oncology, University of Texas Southwestern Medical Center, Dallas, TX, USA*

DAVID J. CHEN • *Division of Molecular Radiation Biology, Department of Radiation Oncology, University of Texas Southwestern Medical Center, Dallas, TX, USA*

MARTA CIPINSKA • *Institute of Molecular Biology, Mainz, Germany*

JONATHAN CLARK • *Eurofins Pharma Discovery Services UK Limited, Gemini Crescent, Dundee Technology Park, Dundee, UK*

PATRICK CONCANNON • *Genetics Institute, University of Florida, Gainesville, FL, USA; Department of Pathology, Immunology, and Laboratory Medicine, University of Florida, Gainesville, FL, USA*

ZBIGNIEW DARZYNKIEWICZ • *Department of Pathology, Brander Cancer Research Institute, New York Medical College, Valhalla, NY, USA*

ANTHONY J. DAVIS • *Division of Molecular Radiation Biology, Department of Radiation Oncology, University of Texas Southwestern Medical Center, Dallas, TX, USA*

FENGXIA DU • *Cancer and Epigenetic Group, Key Laboratory of Genomic and Precision Medicine, China Gastrointestinal Cancer Research Center, Beijing Institute of Genomics, Chinese Academy of Sciences, Beijing, China*

CATHERINE DUBOIS D'ENGHIEN • *Service de Génétique, Pôle de Médecine diagnostique et théranostique, Institut Curie, Paris, France*

MASATO ENARI • *Division of Refractory and Advanced Cancer, National Cancer Center Research Institute, Tokyo, Japan*

FRANCOIS FÉRON • *Aix Marseille Université, CNRS, NICN, UMR7259, Marseille, France; APHM, Centre d'Investigations Cliniques en Biothérapie, CIC-BT 510, Marseille, France*

ALICE FIEVET • *Inserm U830, Institut Curie - Section de Recherche, Paris 75248 France; Service de Génétique, Pôle de Médecine diagnostique et théranostique, Institut Curie, Paris, France*

Jorge Garcia • *Department of Pathology, Brander Cancer Research Institute, New York Medical College, Valhalla, NY, USA*
Richard A. Gatti • *Department of Pathology and Laboratory Medicine, David Geffen School of Medicine at UCLA, Los Angeles, CA, USA*
Aaron A. Goodarzi • *Robson DNA Science Centre, Arnie Charbonneau Cancer Institute, Department of Biochemistry and Molecular Biology and Department of Oncology, Cumming School of Medicine, University of Calgary, Calgar, AB, Canada*
Mark E. Graham • *Children's Medical Research Institute, University of Sydney, Westmead, NSW, Australia*
H. Dorota Halicka • *Department of Pathology, Brander Cancer Research Institute, New York Medical College, Valhalla, NY, USA*
Janelle L. Harris • *QIMR Berghofer Medical Research Institute, Brisbane, QLD, Australia*
Simon Hawdon • *Eurofins Pharma Discovery Services UK Limited, Gemini Crescent, Dundee Technology Park, Dundee, UK*
Jennifer Hill • *Eurofins Pharma Discovery Services UK Limited, Gemini Crescent, Dundee Technology Park, Dundee, UK*
Hailiang Hu • *Department of Pathology and Laboratory Medicine, David Geffen School of Medicine at UCLA, Los Angeles, CA, USA; Department of Pathology, Duke University School of Medicine, Durham, NC, USA*
Richard G. Ivey • *Fred Hutchinson Cancer Research Center, Seattle, WA, USA*
Burkhard Jakob • *GSI Helmholtzzentrum für Schwerionenforschung GmbH, Biophysik, Darmstadt, Germany*
Jacob J. Kennedy • *Fred Hutchinson Cancer Research Center, Seattle, WA, USA*
Kum Kum Khanna • *QIMR Berghofer Medical Research Institute, Brisbane, QLD, Australia*
Amanda W. Kijas • *University of Queensland Centre for Clinical Research, University of Queensland, Herston, QLD, Australia; Australian Institute for Bioengineering and Nanotechnology, University of Queensland, St Lucia, QLD, Australia*
Eiji Kinoshita • *Department of Functional Molecular Science, Institute of Biomedical & Health Sciences, Hiroshima University, Hiroshima, Japan*
Emiko Kinoshita-Kikuta • *Department of Functional Molecular Science, Institute of Biomedical & Health Sciences, Hiroshima University, Hiroshima, Japan*
Karolin Klement • *Robson DNA Science Centre, Arnie Charbonneau Cancer Institute, Department of Biochemistry and Molecular Biology and Department of Oncology, Cumming School of Medicine, University of Calgary, Calgar, AB, Canada*
Tohru Koike • *Department of Functional Molecular Science, Institute of Biomedical & Health Sciences, Hiroshima University, Hiroshima, Japan*
Sergei V. Kozlov • *University of Queensland Centre for Clinical Research (UQCCR), University of Queensland, Herston, Brisbane, QLD Australia*
Martin F. Lavin • *University of Queensland Centre for Clinical Research (UQCCR), University of Queensland, Herston, Brisbane, QLD, Australia*
Jiangwei Li • *Department of Pathology, Brander Cancer Research Institute, New York Medical College, Valhalla, NY, USA*
Yi Chieh Lim • *QIMR Berghofer Medical Research Institute, Brisbane, QLD, Australia*
Alan Mackay-Sim • *Eskitis Institute for Drug Discovery, Griffith University, Nathan, QLD, Australia*
Zixu Mao • *Department of Pharmacology, Emory University School of Medicine, Atlanta, GA, USA*

Jurgen A. Marteijn • *Department of Molecular Genetics, Cancer Genomics Netherlands, Erasmus University Medical Center, Rotterdam, The Netherlands*
Yuko Matsushima-Hibiya • *Division of Refractory and Advanced Cancer, National Cancer Center Research Institute, Tokyo, Japan*
Elisabeth Mian • *Department of Obstetrics and Gynaecology, University of Ulm, Ulm, Germany*
Makoto Miyazaki • *Division of Refractory and Advanced Cancer, National Cancer Center Research Institute, Tokyo, Japan; Department of Medical Genome Sciences, Laboratory of Tumor Cell Biology, Graduate School of Frontier Sciences, The University of Tokyo, Tokyo, Japan*
Shareef Nahas • *Department of Pathology and Laboratory Medicine, David Geffen School of Medicine at UCLA, Los Angeles, CA, USA*
Sam Nayler • *Australian Institute for Bioengineering and Nanotechnology (AIBN), The University of Queensland, Brisbane, QLD, Australia*
Shyam Nyati • *Center for Molecular Imaging, University of Michigan, Ann Arbor, MI, USA; Department of Radiation Oncology, University of Michigan, Ann Arbor, MI, USA*
Ryo Otomo • *Division of Refractory and Advanced Cancer, National Cancer Center Research Institute, Tokyo, Japan*
Amanda G. Paulovich • *Fred Hutchinson Cancer Research Center, Seattle, WA, USA*
Chris Perry • *Department of Otolaryngology Head and Neck Surgery, Princess Alexandra Hospital, Woolloongabba, Brisbane, QLD, Australia*
Rossana Piccinno • *Institute of Molecular Biology, Mainz, Germany*
Andrew Plater • *Eurofins Pharma Discovery Services UK Limited, Gemini Crescent, Dundee Technology Park, Dundee, UK*
Alnawaz Rehemtulla • *Center for Molecular Imaging, University of Michigan, Ann Arbor, MI, USA; Department of Radiation Oncology, University of Michigan, Ann Arbor, MI, USA*
Guillaume Rieunier • *Department of Oncology, University of Oxford, Oxford, UK; Inserm U830, Institut Curie - Section de Recherche, Paris, France*
Vassilis Roukos • *Institute of Molecular Biology, Mainz, Germany*
Brian Dale Ross • *Center for Molecular Imaging, University of Michigan, Ann Arbor, MI, USA; Department of Radiology, University of Michigan, Ann Arbor, MI, USA*
Regine M. Schoenherr • *Fred Hutchinson Cancer Research Center, Seattle, WA, USA*
Hua She • *Department of Pharmacology, Emory University School of Medicine, Atlanta, GA, USA*
Yosef Shiloh • *The David and Inez Myers Laboratory for Cancer Research, Department of Human Molecular Genetics and Biochemistry, Sackler School of Medicine, Tel Aviv University, Tel Aviv, Israel*
Marc-Henri Stern • *Inserm U830, Institut Curie - Section de Recherche, Paris, France*
Romal Stewart • *University of Queensland Centre for Clinical Research, Brisbane, QLD, Australia*
Dominique Stoppa-Lyonnet • *Inserm U830, Institut Curie - Section de Recherche, Paris, France; Service de Génétique, Pôle de Médecine diagnostique et théranostique, Institut Curie, Paris, France*
Yingli Sun • *Cancer and Epigenetic Group, Key Laboratory of Genomic and Precision Medicine, China Gastrointestinal Cancer Research Center, Beijing Institute of Genomics, Chinese Academy of Sciences, Beijing, China*

RATNESWARY SUTHARSAN • *Eskitis Institute for Drug Discovery, Griffith University, Nathan, QLD, Australia; Griffith Institute for Drug Discovery (GRIDD), Griffith University, Nathan, QLD, Australia*
KEIJI SUZUKI • *Atomic Bomb Disease Institute, Nagasaki University Graduate School of Biomedical Sciences, Nagasaki, Japan*
EFRAT TAL • *The David and Inez Myers Laboratory for Cancer Research, Department of Human Molecular Genetics and Biochemistry, Sackler School of Medicine, Tel Aviv University, Tel Aviv, Israel*
GISELA TAUCHER-SCHOLZ • *GSI Helmholtzzentrum für Schwerionenforschung GmbH, Biophysik, Darmstadt, Germany*
MARIA TRESINI • *Department of Molecular Genetics, Cancer Genomics Netherlands, Erasmus University Medical Center, Rotterdam, The Netherlands*
KRISTOFFER VALERIE • *Department of Radiation Oncology and the Massey Cancer Center, Virginia Commonwealth University, Richmond, VA, USA*
WIM VERMEULEN • *Department of Molecular Genetics, Cancer Genomics Netherlands, Erasmus University Medical Center, Rotterdam, The Netherlands*
ASHLEY J. WAARDENBERG • *Children's Medical Research Institute, University of Sydney, Westmead, NSW, Australia*
SEBASTIAN A. WAGNER • *Department of Medicine, Hematology/Oncology, Goethe University, Frankfurt, Germany*
GAUTAM WALI • *Eskitis Institute for Drug Discovery, Griffith University, Nathan, QLD, Australia*
SHIH-YA WANG • *Division of Molecular Radiation Biology, Department of Radiation Oncology, University of Texas Southwestern Medical Center, Dallas, TX, USA*
JIE WEN • *Shanghai ChemPartner Co., Ltd, Shanghai, China*
JEFFREY R. WHITEAKER • *Fred Hutchinson Cancer Research Center, Seattle, WA, USA*
ADRIAN WIEGMANS • *QIMR Berghofer Medical Research Institute, Brisbane, QLD, Australia*
LISA WIESMÜLLER • *Department of Obstetrics and Gynaecology, The University of Ulm, Ulm, Germany*
ERNST WOLVETANG • *Australian Institute for Bioengineering and Nanotechnology (AIBN), The University of Queensland, Brisbane, QLD, Australia*
GRANT YOUNG • *Department of Radiation Oncology, University of Michigan, Ann Arbor, MI, USA*
HONG ZHAO • *Department of Pathology, Brander Cancer Research Institute, New York Medical College, Valhalla, NY, USA*
LEI ZHAO • *Fred Hutchinson Cancer Research Center, Seattle, WA, USA*

Chapter 1

Assaying Radiosensitivity of Ataxia-Telangiectasia

Hailiang Hu, Shareef Nahas, and Richard A. Gatti

Abstract

Ataxia-Telangiectasia (A-T) is a prototypical genomic instability disorder with multi-organ deficiency and it is caused by the defective function of a single gene, *ATM* (Ataxia-Telangiectasia Mutated). Radiosensitivity, among the pleiotropic symptoms of A-T, reflects the basic physiological functions of ATM protein in the double strand break (DSB)-induced DNA damage response (DDR) and also restrains A-T patients from the conventional radiation therapy for their lymphoid malignancy. In this chapter, we describe two methods that have been developed in our lab to assess the radiosensitivity of A-T patients: (1) Colony Survival Assay (CSA) and (2) Flow Cytometry of phospho-SMC1 (FC-pSMC1). The establishment of these more rapid and reliable functional assays to measure the radiosensitivity, exemplified by A-T, would facilitate the diagnosis of other genomic instability genetic disorders as well as help the treatment options for most radiosensitive patients.

Key words DNA damage response, Colony survival, Flow cytometry, Whole blood, Lymphoblastoid cells, Ionizing radiation

1 Introduction

A-T is an autosomal-recessive genetic disorder characterized by progressive cerebellar neurodegeneration, immunodeficiency, radiosensitivity, and cancer susceptibility [1, 2]. Extended phenotypes have been documented for A-T patients as well, including sterility, premature aging, increased risk of metabolic syndromes, such as diabetes and cardiovascular diseases [3]. However, due to the early death of A-T patients, the extended phenotypes that are usually of slow onset are not considered primary characteristics for A-T. To diagnose A-T disease, clinical manifestations such as characteristic ataxia and ocular telangiectasia are first sought and laboratory testing to detect increased serum α-fetoprotein level, lack of cellular ATM protein level and kinase activity, and increased radiosensitivity confirm the clinical diagnosis [4, 5]. To date, no curable treatment is available for A-T.

The pleiotropic symptoms of A-T are caused by the loss of function of a single gene, *ATM*, which encodes a predominantly

Sergei V. Kozlov (ed.), *ATM Kinase: Methods and Protocols*, Methods in Molecular Biology, vol. 1599,
DOI 10.1007/978-1-4939-6955-5_1, © Springer Science+Business Media LLC 2017

nuclear protein kinase with a critical role in the double strand break (DSB)-induced DNA damage response (DDR) [6–8]. The nuclear function of ATM protein in regulating DDR has been extensively studied and is thought to be responsible for immunodeficiency, radiosensitivity, and cancer susceptibility phenotypes of A-T, but its role in cerebellar neurodegeneration and other extended phenotypes remains to be elucidated. ATM is also found in the cytoplasm of many types of cells including neurons, suggesting other cellular functions beyond nuclear DDR, which is also corroborated with a finding that ATM can be activated independently of DNA damage through oxidation [9].

Radiosensitivity of A-T patients to lower dose of ionizing radiation reflects the role of ATM in DNA repair and processing [10, 11]. ATM is a large protein (370 kD) with serine/threonine kinase activity and can be activated by DNA DSBs to coordinate cell cycle checkpoints during DNA repair, recognize and repair the broken ends of DNA, or activate apoptotic pathways in cells that have been damaged beyond repair [12]. In the DNA repair process, ATM can be recruited by MRN (MRE11-RAD50-NBS1) to the DSB sites and activated by autophosphorylation of ATM kinase to sequentially phosphorylate H2AX to activate its downstream repair pathways: non-homologous end joining (NHEJ) and homologous recombination repair (HRR). Many important proteins that participate in the DNA repair, such as 53BP1, CtIP, and BRCA1, are recruited to the damage sites by γ-H2AX (phospho-H2AX) and its binding partner MDC1. Therefore, formation of foci of these DNA repair proteins has been used to measure the DSBs and their phosphorylation kinetics has been used to assess the DNA repair process, which can be used to estimate the radiosensitivity of this group of patients.

Our lab has developed a variety of laboratory tests to diagnose A-T patients. In this chapter, we describe two assays to assess the radiosensitivity of A-T: (1) CSA, the Colony Survival Assay that identifies virtually all patients with A-T by measuring the radiosensitivity of Epstein-Barr virus (EBV)-transformed lymphoblastoid cells [13, 14]; and (2) FC-pSMC1, a flow cytometry (FC)-based ATM kinase assay that measures ATM-dependent phosphorylation of structural maintenance of chromosomes 1 (SMC1) post-DNA damage in patient-derived peripheral blood lymphocytes (PBLs) [15]. These protocols can also be extended to diagnose other DNA repair deficiency disorders and therefore have significant clinical impacts.

2 Materials

2.1 CSA: Equipment and Reagents

2.1.1 Isolation of Peripheral Blood Lymphocytes (PBLs)

1. Tabletop Centrifuge (model Allegra™ 6R) with CH-3.8/GH-3.8A swing bucket rotor (Beckman Coulter).
2. 15 mL conical centrifuge tubes.
3. Ficoll-Paque Solution.
4. RPMI 1640 medium, stored at 4 °C.

5. Fetal Bovine Serum, stored at −20 °C.
6. Penicillin-Streptomycin-Glutamine (PSG), stored at −20 °C.
7. 1× Phosphate-buffered Saline (PBS), pH 7.4.
8. DMSO.
9. Freeze media: RPMI 1640 supplemented with 50% FBS and 10% DMSO.
10. 2 mL Cryovial tubes.
11. Hemocytometer.
12. −20 and −70 °C Freezers.
13. Liquid nitrogen tank.

2.1.2 EBV Transformation

1. B95–8 cells (ATCC®CRL-1612) for producing Epstein-Barr virus.
2. 0.2 μm Filter.
3. 24-well flat-bottom tissue culture plates.
4. CO_2 incubator.

2.1.3 Assay Procedure (CSA)

1. Irradiator (MARK-I/Cs-137).
2. 96-well flat-bottom tissue culture plates.
3. 37 °C water bath.
4. MTT (3-[4,5-Dimethylthiazol-2yl]-2,5-diphenyltetrazolium), stored at 4 °C.
5. Phase contrast microscope.

2.2 FC-pSMC1: Equipment and Reagents

2.2.1 Isolation of Peripheral Blood Lymphocytes (PBLs)

The same equipment and reagents as Subheading 2.1.1

2.2.2 Assay Procedure (FC-pSMC1)

1. Irradiator (MARK-I/Cs-137).
2. CO_2 incubator.
3. Fixation Reagent A (Fix & PERM® Cell Fixation &Permeabilization, Thermo Fisher).
4. Permeabilization Reagent B (Fix & PERM® Cell Fixation &Permeabilization, Thermo Fisher).
5. Methanol (stored at −20 °C).
6. 4 °C Refrigerator.
7. Sodium Azide (1% solution).
8. Vortex Mixer.
9. Anti-pS966-SMC1 (Novus Biologicals NB100–206).

10. Anti-Rabbit-FITC (Jackson ImmunoResearch Laboratories).
11. 16% Formaldehyde (w/v), Methanol-free.
12. 5 mL Falcon polypropylene round-bottom tubes.
13. BD FACSCalibur™ cytometer (BD Biosciences).
14. Cell Quest Software.

3 Methods

3.1 CSA

This assay measures the radiosensitivity or colony-survival fraction (CSF) of EBV-transformed lymphoblastoid cells post-irradiation. Starting with a heparinized blood specimen, the lymphocytes are Ficoll-separated and infected with Epstein-Barr virus. After 4–6 weeks, they begin to grow spontaneously in vitro. These cells (in two concentrations: 50 and 100 cells/well) are then plated into four 96-well tissue culture plates. Two of the plates are irradiated with 1 Gy of ionizing radiation (IR); the other two are not irradiated. All four plates remain in a CO_2 incubator for 10–14 days, after which the number of wells with clumps of at least 32 cells (under a phase contrast microscope, 32 cells represents 5 cell divisions from a single cell) are scored as positive. This result is compared to the nonirradiated plate and the Survival Fraction is calculated.

3.1.1 Isolation of PBLs

1. Invert the whole blood sample-containing green-topped tube gently three times and centrifuge at 920 × *g* for 15 min at RT with "no brake" setting of tabletop centrifuge.
2. Remove the plasma layer, transfer the buffy-coat layer into 5 mL 1× PBS in a 15 mL conical centrifuge tube, and mix by pipetting.
3. Layer the mixture on the top of 4 mL Ficoll-Paque solution in a separate 15 mL centrifuge tube.
4. Centrifuge at 1540 × *g* for 15 min at RT with the "no brake" setting.
5. Discard the top layer, carefully transfer the middle layer (white interphase) to 10 mL 1× PBS in a 50 mL conical tube and mix the cell suspension by pipetting. (For FC-pSMC1, proceed to Subheading 3.2.2).
6. Divide evenly into two 15 mL conical centrifuge tubes and centrifuge at 920 × *g* for 5 min at RT to recover the pellet.
7. Use one tube (>4–8 million cells) for EBV transformation (*see* below Subheading 3.1.2).
8. Resuspend the cell pellet from the other tube with 1.5 mL of Freeze media and transfer to 2 mL cryovial tube (*see* **Note 1**).
9. Freeze cells by lowering the temperature gradually (*see* **Note 2**).
10. Store cells in the liquid nitrogen tank in the next morning.

3.1.2 EBV Transformation to Establish Immortalized Lymphoblastoid Cell Lines

1. Grow EBV-producing B95–8 cells in RPMI containing 15% FBS to a density of 0.5 × 10^6 cells/mL.
2. Centrifuge the B95–8 cells at 250 × *g* for 5 min at RT.
3. Collect the supernatant (containing virus) and pass it through a 0.2 μm filter to remove all cells.
4. Aliquot the EBV filtrate into 2.0 mL cryovial tube and store at −70 °C for future use (*see* **Note 3**) or immediately use as follows:
5. Resuspend the one-half of the cell pellet (described in **step 7** of Subheading 3.1.1) in 2.6 mL RPMI 1640 media containing 15% FBS and 1% PSG.
6. Aliquot 0.6 mL of cell suspension per well into 4 wells of a 24-well plate.
7. Add 0.5 mL of EBV-supernatant (either fresh one from **step 13** or thaw −70 °C stored EBV-supernatant at 37 °C water bath) to each well.
8. Incubate at 37 °C with 5% CO_2 for 1–3 weeks and monitor cell growth under a phase contrast microscope.
9. Transfer to a 25 mL flask or 75 mL flask when cells begin to form large clumps and add 1–5 mL fresh medium (RPMI 1640 supplemented with 15% FBS and 1%PSG) when the culture color becomes yellow.
10. Count cells with a hemocytometer every other day after 1 week and freeze some of the transformed cells (5–10 million cells/vial) when the number of cells reaches 20 × 10^6.
11. In normal situation, 90% of sample cells can be immortalized within 1–2 months and the patient-derived lymphoblastoid cell lines are established (*see* **Note 4**).

3.1.3 Defrost Lymphoblastoid Cell Lines (LCLs)

1. Remove cryovial with LCLs from liquid nitrogen tank.
2. Place in a 37 °C water bath for 1 min (*see* **Note 5**).
3. Remove from the water bath when only a small chunk of ice remains in the tube.
4. Transfer cells to a 15 mL conical tube with 8 mL of RMPI 1640 complete media already in the tube.
5. Centrifuge at 250 × *g* for 5 min at RT.
6. Aspirate supernatant and resuspend the cell pellet in complete media and count cells to make a concentration of 0.2–0.5 × 10^6/mL in 10 mL complete media.
7. Transfer to 25 mL flask and put cells into an incubator to grow to 1.0 × 10^6/mL.

3.1.4 Assay Procedure for CSA

1. Centrifuge cells at 250 × *g* for 10 min.
2. Pass the supernatant through a 0.22 μm filter and collect as "conditioned medium."

3. Resuspend the cell pellets in 10 mL combined media (1 volume of conditioned media and 1 volume of fresh media) and make two cell densities with combined media at 500 cells/mL and 1000 cells/mL (*see* **Note 6**).
4. Aliquot 100 μL of the diluted cells (50 cells/well or 100 cells/well) into two 96-well plates and duplicate.
5. Irradiate one plate of each concentration at 1 Gy with MARK-I/Cs-137 irradiator (another duplicate plate is not irradiated and used as a control).
6. Incubate all four plates at 37 °C incubator with 5% CO_2 for 10–14 days to allow colony formation (*see* **Note 7**).

3.1.5 Colony Counting, Calculation, and Interpretation

1. Add 100 μL 1.0 mg/mL solution of MTT in 1× PBS into each well.
2. Incubate for 2–4 h at 37 °C.
3. Score each well for all four plates under a regular microscope. A well is considered positive if it contains even a single colony with >32 cells (*see* **Note 8**).
4. Calculate the colony formation efficiency (CFE):

   $CFE = -\ln[(N - C)/N]/W$

   N = number of wells.

   C = number of positive wells.

   W = seeded number of cells per well.
5. The survival fraction is calculated as the ratio of CFE% = CFEi/CFEc (CFEi: the CFE of irradiated plates and CFEc: CFE of the control plates).

The mean survival fraction for A-T patients is 13.1% ± 7.2% compared with 50.1% ± 13.5% for healthy control patients. Therefore, based on our previous studies comparing 104 bona fide A-T LCLs and 29 wild-type LCLs (i.e., phenotypically healthy), LCLs with a survival ≤21% are interpreted as radiosensitive and those with a survival ≥37% were considered normal, while intermediate radiosensitivity ranges between 21% and 37% (Fig. 1a). By using this assay, the radiosensitivity of several other genetic diseases was also analyzed (Fig. 1b) (*see* **Note 9**).

3.2 FC-pSMC1 Assay

This assay is a flow cytometry (FC)-based ATM kinase assay that measures ATM- dependent phosphorylation of Structural Maintenance of Chromosomes 1 protein (SMC1) following DNA damage using patient-derived PBLs, which can reliably distinguish A-T heterozygotes and homozygotes from controls.

3.2.1 Isolation of PBLs

Follow **steps 1–5** in Subheading 3.1.1 to isolate peripheral blood lymphocytes and continue to the next Subheading 3.2.2.

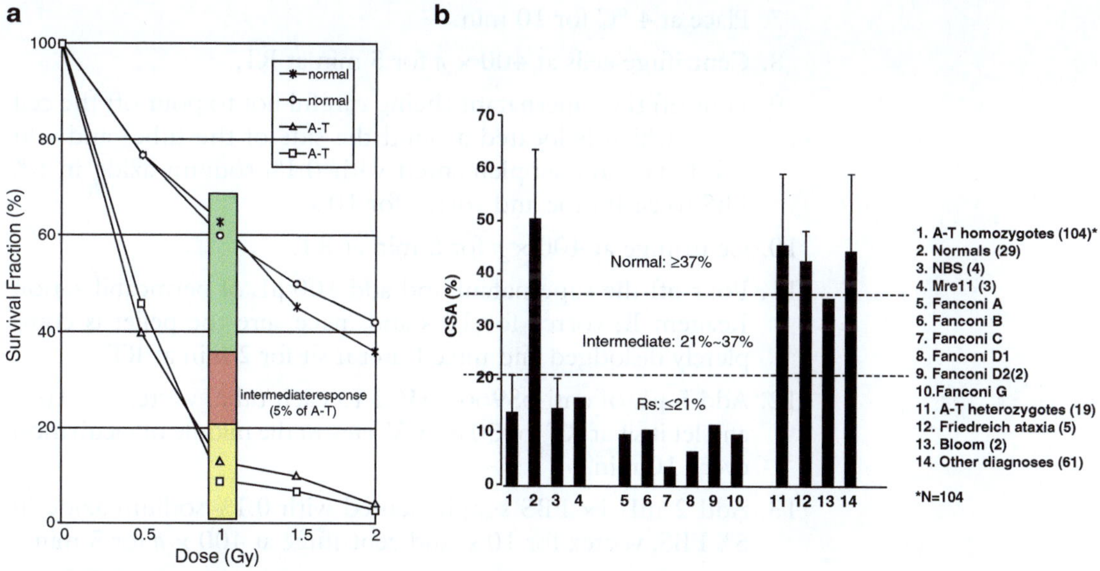

Fig. 1 (**a**) Dose-response curves for normal controls and A-T patients. LCLs were established with EBV-transformation and irradiated with various doses (0.5, 1.0, 1.5, and 2.0 Gy). The dose of 1.0 Gy was selected and used for testing 104 A-T LCLs and 29 normal controls to establish a radiosensitivity range: Normal range (*green*, ≥37%); intermediate range (*red*, 21%–37%); and radiosensitive range (*yellow*, ≤21%). (**b**) Mean colony survival fractions (μSF ± SD) after 1.0-Gy irradiation to LCLs of Normals (*N* = 29), A-T homozygotes (*N* = 104), A-T heterozygotes (*N* = 19), and AT-related disorders. LCLs from four patients with Nijmegen breakage syndrome (NBS) had the same degree of radiosensitivity as A-T cells (13 ± 9%). LCLs from seven patients with Fanconi"s anemia (groups A, B, C, D1, D2, and G) were radiosensitive and three patients with Mre11 deficiency were also radiosensitive (11% ± 6%). LCLs from five patients with Friedreich ataxia were not radiosensitive by this testing. LCLs from two patients with Bloom syndrome had slightly decreased colony survival fractions although studies have suggested that they are of intermediate radiosensitivity. LCLs from 19 A-T heterozygotes showed normal. The diagnoses of another 61 patients with normal CSA responses are largely undetermined. These patients encompassed a wide variety of neurologic signs and symptoms, such as mental and/or motor retardation and seizures. (Adapted from Sun et al., 2002 J. Pediatrics [14])

3.2.2 Assay Procedure for FC-pSMC1

1. Centrifuge the cells at 920 × *g* for 5 min at RT, resuspend in 10 mL complete media, and evenly split the cells into two 15 mL conical tubes with one labeled "+IR" and another labeled "-IR" as a control.
2. Take both tubes to the MARK-I/Cs-137 Irradiator but only irradiate the "+IR" tube with 10 Gy.
3. Return both tubes to the 37 °C incubator and let sit for 50 min (*see* **Note 10**).
4. Centrifuge at 400 × *g* for 5 min at RT.
5. Pour off supernatant (do not aspirate) and add 100 μL of fixation Reagent A to each tube, vortex 10 s, let it sit for 3 min.
6. Add 2 mL cold methanol, drop-wise, to the tube while vortexing at the same time.

7. Place at 4 °C for 10 min.
8. Centrifuge cells at 400 × *g* for 5 min at RT.
9. Pour off the supernatant (being careful not to pour off the cell pellet which is located around the side of the tube) and add 2 mL 1× PBS supplemented with 0.1% sodium azide in 5% FBS to each tube and vortex for 10 s.
10. Centrifuge at 400 × *g* for 5 min at RT.
11. Pour off the supernatant and add 100 μL of permeabilization Reagent B, vortex for 10 s and make sure the pellet is completely dislodged and mixed. Let it sit for 2 min at RT.
12. Add 5 μL of anti-pS966-SMC1 to each tube, vortex for 10 s, and let it sit at RT for 50 min. Vortex in the middle of incubation every 10 min.
13. Add 2 mL 1× PBS supplemented with 0.1% sodium azide in 5% FBS, vortex for 10 s, and centrifuge at 400 × *g* for 5 min.
14. Pour off supernatant, add 100 μL of 1× PBS supplemented with 3% BSA, vortex, make sure cell pellet is completely dislodged and mixed.
15. Add 3 μL of anti-rabbit FITC to each tube, vortex for 10 s, and let it sit at RT in the dark (covered with Foil) for 45 min. Vortex in the middle of the incubation every 10 min.
16. Add 2 mL 1× PBS supplemented with 0.1% sodium azide in 5% FBS, vortex for 10 s, and centrifuge at 400 × *g* for 5 min at RT.
17. Pour supernatant and resuspend the cell pellet in 400 μL of 2% PFA in 1× PBS.
18. Place tubes at 4 °C until ready to run flow cytometry.

3.2.3 Flow Cytometry Analysis and Result Interpretation

1. Analyze cells by flow cytometry on a BD FACSCalibur™ (BD Biosciences) using forward size scale as X-axis and FITC as Y-axis. Data analysis is done using Cell Quest software.
2. Plot geometric mean (GM) fluorescence intensity (FI) on the x axis using a log scale. The GM FI peak of untreated cells is subtracted from the GM FI peak of treated cells to yield the difference (ΔFI), and the ΔFIs for all samples are normalized against the ΔFI of a healthy daily control (DC) and expressed as a proportion (%DC).

We have tested PBLs from 16 healthy unknowns, 10 obligate A-T heterozygotes, and 6 unrelated A-T homozygotes [15]. As shown in Fig. 2, after 10 Gy IR, we observed an average response of 106.1%DC. By comparison, when we tested fresh PBLs isolated from the 10 obligate A-T heterozygotes, the average response to IR damage was significantly lower than that of healthy unknowns: 37.0% DC vs. 106.1%DC ($P < 0.006$). Responses of both the unknowns and A-T heterozygotes were significantly larger than

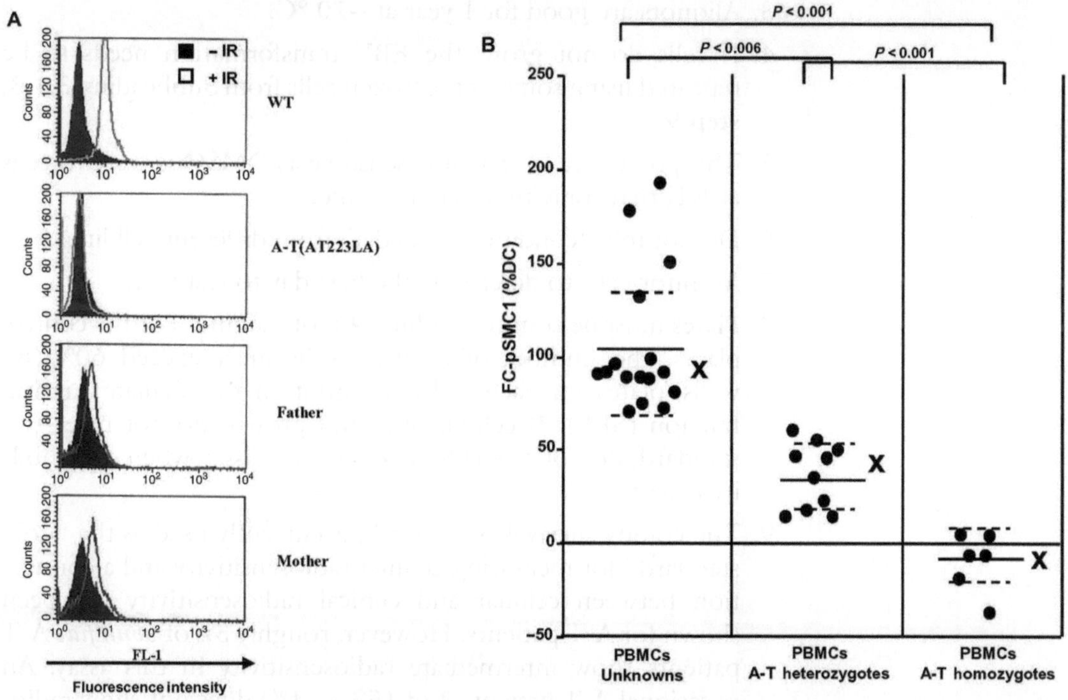

Fig. 2 FC-pSMC1 detection of IR-induced ATM-dependent phosphorylation of SMC1pSer966 using fresh PBLs. (**a**) FC-pSMC1 histograms of PBLs from a healthy unknown (WT), A-T patient (AT223LA), and obligate A-T heterozygotes (father and mother of AT223LA). (**b**) Dot plot of normalized FC-pSMC1 results performed on PBLs from 16 healthy unknowns [106%DC], 10 A-T heterozygotes [37%DC], and 10 A-T homozygotes [−9%DC]. X represents data from PBLs treated with 1.5 μg/mL bleomycin instead of IR. *Solid lines* represent the average normalized ΔFI (%DC), and *dashed lines* represent SDs. P values between genotypes are indicated above each panel. (Adapted from [15]) Shareef A. Nahas, Anthony W. Butch, Liutao Du, Richard A. Gatti Rapid Flow Cytometry–Based Structural Maintenance of Chromosomes 1 (SMC1) Phosphorylation Assay for Identification of Ataxia-Telangiectasia Homozygotes and Heterozygotes Clinical Chemistry Mar 2009, 55 (3) 463–472; Doi: 10.1373/clinchem.2008.107128 (Reproduced with permission from the American Association for Clinical Chemistry)

those of A-T homozygotes: 106.1%DC vs. −8.7%DC ($P < 0.001$), and 37.0%DC vs. −8.7%DC ($P < 0.001$). None of the responses for A-T PBLs fell within the ranges for unknowns or obligate A-T heterozygotes (Fig. 2b). The responses between genotypes were significantly different from each other ($P < 0.001$). We concluded that FC-pSMC1 was able to reliably distinguish between unknowns, A-T heterozygotes, and A-T homozygotes [15] (*see* **Note 11**).

4 Notes

1. Beyond this point, you will not be able to change the label on this tube, which is convenient for tracking, storage, and ship later.
2. Place the tube in −20 °C freezer for 30–60 min and move the tube to −70 °C freezer overnight.

3. Aliquots are good for 1 year at −70 °C.
4. If cells do not grow, the EBV transformation needs to be repeated using some of the frozen cells from Subheading 3.1.3, **step 9**.
5. This procedure is very time sensitive as DMSO is toxic to cells at RT, thaw only two vials at a time.
6. Do not mix "conditioned media" from different cell lines.
7. Monitor cells to determine the best day to test CSA.
8. Plates must be counted within 24 h of staining. For the control plates, the number of positive wells must exceed 60% (60 wells/plate) for each cell concentration to calculate survival fraction (SF%). If cell viability and growth do not meet this standard, do not report SF%. Repeat the assay when cell viability is better.
9. The colony survival assay (CSA) is currently used as the "gold standard" for measuring cellular radiosensitivity and a connection between cellular and clinical radiosensitivity has been shown for A-T patients. However, roughly 5% of *bona fide* A-T patients show intermediate radiosensitivity in this assay. An occasional A-T patient (2 of 153 = <1%) does not show radiosensitivity in this assay, even after repeated testing. By using this assay, several other rare genetic disorders have been shown to have radiosensitivity levels similar to A-T. These include: Nijmegen Breakage Syndrome, Mre11 deficiency, Ligase IV deficiency, Fanconi anemia, and several primary immunodeficiency disorders (Fig. 1b). Therefore, the 1 Gy-irradiated CSA can be clinically used to diagnose the radiosensitive genetic patients, which may facilitate the identification of the underlying genetic factors and help the conventional radio/chemotherapy options for these patients.
10. Alternative: instead of IR, PBLs can be treated with 1.5 μg/mL Bleomycin for 2 h in 37 °C incubator.
11. The CSA requires the establishment of a lymphoblastoid cell line (LCL) (4–8 weeks), and an additional 2–3 week period to conduct the assay. Thus, the CSA process results in a diagnostic turn-around time of approximately 90 days, which diminishes its diagnostic usefulness for timely radio-therapeutic intervention. On the other hand, A-T heterozygotes are at an increased risk for breast cancer and possibly heart disease and the frequency of heterozygosity is about 1–4% in general US population [4]. Identifying heterozygosity in the absence of a prior affected family member is even more challenging. ATM protein levels are usually 40–50% of normal in heterozygotes but cannot be reliably quantified by immunoblotting or ATM-ELISA from a single peripheral blood sample [16, 17].

Radiosensitivity test of cell lines by CSA from known A-T heterozygotes is usually inconclusive, yielding scores in the normal or intermediate range. The above factors prompted us to seek a rapid assay based on ATM function. The FC-pSMC1 assay could be performed in most clinical laboratories using 1–2 mL blood and it shortens the turnaround time for diagnosing A-T homozygotes from approximately 3 months to approximately 3 h. Thus, the FC-pSMC1 is a reliable, rapid screening assay for ATM homozygosity or heterozygosity and can be used for pre-natal counseling or for screening individuals in large study cohorts for potential ATM heterozygosity, which can then be confirmed by sequencing.

References

1. Gatti RA et al (2001) The pathogenesis of ataxia-telangiectasia. Learning from a Rosetta stone. Clin Rev Allergy Immunol 20(1): 87–108
2. Perlman S, Becker-Catania S, Gatti RA (2003) Ataxia-telangiectasia: diagnosis and treatment. Semin Pediatr Neurol 10(3):173–182
3. Ambrose M, Gatti RA (2013) Pathogenesis of ataxia-telangiectasia: the next generation of ATM functions. Blood 121(20):4036–4045
4. Chun HH, Gatti RA (2004) Ataxia-telangiectasia, an evolving phenotype. DNA Repair (Amst) 3(8–9):1187–1196
5. Lavin MF et al (2007) Current and potential therapeutic strategies for the treatment of ataxia-telangiectasia. Br Med Bull 81-82: 129–147
6. Gatti RA et al (1988) Localization of an ataxia-telangiectasia gene to chromosome 11q22-23. Nature 336(6199):577–580
7. Savitsky K et al (1995) A single ataxia telangiectasia gene with a product similar to PI-3 kinase. Science 268(5218):1749–1753
8. Shiloh Y (2006) The ATM-mediated DNA-damage response: taking shape. Trends Biochem Sci 31(7):402–410
9. Guo Z et al (2010) ATM activation by oxidative stress. Science 330(6003):517–521
10. Jeggo P, Lavin MF (2009) Cellular radiosensitivity: how much better do we understand it? Int J Radiat Biol 85(12):1061–1081
11. Nahas SA, Gatti RA (2009) DNA double strand break repair defects, primary immunodeficiency disorders, and 'radiosensitivity'. Curr Opin Allergy Clin Immunol 9(6):510–516
12. Ciccia A, Elledge SJ (2010) The DNA damage response: making it safe to play with knives. Mol Cell 40(2):179–204
13. Huo YK et al (1994) Radiosensitivity of ataxia-telangiectasia, X-linked agammaglobulinemia, and related syndromes using a modified colony survival assay. Cancer Res 54(10):2544–2547
14. Sun X et al (2002) Early diagnosis of ataxia-telangiectasia using radiosensitivity testing. J Pediatr 140(6):724–731
15. Nahas SA et al (2009) Rapid flow cytometry-based structural maintenance of chromosomes 1 (SMC1) phosphorylation assay for identification of ataxia-telangiectasia homozygotes and heterozygotes. Clin Chem 55(3):463–472
16. Butch AW et al (2004) Immunoassay to measure ataxia-telangiectasia mutated protein in cellular lysates. Clin Chem 50(12):2302–2308
17. Chun HH et al (2003) Improved diagnostic testing for ataxia-telangiectasia by immunoblotting of nuclear lysates for ATM protein expression. Mol Genet Metab 80(4):437–443

Chapter 2

Assaying for Radioresistant DNA Synthesis, the Hallmark Feature of the Intra-S-Phase Checkpoint Using a DNA Fiber Technique

Amanda W. Kijas and Martin F. Lavin

Abstract

During S-phase the cell replicates its DNA which is critical to maintaining the integrity of the genome and cell survival amidst damaging events. The cell is equipped with a series of checkpoints to slow progress throughout the cycle and facilitate DNA repair. Ataxia telangiectasia mutated (ATM), defective in the human genetic disorder ataxia-telangiectasia (A-T), is the key to initiating a signaling cascade activating the intra-S-phase checkpoint. This was first identified in A-T cells as radioresistant DNA synthesis using ^{14}C thymidine and ^{3}H thymidine to pulse label replicating cells before and after damage. This technique has been superseded now by direct labeling that distinguishes DNA replication initiations from ongoing sites of replication which are the target for the intra-S-phase checkpoint. Here, we outline how sites of replication are pulse labeled with two different thymidine analogs before and after damage. The DNA is then stretched out as fibers for immunolabeling to enable visual distinction and counting of ongoing replication forks from new initiations. It is this extent of new initiations that is used to detect the intra-S-phase checkpoint after DNA damage.

Key words Intra-S-phase checkpoint, Radioresistant DNA synthesis, DNA fibers, Replication

1 Introduction

During S-phase cells replicate their genome, creating duplicate sister chromatids in the preparation for cell division. Mammalian cells replicate their genome in a coordinated fashion firing multiple replicons of varying sizes, 20–400 kb, throughout S-phase. These replication forks move bidirectionally at about 2–3 kb per minute, depending on the stage of S-phase [1].

To preserve the integrity of this replication process cells have mechanisms to detect aberrant or unreplicated DNA and respond in a coordinated manner to initiate S-phase checkpoints and repair, or in more extreme cases of damage, initiate cell death. The intra-S-phase checkpoint induced by double strand breaks (DSB), signals

Sergei V. Kozlov (ed.), *ATM Kinase: Methods and Protocols*, Methods in Molecular Biology, vol. 1599,
DOI 10.1007/978-1-4939-6955-5_2, © Springer Science+Business Media LLC 2017

to halt new replicon initiations, distinct from the other S-phase checkpoints that affect actively progressing forks [2].

Defects in this intra-S-phase checkpoint were first discovered in ataxia telangiectasia (A-T) patient cells that were unable to reduce the rate of replication after exposure to ionizing radiation and was termed, radioresistant DNA synthesis, RDS [3, 4]. These original experiments utilized two forms of radiolabeled thymidine ^{14}C and ^{3}H to consecutively pulse label DNA synthesis pre and post DSB-induced damage. This yielded a measure of total synthesis post damage, thus only a crude measure for new initiations. Since these initial observations linking mutant ATM to defects in the cell cycle, two members of the MRE11-RAD50-NBN (MRN) complex, MRE11 and NBN, were subsequently identified as key in the activation of the intra-S-phase checkpoint [5, 6]. It was demonstrated later that the key activation events involve a signaling cascade through which cyclin-dependent kinases and cyclins become activated to inhibit progress through the cell cycle [7]. Since these early reports, a RAD50-deficient patient was identified and the patient's fibroblasts also displayed RDS [8, 9]. However, in that case the methodology used to determine the effect of radiation-induced DSB on the level of inhibition on new origin firing compared to control cells was a DNA fiber assay that directly visualizes new initiations after damage. This same fiber assay-based methodology has since been applied to A-T lymphoblastoid cells that reported the effects of multiple autophosphorylation sites of ATM in the activation of the intra-S-phase checkpoint [10], and more recently has also been applied in the characterization of A-T stem cells [11, 12].

DNA fibers have been widely used for investigations of replication, with pioneering work by Cairns [13] and Huberman and Riggs [14] who used autoradiography to visualize [3H]thymidine-labeled DNA, showing that it can be stretched to a resolution comparable to native B-DNA at 0.34 μM/kb [15]. These techniques have since been used to utilize thymidine analogs in combination with fluorescently labeled antibodies where they have also been applied to high-resolution gene mapping [16].

Thymidine nucleotide analogs such as 5-Chloro-2′-deoxyuridine and 5-Iodo-2′-deoxyuridine are transported into the cell and then incorporated in an uninhibited manner into newly synthesized DNA, enabling direct pulse labeling of actively replicating sites in the genome. The subsequent ability to differentially detect these analogs by immunofluorescence in DNA fiber spreads allowing high-resolution analysis of replication fork speed, termination, as well as new replicon initiations. These techniques have been applied to investigate replication dynamics after DNA damage in cells where the proportion of new initiations (shown as red-only tracks, Fig. 1a) to ongoing tracks (shown as green and red) before and after damage was determined. The two early studies used two

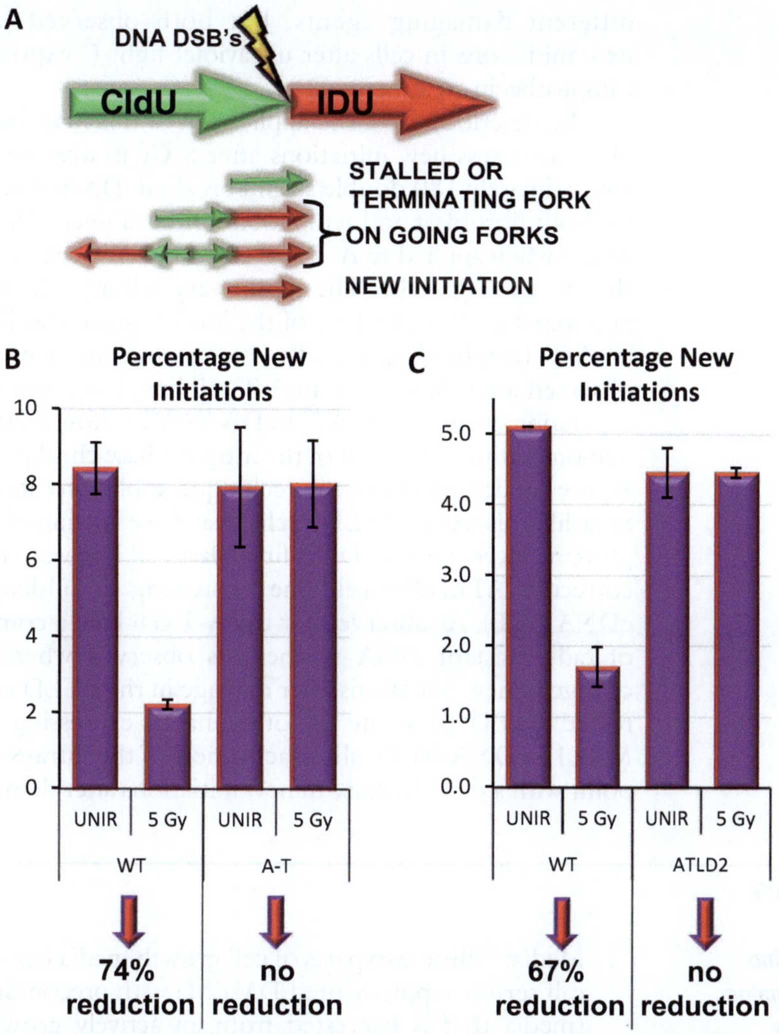

Fig. 1 (**a**) Replication forks are consecutively pulse labeled with 5-Chloro-2′-deoxyuridine (CldU) and 5-Iodo-2′-deoxyuridine (IDU). Double strand break (DSB) inducing agent introduced at the end of CldU labeling enables the subsequent scoring of ongoing forks (*green* and *red tracks*) from new initiations (*red*-only tracks). (**b**) A-T lymphoblastoid cells (A-T) and corrected lymphoblastoid cells (WT) expressing a wild-type ATM cDNA were consecutively labeled with CldU and IDU. Cells were either left as unirradiated (UNIR) or irradiated with 5 Gy at the end of the CldU labeling (*see* Fig. 1a). Cells were then treated as outlined in the lymphoblastoid method below to spread the DNA fibers and immunolabel for the thymidine analogs. The proportion of ongoing (*green* and *red* tracks) and new initiations (*red*-only tracks) was scored. The percentage of new initiations was calculated as the number of new initiations/(ongoing + new initiations) × 100. A total of over 600 forks were scored for each treatment for each cell line. The standard deviation of the mean is shown, $n = 3$. (**c**) ATLD2 fibroblasts (ATLD2) and corrected fibroblasts (WT) expressing a wild-type MRE11 cDNA were consecutively labeled with Cldu and Idu. Cells were either left as unirradiated (UNIR) or irradiated with 5 Gy at the end of the Cldu labeling (*see* Fig. 1a). Cells were then treated as outlined in the fibroblast method below to spread the DNA fibers and immunolabel for the thymidine analogs. The proportion of ongoing (*green* and *red* tracks) and new initiations (*red*-only tracks) was scored. The percentage of new initiations was calculated as the number of new initiations/(ongoing + new initiations) × 100. A total of over 460 forks were scored for each treatment for each cell line. The standard deviation of the mean is shown, $n = 2$

different damaging agents, but both observed inhibition of new initiations in cells after ultraviolet light C exposure [17] and camptothecin treatments [18].

We describe below the application of this fiber-based methodology to assess new initiations after 5 Gy irradiation that induces approximately 180 double strand breaks in DNA. This is described for both fibroblast and lymphoblastoid cell lines. The results generated when applied to A-T and corrected A-T cells as a control are shown in Fig. 1b. The data very clearly demonstrate the radioresistant-based nature of the intra-S-phase check point failure in A-T lymphoblastoid cells. No change in new initiations is observed after damage in an A-T cell line, but corrected A-T cells expressing a full-length ATM cDNA (WT) show a 70% reduction, demonstrating activation of the intra-S-phase checkpoint (Fig. 1b). Figure 1c demonstrates the technique applied to ataxia telangiectasia like disorder (ATLD) cells that have no functional MRE11 protein. Here, an ATLD2ht fibroblast cell line was compared to a corrected ATLD2ht cell line expressing a full-length MRE11 cDNA (WT). As observed for the A-T cell line, a comparable level of radioresistant DNA synthesis is observed where there is no change in new initiations after damage in the ATLD cells. The corrected ATLD cells, on the other hand, expressing a full-length MRE11 cDNA (WT), show activation of the intra-S-phase checkpoint with a 67% decrease in new initiations after damage (Fig. 1c).

2 Materials

2.1 Thymidine Nucleotide Analog Incorporation

1. Media: Dilute two parts of cell growth media (usually 12% fetal calf serum supplemented DMEM) with preconditioned media (media that is harvested from an actively growing culture). This is to be used for all media containing steps.
2. Prepare aliquots of 2 mM 5-Iodo-2′-deoxyuridine (IDU), MW = 354.1. Preheat 40 mL sterile high-purity water to 60 °C. Add 28.34 mg IDU, vortex vigorously, then wrap in foil, and place on a rotating platform to dissolve. Make 1 mL aliquots once dissolved, store at −20 °C.
3. Prepare aliquots of 2 mM 5-Chloro-2′-deoxyuridine (CldU), MW = 262.6. Preheat 40 mL sterile, high-purity water to 60 °C. Add 21.01 mg CldU, vortex to dissolve, then make 0.5 mL aliquots and store at −20 °C.
4. Optional: Prepare 10 mM Thymidine aliquots.
5. Versene.
6. Trypsin-EDTA.
7. PBS.

8. TBST.
9. Trypan blue (for cell counting).
10. Plasticware for tissue culture (6-well plates, 1.5 mL tubes, 5 and 10 mL pipettes, 1 mL pipette tips).

2.2 Preparing DNA Fibers

1. Menzel–Glaser Superfrost plus or ULTRA plus, positively charged slides.
2. Lysis solution: 0.5% sodium dodecyl sulfate, 200 mM Tris–HCL pH 7.4 and 50 mM EDTA.
3. 15–20° ramp and a large glass plate (capable of holding multiple slides). We made one by marking of 20° with protractor on a small styrofoam box. Then cut with razor blade and used an old gel casting glass plate about 20 × 20 cm.
4. Vertical glass staining jars/glass coplin jars. Vertical eight slide jars are good size.
5. Ice-cold methanol/acetic acid (3:1).
6. Lint-free paper towel.

2.3 Immunolabeling of Incorporated CldU and IDU at Sites of DNA Synthesis

1. Vertical glass staining jars/glass coplin jars. (Vertical eight slide jars are good size).
2. 50 × 22 mm cover slips for mounting slides.
3. 2.5 M HCL.
4. PBS.
5. PBS containing 0.01% Tween 20 (0.01% PBST).
6. PBS containing 0.05% Tween 20 (0.05% PBST).
7. Blocking buffer (10% BSA in 0.01% PBST).
8. Antibody dilution buffer, 1% BSA in 0.01% PBST.
9. High stringency wash buffer (1 L: 29.2 g sodium chloride, 4.44 g Tris–HCL adjust to pH 8 with NaOH and then add Tween20 to 0.5% final concentration).
10. Flat plastic tray for washing and staining glass slides.
11. Parafilm cut to the size of the slide (excluding super frosted region).
12. Rat monoclonal anti-bromodeoxyuridine antibody (Rat monoclonal [BU1/75] catalog number Ab6326, Abcam).
13. Goat anti-rat Alexa488 conjugated secondary antibody.
14. Mouse monoclonal anti-bromodeoxyuridine antibody (Anti-BrdU (B44), catalog number 347580, Becton Dickinson).
15. Goat anti-mouse Alexa594 conjugated secondary antibody.
16. Mounting media.

3 Methods

3.1 For Fibroblast Cell Lines

3.1.1 Seeding of Cells

Cells are split 1–3 days before the experiment such that they will be sub-confluent on the day of experiment when you will need at least 1×10^5 cells. Seed to generous media volume (5 mL for 6-well plate) as this will be reused during the labeling experiment as preconditioned media.

3.1.2 Thymidine Nucleotide Analog Incorporation

1. On the morning of the experiment collect excess media from plated cells (about 2.5 mL/well). Dilute this preconditioned media (1/3) with fresh media (2/3) and then place in an incubator to warm to 37 °C and equilibrate to 5% CO_2, so lid needs to be loose or use small filter toped flask. This is to be used for all subsequent media containing steps.
2. Defrost aliquots of CldU (2 mM) and IDU (2 mM), thymidine (10 mM). Prepare 2 mL per well labeling media of each 20 μM CldU and 80 μM IDU, as well as 1 mL per well of each 100 μM thymidine and rinse media.
3. Label cells consecutively with 20 μM CldU media for 20 min including a mock or 5 Gy treatment just prior to the removal of CldU (*see* **Note 1**).
4. Remove all media and rinse consecutively with 100 μM thymidine media (1 mL) and then rinse media (1 mL), removing all residual media between each addition. This can be replaced with just two washes with rinse media, the presence of thymidine does not seem essential for the change over between CldU and IDU.
5. Add 80 μM IDU media for a further 20 min. As the time after CldU removal and IDU addition must be kept to a minimum, the thymidine rinse can be removed prior to performing one 2 mL media rinse, which saves time when processing multiple samples.
6. Remove media, rinse twice with versene, then add trypsin-EDTA in small volume (about 600 μL per well) until cells detach fully with a gentle tap.
7. Add cold fetal calf serum or fetal bovine serum to neutralize trypsin and pellet cells in 1.5 mL Eppendorf tubes (spinning at 400 × *g* for 5 min at 4 °C).
8. Remove trypsin-FCS and rinse cells once with ice-cold PBS.
9. Resuspend cell pellet with 50 μL ice-cold PBS. Leave cells on ice and perform trypan blue cell count using hemocytometer. Adjust volume to approximately $1–2 \times 10^6$ cells per mL with ice cold PBS.

3.1.3 Preparing DNA Fibers

DNA fibers are made following the approach by Parra and Windle [16] where labeled DNA fibers are stretched out by gravity as the DNA solution slides down a charged slide.

1. Label slides and set onto clean dust-free flat bench top (check with level).
2. Place 2 μL of cells (1–2 × 10^6 cells/mL) at the top of a Menzel–Glaser Superfrost plus or ULTRA plus (positively charged) slide, as shown in Fig. 2a (*see* **Note 2**).
3. Add 8 μL of lysis solution to the cell droplet sitting at the top of the slide and gently pipette up and down once to ensure mixing (Fig. 2a). Allow the cells to lyse in the droplet for 8–10 min without disturbing the slides (*see* **Note 3**).

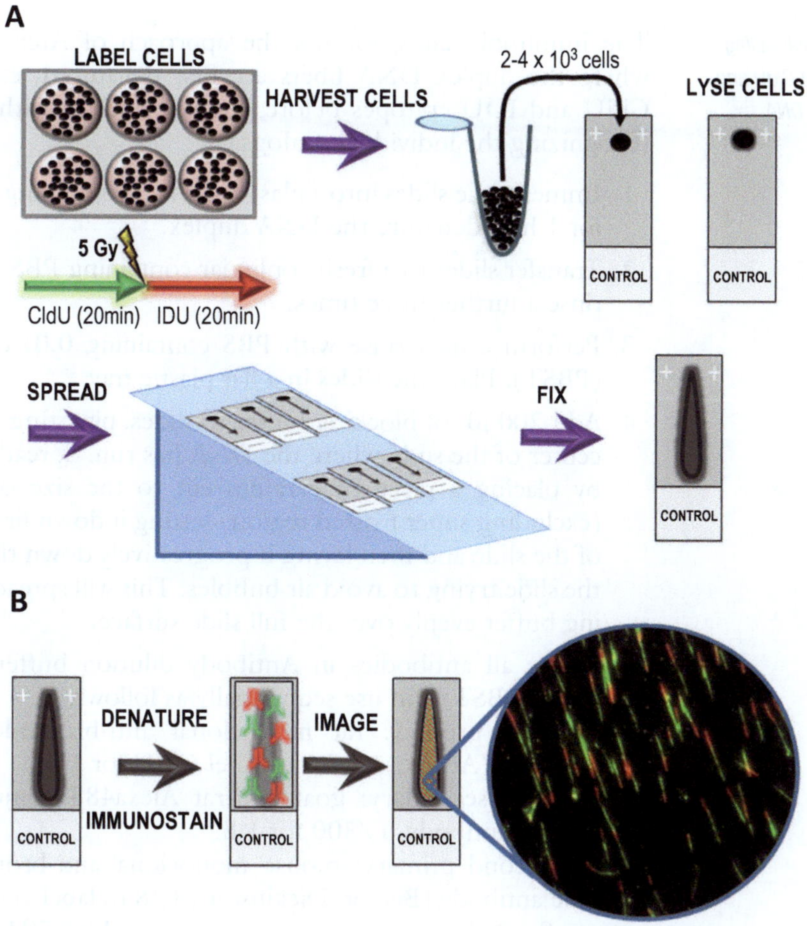

Fig. 2 **(a)** Cells are seeded to subconfluent density and replication forks consecutively pulse labeled with CldU and IDU. Cells were then harvested and lysed using SDS on positively charged slides before spreading by gravity at 15–20°. Stretched DNA fibers were then fixed in methanol acetic acid before storage. (**b**) DNA fibers were denatured in HCL, before sequential immunolabeling with antibodies recognizing the two thymidine analogs, CldU and IDU. Ongoing forks and new initiations were imaged using fluorescence microscopy

4. Pick up slides carefully so as not to disturb the droplet and place onto a large glass plate (capable of holding multiple slides), tilted to 15–20°. This will allow gravity to slowly stretch out the labeled DNA into fibers down the length of the slide (Fig. 2a). Keep the droplets running roughly down the center of the slide to make subsequent immunolabeling and imaging more straightforward, knowing where to expect the DNA fibers to be.
5. Leave slides to air dry and then fully immerse into glass coplin jar containing ice-cold methanol/acetic acid (3:1) for 10 min (*see* **Note 4**).
6. Remove slides, draining well before placing DNA side up onto lint-free paper towel in fume hood. Store the slides at 4 °C for at least 1 day (to age the DNA) before immunolabeling.

3.1.4 Immunolabeling of Incorporated CldU and IDU at Sites of DNA Synthesis

The immunolabeling follows the approach of Aten et al. [19] where the duplex DNA fibers are first denatured, exposing the CldU and IDU epitopes before immunolabeling with antibodies recognizing the individual analogs.

1. Immerse the slides into a glass coplin jar containing 2.5 M HCl for 1 h to denature the DNA duplex.
2. Transfer slides to a fresh coplin jar containing PBS, drain, then rinse a further three times.
3. Perform a final rinse with PBS containing 0.01% Tween 20 (PBST). Place the slides in a flat plastic tray.
4. Add 200 μL of blocking buffer to slides, pipetting it down the center of the slide where the DNA has run. Spread this evenly by placing a piece of parafilm cut to the size of the slide, (excluding super frosted region) setting it down first at the top of the slide and then laying it progressively down the length of the slide trying to avoid air bubbles. This will spread the blocking buffer evenly over the full slide surface.
5. Dilute all antibodies in Antibody dilution buffer (1% BSA, 0.01% PBST) and use sequentially as follows:

   First primary: rat monoclonal anti-bromodeoxyuridine antibody (Abcam), 1/70 to label CldU for 1.5 h.

   First secondary: goat anti-rat Alexa488 conjugated secondary antibody, 1/300 for 1 h.

   Second primary: mouse monoclonal anti-bromodeoxyuri dine antibody (Becton Dickinson), 1/8 to label IDU for 1 h.

   Second secondary: goat anti-mouse Alexa594 conjugated secondary highly cross absorbed antibody, 1/300 for 1 h.
6. Prepare 90 μL of each antibody dilution for each slide you are labeling (*see* **Note 5**). Pipette antibody dilution down the center of the slide where the DNA has run. Spread this evenly

over the slide surface by placing a piece of parafilm cut to the size of the slide, (excluding super frosted region) setting it down first at the top of the slide and then laying it progressively down the length of the slide trying to avoid air bubbles. This will spread the antibody evenly over the full slide surface.

7. Between antibodies rinse once and wash 2 × 10 min in 0.05% PBST, with the exception after second primary (mouse anti-bromodeoxyuridine antibody) where a high stringency wash is used for 9 min (no longer or may experience loss of signal).

3.1.5 Mounting and Imaging of Labeled DNA Fibers

1. Perform a final rinse in PBS and drain slides well. For each slide dry the back of the slide and then place long edge down on lint-free tissue to drain.
2. Try to dot mounting media down the center of the slide (DNA track). Using forceps and finger slowly lower 50 × 22 mm cover slip down the length of the slide.
3. Use fluorescent microscope to follow labeled DNA tracks scanning down the slide, using 594 channel as the signal is usually stronger for IDU staining. You will also need a 488-channel filter for CldU detection. Magnification: 63× and or 40× lenses.
4. Score the proportion of ongoing (green and red tracks) and new initiations (red-only tracks). The percentage of new initiations is calculated as the number of new initiations/(ongoing + new initiations) × 100 for each treatment.

3.2 METHODS for Lymphoblastoid Cultures

3.2.1 Dilution of LCL Cultures

1. Cells are split 1 day before experiment to densities of about 3–4 × 10^5 cells/mL such that they will be at about 6–8 × 10^5 cells/mL on the day of experiment, when you will need 200 μL of cells per treatment.

3.2.2 Nucleotide Analog Incorporation

1. On the morning of the experiment collect excess media from diluted cells (preconditioned media). Dilute this preconditioned media (1/3) with fresh media (2/3) and then place in an incubator to warm to 37 °C and equilibrate to 5% CO_2, so the lid needs to be loose or use a small filter toped flask. This is to be used for the dilution of the IDU and you will need 5 mL per sample.
2. Defrost aliquots of CldU (2 mM) and IDU (2 mM). Prepare 100 μM IDU stock (5 mL per sample) in media and put to warm at 37 °C, and equilibrate to 5% CO_2. Also put versene on ice.

3. Label cells consecutively with 20 μM CldU (2 μL of 2 mM stock per 200 μL of cells for each sample) for 20 min at 37 °C and 5% CO_2, including a mock or 5 Gy treatment during the last minutes of CldU labeling. Then add 5 mL of 100 μM IDU for a further 20 min (at 37 °C, 5% CO_2).
4. Add 5 mL ice-cold versene supplemented with 10 mM EDTA pH 8, pellet cells by centrifugation, and rinse once with ice-cold versene.
5. Resuspend cell pellet in 50 μL of ice-cold PBS. Leave cells on ice while doing cell counts using trypan blue and hemocytometer. Adjust volume to be approx. 1–2 × 10^6 cells per mL using ice cold PBS.

3.2.3 Preparing DNA Fibers

Treat the same as outlined above for fibroblasts.

3.2.4 Immunolabeling of Incorporated CldU and IDU at Sites of DNA Synthesis

Treat the same as outlined above for fibroblasts.

3.2.5 Imaging of Labeled DNA Fibers

Treat the same as outlined above for fibroblasts.

4 Notes

1. Take into account your irradiator location and output for timing of CldU labeling. I routinely start irradiating at 15 min, to give me time to complete irradiation and return to lab within 20 min CldU labeling.
2. The number of cells needs to be optimized depending on the percentage you expect to be in S phase. If you are using rapidly dividing cells, you will need to supplement the labeled cells with unlabeled cells 1:1, you should not need to do this with control and A-T cells.
3. The time varies greatly depending on air conditioning and temperature (affecting droplet drying time), and cell number (affecting the viscosity of the droplet, the more DNA the more viscous). Plan to run several extra slides to get this right. The droplet should migrate slowly, if it migrates too fast, leave the droplet to dry a couple of minutes longer before tilting to 15–20°. If droplet does not reach the bottom of the slide or migrates too slowly (a few minutes), then reduce the droplet cell lysis time by a minute or two, or increase lysis solution to 10 μL.
4. If the DNA track left by the droplet has dried but a droplet at the base of the slide is still sitting there, the corner of a lint-free tissue

can be used to touch into the base of droplet (superfrosted end); this sucks of some of the excess fluid, to enable faster drying and stops the formation of a crusty region at the bottom.

5. If using other primary antibodies raised against these thymidine analogs, you will have to confirm their specificity.

References

1. Hyrien O (2015) Peaks cloaked in the mist: the landscape of mammalian replication origins. J Cell Biol 208:147–160
2. Bartek J, Lukas C, Lukas J (2004) Checking on DNA damage in S phase. Nat Rev Mol Cell Biol 5:792–804
3. Houldsworth J, Lavin MF (1980) Effect of ionizing radiation on DNA synthesis in ataxia telangiectasia cells. Nucleic Acids Res 8:3709–3720
4. Painter RB, Young BR (1980) Radiosensitivity in ataxia-telangiectasia: a new explanation. Proc Natl Acad Sci U S A 77:7315–7317
5. Stewart GS et al (1999) The DNA double-strand break repair gene hMRE11 is mutated in individuals with an ataxia-telangiectasia-like disorder. Cell 99:577–587
6. Taalman RD, Jaspers NG, Scheres JM, de Wit J, Hustinx TW (1983) Hypersensitivity to ionizing radiation, in vitro, in a new chromosomal breakage disorder, the Nijmegen breakage syndrome. Mutat Res 112:23–32
7. Bartek J, Lukas J (2003) Chkl and Chk2 kinases in checkpoint control and cancer. Cancer Cell 3:421–429
8. Waltes R et al (2009) Human RAD50 deficiency in a Nijmegen breakage syndrome-like disorder. Am J Hum Genet 84:605–616
9. Gatei M et al (2011) ATM protein-dependent phosphorylation of Rad50 protein regulates DNA repair and cell cycle control. J Biol Chem 286:31542–31556
10. Kozlov SV et al (2011) Autophosphorylation and ATM activation: additional sites add to the complexity. J Biol Chem 286:9107–9119
11. Stewart R et al (2013) A patient-derived olfactory stem cell disease model for ataxia-telangiectasia. Hum Mol Genet 22: 2495–2509
12. Nayler S et al (2012) Induced pluripotent stem cells from ataxia-telangiectasia recapitulate the cellular phenotype. Stem Cells Transl Med 1:523–535
13. Cairns J (1963) The bacterial chromosome and its manner of replication as seen by autoradiography. J Mol Biol 6:208–213
14. Huberman JA, Riggs AD (1968) On the mechanism of DNA replication in mammalian chromosomes. J Mol Biol 32:327–341
15. Watson JD, Crick FH (1953) Molecular structure of nucleic acids; a structure for deoxyribose nucleic acid. Nature 171:737–738
16. Parra I, Windle B (1993) High resolution visual mapping of stretched DNA by fluorescent hybridization. Nat Genet 5:17–21
17. Merrick CJ, Jackson D, Diffley JF (2004) Visualization of altered replication dynamics after DNA damage in human cells. J Biol Chem 279:20067–20075
18. Seiler JA, Conti C, Syed A, Aladjem MI, Pommier Y (2007) The intra-S-phase checkpoint affects both DNA replication initiation and elongation: single-cell and -DNA fiber analyses. Mol Cell Biol 27:5806–5818
19. Aten JA, Bakker PJ, Stap J, Boschman GA, Veenhof CH (1992) DNA double labelling with IdUrd and CldUrd for spatial and temporal analysis of cell proliferation and DNA replication. Histochem J 24:251–259

Chapter 3

ATM Gene Mutation Detection Techniques and Functional Analysis

Guillaume Rieunier, Catherine Dubois D'Enghien, Alice Fievet, Dorine Bellanger, Dominique Stoppa-Lyonnet, and Marc-Henri Stern

Abstract

Ataxia Telangiectasia (A-T) is caused by biallelic inactivation of the Ataxia Telangiectasia Mutated (*ATM*) gene, due to nonsense or missense mutations, small insertions/deletions (indels), splicing alterations, and large genomic rearrangements. After establishing A-T clinical diagnosis, a molecular confirmation is needed, based on the detection of one of these loss-of-function mutations in at least one allele. In most cases, the pathogenicity of the detected mutations is sufficient to make a definitive diagnosis. More rarely, mutations of unknown consequences are identified and direct biological analyses are required to establish their pathogenic characters. In such cases, complementary analyses of ATM expression, localization, and activity allow fine characterization of these mutations and facilitate A-T diagnosis. Here, we present genetic and biochemical protocols currently used in the laboratory that have proven to be highly accurate, reproducible, and quantitative. We also provide additional discussion on the critical points of the techniques presented here.

Key words DNA sanger sequencing, MLPA, Functional assays

1 Introduction

Ataxia-Telangiectasia (A-T) is an autosomal-recessive disease. After *ATM* gene location by linkage study on the 11q22-23 region, the Shiloh group cloned it and determined its sole responsibility in the occurrence of A-T [1]. Later, *ATM* was described as a 150 kb gene with 62 coding exons, its huge size making mutations detection very difficult. *ATM* encodes a 350 kDa protein kinase (3056 amino acids) with a catalytic domain resembling that of phosphatidylinositol-3 kinases (PI3K) [2], making assessment of its activity possible.

More than 400 *ATM* mutations have been shown deleterious throughout the whole gene without over-representation at protein functional domains [3]. A few founder mutations have been found in various countries like the Norwegian (3245ATC>TGAT), the Amish (1563delAG), or the Sardinian (3894insT) mutations [4–6],

Sergei V. Kozlov (ed.), *ATM Kinase: Methods and Protocols*, Methods in Molecular Biology, vol. 1599,
DOI 10.1007/978-1-4939-6955-5_3,

but they represent only a small fraction of the mutation spectrum, and do not facilitate mutation detection. The identification of *ATM* mutations is performed using direct Sanger sequencing of genomic DNA (Next Generation Sequencing being still under validation in our laboratory) by testing simultaneously the 32 most mutated amplicons and then the 29 less mutated ones. When no or one point mutation is identified, search for large gene rearrangements is performed with semiquantitative PCR by multiple ligation of probe amplification (MLPA). In the vast majority of cases, nonsense mutations, small insertions/deletions, large genomic rearrangements, and splicing alterations are detected, leading to the total loss of ATM protein. A few patients present missense mutations with unknown consequences for the protein. To assess the pathogenicity of these mutations, ATM level, localization, and kinase activity can be evaluated after ATM activation.

After DNA damage, ATM phosphorylates a thousand targets, activating cell cycle checkpoints, DNA repair, metabolic and senescence pathways. Taking advantage of the number of informative targets, A-T molecular diagnosis can be refined through the evaluation of the ATM targets phospho-KAP1 Ser824, phospho-Chk2 Thr68, and phospho-p53 Ser15 by western blot after ionizing radiation [7]. ATM total level and cyto-nuclear repartition, quantified by western blot after cyto-nuclear extraction, also give information on ATM deficiency for nuclear and cytoplasmic functions [8].

In this chapter, we introduce protocols to sequence and search for large gene rearrangements in *ATM*, as well as functional tests to assess ATM kinase function, which are currently used with success in our laboratory and have proven to be highly accurate, reproducible, and quantitative. Additional critical points are given here to troubleshoot and enhance the techniques described here.

2 Materials

2.1 DNA Extraction from Blood Samples

1. QuickGene-610L extraction system (Fujifilm).
2. QuickGene DNA whole blood kit (Fujifilm).
3. Heating block.
4. Spectrophotometer.

2.2 PCR

1. ATM primers 20 μM (Table 1).
2. Taq polymerase 5 U/μL.
3. Premixed 10 mM dNTPs.
4. DMSO.
5. 15 mM $MgCl_2$.
6. 96-well PCR plate.

Table 1
ATM primers

Primer pair name	Primer sequence 1 (5′>3′)	Primer sequence 2 (5′>3′)	chr	genomic position hg19 UCSC In-Silico PCR	Amplicon size	Targeted region for primer choice
ATM0405Abis_and_ATM0405Bbis	ATGTGCCTCTAATTGTACAG	TGCCAAATTCATATGCAAGG	11	chr11:108098180-108098754	575bp	5'UTR-172-IVS05+139
ATM06A_and_ATM06B	GATGGCATGAACAGCTTTTG	CGACAGTAATCTGTTAAGCC	11	chr11:108099826-108100099	274bp	IVS05-79-IVS06+49
ATM07A_and_ATM07B	TAGTTGCCATTCCAAGTGTC	GACAGAGTGCTTTCTTTGGTG	11	chr11:108106327-108106633	307bp	IVS06-70-IVS07+72
ATM08ANEW_and_ATM08BNEW	CTTTTTCTGTATGGGATTATGG	TTACTGAGTCTAAAACATGGTC	11	chr11:108114578-108114921	344bp	IVS07-102-IVS08+76
ATM09A_and_ATM09B	CCACTTCATAACTGTTCAGTTTG	GTCATATCCTCCTAAAGAACAC	11	chr11:108115440-108115845	406bp	IVS08-75-IVS09+92
ATM10A_and_ATM10B	CAGCTGAATAATTTTGTGGG	TAGGCTTTTTGTGAGAACAC	11	chr11:108117611-108117956	346bp	IVS09-80-IVS10+102
ATM11AATMex11A_and_ATM11Bnew	TGATACGAGATCGTGCTGTTC	GGATTCCACTGAAAGTTTTCTG	11	chr11:108119558-108119908	351bp	IVS10-102-IVS11+79
ATM12A_and_ATM12B	GTTTGTTAATGTGATGGAATAG	TGTGTTTATCTGTAAGTCAG	11	chr11:108121375-108121841	467bp	IVS11-53-IVS12+42
ATM13A_and_ATM13B	GATAGATAAAGTCTTTGCCC	ATCAAATAAGTGGAGAGAGC	11	chr11:108122500-108122828	329bp	IVS12-64-IVS13+70
ATM14A_and_ATM14B	TCTTTACATGGCTTTTGGTC	ATTAAGATGCAGCTACTACC	11	chr11:108123492-108123730	239bp	IVS13-52-IVS14+91

(continued)

Table 1
(continued)

Primer pair name	Primer sequence 1 (5′>3′)	Primer sequence 2 (5′>3′)	chr	genomic position hg19 UCSC In-Silico PCR	Amplicon size	Targeted region for primer choice
ATM15A_and_ATM15B	AAGGCAAAGCATTAGGTACTTG	TTCTCCTTCCTAACAGTTTACC	11	chr11:108124487-108124835	349bp	IVS14-54-IVS15+69
ATM16A_and_ATM16B	TTAGAGCTATCCAGGATATG	AAAGAGAAAGGGTTAACCTG	11	chr11:108126796-108127184	389bp	IVS15-146-IVS16+117
ATM17A_and_ATM17B	TGTCCAAGATCAAAGTACAC	TGACAGAGAAAGATCCTATC	11	chr11:108128123-108128386	264bp	IVS16-85-IVS17+53
ATM18A_and_ATM18B	ATAGAGAAAACACTGTCTGC	AGCTATATGTTGTGAGATGC	11	chr11:108129636-108129905	270bp	IVS17-77-IVS18+103
ATM19A_and_ATM19B	ATTTTGACTACAGCATGCTC	ATCAATGAGGCCTCTTATAC	11	chr11:108137834-108138144	311bp	IVS18-64-IVS19+75
ATM20ABis_and_ATM20BBis	TGAAGAGGAGGAAATTTGAGTT	TCTTCAAAGACACCATGTGATTC	11	chr11:108139052-108139434	383bp	IVS19-85-IVS20+98
ATM21A_and_ATM21B	ATGATTTGTGGATAAACCTG	TAAGCTTAACAGAACACATC	11	chr11:108141746-108141944	199bp	IVS20-45-IVS21+71
ATM22A_and_ATM22B	AACTGATGTGTTCTGTTAAGC	ACAAAACTTGCATTCGTATCC	11	chr11:108141921-108142200	280bp	IVS21-57-IVS22+67
ATM23A_and_ATM23B	TGAGTGCTTTTATCAGAATG	CATTAACAAACAAAGACTGC	11	chr11:108143205-108143401	197bp	IVS22-54-IVS23+67
ATM24A_and_ATM24B	GCAGTCTTTGTTTGTTAATGAG	TGAGCTGTTACTATGTAAGAC	11	chr11:108143382-108143654	273bp	IVS23-67-IVS24+75
ATM25A_and_ATM25B	ATATGCTTTGGAAAGTAGGG	CTCTTATACCAAACTTGGTG	11	chr11:108150141-108150420	280bp	IVS24-77-IVS25+85
ATM26A_and_ATM26B	GTAGGAGGTTTCTGAAATGTG	TGCTTTCCAATATTCCCAACC	11	chr11:108151602-108152037	436bp	IVS25-120-IVS26+142

ATM27A_and_ATM27B	ATGTTGTTTCTAGGTCCTAC	AGGGGACTTGCTAAGTATTG	11	chr11:108153339-108153677	339bp	IVS26-98-IVS27+71
ATM28A_and_ATM28B	TATGATACTTTAATGCTGATGG	GGTTATATCTCATATCATTCAG	11	chr11:108154845-108155255	411bp	IVS27-109-IVS28+55
ATM29A_and_ATM29B	CAGTCATCGAATACTTTTGG	TAGACATTGAAGGTGTCAAC	11	chr11:108158222-108158525	304bp	IVS28-105-IVS29+83
ATM30A_and_ATM30Bbis	AAGTTCACTGGTCTATGAAC	CCTCTTTAAGATGTATTTACAAAG	11	chr11:108159637-108159911	275bp	IVS29-67-IVS30+81
ATM31A_and_ATM31B	GTGTATTTATTGTAGCCGAG	AGGAAGAACAGGATAGAAAG	11	chr11:108160266-108160582	317bp	IVS30-63-IVS31+54
ATM32A_and_ATM32B	AACAAAAGGACTTCTGAATG	AACACTCAAATCCTTCTAAC	11	chr11:108163260-108163590	331bp	IVS31-86-IVS32+70
ATM33A_and_ATM33B	CAGTAAGTTTTGTTGGCTTAC	TTACTGCTAGAGCATTACAG	11	chr11:108163969-108164286	318bp	IVS32-71-IVS33+82
ATM34A_and_ATM34B	TCAGTGTCTATAAATGGCAC	AATGAAACCAAGAGCAAGAC	11	chr11:108165550-108165869	320bp	IVS33-104-IVS34+83
ATM35A_and_ATM35B	CAAAAGTGTTGTCTTCATGC	ATCTGTCCTATATGTGATCC	11	chr11:108167960-108168171	212bp	IVS34-54-IVS35+62
ATM36A_and_ATM36B	TCGGCCTTAAGGTTAATTCTTG	GGACAAAGCTTTAGTTACTGAG	11	chr11:108170318-108170711	394bp	IVS35-123-IVS36+99
ATM37A_and_ATM37B	GGAGGTTAACATTCATCAAG	GCAATTTAACAGTCATGACC	11	chr11:108172309-108172599	291bp	IVS36-66-IVS37+83
ATM38A_and_ATM38B	TGTGTAGGAAAGGTACAATG	GCACTCTTTAGATAAACAGG	11	chr11:108173515-108173817	303bp	IVS37-65-IVS38+61
ATM39A_and_ATM39B	CAGTATGTTGAGTTTATGGCAG	AGTGGGATTCCATCTTAAATCC	11	chr11:108175260-108175655	396bp	IVS38-142-IVS39+76
ATM40A_and_ATM40B	AAGGAAGAAGGTGTGTAAGC	GCCGATAGTTAACAAGTTAC	11	chr11:108178563-108178773	211bp	IVS39-61-IVS40+62

(continued)

Table 1
(continued)

Primer pair name	Primer sequence 1 (5′>3′)	Primer sequence 2 (5′>3′)	chr	genomic position hg19 UCSC In-Silico PCR	Amplicon size	Targeted region for primer choice
ATM41A_and_ATM41B	CAGTCACTACCATTGTATTC	TACCCTTATTGAGACAATGC	11	chr11:108180795-108181103	309bp	IVS40-92-IVS41+61
ATM42A_and_ATM42B	GTATATGTATTCAGGAGCTTCC	GCATCTGTACAGTGTCTATAAC	11	chr11:108183072-108183306	235bp	IVS41-66-IVS42+81
ATM43Anew_and_ATM43Bnew	TAGAGTTGGGAGTTACATATTG	TTTACACACACATAACTCCTTC	11	chr11:108186446-108186704	259bp	IVS42-104-IVS43+66
ATM44A_and_ATM44B	AAACAACGGTATAGTAATTCTG	AACATACTGAAATAACCTCAGC	11	chr11:108186656-108186905	250bp	IVS43-82-IVS44+65
ATM45A_and_ATM45B	CTTTGCTGTTTTTTCTCTGG	TGTTGTTTAGAATGAGGAGAG	11	chr11:108188024-108188304	281bp	IVS44-76-IVS45+56
ATM46A_and_ATM46B	CATGTATATCTTAGGGTTCTG	CAAGTTTTCAGAAAAGAAGCC	11	chr11:108190623-108190852	230bp	IVS45-58-IVS46+67
ATM47A_and_ATM47B	ATTTCCCTGAAAACCTCTTC	ATTGGTAACAGAAAAGCTGC	11	chr11:108191972-108192201	230bp	IVS46-56-IVS47+54
ATM48A_and_ATM48B	CATTTCTCTTGCTTACATGAAC	TAATAAAAGGAAAGTCAAGAGG	11	chr11:108195996-108196323	328bp	IVS47-41-IVS48+52
ATM49Abis_and_ATM49Bbis	TCAATGAATGGTAGTTGCTG	TAAATTACTAATTTCAAGGCTC	11	chr11:108196688-108197065	378bp	IVS48-97-IVS49+113
ATM50A_and_ATM50B	TACATGAAGGGCAGTTGGGT	CCCTAGAGACTATACAGAGC	11	chr11:108198303-108198635	333bp	IVS49-69-IVS50+150
ATM51A_and_ATM51B	GTGTATTACCTTAATTTGAGTG	TTAGAGCTCAAAACTATATCAG	11	chr11:108199666-108200076	411bp	IVS50-81-IVS51+111
ATM52A_and_ATM52B	TTCCCTGGGATAAAAACCCAAC	GGTATACACGATTCCTGACATC	11	chr11:108200828-108201231	404bp	IVS51-113-IVS52+83

ATM53A_and_ATM53B	AACTTACTTGCTTAGATGTG	GTATTTCCATTTCTTAGAGG	11	chr11:108202098-108202386	289bp	IVS52-73-IVS53+102
ATM54AEX54A_and_ATM54Bnew	ACATGCAGGCATACACGCTCT	CCAGCCTTGAACCGATTTTAGATGG	11	chr11:108202474-108202879	406bp	IVS53-132-IVS54+115
ATM55Anew_and_ATM55Bnew	TCCATTGTCTAGATTTGTGC	ATCTCTACAGAGAGTAACAC	11	chr11:108203377-108203685	309bp	IVS54-112-IVS55+58
ATM56A_and_ATM56B	TTGACCTTCAATGCTGTTC	TTGAGTATTTTAGGGCTTGG	11	chr11:108204555-108204871	317bp	IVS55-58-IVS56+176
ATM57Anew_and_ATM57Bnew	ATCACATCGTCATTTGTTTCTC	TTCTACTCTACAAATCTTCCTC	11	chr11:108205580-108205972	393bp	IVS56-116-IVS57+136
ATM58A_and_ATM58B	TTGCTATTCTCAGATGACTCTG	CCAAAGCATTATGAATATGGGC	11	chr11:108206495-108206783	289bp	IVS57-77-IVS58+95
ATM59A_and_ATM59B	GCTGAATGATCATCAAATGC	AGCTGTCAGCTTTAATAAGC	11	chr11:108213881-108214163	283bp	IVS58-68-IVS59+65
ATM60A_and_ATM60B	TTATTGCCCCTATATCTGTC	CAGAATGTAGAAAAAGTGCTG	11	chr11:108216370-108216723	354bp	IVS59-100-IVS60+88
ATM61Anew_and_ATM61Bnew	ACCAAGTCAGTGGTCTTAATTG	CAAACAACATTCCATGATGACC	11	chr11:108217915-108218158	244bp	IVS60-91-IVS61+66
ATM62A_and_ATM62B	GGTAACATGTGGTTTCTTGCC	CCCAGCCCATGTAATTTTGAC	11	chr11:108224392-108224667	276bp	IVS61-101-IVS62+60
ATM63A_and_ATM63B	ATAGGCTCAGCATACTACAC	TGACGAGATACACAGTCTAC	11	chr11:108225440-108225664	225bp	IVS62-98-IVS63+63
ATM64A_and_ATM64B	CTGGTTCTACTGTTTCTAAG	TGAACAGTTTAAAGGCCTTGG	11	chr11:108235736-108236019	284bp	IVS63-73-IVS64+74
ATM65A_and_ATM65B	ATTCCCAAGGCCTTTAAACTG	GGCAGGTTAAAAATAAAGGC	11	chr11:108235995-108236306	312bp	IVS64-57-3'UTR+71

7. Multichannel pipet.
8. Thermal cycler for PCR.

2.3 Purification of PCR Products

1. 1.5% Agarose supplemented with Syber Safe 1×.
2. Purification plate (NucleoFast, Macherey-Nagel).
3. Vacuum Pump.

2.4 Sequencing

1. BigDye Terminator v1.1 Cycle Sequencing Kit.
2. 3.2 μM Forward or Reverse sequencing primers (Table 1).

2.5 Large-Sized Deletion Detection by MLPA

1. SALSA MLPA P041-A2 ATM/P042-A2 ATM probemix (MRC-Holland).
2. SALSA MLPA EK5 reagent kit (MRC-Holland).
3. Hi-Di™ formamide.
4. Thermal cyclers for PCR.
5. 96-well Microplate "Microamp Optical."
6. 0.2 mL PCR tubes.
7. Size standard HD500 (ThermoFisher Scientific).
8. Capillary sequencer.

2.6 Lymphoblastoid Cell Lines Culture

1. RPMI 1640 medium supplemented with 10% decomplemented fetal bovine serum (FBS) and 1% Penicillin-Streptomycin. Store at 4 °C.
2. 15 or 50 mL conical centrifuge tubes.
3. Cell culture flasks with filter caps.
4. Pipettes.
5. Filter tips.
6. Pipettors.
7. 5% CO_2 tissue culture incubator.
8. Centrifuge.
9. Laminar flow hoods (PSM type II).
10. Cell counter.

2.7 Protein Extraction, Quantification, Western Blotting, and Immunodetection

1. Ionizing radiation apparatus (γ-ray, Cesium-137).
2. Camptothecin.
3. Protein extraction buffer RIPA: 50 mM Tris–HCl, pH 7.5, 150 mM NaCl, 1 mM EDTA, 1% NP40, 0.25% sodium deoxycholate, 0.1% sodium dodecyl sulfate (SDS), supplemented with NaF, Na_3VO_4, phenylmethanesulfonylfluoride (PMSF), and a cocktail of protease inhibitors (cOmplete©, Roche).
4. Vortex.

5. Refrigerated microcentrifuge.
6. BCA Protein Assay Kit with Albumin Standard.
7. 0.2 mL 8-strip PCR Tubes with domed caps.
8. 37 °C incubator.
9. UV/Visible spectrophotometer.
10. 500 mM dithiothreitol (DTT).
11. LDS Sample Loading Buffer (4×).
12. Precast polyacrylamide gradient (4–15%) gels.
13. Gel Saver gel loading tips.
14. Molecular weight markers.
15. Tris/Glycine/SDS electrophoresis running buffer (25 mM Tris, 192 mM glycine, 0.1% SDS).
16. Electrophoresis apparatus and electrophoresis power supply.
17. iBlot® Dry Blotting transfer apparatus.
18. iBlot® transfer stack with nitrocellulose membrane.
19. Ponceau S staining solution.
20. TBS-0.1% Tween buffer.
21. Odyssey blocking buffer (LI-COR Biosciences).
22. Primary antibodies: rabbit anti-ATM (Novus Biologicals; # NB110-55475), rabbit anti-phospho-CHK2 (Thr68), mouse anti-CHK2 (Cell Signaling Technology; #2661 and #3440, respectively), rabbit anti-phospho-KAP1 (Ser824), and anti-KAP1 (Bethyl laboratories; #A300-767A and #A300-275A, respectively), rabbit anti-phospho-p53 (Ser15; Abcam; #ab1431), mouse anti-p53 (NeoMarkers; #MS-187-P0), and mouse anti-β-actin (Sigma; #A2228).
23. Roller mixer.
24. Secondary antibodies anti-mouse and anti-rabbit IRDYE 800CW/680LT (LI-COR Biosciences).
25. Odyssey Quantitative Fluorescence Imaging System (LI-COR Biosciences).

2.8 Cytoplasmic and Nuclear Protein Extraction

1. NE-PER kit Nuclear and Cytoplasmic extraction reagents (Thermo Scientific Pierce).
2. NaF 1 M stock (100×) solution.
3. Na_3VO_4 200 mM stock (100×) solution.
4. Phenylmethanesulfonylfluoride (PMSF) 100 mM stock (100×) solution in isopropanol.
5. Cocktail of protease inhibitors.
6. Vortex.
7. Refrigerated microcentrifuge.

3 Methods

3.1 DNA Extraction

Genomic DNA is extracted using the whole blood DNA isolation kit (QuickGene, Fujifilm) and extraction system (QuickGene-610 L, Fujifilm).

1. Reconstitute protease at room temperature 30–40 min before use.
2. Put 300 μL protease in a 15 mL tube.
3. Add 2 mL of homogenized whole blood.
4. Add 2.5 mL of lysis buffer.
5. Homogenize with 10 up-down and then vortex for 15 s.
6. Incubate for 10 min at 56 °C.
7. Add 2.5 mL 100% ethanol.
8. Homogenize (10 up-downs).
9. Vortex 15 s.
10. Put the lysate in the extraction column.
11. Place the extraction column in the extraction system and start isolation program (12 min).
12. Quantify DNA concentration using the spectrophotometer and dilute at 50 ng/μL.
13. Store DNA at −20 °C.

3.2 PCR Amplification, Purification of PCR Products and Direct Sequencing

3.2.1 Primer Plating

All ATM fragments are amplified with the same PCR conditions. Primers plating allows direct sequencing automation. Due to the large size of ATM, amplicons are studied in two steps, first the 32 most commonly mutated amplicons and then the 29 least commonly mutated amplicons (Fig. 1). Sequencing is performed on one DNA strand and identified mutations are checked on the other strand.

PCR amplification: Pool 20 μM forward and reverse primers (Table 1) in a 96 plate as follows:

Direct sequencing: Pool 3.2 μM forward and reverse primers in a 96 plate as follows:
Plates can be stored at −20 °C.

3.2.2 PCR

All ATM fragments are amplified under standard conditions: 1.5 mM $MgCl_2$ in 2× Taq. All the steps must be performed on ice.

1. For single DNA sample, prepare 32 × 200 μL microtubes for **step 1** and 29 for **step 2**.
2. Prepare the mix (Fig. 2a) without the DNA in the first place.
3. Add 0.8 μL of pooled primers (20 μM) in each microtube using a multichannel pipet.

a

Step 1:			
ATM04+05 A+B	ATM24 A+B	ATM54 A+Bnew	ATM10 A+B
ATM06 A+B	ATM27 A+B	ATM55 Anew+Bnew	ATM15 A+B
ATM07 A+B	ATM28 A+B	ATM57 A+B	ATM16 A+B
ATM09 A+B	ATM39 A+B	ATM58 A+B	ATM22 A+B
ATM12 A+B	ATM41 A+B	ATM59 A+B	ATM26 A+B
ATM18 A+B	ATM48 A+B	ATM60 A+B	ATM31 A+B
ATM20 A+B	ATM52 A+B	ATM62 A+B	ATM46 A+B
ATM21 A+B	ATM64 A+B	ATM65 A+B	ATM56 A+B

Step 2:			
ATM08 Anew+B	ATM11 A+B	ATM36 A+B	ATM49 Abis+Bbis
ATM13 A+B	ATM19 A+B	ATM37 A+B	ATM51 A+B
ATM14 A+B	ATM23 A+B	ATM40 A+B	ATM53 A+B
ATM17 A+B	ATM25 A+B	ATM42 A+B	ATM61 Anew+Bnew
ATM29 A+B	ATM30 A+B	ATM43 Anew+Bnew	ATM63 A+B
ATM35 A+B	ATM32 A+B	ATM44 A+B	
ATM38 A+B	ATM33 A+B	ATM45 A+B	
ATM50 A+B	ATM34 A+B	ATM47 A+B	

b

Step 1:			
ATM04+05 B	ATM24 B	ATM54 A	ATM10 A
ATM06 A	ATM27 B	ATM55 Anew	ATM15 A
ATM07 A	ATM28 A	ATM57 A	ATM16 A
ATM09 A	ATM39 A	ATM58 A	ATM22 B
ATM12 B	ATM41 A	ATM59 A	ATM26 B
ATM18 B	ATM48 A	ATM60 A	ATM31 A
ATM20 B	ATM52 A	ATM62 A	ATM46 B
ATM21 B	ATM64 A	ATM65 B	ATM56 B

Fig. 1 (**a**) PCR amplification. (**b**) Direct sequencing

Step 2:			
ATM08 B	ATM11 B	ATM36 A	ATM49 Bbis
ATM13 A	ATM19 A	ATM37 A	ATM51 A
ATM14 B	ATM23 B	ATM40 A	ATM53 A
ATM17 A	ATM25 A	ATM42 A	ATM61 Anew
ATM29 A	ATM30 A	ATM43 Anew	ATM63 A
ATM35 A	ATM32 A	ATM44 A	ATM04+05 A
ATM38 A	ATM33 A	ATM45 A	
ATM50 A	ATM34 A	ATM47 A	

Fig. 1 (continued)

a

Component	Stock concentration	Final concentration	Volume (final volume: 25μL)
H_2O	-	-	629
Buffer + $MgCl_2$	10X + 15 mM	1X + 1.5 mM	85
dNTP	10 mM	0.8 mM	68
Taq	5 U/μL	0.05 U/μL	8.5
Add DNA at the end			
DNA	25ng/μL		68
Distribute 24μL			

b

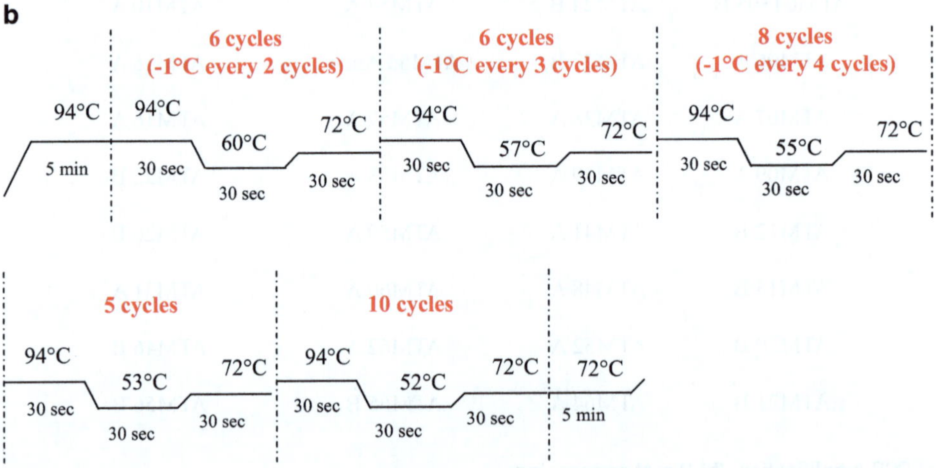

Fig. 2 (a) Standard mix preparation for 34 reactions with 2× Taq. **(b)** Touchdown PCR program prior to sequencing

4. Add the volume of DNA indicated in the mix (25 ng/μL).
5. Vortex.
6. Distribute 24 μL of mix.
   The PCR program is a Touchdown TD60 as shown in Fig. 2b.
7. Check amplification by loading a fraction of the PCR product on an agarose 1.5% gel supplemented with Syber Safe 1×.
8. Purify PCR products with purification plate (NucleoFast, Macherey-Nagel) on a vacuum pump (10 min at 22.5 mm Hg).
9. Resuspend purified PCR products in 25 μL H_2O.
10. Sequence the purified PCR products with the BigDye Terminator v1.1 Cycle Sequencing Kit. Mix 4 μL of Big Dye Terminator V1.1, 0.5 μL of 3.2 μM Primer and 1.5 to 5.5 μL* (*depending on the quantification) of purified PCR product. Adjust to 10 μL with water if necessary. The cycling conditions are as follows: Cycle for 25 cycles of 96 °C for 10 s, 50 °C for 5 s, and 60 °C for 2 min followed by incubation at 4 °C.
11. Sanger sequencing analysis is performed on Sequence Scanner v2.0 software. Results are analyzed with the SeqScape® software and sequences are printed with the FinchTV software.
12. **Mutation analysis**
    Mutation pathogenicity is assessed with the average Grantham difference, Align GVGD and the PolyPhen and SIFT scores, all included in the routinely used package Alamut (http://www.interactive-biosoftware.com/alamut-visual/). Effects on splicing are predicted using MaxEntScan (http://genes.mit.edu/burgelab/maxent/Xmaxentscan_scoreseq.html), NNSPLICE (http://omictools.com/nnsplice-s10697.html), and GeneSplicer (http://www.cbcb.umd.edu/software/GeneSplicer/gene_spl.shtml) prediction software.
13. Molecular validations for splicing alteration are performed on cDNA using puromycin (*see* **Note 1**). Such validations are particularly useful to identify deep intronic mutations and to validate splicing alteration predicted by Alamut.

3.3 Large-Sized ATM Rearrangement Detection by MLPA

Detection of rearrangements is performed with MLPA P041-A2 ATM/P042-A2 ATM probemix kits (MRC-Holland). Volumes are given for one reaction (*see* **Note 2**).

3.3.1 Denaturing

1. Put 200 ng of DNA in each well of a microplate 96-well "Microamp Optical" (4 μL of DNA diluted at 50 ng/μL) and complete with 1 μL of TE 10:1.
2. Centrifuge the microplate 5 s at 120 × *g*.
3. Denature the samples for 20 min at 98 °C using a thermal cycler.

4. Cool the reaction down to 25 °C before opening the lid.
5. Put the plate at room temperature.
6. Spin-down the plate.

3.3.2 Hybridization

1. Prepare Hybridization Master Mix: 1.5 μL MLPA buffer +1.5 μL probe mix per sample (vortex each stock tube before opening).
2. Add 3 μL of Hybridization Master Mix.
3. Centrifuge the plate 5 s at 1000 rpm.
4. Incubate for 1 min at 95 °C.
5. Incubate for 16–20 h at 60 °C.

3.3.3 Ligation

1. Prepare Ligase Master Mix: 3 μL of ligase-65 buffer A + 3 μL of ligase-65 buffer B + 25 μL H_2O. Vortex.
2. Add 1 μL of ligase-65 to the Ligase Master Mix. Mix by pipetting.
3. Stop the hybridization program (16–20 h at 60 °C) and cool the reaction down to 54 °C.
4. Put the plate in the thermal cycler.
5. Distribute 32 μL of Ligase Master Mix per well using a multichannel pipet. Mix by pipetting (avoid bubbles).
6. Incubate for 15 min at 54 °C and then for 5 min at 98 °C (enzyme inactivation). Leave at room temperature.

3.3.4 PCR

1. Warm the polymerase for 10 s by hand.
2. Prepare the Polymerase Master Mix on ice: 7.5 μL H_2O + 2 μL SALSA-PCR primer +0.5 μL SALSA polymerase.
3. Vortex.
4. Add 10 μL of Polymerase Master Mix at room temperature using automatic pipet.
5. Put immediately on the thermal cycler pre-warmed at 60 °C. The cycling conditions are as follows: Cycle for 35 cycles of 95 °C for 30 s, 60 °C for 30 s, and 72 °C for 60 s followed by 72 °C for 20 min and incubation at 15 °C.
6. Genotype the PCR products: migration of 2 μL the amplified fragments on a capillary sequencer, using the size standard HD500.
7. Analyze with the GeneMapper software (*see* **Note 3**).

3.4 Lymphoblastoid Cell Lines (LCL) Culture

We found the utilization of Epstein-Barr immortalized LCL (*see* **Note 4**) particularly useful to demonstrate the pathogenicity of mutations for complex diagnosis (unknown variant, deep intronic mutations). Feed the cells with fresh medium every 3–4 days to a cell density of 0.8×10^6 cells/mL and split into new cultures as required.

3.5 Protein Extraction, Quantification, Western Blotting, and Immunodetection

1. Put 5 × 10^6 cells in 15 mL tubes and use a polystyrene box to transport them to the ionizing radiation apparatus.
2. Irradiate cells (5 Gy) and incubate for 45 min at 37 °C.
3. Centrifuge the cells at 500 × *g* for 5 min at 4 °C. Discard the supernatant.
4. Wash twice in ice-cold TBS. Resuspend the dry pellet in 60 µL of RIPA extraction buffer. Vortex. Keep on ice for 10 min. Centrifuge the lysed cells at 15,000 × *g* for 15 min at 4 °C in a microcentrifuge.
5. Collect supernatant in prechilled tubes. Keep at −20 °C for short-term uses and −80 °C for long-term uses.
6. Prepare bovine serum albumin (BSA) standards: 25, 125, 250, 500, 750, 1000, 1500, 2000 µg/mL.
7. Prepare 100 µL of ice-cold BCA reagent (1 µL of component B for 50 µL of component A in BCA Protein Assay Kit) per sample to quantify.
8. Add 5 µL of proteins for samples and 1 µL of BSA for the standard curve.
9. Mix and incubate for 10–15 min at 37 °C.
10. Measure OD with a spectrophotometer.
11. Estimate the protein concentration.
12. Denature 50 µg of protein supplemented with 50 mM DTT and LDS loading buffer 4× by boiling them for 5 min at 95 °C. Vortex, spin down, and load on the polyacrylamide gradient (4–15%) gel.
13. Run the gel for 45 min at 180 V, in Tris/Glycine/SDS running buffer.
14. Transfer the gel using iBlot transfer stack on the iBlot transfer apparatus for 13 min on program 3.
15. Incubate the membrane on Ponceau S staining solution for 5 min for quality control. Wash 3 min in TBS-0.1% Tween buffer.
16. Block for 45 min in Odyssey blocking buffer at room temperature on a roller mixer.
17. Incubate overnight at 4 °C on a roller mixer with primary antibodies diluted at 1/1000 in Odyssey blocking buffer supplemented with Tween 0.1%.
18. The membrane is probed with:
    - Day 1: anti-phospho-KAP1, anti-phospho-CHK2, anti-phospho-p53, anti-ATM.
    - Day 2: anti-KAP1, anti-CHK2, anti-p53, and anti-β-actin (Fig. 3).
19. Wash three times for 5 min with TBS-0.1% Tween buffer.

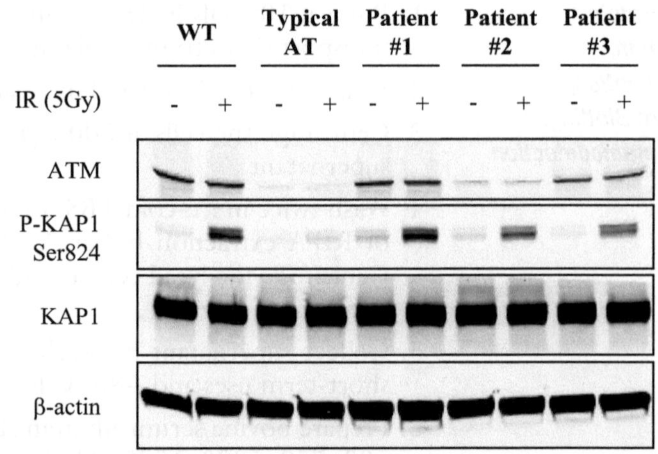

Fig. 3 Functional ATM analysis. Western blot analysis after ionizing radiation (IR) of lymphoblastoid cell lines from control (WT), typical ataxia telangiectasia patient (Typical AT), and 3 variant AT patients #1, 2, and 3

20. Incubate for 1 h at room temperature on a roller mixer with IRDYE 800CW secondary antibodies diluted at 1/10000 in Odyssey blocking buffer supplemented with 0.1% Tween.
21. Wash three times for 5 min with TBS-0.1% Tween buffer.
22. Dry the membrane with Whatman paper.
23. Scan the membrane and quantify band intensities using Odyssey Quantitative Fluorescence Imaging System.

3.6 Cytoplasmic and Nuclear Protein Extraction

Cytoplasmic-nuclear protein extraction is performed based on an improved version of the NE-PER protocol.

1. Centrifuge 5×10^6 lymphoblastoid cell lines at $500 \times g$ for 5 min.
2. Wash cells gently with TBS.
3. Centrifuge at $500 \times g$ for 3 min.
4. Discard supernatant and let the pellet as dry as possible.
5. Add 50 μL of ice-cold CER I buffer, supplemented with protease and phosphatase inhibitors, to the cell pellet.
6. Vortex the tube vigorously at maximum speed for 15 s.
7. Incubate the tube on ice for 10 min.
8. Add 2.75 μL of ice-cold CER II buffer to the tube.
9. Vortex the tube vigorously at maximum speed for 5 s.
10. Incubate the tube on ice for 1 min.
11. Vortex the tube vigorously at maximum speed for 5 s.

12. Centrifuge at maximum speed (15,000 × *g*) for 5 min at 4 °C.
13. Immediately transfer the supernatant (cytoplasmic extract) to a prechilled tube (*see* **Note 5**) and place it at −80 °C or use immediately.
14. Suspend vigorously the insoluble pellet fraction (nuclei) in 25 μL of ice-cold NER buffer, supplemented with protease and phosphatase inhibitors.
15. Vortex the tube vigorously at maximum speed for 1 min.
16. Place the tube on ice for 10 min and repeat the 1 min vortex every 10 min for 1 h.
17. Centrifuge at maximum speed (15,000 × *g*) for 10 min at 4 °C.
18. Immediately transfer the supernatant (nuclear extract) to a prechilled tube and place it at −80 °C or use immediately.

4 Notes

1. Puromycin treatment inhibits translation enabling the analysis of transcripts (with premature stop codons for example) that would be otherwise degraded by the nonsense-mediated mRNA decay (NMD) pathway. Puromycin treatment will determine the potential effects of a mutation on splicing errors by increasing the level of these splicing errors.
2. For quality controls, amplify also wild-type, mutated and negative samples.
3. Genemapper analysis: A sizing curve is generated on GeneMapper using the standard fragments (calibration). The sizes of the other fragments are then determined by linear regression fit.
4. Other laboratories traditionally work on primary fibroblast cell lines established from skin biopsies. The determination of ATM protein alterations cannot be done on Peripheral Blood Lymphocytes as its level is very low in resting lymphocytes.
5. Retrieve the maximum part of the supernatant but leave a few drops in the first place. After transferring to a prechilled tube, discard the last drops and let the pellet as dry as possible.

Acknowledgments

G.R. is supported by grants from Ligue Nationale Contre le Cancer and Fondation pour la Recherche Médicale n°FDT20130928274. This work was supported in part by a grant of the Fondation AT EUROPE. We thank Genethon for access to cell lines, and the Institut Curie sequencing facility.

References

1. Savitsky K, Bar-Shira A, Gilad S, Rotman G, Ziv Y, Vanagaite L, Tagle DA, Smith S, Uziel T, Sfez S, Ashkenazi M, Pecker I, Frydman M, Harnik R, Patanjali SR, Simmons A, Clines GA, Sartiel A, Gatti RA, Chessa L, Sanal O, Lavin MF, Jaspers NG, Taylor AM, Arlett CF, Miki T, Weissman SM, Lovett M, Collins FS, Shiloh Y (1995) A single ataxia telangiectasia gene with a product similar to PI-3 kinase. Science 268(5218):1749–1753
2. Savitsky K, Sfez S, Tagle DA, Ziv Y, Sartiel A, Collins FS, Shiloh Y, Rotman G (1995) The complete sequence of the coding region of the ATM gene reveals similarity to cell cycle regulators in different species. Hum Mol Genet 4(11):2025–2032
3. Mitui M, Campbell C, Coutinho G, Sun X, Lai CH, Thorstenson Y, Castellvi-Bel S, Fernandez L, Monros E, Carvalho BT, Porras O, Fontan G, Gatti RA (2003) Independent mutational events are rare in the ATM gene: haplotype prescreening enhances mutation detection rate. Hum Mutat 22(1):43–50. doi:10.1002/humu.10232
4. Telatar M, Teraoka S, Wang Z, Chun HH, Liang T, Castellvi-Bel S, Udar N, Borresen-Dale AL, Chessa L, Bernatowska-Matuszkiewicz E, Porras O, Watanabe M, Junker A, Concannon P, Gatti RA (1998) Ataxia-telangiectasia: identification and detection of founder-effect mutations in the ATM gene in ethnic populations. Am J Hum Genet 62(1):86–97. doi:10.1086/301673
5. Mitui M, Bernatowska E, Pietrucha B, Piotrowska-Jastrzebska J, Eng L, Nahas S, Teraoka S, Sholty G, Purayidom A, Concannon P, Gatti RA (2005) ATM gene founder haplotypes and associated mutations in Polish families with ataxia-telangiectasia. Ann Hum Genet 69(Pt 6):657–664. doi:10.1111/j.1529-8817.2005.00199.x
6. Cavalieri S, Funaro A, Porcedda P, Turinetto V, Migone N, Gatti RA, Brusco A (2006) ATM mutations in Italian families with ataxia telangiectasia include two distinct large genomic deletions. Hum Mutat 27(10):1061. doi:10.1002/humu.9454
7. Meneret A, Ahmar-Beaugendre Y, Rieunier G, Mahlaoui N, Gaymard B, Apartis E, Tranchant C, Rivaud-Pechoux S, Degos B, Benyahia B, Suarez F, Maisonobe T, Koenig M, Durr A, Stern MH, Dubois d'Enghien C, Fischer A, Vidailhet M, Stoppa-Lyonnet D, Grabli D, Anheim M (2014) The pleiotropic movement disorders phenotype of adult ataxia-telangiectasia. Neurology 83(12):1087–1095. doi:10.1212/WNL.0000000000000794
8. Jacquemin V, Rieunier G, Jacob S, Bellanger D, d'Enghien CD, Lauge A, Stoppa-Lyonnet D, Stern MH (2012) Underexpression and abnormal localization of ATM products in ataxia telangiectasia patients bearing ATM missense mutations. Eur J Hum Genet 20(3):305–312. doi:10.1038/ejhg.2011.196

Chapter 4

An HTRF® Assay for the Protein Kinase ATM

Phillip Adams, Jonathan Clark, Simon Hawdon, Jennifer Hill, and Andrew Plater

Abstract

Ataxia telangiectasia mutated (ATM) is a serine/threonine kinase that plays a key role in the regulation of DNA damage pathways and checkpoint arrest. In recent years, there has been growing interest in ATM as a therapeutic target due to its association with cancer cell survival following genotoxic stress such as radio- and chemotherapy. Large-scale targeted drug screening campaigns have been hampered, however, by technical issues associated with the production of sufficient quantities of purified ATM and the availability of a suitable high-throughput assay. Using a purified, functionally active recombinant ATM and one of its physiological substrates, p53, we have developed an in vitro FRET-based activity assay that is suitable for high-throughput drug screening.

Key words ATM, Kinase, Therapeutic, Cancer, p53, FRET, Assay

1 Introduction

Ataxia telangiectasia mutated (ATM) is a member of the phosphoinositide 3-kinase (PI3K)-related protein kinase (PIKK) family that includes DNA-dependent protein kinase catalytic subunit (DNA-PKc), ataxia telangiectasia and rad3-related (ATR), mammalian target of rapamycin (mTOR) and suppressor of morphogenesis in genitalia (SMG1). Through orchestrated phosphorylation events, these enzymes control a host of cell processes including cell survival, proliferation, metabolism, differentiation, motility, nonsense-mediated mRNA decay, and the cellular DNA damage response (DDR) [1, 2]. A sixth member of the family, transformation/transcription-associated protein (TRRAP), is a component of various histone acetyltransferase enzymes, but has no kinase activity [2].

Specific mutations in the ATM gene have been found to be the cause of the rare autosomal-recessive disorder ataxia–telangiectasia (A–T), which is characterized by cerebellar degeneration, immunodeficiency, and an increased risk of cancer [3]. Cells from A-T

Sergei V. Kozlov (ed.), *ATM Kinase: Methods and Protocols*, Methods in Molecular Biology, vol. 1599,
DOI 10.1007/978-1-4939-6955-5_4, © Springer Science+Business Media LLC 2017

patients have defects in DDR pathways, hypersensitivity to ionizing radiation (IR), and an increased frequency of chromosome breakage [4]. Their responses to base-damaging agents and ultraviolet radiation are normal, however, indicating that ATM is activated specifically in response to IR and double-stranded DNA breaks (DSBs). Genotoxic chemical stress, ionizing radiation and normal DNA replication errors are all causes of DSBs and result in the activation of ATM, which responds by phosphorylating a number of proteins involved in cell cycle checkpoint arrest, DNA repair, and apoptosis. One such target protein, the tumor-suppressor p53, plays a critical role in regulating cell cycle progression or apoptosis after DNA damage. Following DSBs, p53 is phosphorylated by ATM and subsequently orchestrates a myriad of pleiotropic processes to repair cellular damage or promote cell death [5]. One crucial role of p53 is the activation of the G1/S cell cycle checkpoint through up-regulation of the CDK inhibitor p21, which ensures that DNA replication and cell division is halted in the presence of DSBs [6].

Disruptions in DDR pathways are common in many tumor types, and although the perturbation of these mechanisms gives cancer cells a survival and growth advantage, it leaves them potentially vulnerable to attack on remaining DNA repair mechanisms [7]. In light of this, there is a growing interest in DDR regulators as targets for drug discovery and a number of studies have shown that inhibition of specific DDR elements can sensitize cancer cells to genotoxic stress [8, 9]. ATM was first recognized as an attractive target for therapeutic intervention due to the hypersensitivity of A-T cells to IR. Since then, therapeutics designed to disable ATM during chemo- or radiotherapy have been investigated [10]. Generating sufficient quantities of purified enzyme for extensive drug discovery purposes, however, has proven difficult and until very recently, no suitable purified protein has been available for large-scale in vitro screening projects. To address this, we have developed a recombinant protein production process in a human cell line that generates significant amounts of a 352 kDa, full-length, human ATM that is capable of phosphorylating its native substrate p53. A recombinant full-length p53 protein expressed in *E. coli* provides an appropriate substrate for kinase activity and these purified proteins have been used to develop a high-throughput ATM assay for drug discovery purposes.

Here, we present an in vitro FRET-based kinase assay measuring phosphorylation of p53 on Ser-15 by ATM (*see* **Notes 1–4**). Figure 1 shows the architecture formed when the HTRF antibodies are added to the stopped reaction.

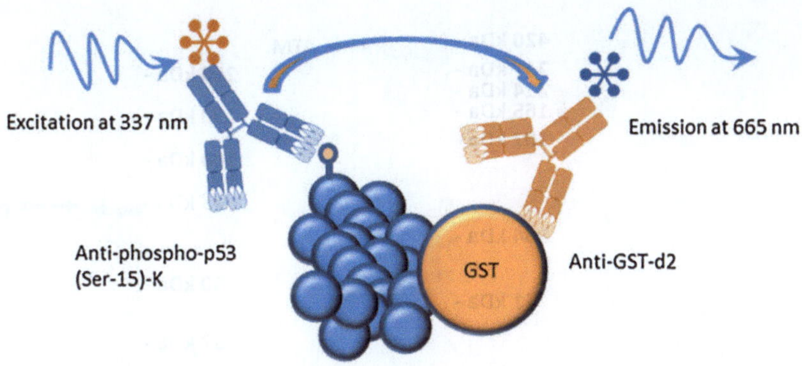

Fig. 1 Schematic of the HTRF architecture. The HTRF signal is generated when the phosphorylated p53 product recruits the donor and acceptor antibodies, bringing them into close proximity. Excitation of the donor antibody (Europium-labeled anti-phospho-53) at 337 nm results in resonance energy transfer, allowing detection of emitted light at 665 nm from the anti-GST-d2 antibody

2 Materials

1. Recombinant purified ATM kinase (Eurofins catalogue number 14-933, aliquot and store at −70 °C; Fig. 2).
2. p53 substrate (GST-Myc-p53-(hu,FL), (Eurofins catalogue number 14-952, aliquot and store at −70 °C; Fig. 2).
3. 100 mM Magnesium acetate stock in water (store refrigerated).
4. 100 mM Manganese chloride stock in water (store refrigerated).
5. 1 M Dithiothreitol (DTT) stock in water (store frozen).
6. 100 mg/mL Bovine serum albumin (BSA) prepared in water. This is a 10% (w/v) solution, and can be stored in the −20 °C freezer.
7. 500 mM EDTA (store refrigerated) (*see* **Note 5**).
8. 10 mM Adenosine 5′-triphosphate (ATP) prepared in 50 mM HEPES pH 7.4 (store frozen at −20 °C) (*see* **Note 6**).
9. Reaction buffer: 25 mM HEPES pH 8.0, 0.01% Brij-35, 1% glycerol (can be stored refrigerated).
10. Enzyme dilution buffer: 25 mM HEPES pH 8.0, 0.01% Brij-35, 1% glycerol, 5 mM DTT, 1 mg/mL BSA. The DTT and BSA are added fresh. *See* **step 1** of Subheading 3.1 for the recipe.
11. Stop solution: 12.5 mM MOPS pH 8.0, 0.005% Brij-35, 0.5% glycerol, 250 mM EDTA (This is a 50:50 mix of reaction buffer and 500 mM EDTA).

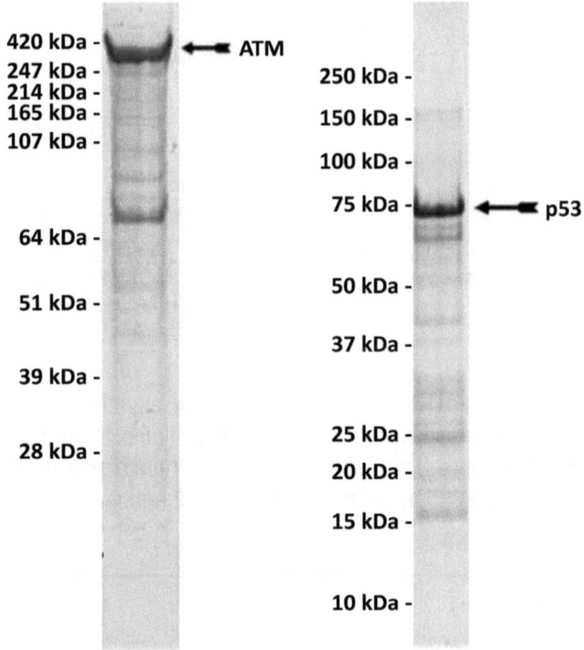

Fig. 2 Purified Recombinant ATM and p53. The purified ATM enzyme and p53 substrate were analyzed by SDS-PAGE and coomassie blue staining. The recombinant ATM was recovered at approximately 70% purity with a yield of 1–2 mg/L. A major contaminant migrating at approximately 70 kDa on the gel was identified as Hsp70 by mass spectrometry. The recombinant p53 is recovered at approximately 75% purity with a yield of 1 mg/L

12. Antibody dilution buffer (4×): 200 mM HEPES pH 7.0, 600 mM NaCl, 0.4% sodium cholate, 0.04% Tween 20, 0.05% sodium azide (store refrigerated).
13. Anti-phospho-p53 (Ser-15)-K (CisBio 61P08KAE): supplied by Cisbio at a concentration of 0.1 mg/mL. Dilute in the antibody dilution buffer to 0.840 μg/mL (8.4 μL of antibody, 250 μL antibody dilution buffer, 741.6 μL water). This can be stored at −70 °C.
14. Anti-GST-d2 (CisBio 61GSTDLA): supplied by Cisbio at a concentration of 0.25 mg/mL. Dilute in the antibody dilution buffer to 50 μg/mL (200 μL of antibody, 250 μL antibody dilution buffer, 550 μL water). This can be stored at −70 °C.
15. Detection buffer: 50 mM HEPES pH 7.0, 150 mM NaCl, 267 mM KF, 0.1% sodium cholate, 0.01% Tween 20, 0.0125% sodium azide (store refrigerated).
16. Detection solution: Detection buffer containing anti-phospho-p53 (Ser-15)-K (CisBio 61P08KAE) and anti-GST-d2 (CisBio 61GSTDLA). The final concentrations of antibodies in the stopped assay are 0.014 μg/mL and 0.833 μg/mL for

anti-phospho-p53 (Ser-15)-K and anti-GST-d2 respectively. This solution should be prepared on the day of assay—*see* **step 4** of Subheading 3.1 for the preparation.

17. 4× ATP solution: (X μM ATP (concentration as required), 20 mM manganese chloride, 20 mM magnesium acetate). This solution should be prepared on the day of assay—*see* **step 3** of Subheading 3.1 for the preparation.
18. Microtitre plates (black 384-well, e.g., Corning Costar 3573).
19. Polypropylene tubes for preparing working stocks of reagents and reaction mixes.
20. Microplate reader fitted with optics appropriate for reading HTRF®.

3 Methods

The volumes added per well detailed below are suitable for a 384-well plate using a 20 μL reaction volume (30 μL stopped volume).

All working solutions should be prepared in advance and stored on ice until needed. The dilution buffer is readily prepared from the reaction buffer with the addition of BSA and DTT. It is recommended that the ATP solution is prepared from its constituent parts (ATP, magnesium acetate, and manganese chloride) on the day of assay.

Freeze-thaw cycles of the kinase should be minimized. Thawing of the kinase should be rapid (vial held in hand until thawed), and refreezing should be achieved by snap-freezing in liquid nitrogen. To reduce the number of freeze-thaw cycles, the enzyme should be aliquoted following the initial thawing.

3.1 Preparation of Working Solutions

1. Prepare the enzyme dilution buffer by adding 5 μL of 1 M DTT and 10 μL 10% (w/v) BSA per mL of reaction buffer. For a single 384-well plate, 1 mL of enzyme in dilution buffer is required.
2. Prepare the stop solution by mixing equal volumes of reaction buffer and 500 mM EDTA.
3. Prepare the 4× ATP solution by adding 500 μL each of 100 mM magnesium acetate and 100 mM manganese chloride to 1.49 mL of deionized water. Add 10 μL of 10 mM ATP. This creates 2.5 mL of 400 μM ATP, 20 mM manganese chloride, and 20 mM magnesium acetate. For other concentrations of ATP, adjust the volume of water and ATP to suit. The ATP solution makes up 25% of the total assay volume, so 4× final assay concentrations are required. Titrations of ATP should be prepared in a solution of 20 mM manganese chloride and

20 mM magnesium acetate (this is to keep the concentration of divalent metal ions constant).

4. Prepare the detection solution by mixing the detection buffer with the anti-GST-d2 and anti- phospho-p53 (Ser-15)-K antibodies. To prepare 1 mL of detection solution, add 100 μL of each antibody solution (as prepared in Subheading 2, **steps 13** and **14**) to 800 μL of detection buffer. This reagent should be prepared on the day of assay. This is a 6× solution. 5 μL of this solution is added to the stopped assay.
5. Dilute the enzyme in dilution buffer to the desired concentration(s). The enzyme solution makes up 12.5% of the total assay volume, so 8× final assay concentrations need to be prepared. Titrations of the enzyme should be prepared in the enzyme dilution buffer.
6. Dilute the p53 substrate in reaction buffer to the desired concentration(s). The substrate solution makes up 12.5% of the total assay volume, so 8× final assay concentrations need to be prepared. Titrations of p53 should be prepared in reaction buffer.

3.2 Assay Procedure

1. Add 10 μL of reaction buffer per assay to all wells to be used.
2. Add 2.5 μL of p53 substrate diluted in reaction buffer to all wells.
3. Add 2.5 μL (typically 0–50 ng) of ATM prepared in enzyme dilution buffer. For blanks, either add 2.5 μL enzyme dilution buffer, or add Stop solution before the addition of ATP (*see* **Note** 7).
4. Add 5 μL of the 4× ATP mixture to all wells to initiate the reaction (*see* **Note 8**).
5. Incubate for 30 min at room temperature.
6. Add 5 μL of the stop solution to all wells to terminate the assay.
7. Add 5 μL of the detection solution to all wells.
8. Following the termination of the reaction and addition of the detection solution, it is recommended that the plate is sealed and read after an overnight incubation in the fridge.
9. Measure HTRF® ratio on an appropriate microplate reader according to the following parameters:

Excitation	337 nm
Emission	665 and 620 nm
Counting delay	50 μs
Integration time	400 μs
No. flashes per well	200

The above settings are suitable for a BMG PHERAstar. Refer to your instrument manufacturer for further guidance on measurement parameters recommended for HTRF®.

HTRF® ratio is calculated as follows:

$$HTRF\ Ratio = \left(\frac{Emission\ at\ 665nm}{Emission\ at\ 620nm}\right) \times 10{,}000$$

The ATM, ATP, and substrate should be titrated to determine the optimal concentrations required for subsequent experiments (e.g., testing potential inhibitors) (*see* **Note 9**). For the purposes of screening potential inhibitors, the concentration of ATM chosen should be on the upslope of the reaction velocity curve (the linear phase). At higher enzyme concentrations, the signal in the assay does not change. This could be due to the higher enzyme concentrations taking the assay out of the initial linear range (e.g., by substrate depletion) or due to the limits of the dynamic range of detection of the assay being reached.

Once the appropriate concentrations have been determined, the preparation of a reaction mix containing ATM and substrate in buffer facilitates the assay build, which can be reduced to fewer steps. In this case, ATM is diluted to 8× the desired final assay concentration in dilution buffer, and then mixed with the substrate diluted in reaction buffer (i.e. the solutions referred to in **steps 1–3** in the assay procedure are mixed to give a single solution). This solution is then added to all test wells (15 μL per well) before starting the assay with the addition of 5 μL ATP (**step 4** of the assay procedure).

Figures 3, 4 and 5 show titrations of ATM, p53, and ATP respectively using this assay format. Titrations of enzyme and substrate define the basic working concentrations required for assay applications, for example screening compounds. These titrations should mimic the conditions of the assay to be used in terms of incubation time and temperature. A compound screen conducted at 10 μM ATP should be preceded by an enzyme titration at that concentration. Figure 6a, b show the assay used to screen some commonly used protein and lipid kinase inhibitors [11–14]. Figure 7 shows the dose-response for some of the lipid kinase inhibitors, including the ATM inhibitor KU55933 [16] (*see* **Note 10**). Table 1 shows the EC_{50} values determined using the HTRF® assay and data generated using an orthogonal ELISA (*see* **Note 3**), alongside those determined by other laboratories [13].

4 Notes

1. The choice of assay for any given enzyme is determined by a variety of factors, and these will include cost, availability of reagents, personal preference (i.e. what the operator has experience with), and whether or not equipment is available for the readout. Different assay platforms have different advantages and disadvantages, whether they suffer from compound

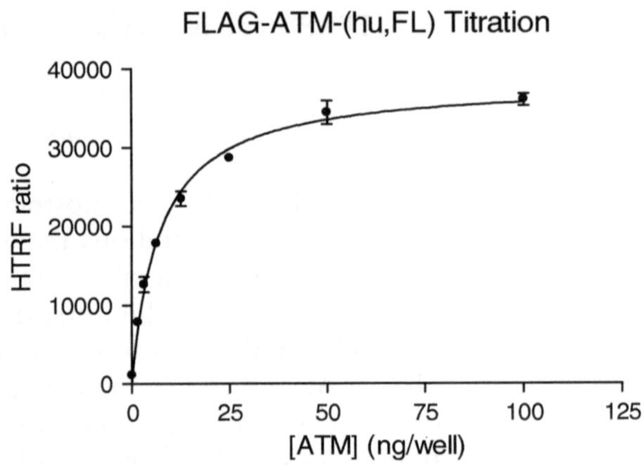

Fig. 3 ATM Titration. ATM was titrated and tested in the HTRF® assay using 30 nM p53 substrate and 100 μM ATP. The assay was conducted for 30 min at room temperature before stopping and addition of detection solution. The plate was sealed and read following an overnight incubation in the fridge. The data were fitted to the Michaelis-Menten equation using GraphPad Prism®. The concentration of ATM chosen for subsequent experiments was 20 ng/well. This is on the upslope of the velocity curve (i.e., part of the linear phase). At higher enzyme concentrations the signal in the assay does not change. This could be due to the higher enzyme concentrations taking the assay out of the initial linear range (e.g., by substrate depletion) or due to the limits of the dynamic range of detection of the assay being reached. For these reasons, it is important that the enzyme is titrated under the assay conditions to be employed in downstream experiments, and a concentration selected which is on the upslope (linear phase) of the reaction velocity curve (*see* **Note 9**)

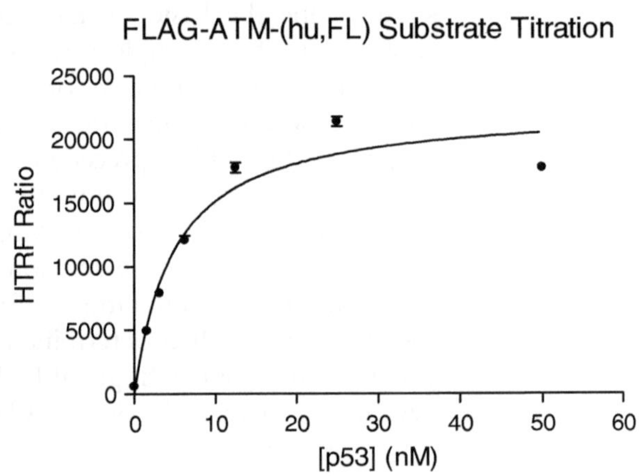

Fig. 4 p53 Titration. The substrate p53 was titrated at fixed ATM (20 ng per test point) and 100 μM ATP. The assay was conducted for 30 min at room temperature before stopping and the addition of detection solution. The plate was sealed and read following an overnight incubation in the fridge. The data were fitted to the Michaelis-Menten equation using GraphPad Prism®. From this, the concentration of p53 chosen for screening potential inhibitors was 30 nM (*see* **Note 11**). This concentration is on the plateau (saturating concentration), which is appropriate when screening for ATP competitors

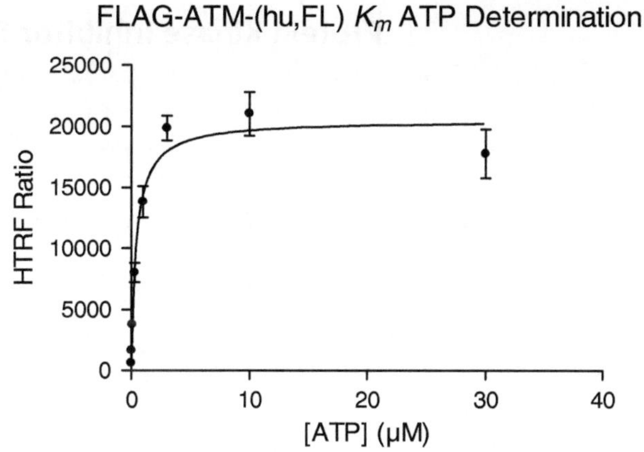

Fig. 5 ATP Titration. ATP was titrated to estimate the apparent *Km*, using a fixed ATM concentration (20 ng per test point) and 30 nM p53 substrate. The assay was conducted for 30 min at room temperature before stopping and the addition of detection solution. The plate was sealed and read following an overnight incubation in the fridge. The data were fitted to the Michaelis-Menten equation using GraphPad Prism®. The apparent *Km* for ATP under these conditions is 0.5 μM

interference, or are low throughput. While the homogeneous HTRF® assay was our end-point for ATM, the assay development began with a western blot using an unconjugated form of the antibody, coupled to an anti-mouse-HRP detection. The antibody used for the detection in this ATM assay is against the phosphorylated Ser-15 of p53, and is available as a europium conjugate from CisBio, but is also available from Cell Signalling Technologies (catalogue number 9286) in an unconjugated form. The western blot approach reduced complexity to allow us to verify that we did indeed have an active ATM, and that it was phosphorylating p53 as expected. One additional important point here is that we were also able to verify that the p53 we were using did not have endogenous phosphorylation at the site of interest. The version of p53 used here is produced in a bacterial host. Other p53 constructs we trialed were produced in insect cells, and did have low levels of endogenous phosphorylation at this site. While these preparations did work, there was a slightly higher background in the assays. The western blot was a useful tool in the early stages of development and remains a good trouble-shooting aid.

2. The assay can also be set up as an ELISA, using an anti-Myc antibody (Merck-Millipore 05-724) to coat the wells of an ELISA plate. This format was also employed before the HTRF® assay was developed fully, and can be used as a hit confirmation assay, or in instances where compound may interfere strongly in the HTRF® format, or where laboratories do not have access to HTRF® readers.

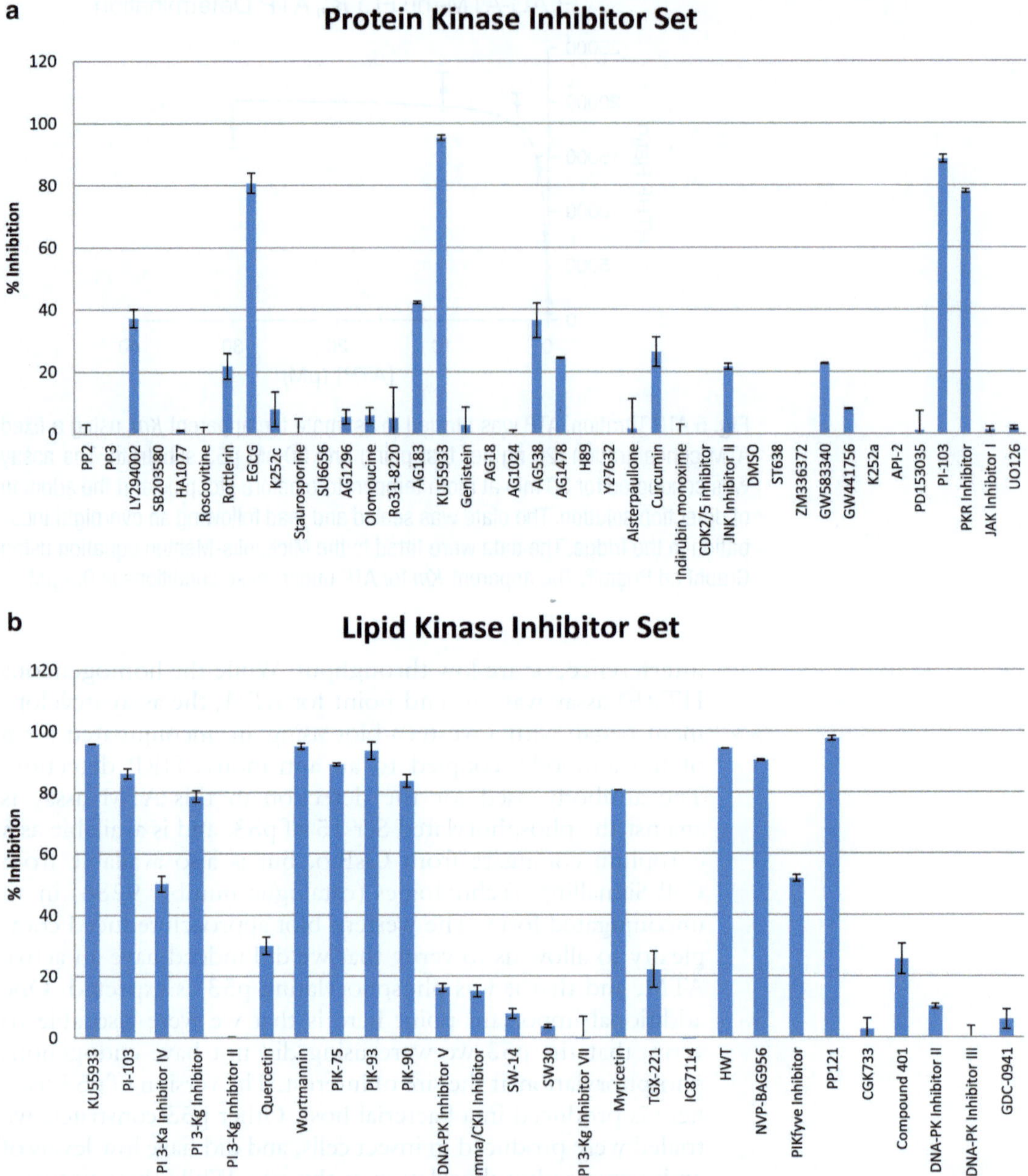

Fig. 6 Reference Inhibitor Screening. A set of commonly used protein kinase inhibitors (*panel a*) and a set of commonly used lipid kinase inhibitors (*panel b*) were tested in the ATM HTRF® assay. Assays were conducted as described in the methods using 20 ng of ATM, 30 nM p53, and 10 μM ATP (*see* **Note 13**)

3. In the ELISA, the reaction and dilution buffers are the same as those used in the HTRF® assay. ATM, p53, and ATP should be titrated as noted above. Assays may be conducted according to the steps for the HTRF® format, but once stopped (**step 6** of the

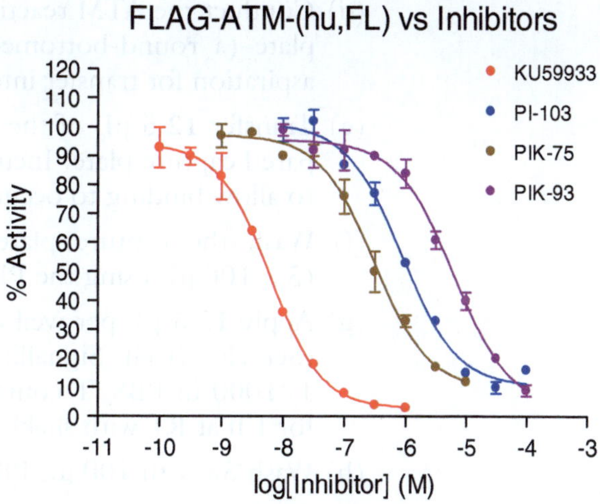

Fig. 7 EC_{50} determinations. Selected inhibitors from the single concentration screen shown in Fig. 7 were tested in dose-response, using the same assay conditions as employed previously. Data were converted to % activity by comparison to the control (uninhibited reaction) and plotted against the log (M) concentration of inhibitor (*see* **Note 10**)

Table 1
EC_{50} values

	EC_{50} (nM)		
Compound	ELISA	HTRF	Lit.
KU-55933	7	6	5
PI-103	451	667	920
PIK-75	113	184	490
PIK-93	3769	3543	2300

The dose-response curves shown in Fig. 7 were generated by fitting the inhibition data to a four-parameter logistic using GraphPad Prism®. The EC_{50} values determined from the fit, in nanomolar, are shown.

above assay procedure) the protocols deviate. The ATM reactions are conducted in a microplate separate to the capture plate.

(a) Coat a suitable microplate (e.g. Corning Costar 3693) with anti-Myc antibody (diluted 1/250 in phosphate-buffered saline (PBS)). For the half-area white plate, use 25 μL per well to coat. Allow at least 1 h to coat, but preferably seal the plate and leave overnight in the fridge.

(b) Discard the antibody solution and block with 100 μL 10% (w/v) BSA in PBS. Shake for 1 h.

(c) Wash the capture plate three times with PBS containing 0.01% (v/v) Tween 20 (PBS/T).

(d) Conduct the ATM reactions required in a separate microplate (a round-bottomed polypropylene plate facilitates aspiration for transfer into the capture plate).

(e) Transfer 12.5 μL of the stopped reaction to the pre-prepared capture plate. Incubate, with shaking, for 1 h at RT to allow binding to occur.

(f) Wash the capture plate to remove unbound reaction (3 × 100 μL using the PBS/T solution).

(g) Apply 12.5 μL per well of biotinylated anti-phospho-p53 (Ser-15) (Cell Signalling Technologies 4030) diluted 1/1000 in PBS/T containing 10% (w/v) BSA. Incubate for 1 h at RT with shaking.

(h) Wash 3× with 100 μL PBS/T.

(i) Apply streptavidin conjugated horse-radish peroxidase (GE Healthcare RPN1231, diluted 1/5000 in PBS/T containing 10% (w/v) BSA). Incubate for 30 min at room temperature.

(j) Wash 3× with 100 μL PBS/T.

(k) Add 20 μL per well of SuperSignal ELISA Pico Chemiluminescent Substrate (Thermo Scientific 37070) and read on a luminometer.

The volumes stated are for using a half-area 96-well plate, so scale-up may be required if a standard well ELISA plate is used. A 2 min. spin at 1000 rpm (164 × *g*) after each addition of coating antibody, stopped reaction, primary antibody, and streptavidin-HRP is recommended to ensure solutions are at the bottom of the well.

4. The assay described here is suitable for other PIKK enzymes (ATR and DNA-PK) with some minor modifications to the divalent cations used. For DNA-PK, use only 10 mM magnesium acetate as the divalent counter-ion for ATP, and for ATR, use only 10 mM manganese chloride.

5. To prepare EDTA to a concentration of 500 mM, add the EDTA to stirred water in small amounts, bringing the pH back up to 8.0 using sodium hydroxide to facilitate the dissolution. Once the EDTA is dissolved adjust volume with water to create the 500 mM stock.

6. The quality of reagents is clearly of importance to any assay. However, their preparation and shelf-life is also to be considered. When preparing stocks of ATP, it is important to buffer the solution, and to avoid extremes of pH during the preparation. Long-term storage in the fridge is to be avoided, but the solutions may be stored in the freezer at −20 °C. These measures are to avoid hydrolysis of the tri-phosphate into di- or monophosphates. Manganese chloride stocks should also be replaced frequently.

7. The inhibitor experiments described above all use a pre-stop to generate the blank signal. This is simply the stop solution being added before the ATP. For operational reasons, a compound blank could be used instead. This would allow the same reaction mix to be added to all wells and would simplify the start and stop steps. In a titration of enzyme, 0 enzyme as a blank is the preferred option.
8. During the assay build process, it can be beneficial to spin the assay plates to ensure that all the reagents are mixed together at the bottom of the well. This can help to improve replicate data. A spin of 1000 rpm (164 × *g*) for 2 min is sufficient to do this. Spin the plate after the addition of reaction mix containing enzyme, substrate and buffer, after the addition of ATP, and after the addition of stop and detection solutions.
9. Different preparations of recombinant proteins may have different concentrations, purities, and activities. Any new preparation should be titrated prior to its use in order to verify activity, and to determine optimal concentrations for the assay.
10. The availability of a characterized inhibitor for ATM was invaluable. The KuDOS Pharmaceuticals molecule KU55933 [15] has been tested by a number of groups against ATM, is commercially available, and is a good tool to use when establishing the ATM assay. This compound performed comparably in the HTRF® and ELISA formats. It should be noted that different assay technologies may generate different responses from the same compound set for the same target [16, 17].
11. The concentration of p53 in the HTRF® assay is critical to the assay: too much reduces the signal notably (hook effect). Problems with low signal should therefore include p53 concentration as part of the investigation. The western blot and ELISA can tolerate a wider range of p53 concentrations.
12. The quality or robustness of the assay can be assessed using assay variability ratios such as the Z′ measurement [18]. This assay routinely generates Z′ scores of >0.7. The Z′ can be calculated as follows:

$$1 - \frac{3x\left(SD\left(Blank\right) + SD\left(Ctrl\right)\right)}{\left(mean\left(Ctrl\right) - mean\left(Blank\right)\right)}$$

where SD is the standard deviation, Ctrl is the control (uninhibited signal), and Blank is the signal from the pre-stopped wells.

13. The compound data that is shown here uses compounds at 50× final assay concentration in 100% DMSO. 0.4 μL compound is added to the test wells, using a TTP Labtech Mosquito. Volume from the reaction buffer is sacrificed to

maintain the 20 μL assay volume. The differences in final buffer composition due to this are small, and not considered to be an issue. The final DMSO concentration is 2%.

14. The phosphorylation state of ATM is linked to its activity [19]. The recombinant form produced in the mammalian expression system used here has been shown by western blot to have some phosphorylation on Ser-1981.

References

1. Meyn MS (1995) Ataxia-telangiectasia and cellular responses to DNA damage. Cancer Res 55:5991–6001
2. Lempiäinen H, Halazonetis TD (2009) Emerging common themes in regulation of PIKKs and PI3Ks. EMBO J 28:3067–3073
3. Swift M, Morrell D, Massey RB, Chase CL (1991) Incidence of cancer in 161 families affected by ataxia-telangiectasia. N Engl J Med 325:1831–1836
4. Shiloh Y, Ziv Y (2013) The ATM protein kinase: regulating the cellular response to genotoxic stress, and more. Nat Rev Mol Cell Biol 14:197–210
5. Shiloh Y (2003) ATM and related protein kinases: safeguarding genome integrity. Nat Rev Cancer 3:155–168
6. Kastan MB, Lim DS (2000) The many substrates and functions of ATM. Nat Rev Mol Cell Biol 1:179–186
7. Weber AM, Ryan AJ (2015) ATM and ATR as therapeutic targets in cancer. Pharmacol Ther 149:124–138
8. Zhao Y, Thomas HD, Batey MA, Cowell IG, Richardson CJ, Griffin RJ, Calvert AH, Newell DR, Smith GC, Curtin NJ (2006) Preclinical evaluation of a potent novel DNA-dependent protein kinase inhibitor NU7441. Cancer Res 66:5354–5362
9. Rainey MD, Charlton ME, Stanton RV, Kastan MB (2008) Transient inhibition of ATM kinase is sufficient to enhance cellular sensitivity to ionizing radiation. Cancer Res 68:7466–7474
10. Cremona CA, Behrens A (2014) ATM signalling and cancer. Oncogene 33:3351–3360
11. Davies SP, Reddy H, Caivano M, Cohen P (2000) Specificity and mechanism of action of some commonly used protein kinase inhibitors. Biochem J 351:95–105
12. Bain J, McLauchlan H, Elliott M, Cohen P (2003) The specificities of protein kinase inhibitors: an update. Biochem J 371:199–204
13. Knight ZA, Gonzalez B, Feldman ME, Zunder ER, Goldenberg DD, Williams O, Loewith R, Stokoe D, Balla A, Toth B, Balla T, Weiss WA, Williams RL, Shokat KM (2006) A pharmacological map of the PI3-K family defines a role for p110alpha in insulin signaling. Cell 125: 733–747
14. Bain J, Plater L, Elliott M, Shpiro N, Hastie CJ, McLauchlan H, Klevernic I, Arthur JS, Alessi DR, Cohen P (2007) The selectivity of protein kinase inhibitors: a further update. Biochem J 408:297–315
15. Hickson I, Zhao Y, Richardson CJ, Green SJ, Martin NM, Orr AI, Reaper PM, Jackson SP, Curtin NJ, Smith GC (2004) Identification and characterization of a novel and specific inhibitor of the ataxia-telangiectasia mutated kinase ATM. Cancer Res 64:9152–9159
16. Wu X, Sills MA, Zhang JH (2005) Further comparison of primary hit identification by different assay technologies and effects of assay measurement variability. J Biomol Screen 10:581–589
17. Sills MA, Weiss D, Pham Q, Schweitzer R, Wu X, Wu JJ (2002) Comparison of assay technologies for a tyrosine kinase assay generates different results in high throughput screening. J Biomol Screen 7:191–214
18. Zhang JH, Chung TD (1999) Oldenburg KR (1999) A simple statistical parameter for use in evaluation and validation of high throughput screening assays. J Biomol Screen 4:67–73
19. Bakkenist CJ, Kastan MB (2003) DNA damage activates ATM through intermolecular autophosphorylation and dimer dissociation. Nature 421:499–506

Chapter 5

ATM Kinase Inhibitors: HTS Cellular Imaging Assay Using Cellomics™ ArrayScan VTI Platform

Catherine Bardelle and Joanna Boros

Abstract

Small molecule inhibitors of the ATM pathway could represent a promising opportunity for cancer therapy, working either by enhancing the clinical efficacy of radiotherapy and existing chemotherapies or by synthetic lethality-based mechanisms. In this chapter, we describe a high-throughput, high-content imaging assay monitoring levels of ATM phosphorylation at Serine 1981 following induction of DNA damage by ionizing radiation.

Key words DNA damage, High content, Cellular assay, High-throughput screen, Ionizing radiation, ATM phosphorylation, Kinase inhibitors

1 Introduction

High-content screening or HCS has increased significantly in the last decade as quite a few integrated instrumentation platforms were released. HCS refers to any method where simultaneous acquisition of several independent parameters is used to analyze whole cells or components of cells. The ability to study many features at once is what gives HCS such a great power and makes it an attractive approach. HCS can be used throughout the drug discovery process from the identification or validation of relevant drug targets all the way to the analysis of clinical data. Often, compounds identified by other approaches can then be clustered into cellular phenotypes based on known drugs with fully characterized mode of actions. One of the technical challenges associated with HCS is the storage of the sheer size of the collected data running in the terabytes level.

HCS can be used to study many cellular processes such as protein expression, protein phosphorylation, cell cycle, receptor ligand binding, or cell proliferation. The strength here is that HCS enables both target-based and phenotypic approaches. Protein translocation, changes in cellular protein levels, posttranslational modifications,

Sergei V. Kozlov (ed.), *ATM Kinase: Methods and Protocols*, Methods in Molecular Biology, vol. 1599,
DOI 10.1007/978-1-4939-6955-5_5, © Springer Science+Business Media LLC 2017

cells and organelles morphology changes, etc. can be visualized. HCS can provide valuable insights into toxicity and mode of action, therefore enabling the selection of the most promising compounds from a screen. HCS-based strategy has become an integral member of the variety of approaches available for an HTS campaign.

Ataxia-telangiectasia mutated (ATM) is one of the major regulators of the DNA-damage response (DDR) and plays a key role in regulating the cellular response to ionizing radiation [1–3]. ATM, together with ATM and RAD3-related (ATR) and DNA-dependent protein kinase (DNA-PK), belongs to a closely related protein family known as the phosphoinositide-3-kinase-related protein kinases (PIKKs) [4–6]. They all have a very similar domain architecture and a high sequence homology, especially in the C-terminal kinase, hence the strong preference for phosphorylation of Ser and Thr residues that are followed by Gln. Even though the PIKKs act in distinct pathways, and are activated by different types of DNA damage (ATR primarily through single strand breaks and collapse of replication forks, whereas ATM and DNA-PK upon double strand breaks) they share many downstream substrates such as CHK1, CHK2, p53, γH2AX and KAP1.

ATM was first mentioned in 1995, showing that mutations in the ATM gene are responsible for the multisystem disorder ataxia-telangiectasia (A-T), characterized by neurodegeneration, telangiectasia, immunodeficiency, predisposition to cancer, and sensitivity to ionizing radiation (IR) [7]. It was mostly this hypersensitivity of ATM-defective cells to IR that put ATM in the spotlight as a therapeutic target for cancer therapy.

First studies aimed to test the effects of ATM inhibition on radiosensitivity have mainly used relatively nonspecific PIKK and PI3K inhibitors: wortmannin and caffeine [8, 9]. In the last decade, however, more specific inhibitors have been identified. The first selective inhibitor, KU55933, was identified through an in vitro kinase assay screen by KuDOS [10]. Soon after an analog was developed, KU60019, which was tenfold more effective than KU-55933 at blocking radiation-induced phosphorylation of key ATM targets in human glioma cells [11]. However, both of those inhibitors have not been tested in clinical trials due to either poor bio-availability or high lipophilicity. Recently, another ATM inhibitor from the same class of compounds, KU-59403, has been described [12]. It has even better potency over KU-55933, good solubility, and bio-availability, and it is the first compound to show, without major toxicity, significant chemosensitization in in vivo models of human cancer. CP-466722 is another available transient ATM inhibitor for research use, which was also identified through an in vitro kinase assay [13].

Currently, there are no ATM inhibitors in clinical development. However, the data provided through various in vitro experiments shows great potential of having such inhibitors. In addition, continuous improvements in chemical structures and high-throughput screens of chemical libraries led to the identification in the last decade of specific ATM inhibitors. Since all currently available ATM inhibitors were discovered through in vitro kinase assays, development of different types of screening assays, especially cellular assays, is important for further development of inhibitors with improved properties for in vivo and clinical use.

There are many commercial kits measuring the activity of ATM downsteam targets (phospho-CHK2, phospho-p53, phospho-4E-BP1, γH2AX). These are available in a variety of formats such as ELISA, SureFire®, flow cytometry, reporter assays and can be adapted for high-throughput and drug discovery applications. Most published imaging assays for DNA Damage and Repair (DDR) follow histone H2AX phosphorylation on Ser139 (γH2AX). γH2AX is a sensitive marker for DNA double strand breaks, but it is not selective for ATM activation only. Upon DSBs formation one or more of the PI3K-like kinases, such as ATM, ATR, or DNA-PK, are activated and phosphorylate H2AX among other relevant substrates. A high-content method was published [14] in 2011 looking at γH2AX foci formation following chemical-induced DNA damage to identify inhibitors of the DDR pathways. The assay described below is specific for ATM inhibition as it measures direct effects of inhibitors on the autophosphorylation site on Ser1981. Several publications showed that phosphorylation of Ser1981 is not DNA damage specific in itself in the ATM function [15, 16]; however, the combination with ionizing radiation, and therefore introduction of DNA damage, links it to DDR and gives it that specificity.

Protocols for HCS follow a very similar scheme and consist of four parts. First the assay plates have to be prepared. This involves plating cells, treatment with inhibitors, activators, and/or appropriate stimulus. Then cells are fixed and stained with a probe or antibody. The third step is image acquisition and finally image analysis. On the Arrayscan the images are acquired and analyzed simultaneously. Once good quality images are acquired, they are stored and can be reanalyzed with modified or new parameters. Furthermore, images from specific plates and wells can be retrieved and viewed in more depth. A detailed flow chart of the overall process is shown in Fig. 1.

2 Materials

1. Assay media: MEM, 10% fetal calf serum, 2 mM L-glutamine.
2. Dulbecco's PBS.
3. Fixative solution: 3.7% Formaldehyde in PBS.
4. Permeabilization Solution: 0.1% Triton- X100 in PBS.

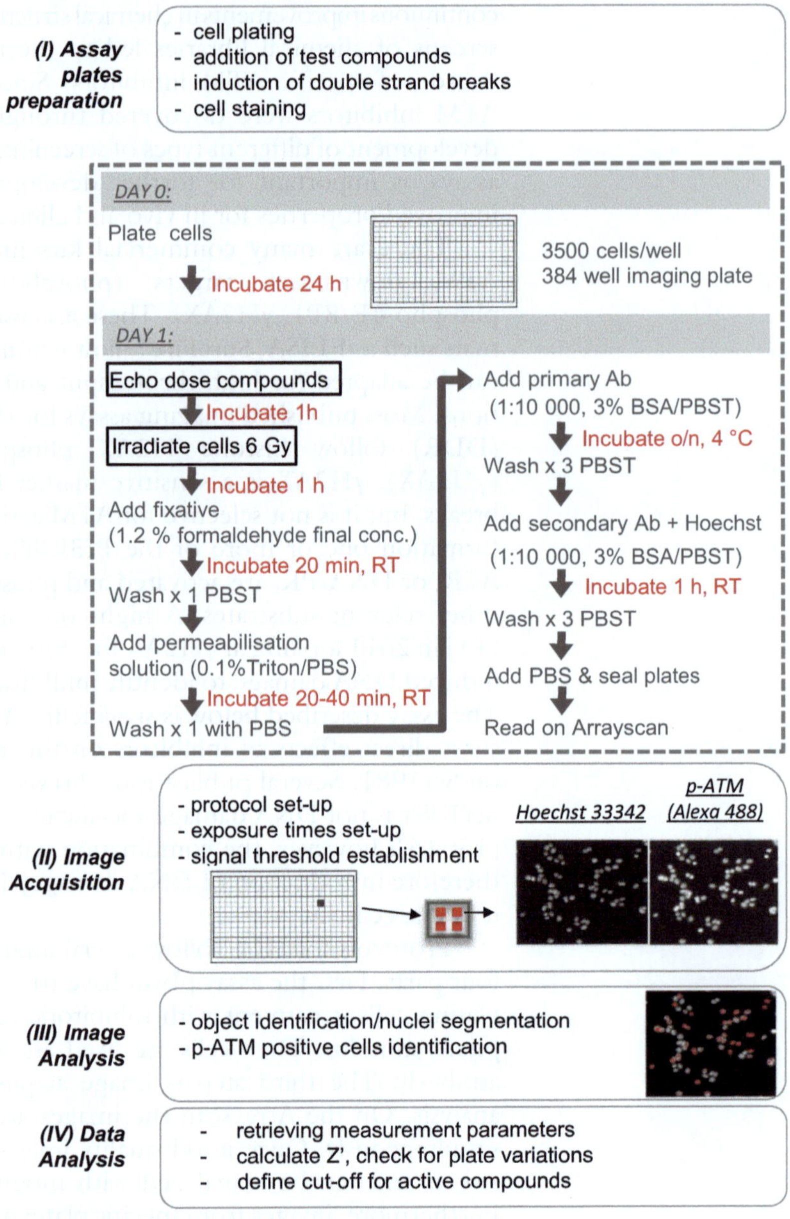

Fig. 1 Schematic representation of the assay

5. Antibody buffer (primary and secondary antibodies): 3% BSA in PBS, 0.05% Tween-20.
6. Wash buffer: PBS, 0.05% Tween-20.
7. Primary antibody: ATM pSer1981 mouse monoclonal antibody, Clone 10H11. E12 (Merck Millipore).
8. Hoechst 33342 nucleic acid stain.
9. Secondary antibody: Alexa Fluor® 488 goat anti-mouse IgG (H + L) (Life technologies).

10. DMSO.
11. A suitable compound that can be used as an inhibitor control such as KU-60019 (Tocris).
12. Black plate seal (Thermo Scientific).
13. 384-well imaging tissue culture-treated plates (Costar #3712 or equivalent).
14. Echo source plates, Labcyte #P-05525-CV1 or equivalent.
15. Gamma cell 3000 irradiator (Best Theratronics). Alternatively, an X-Ray system such as the X-RAD320 (Precision X-Ray) can be used.
16. Biotek Microplate washer (ELx405) or other suitable plate washers.
17. Echo Liquid Handler (Labcyte).
18. Cellomics™ ArrayScan VTI HCS Reader or other imaging platforms such as CellInsight (ThermoScientific™).
19. Multidrop™ Combi dispenser or other liquid dispensers such as Wellmate® (ThermoScientific™).
20. Data analysis software such as Genedata Screener® (Genedata).

3 Methods

Here, we describe a high-content imaging assay to measure ATM activity that has been developed and used in HTS to identify ATM inhibitors [17]. In the assay we directly measure levels of ATM phosphorylation after induction of double strand breaks. The protocol presented below can be modified to further multiplex the assay and include other endpoints. Furthermore, the analysis of acquired images can be as well extended to provide a more comprehensive phenotypic characterization of cells and the effects of drugs on cell survival, proliferation, or apoptosis. This can be done by extracting additional data for the nuclear morphology and nuclear signal intensity. The method we describe here can be used to test compounds in both single concentration and dose response format.

3.1 Assay Plates Preparation

3.1.1 Cell Plating

The seeding and treatment protocol described below was optimized for HT29 cells in a 384-well plate format. For other cell lines the plating protocol as well as timing of the compound treatment will have to be determined empirically. For 96 or 1536-well plates, cell seeding density will have to be optimized to ensure cells are in logarithmic growth phase during compound treatment.

1. Aspirate media and wash cells once with PBS.
2. Add 2 mL pre-warmed Trypsin and incubate for 2–4 min at 37 °C (longer incubation times should be avoided due to the risk of cell damage and cell clumping).

3. Transfer cells to a sterile tube containing pre-warmed assay media and centrifuge for 3 min at 300 × *g*. Discard supernatant and gently resuspend cells in 10 mL of assay media.
4. Syringe the cells suspension 3× using a wide bore needle (18G) and count cells. This step ensures cells are well separated as this is crucial for image analysis using the cell identification algorithm.
5. Dilute cells to 8.75 × 10^4 cells/mL.
6. Dispense cells into 384-well assay plates at 40 μL per well (3500 cells/well) using a Multidrop combi dispenser (*see* **Note 1**).
7. Leave plates for ~10 min at room temperature before placing them in the 37 °C incubator. With some cell lines this step helps in having cells distributed more evenly across the well.
8. Place plates into a rotating incubator and incubate overnight at 37 °C, 5% CO_2 with high humidity (*see* **Note 2**).

3.2 Compounds Addition and Irradiation

1. All compounds are prepared at 10 mM concentration in DMSO in the Labcyte Echo source plates, and an Echo Liquid Handler (Labcyte) is used to dispense compounds (120 nl) into plates. The final DMSO concentration in the well is 0.3% (*see* **Note 3**).

   Include on every plate Max and Min controls to monitor plate-to-plate reproducibility. The average signal for Max and Min controls will also be used for calculating % effect. Hence, it is important, when designing your assay, to choose proper negative and positive controls to minimize the amount of false-positive and/or false-negative results. Use DMSO (final concentration 0.3%) as Max control and 10 μM KU-60019 inhibitor as Min control.
2. Once dosing is complete, incubate plates for 1 h at 37 °C with 5% CO_2.
3. Irradiate cells with 6 Gy and return plates to the incubator for 1 h for recovery.

3.3 Staining

All steps described below are adapted for automation and processing multiple plates simultaneously. Addition of all reagents is done using Multidrop Combi liquid dispensers and the washing/aspiration steps using BioTek Microplate washer. For lower throughput plate processing (especially 96-well plates) reagents can be added with a multichannel pipette and washes done manually. For 384-well plates, however, this is not recommended as it may result in high signal variability from well to well and high background due to insufficient washes.

1. After 1 h take plates out from the incubator and add 20 μL/well of fixative solution using a Multidrop Combi liquid dispenser in a fume hood (final formaldehyde concentration 1.2%) (*see* **Note 4**).

2. Incubate for 20 min at RT.
3. Aspirate the media/fixative solution using a BioTek Microplate washer and wash plates once with 80 μL/well of PBS (*see* **Note 5**).
4. Aspirate the PBS and add 20 μL/well of permeabilization solution and incubate for 20 min at RT.
5. Aspirate the permeabilization solution and wash plates once with 80 μl/well of PBS.
6. Dilute the ATM pSer1981 antibody 1/10000 in antibody buffer (3% BSA in PBS, 0.05% Tween) and add 20 μL/well.
7. Incubate plates with primary antibody overnight at 4 °C.
8. Wash plates three times with wash buffer using BioTek Microplate washer. After final wash aspirate the wash buffer.
9. Dilute the secondary antibody (Alexa Fluor® 488 goat anti-mouse IgG) 1/500 and Hoechst 33342-1/10000 in the antibody buffer (3% BSA in PBS + 0.05% Tween).
10. Add 20 μL/well of the antibody/Hoechst 33342 solution and incubate for 1 h at RT. Protect from direct light.
11. Wash plates three times with wash buffer using BioTek Microplate washer. After final wash aspirate the wash buffer.
12. Add 80 μL/well of PBS to all plates and cover with black plate seal.

    Make sure there are no air bubbles in the wells before sealing the plate.

    If needed store plates at 4 °C before imaging (*see* **Note 6**).

3.4 Image Acquisition

The Cellomics™ ArrayScan VTI HCS Reader is an automated epifluorescent microscope system designed for HCS, which enables real-time image acquisition and analysis. The instrument is designed to work with validated image analysis modules called- BioApplications. Those are validated algorithms that not only convert images into numeric data, but as well indicate which input parameters need to be defined to measure specific biological phenomenon (for the complete list of available BioApplications, please refer to the ThermoFisher™ Scientific High-Content Screening web page).

3.4.1 ArrayScan Assay Protocol Setup

The BioApplication used for this assay is called Compartmental Analysis, which is designed to measure intracellular signal intensity changes between different subcellular compartments. Channels and filter settings used for this experiment are shown in Fig. 2. For setting up image acquisition, algorithm and measurement parameters follow the procedure below:

1. Select the XF53 filter and 10× objective.
2. Use a 2-color setup to analyze Hoechst nuclear staining together with 488 nm secondary antibody.

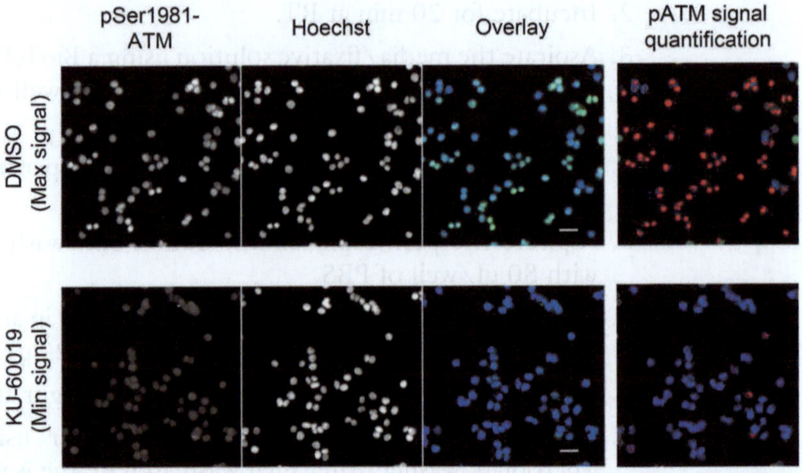

Fig. 2 Sample images for positive (DMSO-treated) and negative (10 μM Ku-60019-treated) phospho-ATM signal taken with the Cellomics™ ArrayScan VTI HCS Reader using 10 × objective. Starting from the left are the two single channel images and the overlay. In the overlay image Hoechst staining is depicted in *blue* and phospho-ATM in *green*, the *scale bars* indicate 80 μm. The *right panel* is the visual representation of the phospho-ATM signal quantification using the Compartmental Analysis algorithm and measuring the signal intensity above a set threshold inside the Circ area. *Red color* indicates the positive signal

Channel	1	2
Stain	Hoechst 33342	Alexa 488
Cell Parameter	Nuclear stain, cell count	Phospho-ATM intensity
Filter	XF93 Hoechst 33342	XF93 FITC
Excitation/Emission filter wavelengths	365±25 nm - 535±20 nm	475±20 nm - 535±20 nm

Fig. 3 Emission and excitation filter settings for the ArrayScan VTI Channel 1 and 2 (for nuclear stain and phospho-ATM signal detection respectively)

3. Scan using the Compartmental Analysis algorithm.
4. Parameters of interest:
   - MEAN_CircSpotAvgIntenCh2.
   - Valid Object Count.
5. Acquire four fields of view per well.
   - Exposures for each channel are established using wells with Max and Min phospho-ATM signals and then fixed for the entire batch (Fig. 3).

- Parameters to check:
  - Channel 1: Exposure, segmentation, threshold.
  - Channel 2: Exposure, threshold.

3.4.2 Basic Assay Parameters

- **Fixed Threshold Object identification methods for both channels:** The fixed threshold method sets an intensity threshold independently of the image data. When the threshold is set, only pixels with intensities above this threshold will be considered as belonging to these objects and retained for the analysis specific to this channel.
- **Background correction:** prior to image analysis, the noncellular background can be computed and subtracted from the image separately in each channel if desired. Setting the value to −1 lets the software decide on the value needed for optimal background correction.
- **Object segmentation:** this is a very important parameter and needs to be checked regularly during each plate batch run. Usually, the software algorithm can and will identify individual objects, but if the cells are densely packed or clumped then this might not be performed correctly. One will then need to test which value, positive or negative, is better for setting the correct nuclear mask. Negative value segments the nuclei based on the peak intensity method, while a positive value uses the geometric one. Too small a value will cause excessive segmentation, whereas too big a value results in incorrect individual objects resolution.
- **Object type:** this parameter allows us to specify whether the objects of interest are brighter or darker than the background of the image. Setting it to 0 considers bright areas on a dark background as potential objects.
- **Reject Border Objects:** set to 1 (this will ensure that the objects touching the edges of the image are excluded from analysis).
- **Object Area:** measures the size of nucleus in Channel 1 (important to properly mark the nucleus), as the nucleus region will be used to determine the area to measure phospho-ATM signal (Circ area) in Channel 2.
- **Compartmental Analysis**: for this analysis one needs to adjust cellular regions, Circ and Ring, in Channel 2. The Circ region is set based on the primary object in Channel 1.
- **Circ modifier:** this parameter is used to adjust the Circ region and is equal to the number of pixels that the primary object (nucleus) is dilated or eroded to define the mask. Setting it to a negative value means the Circ region will be slightly smaller than the perimeter of the primary object.

- **SpotKernalRadius:** the software application first identifies all the spots in each of the dependent channels used and then identifies those that are Circ Spots. Spots in cells are usually surrounded by background cellular fluorescence and the area covered by a Spot has to be distinguished from this background. This is controlled by the "SpotKernelRadius" parameter. The size of this parameter is similar to the maximum size, in pixels, of the Spots to be detected. The higher the intensity threshold is set, the more stringent the criteria are for a pixel to be identified as a Spot's and for its intensity to be used in the Spot's intensity computation.
- **CircSpotAvgIntensCh2:** this is the average intensity of all spot pixels with Channel 2 Circ mask.

3.4.3 ArrayScan Multi/Plate Run Protocol Setup

1. If stored at 4 °C, equilibrate plates first to room temperature prior to proceeding with the reading.
2. Load the first plate into the ArrayScan VTI instrument.
3. Select the established assay protocol.
4. Select the correct plate format.
5. Start with Channel 1 (Nuclei):
   (a) Select a Max control well.
   (b) Click on "Acquire image," click "Autofocus," and then click "Auto Expose."
   (c) Set the exposure time so that % saturation is between ~25 and 50% and nuclei look white on a black background. Fix this value.
6. Go to Channel 2.
   (a) Select "Acquire image set."
   (b) Set exposure time so % saturation is approximately 40–50%.
7. Return to Channel 1 and run the algorithm.
   (a) Blue masks = Selected Object (Nuclei).
   (b) Orange masks = Rejected Object.
8. The blue masks should fit neatly around the nuclei. The Object Identification Fixed Threshold value has to be adjusted if the masks are too small or too large.
9. Once Channel 1 algorithm is optimized look at Channel 2.
10. Select "Identify Objects" and on the image click on the objects with strong phospho-ATM staining (those will be cells with red color). Values for MeanCircSpotAvgIntensity will be displayed on the right hand side. Use these values to set the threshold. Run the algorithm to check what is being picked up. Check the values by double clicking on the image: these should be about 1000–1200. Repeat the same procedure on a Min well, the values here should be ~100–150.

11. Confirm that the threshold settings and Object Identification/Segmentation parameters are correct by running the algorithm on few more randomly selected Max and Min control wells.
12. Once you optimize the parameters, save them and start a run for the entire batch of plates.

3.4.4 Viewing and Exporting Data from the ArrayScan

Use vHCS View to export the data.

Export the valid object count (cell number) and MEAN_CircSpotAvgIntenCh2 measured values (phospho-ATM signal) into Excel spreadsheet and save as csv files.

3.5 Data Analysis

There are some important points to consider when interpreting the assay results. When plates or wells do not pass the quality control criteria, they need to be excluded from the analysis.

1. Ensure that Controls and Plate Medians data pass acceptable criteria:
   (a) Z prime ≥0.5.
   (b) Max and Min controls are at expected levels and perform consistently throughout the run.
   (c) Plate medians do not significantly shift throughout the run.
   (d) Also check that the assay plates exhibit similar levels of assay noise, and if patterns exist determine what level of impact, if any, they are having on the results.
   (e) Examine raw data in plate format for patterns (i.e., dispensing patters from a Multidrop Combi, edge effects).
   (f) Look for plate to plate patterns indicative of blocked tips, e.g., Multidrop Combi will have strip pattern across rows or columns depending on the dispensing direction, BioTek Microplate washer will show a pattern of two adjustment wells with lower signal if washes were not sufficient.
2. Calculate the percent effect for each compound as follows: %effect = 100 × [1−(test compound—median min control)/(median Max control-median min control)].
3. Set a cutoff for active compounds (i.e., any compound that exhibits greater average percent activity than the cutoff parameter is declared active). This is usually done based on the average effect of the Min controls and three times their standard deviation. A Robust Z-Score normalization can also be used (*see* **Notes 7–9**).

4 Notes

1. If liquid dispensers (Multidrop Combi) are not available, automated pipettes can be used for plating cells. If the latter is used, after plating centrifuge the plates to make sure no media is left

on the side of the wells. This is crucial, if compound addition is done using Echo Liquid Handler, to avoid media spills when the plates are turned upside down during compound addition.

2. Incubating cells on assay plates in the rotating incubator with high humidity greatly reduces edge effects in signal intensity. This is especially important if cells are grown in lower media volumes and in 384 or 1536-well plates. If such an incubator is not available, placing the plates in a 37 °C incubator in a box with pre-wetted paper towels can also be used.
3. If compounds are added to the cells at a higher volume, the final DMSO concentration will increase. Hence, before proceeding with the screen, one needs to perform a DMSO dose response to determine the DMSO tolerance. A similar experiment should be repeated when different cell lines are used for the assay.
4. Fixation and permabilization times are quite critical for good staining, so it is essential not to prolong the incubation times for these two steps when handling multiple plates.
5. For washing steps, addition of solutions should be done gently so as not to detach cells. Also, leaving a thin film of liquid at the bottom of the wells during aspiration steps will reduce cell loss during washes and subsequent solution addition. For cells that are easily detached, using high binding or coated plates (e.g., Poly-d-Lysine) is recommended.
6. For longer storage of plates before image acquisition, sodium azide can be added to a final concentration of 0.02% (w/v) to prevent microbial contamination—storage of longer than 2 weeks is not recommended.
7. Statistical plate parameters are very important for the analysis of any HTS data. First, normalization ensures that results are comparable within and across experiments. When data is normalized the measured signal values are scaled to a common range spanning from 0 to 100, where 0 corresponds to no compound activity and 100 full compound inhibitory activity. The transformation of the activity values to this range is based on a neutral reference (maximum signal) and an inhibitor reference (minimal signal).
8. In an ideal assay, the signal distributions for the maximum signal reference and the minimum signal reference control wells are nicely separated. This is true when the signal difference between the two references is large compared to the signal distribution variability within the two groups. The Z' Factor takes this into account by relating the sum of the standard deviations (SDs) for the reference wells to the signal range given by the difference in their mean values.

$$Z' = 1 - \frac{3 \times SD(Max) + 3 \times SD(Min)}{(mean(Max) - mean(Min))}$$

The Robust Z' Factor is based on the same formula as the Z' Factor, but standard deviations and means are replaced by the robust standard deviations and medians respectively.

9. The Robust Z-score normalization uses a single control well group as a reference distribution. For this normalization, all wells are treated according to the robust Z-score formula, where the standard deviation and mean are replaced by the robust standard deviation and median, respectively.

For a single plate the median and robust standard deviation of the raw values are obtained from the Central Reference well type. For each well, the median is subtracted from its raw value and this is divided by the robust standard deviation to obtain the robust Z-score.

The single plate robust Z-score normalization can be used to normalize out an offset and a multiplicative distortion per plate. It is based, however, on a relatively small number of control wells, typically 8 or 16, and a central value and variation from these wells has to be estimated.

Robust Z-Score = $(x-m)/\text{RSD}$; where x—is the raw data value of the well to be standardized, and m—is the median of the chosen control well group. RSD is the Robust Standard Deviation for the chosen control well group.

References

1. Bakkenist CJ, Kastan MB (2003) DNA damage activates ATM through intermolecular autophosphorylation and dimer dissociation. Nature 421(6922):499–506
2. Kozlov SV, Graham ME, Peng C, Chen P, Robinson PJ, Lavin MF (2006) Involvement of novel autophosphorylation sites in ATM activation. EMBO J 25(15):3504–3514
3. So S, Davis AJ, Chen DJ (2009) Autophosphorylation at serine 1981 stabilizes ATM at DNA damage sites. J Cell Biol 187(7):977–990
4. Nam EA, Zhao R, Glick GG, Bansbach CE, Friedman DB, Cortez D (2011) Thr-1989 phosphorylation is a marker of active ataxia telangiectasia-mutated and Rad3-related (ATR) kinase. J Biol Chem 286(33): 28707–28714
5. Jackson SP (2009) The DNA-damage response: new molecular insights and new approaches to cancer therapy. Biochem Soc Trans 37(Pt 3):483–494
6. Hiom K (2005) DNA repair: how to PIKK a partner. Curr Biol 15(12):R473–R475
7. Savitsky K, Bar-Shira A, Gilad S, Rotman G, Ziv Y, Vanagaite L, Tagle DA, Smith S, Uziel T, Sfez S et al (1995) A single ataxia telangiectasia gene with a product similar to PI-3 kinase. Science 268(5218):1749–1753
8. Zhou BB, Chaturvedi P, Spring K, Scott SP, Johanson RA, Mishra R, Mattern MR, Winkler JD, Khanna KK (2000) Caffeine abolishes the mammalian G(2)/M DNA damage checkpoint by inhibiting ataxia-telangiectasia-mutated kinase activity. J Biol Chem 275(14): 10342–10348
9. Sarkaria JN, Tibbetts RS, Busby EC, Kennedy AP, Hill DE, Abraham RT (1998) Inhibition of phosphoinositide 3-kinase related kinases by the radiosensitizing agent wortmannin. Cancer Res 58(19):4375–4382
10. Hickson I, Zhao Y, Richardson CJ, Green SJ, Martin NM, Orr AI, Reaper PM, Jackson SP, Curtin NJ, Smith GC (2004) Identification

and characterization of a novel and specific inhibitor of the ataxia-telangiectasia mutated kinase ATM. Cancer Res 64(24):9152–9159

11. Golding SE, Rosenberg E, Valerie N, Hussaini I, Frigerio M, Cockcroft XF, Chong WY, Hummersone M, Rigoreau L, Menear KA et al (2009) Improved ATM kinase inhibitor KU-60019 radiosensitizes glioma cells, compromises insulin, AKT and ERK prosurvival signaling, and inhibits migration and invasion. Mol Cancer Ther 8(10):2894–2902
12. Batey MA, Zhao Y, Kyle S, Richardson C, Slade A, Martin NM, Lau A, Newell DR, Curtin NJ (2013) Preclinical evaluation of a novel ATM inhibitor, KU59403, in vitro and in vivo in p53 functional and dysfunctional models of human cancer. Mol Cancer Ther 12(6):959–967
13. Rainey MD, Charlton ME, Stanton RV, Kastan MB (2008) Transient inhibition of ATM kinase is sufficient to enhance cellular sensitivity to ionizing radiation. Cancer Res 68(18):7466–7474
14. Kim S, Jun DH, Kim HJ, Jeong KC, Lee CH (2011) Development of a high-content screening method for chemicals modulating DNA damage response. J Biomol Screen 16(2):259–265
15. Guo Z, Kozlov S, Lavin MF, Person MD, Paull TT (2010) ATM activation by oxidative stress. Science 330(6003):517–521
16. Irarrazabal CE, Liu JC, Burg MB, Ferraris JD (2004) ATM, a DNA damage-inducible kinase, contributes to activation by high NaCl of the transcription factor TonEBP/OREBP. Proc Natl Acad Sci U S A 101(23):8809–8814
17. Bardelle C, Boros J (2012) Development of a high-content high-throughput screening assay for the discovery of ATM signaling inhibitors. J Biomol Screen 17(7):912–920

Chapter 6

Image-Based High Content Screening: Automating the Quantification Process for DNA Damage-Induced Foci

Yi Chieh Lim

Abstract

Visual inspection of cellular activities based on conventional fluorescence microscope is a fundamental tool to study the role of DNA damage response (DDR). In the context of drug discovery where the capture of thousands of images is required across parallel experiments, this presents a challenge to data collection and analysis. Manual scoring is laborious and often reliant on trained personnel to intuit biological meaning through visual reasoning. On the other hand, high content screening combines the automation of microscopy image acquisition and analysis in a single platform to quantify cellular events of interests. The data generated is rapid and accurate, lessening the bias of human interpretation. Herein, this chapter will describe an image-based high content screen approach and the data analysis of Ataxia-Telangiectasia Mutated (ATM) DNA damage-induced foci.

Key words ATM, DNA repair, Nuclear foci, DNA damage foci, DNA damage response, DNA double strand breaks, High content screen, IN cell analyzer, Cancer stem cells, Glioma

1 Introduction

1.1 Benefits of Image-Based High Content Screening

Image-based high-content screening (HCS) comes as a concept to bridge the gap between acquiring large volume of microscopic images and in-depth data analysis. This has been made possible in the last two decades because of the improved auto-focusing system of lenses coupled with a highly sensitive sensor to capture high-resolution images rapidly and accurately. The management and interpretation of gigabyte-sized image data is then facilitated by the integration of computational software, designed with complex algorithms for customizable analysis. This is in contrast to traditional screening methods such as ELISA and other plate-based reporter assays that quantify cellular activities according to the total cell population responses. With these studies, subtle biological variance in expression profiles being identified has been questioned, on the likelihood of perturbation due to the overall measure of total cell population [1]. In the context of pluripotent or cancer

Sergei V. Kozlov (ed.), *ATM Kinase: Methods and Protocols*, Methods in Molecular Biology, vol. 1599,
DOI 10.1007/978-1-4939-6955-5_6, © Springer Science+Business Media LLC 2017

cells where a heterogeneous population exists, the need to segregate biological readings to an individual cell is required for interpretable data. The ability to perform different measurements on a single cell and to extract multiple information from cellular images is one of many conspicuous features and strategies in HCS. The wealth of content in morphometry information such as fluorescence intensity, cell segmentation, subcellular distribution, and other parameters is archived for computational quantification with the variability adjusted to the needs of a particular research aim [2, 3]. The ease of reproducibility while without the necessary prior knowledge in computer programming has gained considerable adoption in the field of DNA damage with a strong focus on microscopic work.

1.2 Early Response of DNA Damage

There are several DNA lesions that can impact the integrity of the genome without compromising cell survival with the exception of DNA double strand breaks (DSBs) [4]. DSBs can originate from a variety of sources including ionizing radiation (IR), alkylating agents, reactive oxygen species (ROS) production, topoisomerase inhibitors, and replication forks collapse [5]. Combating against DSB insults requires DDR intervention. Activation of the DDR is cascade-dependent and relies on sequential signaling to regulate different downstream pathways to determine the feasibility of either restoring DNA fragments in the genome or undergo DNA damage tolerance by erroneous replication [6]. In the event where DNA impairment is beyond functional repair, programmed cell-death will initiate [7]. These cellular fate decisions are tightly regulated in the form of posttranslational modifications (PMT) that are essential for subcellular localization and subsequent cellular activity of many proteins [8]. The serine/threonine protein kinase ATM is a classic principle responder to DNA DSB [9]. During the early events of DNA strand breakage, various proteins are recruited to the site of damage which can be categorized functionally as sensor, transducer, mediator, and effector proteins. For instance, the MRE11, RAD50, and NBS1 (MRN) sensor complex binds to the ends of DSBs and recruits the transducer protein ATM [9]. Activated ATM then enlists phosphorylated MDC1 and γH2AX to the site as mediators. NBS1, MDC1, and γH2AX are crucial elements in creating a positive feedback loop that maintains ATM activity at the site of DNA breaks during DDR. Additional recruitments include 53BP1, BRCA1, TopBP1, and Claspin [10, 11]. Effectors kinases CHK1 and CHK2 are also initiated in the process to allow downstream dissemination of cellular signaling to different pathways [12].

1.3 Measure of ATM Foci in Response to DNA Double Strand Break

Assessing DDR frequently requires microscopic work by immunofluorescence or fluorescent-tagged fusion protein constructs to visualize protein localization. Often, this takes the form of discernible "spots" or "speckles" measuring approximately several hundred nanometers in diameter [13, 14]. Several scientific terms have been given to these structures such as nuclear foci, ionizing radiation-induced foci (IRIF), DNA repair foci, radiation-induced foci (RIF), and DNA damage foci to broadly describe DDR-associated proteins localizing to DNA break sites [13–16]. Active pS1981-ATM foci within the nucleus is one of many well-characterized events that are associated with DSBs and visible within minutes after DNA damage [9, 17]. The count for phosphorylated ATM foci has a direct correlation with the presence of DSB numbers in a linear IR dose-response fashion. It is important to note the presence of phosphorylated ATM foci varies across different cell types because of differences in kinase activity. Where comparison of phosphorylated ATM foci is made, experimental design should be kept according to cell type specificity. Another factor to consider when counting phosphorylated ATM foci is the severity of DNA damage. Measuring DNA damage foci in whole nucleus should be preferentially maintained at IR doses between 0.4 Gy and 6 Gy where statistical quantification is robust and reliable [18, 19]. Saturation of fluorescence intensity and DNA damage foci overlap after a higher dose of IR can perturb the accuracy of measurement. Several studies suggested that the optimum period to measure maximum phosphorylated ATM foci numbers is between 10 min to an hour after DNA damage induction. After this time, the progressive dephosphorylation of ATM leads to the resolution of DNA damage foci. The reduction in foci numbers and fluorescence intensity within the nucleus is a direct consequence of DNA repair. There are several methods to measure phosphorylated ATM foci. The kinetic-based approach focuses on the dynamic change in phosphorylated ATM foci by fixing cultures in a linear time progression (e.g., 1 h, 3 h, 6 h, 12 h, and 24 h). Alternatively, a single time-point comparison can also be achieved. Experimental endpoint, in general, measures residual phosphorylated ATM foci as an indication for DNA repair deficiency when unresolved DNA break sites repair should be completed (e.g., 12–24 h posttreatment). If the study focuses on understanding the activation of ATM kinase in DNA damage foci, measurement of phosphorylated ATM events should be investigated at early time (e.g., 30 min–1 h).

2 Materials

2.1 Plate Coating

1. Matrigel (BD Biosciences, USA).
2. Phosphate-buffered saline (PBS), pH 7.4.
3. 0.22 μM membrane filter.
4. 96-well black optical-bottom plate (Thermo Scientific, USA).

2.2 Treatment and Fixation of Cells

1. Cs-137 irradiation source of gamma rays (MDS Nordion Gammacell irradiator or similar).
2. Dimethyl sulfoxide (DMSO).

3. Extraction buffer: 20 mM NaCl, 5 mM $MgCl_2$, 10 mM PIPES-NaOH pH 6.8, 300 mM sucrose, and 0.5% NP-40. Supplement with cOmplete and phoSTOP inhibitor tablets before use.
4. cOmplete protease inhibitor tablet (Roche Diagnostics, Germany).
5. PhoSTOP inhibitor cocktail tablet (Roche Diagnostics, Germany).
6. Paraformaldehyde (2% PFA): 2 g of PFA in 10 mL of PBS (pH 7.4).

2.3 Immunostaining of Cells

1. Phosphate-buffered saline (PBS), pH 7.4.
2. Triton X-100.
3. Bovine serum albumin (BSA).
4. Blocking solution: 5% BSA in PBS.
5. Antibody diluent solution: 1% BSA in PBS.
6. ^{pS1981}ATM primary antibody (Abcam, UK or Rockland Inc., USA).
7. Alexa488-conjugated secondary antibody raised against primary antibody.
8. 4″,6-diamidino-2-phenylindole (DAPI). Stock concentration at 10 mg/mL in water.

2.4 Automation of ATM Foci Count

1. IN Cell analyzer 2200 (GE Healthcare Life Sciences, USA).
2. IN Cell investigator software (GE Healthcare Life Sciences, USA).

3 Methods

We primarily use patient-derived cancer stem cells, ATCC tumor cells of brain and breast origin, and fibroblast cell lines. Typically, adherent cells grown in serum-free media require the plate coating step. If serum is used, proceed straight to the later section (*see* Subheading 3.2).

3.1 Plate Coating

1. Thaw a 500 μL vial of Matrigel in an ice-water bucket for 30 min.
2. In a tissue culture laminar flow hood, prepare a 5 mL working stock solution of Matrigel (1:100 dilution) in sterile PBS.
3. Sterilize the diluted Matrigel solution using a 0.22 μM membrane filter.
4. Remove the cover lid of the 96-well optical-bottom plate and distribute 50 μL of working Matrigel solution evenly to cover the base of each well.

5. Return the cover lid back to the 96-well optical-bottom plate and store the vessel in a tissue culture incubator at 37 °C for a minimum period of 30 min before use. Avoid touching the bottom of the plate during transfer.
6. Remove the 96-well optical-bottom plate from the tissue culture incubator.
7. In a laminar flow hood, remove the cover lid of the 96-well optical-bottom plate.
8. Invert and gently tap the plate above a stack of paper towels to remove the Matrigel solution.
9. Rinse each well once with 100–200 μL of sterile PBS and repeat the step above. Plate is now ready for use.

3.2 Preparation of Cells

1. Carefully seed 5000–15,000 cells (depending on the size of individual cell) in 100 μL of media solution to each individual well (*see* **Note 1**).
2. Pipette the solution twice to ensure cells are evenly distributed in each well.
3. Return the cover lid back to the 96-well optical-bottom plate and store the vessel overnight in the 37 °C incubator to allow cells adhesion.

3.3 Treatment of Cells

Induction of DNA damage can be introduced in several ways: an irradiator (e.g., Cs-137), radiomimetic compounds (e.g., Bleomycin), topoisomerase inhibitors (e.g., Etoposide, doxorubicin), alkylating agent (e.g., Temozolomide). Apart from the use of an irradiator, chemotherapeutic agents (single or in combination) are often added freshly to the culture media.

1. Dilute compound/s of interest to the desired concentration in pre-warmed media solution at 37 °C (*see* **Note 2**).
2. Remove the 96-well optical-bottom plate from the 37 °C incubator. In the laminar flow hood, remove the cover lid of the 96-well optical-bottom plate. Invert and gently tap the plate above a stack of paper towels to remove media solution.
3. Gently add 100 μL media containing compound/s of interest onto the wall of each well.
4. Return the cover lid back to the 96-well optical-bottom plate and keep the plate in the 37 °C incubator for 1 h.
5. Irradiate plate at desired dose (0.5Gy–6Gy) and return back to 37 °C incubator.

3.4 Fixation of Cells

1. At the desired time, remove the 96-well optical-bottom plate from the tissue culture incubator. In the laminar flow hood, remove the cover lid.

2. Invert and gently blot the 96-well optical-bottom plate on a stack of paper towels to remove media solution.
3. Rinse each well twice with 100 μL of sterile PBS and repeat the step above.
4. Pipette 50 μL of prechilled extraction buffer (containing cOmplete and phoSTOP inhibitor tablets) solution slowly to the side of each well and leave it at room temperature for 5 min (*see* **Note 3**).
5. Discard the supernatant by inverting and gently blotting the 96-well optical-bottom plate on a stack of paper towels.
6. Rinse twice with 100 μL of sterile PBS by slow pipetting onto the wall of each well and repeat the step above.
7. Add 50 μL of 2% PFA (pH 7.2–pH 7.8) in PBS and leave the plate at room temperature for 30 min to fix the cells.
8. Discard the supernatant by inverting and gently blotting the 96-well optical-bottom plate on a stack of paper towels.
9. Rinse twice with 100 μL of sterile PBS by slow pipetting to the wall of each well and repeat the step above.

3.5 Immunostaining of Cells

1. Pipette 50 μL of 0.25% Triton X-100 in PBS gently down the wall of each well and incubate for 5 min at room temperature to permeablize the cells.
2. Discard the supernatant by inverting and gently blotting the 96-well optical-bottom plate on a stack of paper towels.
3. Rinse twice with 100 μL of sterile PBS by slow pipetting onto the wall of each well and repeat the step above.
4. Add 100 μL of blocking solution (5% BSA in PBS) by slow pipetting onto the wall of each well. Allow blocking for 1 h at room temperature.
5. Discard the supernatant by inverting and gently blotting the 96-well optical-bottom plate on a stack of paper towels.
6. Add 25 μL of anti-^{pS1981}ATM antibody in the antibody diluent solution (1% BSA in PBS) at an appropriate dilution of 1:500–1:1000 for each well. Perform the incubation for 1–2 h at room temperature.
7. Discard the supernatant by inverting and gently blotting the 96-well optical-bottom plate on a stack of paper towels.
8. Rinse twice with 100 μL of sterile PBS by slow pipetting to the wall of each well and repeat the step above.
9. Add 25 μL of the appropriate Alexa488 secondary antibody raised against the host species of the primary antibody in the antibody diluent solution (1% BSA in PBS) at 1:250–1:1000 and DAPI (1:10,000 dilution). Incubate samples in the dark for 1–2 h at room temperature.

10. Discard the supernatant by inverting and gently blotting the 96-well optical-bottom plate on a stack of paper towels.
11. Rinse thrice with 100 μL of sterile PBS by slow pipetting to the wall of each well and repeat the step above.
12. Pipette 200 μL of sterile PBS gently to the wall of each well.
13. Wrap the 96-well optical-bottom plate in aluminum foil to prevent photobleaching.

3.6 Image Acquisition and Processing

3.6.1 Image Acquisition of Cells

1. Initialize the IN Cell analyzer 2000 software. In the setup protocol page, select the "eject the plate/slide" icon located at the top of the main screen to reveal the holding stage.
2. Remove the aluminum foil and insert the 96-well optical-bottom plate onto the holding stage of the machine. Select both the "eject the plate/slide" plate icon to close the sliding door and the "illuminator" icon to warm up the lamp.
3. From the "dashboard," follow according to this sequence. Label the experiment under "protocol name." In the "plate/slide" section, choose the "plate series" (catalogue number) for the 96-well optical-bottom plate used. Ensure the image focus is set according to the thickness of the optical bottom by "verify LAF." Lens objective can be selected at "20×" or "60×." Next, go to the "channel settings" and add the first wavelength to "DAPI" as the image channel with an exposure time between 0.01 and 0.1 s. A similar approach is applied to the addition of the second wavelength. Set the channel to "FITC" for detecting nuclear foci. Exposure time ranges between 0.2 and 3 s. Both acquisitions for "DAPI" and "FITC" should be set as "2D–convolution" images (*see* **Note 4**).
4. Under the "focus options," ensure the laser power for image hardware autofocus runs at "20%" while the software autofocus is set at 20 μm depth (default). Perform a prescan by selecting the "camera icon" to ensure images captured are in focus. Once completed, go to the "plate view" section located at the bottom of the main screen and select the "number of fields" required for each well. It is typical to have approximately 8–16 images taken at different areas of the well by using the "distance of *x* and *y* axis" for each acquisition field (*see* **Note 5**).
5. "Save" and "run protocol" to begin high content image acquisition.

3.6.2 Image Partition (Nucleus Mask)

Acquired images often contain autofluorescent bodies that can present an issue of false-positive measurement. Therefore, a computational mask design is essential to exclude autofluorescent counts and quantify nuclear foci within the boundary of the nucleus.

1. Open the IN Cell developer toolbox software. Select the "view/ analyze image stack" from the "operations" column in the main screen and upload the experiment file of interest (e.g., experiment_name.xdce). Image brightness can be adjusted under the "visuals" icon within the "operations" column to reduce the overall background fluorescence accordingly.
2. From the "protocol explorer" icon, select the "target sets" and right-click to create a "new target type." Label the target name as "Nucleus" for the identification of DAPI-stained cells and configure the source image of analysis as "DAPI-DAPI" (*see* **Note 6**).
3. Change the "segmentation" type to "objection" and adjust the "kernel size" index value to 49 with a "sensitivity" rate of 40. Right-click on the "segmentation" and preview the accuracy of the mask overlaying the nucleus. Adjust the "kernel size" value according to the cell-type used.
4. In the "postprocessing" option, the "sieve (binary)" selection offers discrimination of disproportionate (e.g., cell doublets) and apoptotic cells. Ensure the setting of the first "sieve (binary)" area of interest is greater than 50 μm² and include the second "sieve (binary)" option to exclude any nucleus that is less than 200 μm² in size. It is essential to incorporate the "boarder object removal" setting to exclude analysis of cells at the edge of the image with a partially shown nucleus (*see* **Note** 7).

3.6.3 Image Partition (Foci Mask)

1. From the "target sets," create a "new target type" and label the target name as "Nuclear foci." Configure the source image as "FITC-FITC" for the identification of nuclear foci (*see* **Note 6**).
2. Under the "segmentation," change the option to "vesicle." Maintain the detection area of nuclear foci between 0.1 and 0.4 μm for "minimum granule size" and 0.5–0.6 μm for "maximum granule size." Right-click and preview the surface area of the mask overlaying nuclear foci. Adjust the segmentation value of the vesicle mask accordingly to exclude abnormally large nuclear foci and autofluorescent bodies. Set the "scale" to 2 with a "sensitivity value" of 40 to expand the rate of detection while reducing the incidental potential of background nuclear foci. The checkbox for "precise mask" and "use octagonal morphology" should be marked to improve segmentation result for diminutive objects and refine the detection of nuclear foci as octagonal structures rather than actual spherical morphology element, respectively.
3. In the "postprocessing" option, include the "boarder object removal" setting to filter out analysis of nuclear foci at the edge of the image with a partial nucleus captured by the camera.

3.6.4 Image Partition (Foci Intensity Mask)

1. From the "target sets," create a "new target type" and label the target name as "Foci intensity." Configure the source image of analysis as FITC-FITC to differentiate nuclear foci and background fluorescence.
2. Under the "segmentation," choose the "intensity" option. Adjust the "minimum threshold" to an approximate value of >35 to eliminate detection of background fluorescence within the nucleus. Right-click and preview the intensity mask. Verify the overlay setting is detecting mainly nuclear foci and not the background luminance within the nucleus. Adjust the value accordingly. Perform the same approach for "maximum threshold" and restrict the value at <200 to eliminate autofluorescence bodies as part of analysis (*see* **Note 8**).
3. In the "postprocessing" section, the "erosion" option further segments close proximity nuclear foci by reducing the area of the boundary of detection. The "kernel size" is set at a default value of 3.

3.7 Parameter Linkage

1. To execute the correct analysis, information from the earlier settings (*see* Subheadings 3.6.2–3.6.4) must be jointly associated. In the "target linking" section, right-click and create a "one to many" link for the exclusive identification of nuclear foci within the nucleus. The "output target set" label for this link is "ATM foci." Determine the "primary target" as "Nucleus" and the "Nuclear foci" as "secondary target." Ensure only the checkbox for "secondary target within primary target" is selected with a "100% overlap."
2. Perform a second "one to many" target link to distinguish nuclear foci with signal intensity higher than background fluorescence. Ensure the "Nuclear foci" is selected as the "primary target" for analysis while choosing the "secondary target" as "Foci intensity." Label this particular link as "ATM foci intensity." Mark both checkboxes for "primary target within secondary target" with a "100% overlap" and "allow multiple primary targets to share the same secondary targets."
3. The last target link is to assign the analysis of nuclear foci within the nucleus that fluoresces with an intensity higher than background luminance. Select the "compose one to one" and label the "output target set" as "ATM foci cells." The "primary target" source is "ATM foci" and the "match pathway" originates from "ATM foci\Nucleus." Similarly, select the "ATM foci intensity" as the "secondary target" with a "match pathway" that derives from "ATM foci intensity\Nuclear foci." Mark checkboxes of "secondary target within primary targets," "allow multiple primary targets to share the same secondary targets" and "display measures in primary targets" (Fig. 1).

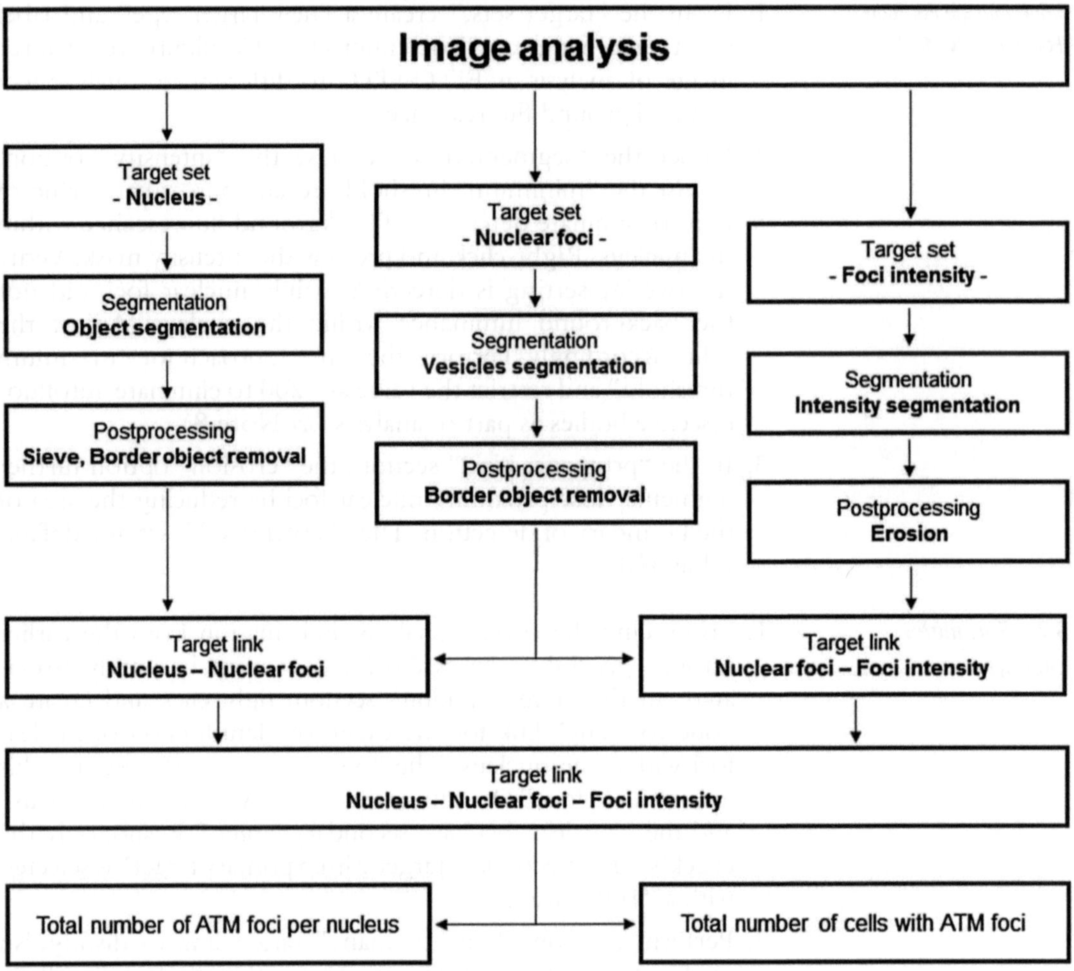

Fig. 1 Analysis flowchart for the quantification of pS1981ATM foci. A brief summary of the workflow to set up an accurate analysis and quantification for nuclear foci using the IN Cell developer software

3.8 DNA Damage Foci Quantification

1. Analysis is performed by generating an "add user defined measure." To calculate the total number of DAPI-stained cells regardless of the presence or absence of DNA damage foci, insert the formula "*[Count<\Nucleus>]*" into the "measure" section for the target set "Nucleus" created earlier (*see* **Note 9**). Label the "measure name" as "Total DAPI-stained cells" (*see* **Note 10**).
2. To calculate the total number of nuclear foci in each DAPI-stained cell, generate an "add user defined measure" for "ATM foci cells" target set. Mark checkbox "advance" and insert the formula "*[Count<\ATM foci cells\ATM foci\All:Nuclear foci>]*" into the "measure" equation section. Label the "measure name" as "ATM foci per cell" (*see* **Note 10**).

3. Analysis can also be adjusted specifically to account for cells that express a certain number of nuclear foci. For instance, an "add user defined measure" to identify only DAPI-stained cells with >5 foci will have the following input "*[Count<\ATM positive cells\ATM foci\All:Nuclear foci >]>=5*" in the "measure" equation section. Label the "measure name" as "Cells with ATM foci" (*see* **Note 10**).
4. Save the protocol and perform a preview run "analyze current field" on both the positive and control images. Adjust individual segmentation setting accordingly if analysis requires a more or less stringent criterion (Fig. 2).

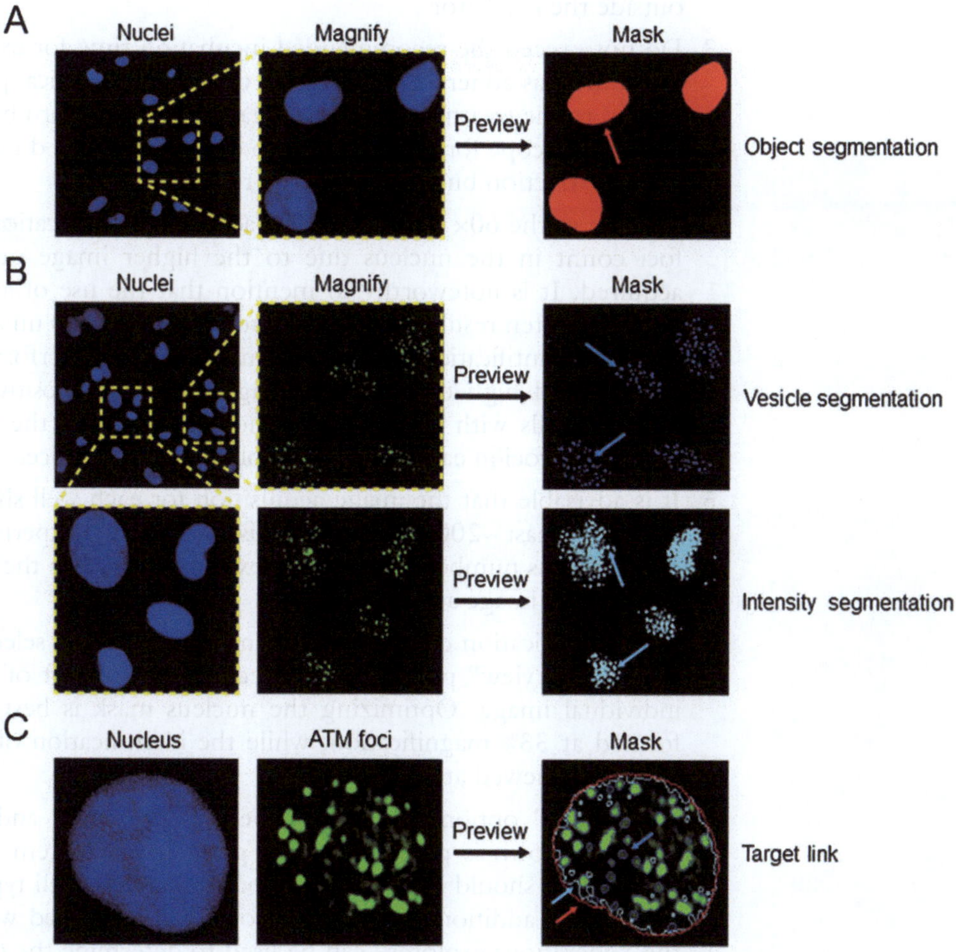

Fig. 2 Image analysis and quantification of pS1981ATM foci in fixed cells. (**a**) The object segmentation mask (*red arrow*) based on cell size to identify DAPI-stained nucleus while excluding apoptotic and "doublet" cells. (**b**) The process then continues to catalogue fluorescence bodies using a vesicle segmentation mask to determine the appropriate size of nuclear foci and an intensity segmentation mask to exclude autofluorescence. (**c**) Target linking of all three independent segmentation identifies specifically nuclear foci in cells

4 Notes

1. Most cells proliferate at different rates and it can be difficult to standardize a starting cell number to examine nuclear foci. A high density cell number of >90% confluency can exert on nutrient exhaustion, leading to reactive oxygen species production. Therefore, it is essential to perform a seeding optimization step. The aim is to achieve a cell density of 50–60% confluency prior to the actual experiment.
2. A negative control should be included. The mock-irradiated culture plate (required to determine the background ATM foci signal) is treated exactly like the sample cultures but left outside the irradiator.
3. Do not exceed the recommended incubation time for extraction buffer as adherent cells can lift off from the optical plate. Attend to the treatment process visually with a standard bright field microscope for the first time to determine if the duration of the extraction buffer needs to be reduced.
4. The use of the 60× objective offers accurate quantification for foci count in the nucleus due to the higher image quality acquired. It is noteworthy to mention that the use of a 20× objective often results in pixelated nuclear foci and is unfavorable for quantification. If the experiment aim is to perform an image-based high content screening to examine positive or negative cells with nuclear foci (and not to count the total number of foci in each cell), a 20× objective will suffice.
5. It is advisable that the image acquisition for each well should achieve at least ~200 cells for analysis. If the initial experiment captured less number of cells than expected, increase the field number for image acquisition.
6. The magnification of the images can be adjusted by selecting the "image view" panel icon located on the top left of each individual image. Optimizing the nucleus mask is best performed at 33% magnification, while the identification of foci should be viewed at 100%.
7. The "sieve" option is cell type-specific dependent and the example shown is performed with glioma neural stem cells. The setting should be adjusted accordingly to the cell type of interest. In addition, the up and down arrow located within the "operations explorer" can be used to determine the order of postprocessing.
8. It is advisable to run the foci intensity preview on a negative control image where cells are stained with both an isotype control and/or secondary antibody-only control. Ensure the minimum threshold value is always higher than the negative control stain.

9. Ensure the view setting for analysis prior to computational quantification is set to "Target table." The tab option is located at the top of the main screen. Also ensure the protocol is regularly saved after changes have been made.
10. The mathematical formula can be found under "grouping" while the masks and linkage parameters generated for analysis are located in "target path." Both extensions are found in the "user define measure" section.

Acknowledgments

This work was supported by grants from the Cancer Council Queensland and the Brain Cancer Discovery Collaborative (BCDC) of Cure Brain Cancer Foundation, Australia.

References

1. Kim K-T, Lee HW, Lee H-O, Kim SC, Seo YJ, Chung W, Eum HH, Nam D-H, Kim J, Joo KM, Park W-Y (2015) Single-cell mRNA sequencing identifies subclonal heterogeneity in anti-cancer drug responses of lung adenocarcinoma cells. Genome Biol 16:127
2. Jones TR, Carpenter AE, Lamprecht MR, Moffat J, Silver SJ, Grenier JK, Castoreno AB, Eggert US, Root DE, Golland P, Sabatini DM (2009) Scoring diverse cellular morphologies in image-based screens with iterative feedback and machine learning. Proc Natl Acad Sci 106(6):1826–1831
3. Martin HL, Adams M, Higgins J, Bond J, Morrison EE, Bell SM, Warriner S, Nelson A, Tomlinson DC (2014) High-content, high-throughput screening for the identification of cytotoxic compounds based on cell morphology and cell proliferation markers. PLoS One 9(2):e88338
4. Rodier F, Coppé J-P, Patil CK, Hoeijmakers WAM, Muñoz DP, Raza SR, Freund A, Campeau E, Davalos AR, Campisi J (2009) Persistent DNA damage signaling triggers senescence-associated inflammatory cytokine secretion. Nat Cell Biol 11(8):973–979
5. Pearl LH, Schierz AC, Ward SE, Al-Lazikani B, Pearl FMG (2015) Therapeutic opportunities within the DNA damage response. Nat Rev Cancer 15:166–180
6. Ziv O, Zeisel A, Mirlas-Neisberg N, Swain U, Nevo R, Ben-Chetrit N, Martelli MP, Rossi R, Schiesser S, Canman CE, Carell T, Geacintov NE, Falini B, Domany E, Livneh Z (2014) Identification of novel DNA-damage tolerance genes reveals regulation of translesion DNA synthesis by nucleophosmin. Nat Commun 5(5437):1–13
7. Ciccia A, Elledge SJ (2010) The DNA damage response: making it safe to play with knives. Mol Cell 40(2):179–204
8. Oberle C, Blattner C (2010) Regulation of the DNA damage response to DSBs by post-translational modifications. Curr Genomics 11(3):184–198
9. Lavin MF (2008) Ataxia-telangiectasia: from a rare disorder to a paradigm for cell signalling and cancer. Nat Rev Mol Cell Biol 9:759–769
10. Lindsey-Boltz LA, Sancar A (2011) Tethering DNA damage checkpoint mediator proteins topoisomerase IIβ-binding protein 1 (TopBP1) and claspin to DNA activates ataxia-telangiectasia mutated and RAD3-related (ATR) phosphorylation of checkpoint kinase 1 (Chk1). J Biol Chem 286(22):19229–19236
11. Wang B, Matsuoka S, Carpenter PB, Elledge SJ (2002) 53BP1, a mediator of the DNA damage checkpoint. Science 298(5597):1435–1438
12. Bartek J, Lukas J (2003) Chk1 and Chk2 kinases in checkpoint control and cancer. Cancer Cell 3:421–429
13. Neumaier T, Swenson J, Pham C, Polyzos A, Lo AT, Yang P, Dyball J, Asaithamby A, Chen DJ, Bissell MJ, Thalhammer S, Costes SV (2012) Evidence for formation of DNA repair centers and dose-response nonlinearity in human cells. Proc Natl Acad Sci 109(2):443–448

14. Takai H, Smogorzewska A, Td L (2003) DNA damage foci at dysfunctional telomeres. Curr Biol 13(17):1549–1556
15. Bao S, Wu Q, McLendon RE, Hao Y, Shi Q, Hjelmeland AB, Dewhirst MW, Bigner DD, Rich JN (2006) Glioma stem cells promote radioresistance by preferential activation of the DNA damage response. Nature 444(7120):756–760. doi:10.1038/nature 05236
16. Lim YC, Roberts TL, Day BW, Stringer BW, Kozlov S, Fazry S, Bruce ZC, Ensbey KS, Walker DG, Boyd AW, Lavin MF (2014) Increased sensitivity to ionizing radiation by targeting the homologous recombination pathway in glioma initiating cells. Mol Oncol 8(8):1603–1615
17. Lavin MF, Kozlov S (2007) ATM activation and DNA damage response. Cell Cycle 6(8):931–942
18. Collis SJ, Schwaninger JM, Ntambi AJ, Keller TW, Nelson WG, Dillehay LE, DeWeese TL (2004) Evasion of early cellular response mechanisms following low level radiation-induced DNA damage. J Biol Chem 279(48):49624–49632
19. Markova E, Schultz N, Belyaev IY (2007) Kinetics and dose-response of residual 53BP1/g-H2AX foci: co-localization, relationship with DSB repair and clonogenic survival. Int J Radiat Biol 83(5):319–329

Chapter 7

Analyzing ATM Function by Electroporation of Endonucleases and Immunofluorescence Microscopy

Keiji Suzuki

Abstract

Ataxia-telangiectasia mutated (ATM) protein, which plays a crucial role in DNA damage checkpoint signaling, is activated by DNA double strand breaks (DSBs) caused by ionizing radiation. While radiation exposure induces various types of DNA break ends, here, we describe a method, which enables creating defined types of DSBs by applying restriction endonucleases and foci analysis by immunofluorescence microscopy. The protocol greatly improves our knowledge on specific roles of ATM function in different DNA repair pathways.

Key words ATM, DNA double strand breaks, DNA damage response, Restriction endonuclease, Phosphorylation, DNA repair

1 Introduction

DNA double strand breaks (DSBs) have been well recognized as the most detrimental DNA damage caused by ionizing radiation. While DSBs are efficiently repaired by intracellular DNA repair pathways, such as non-homologous end-joining (NHEJ) and homologous recombination (HR), un-rejoined or mis-rejoined DNA damage could threaten the integrity of the genome [1–4]. Therefore, the cells have evolved a sophisticated system called DNA damage checkpoints, which is indispensable for keeping the stability of the genome [5, 6]. DNA damage checkpoint pathway requires ATM function [7–10]. ATM kinase is activated by autophosphorylation leading to dissociation of inactive ATM dimers [11–14]. Activated ATM phosphorylates various downstream proteins including those that regulate cell cycle progression, cell death, as well as DNA repair [7–10]. Thereby, ATM plays a critical role in orchestrating DNA damage signaling and DNA damage repair.

Since the earlier studies demonstrating the essential role of ATM function in cell cycle regulation, ionizing radiation has been used to create DSBs [15, 16]. However, deposition of radiation energy

Sergei V. Kozlov (ed.), *ATM Kinase: Methods and Protocols*, Methods in Molecular Biology, vol. 1599, DOI 10.1007/978-1-4939-6955-5_7, © Springer Science+Business Media LLC 2017

results in various types of break end structures, which makes it difficult to define specific roles of ATM in sensing and amending the specific type of broken DNA ends. Here, we developed a method, which enables generation of defined types of DSBs by applying restriction endonucleases (REs) [17]. REs were introduced into the synchronized G0/G1 cells following the previous report [18]. While the ATM function was required for proper activation of DNA damage checkpoint after the treatment of any type of REs, ATM-dependent pathway is required for DNA repair of certain types of termini. Thus, the novel protocol greatly improved our knowledge on specific roles of ATM function in DNA repair pathways.

2 Materials

2.1 Cell Culture

1. Normal human diploid fibroblasts (NHDFs) and AT patient-derived fibroblasts (AT2KY) (*see* ref. [19]).
2. Culture medium: Prepare MEM stock solution (1×) by dissolving 9.4 g of Eagle's MEM powder in 450 mL of Milli-Q water and adding 55 mg/L sodium pyruvate, 4.768 g/L HEPES, and 1.1 g/L $NaHCO_3$. Mix well and adjust pH to 7.2 by adding 1 N HCl. Make up to 500 mL with Milli-Q water. Store at 4 °C. Prepare culture medium by mixing 450 mL of MEM stock solution and 50 mL of FBS (Hyclone, Victoria, Australia). Culture medium can be stored for 1 month at 4 °C.
3. PBS(−): Prepare stock solution (10×) by dissolving 80 g/L NaCl, 28.8 g/L Na_2HPO_4, 2 g/L KCl, and 2 g/L KH_2PO_4 in Milli-Q water. The solution can be stored at room temperature for several months.
4. Trypsin/PBS: Prepare 0.2% (wt/vol) solution by dissolving 2 g/L trypsin powder in PBS(−). The solution can be stored at −20 °C for several months.
5. Cell Culture flasks (T25 with 25 cm^2 growth area).
6. Cell culture tubes (15 mL).

2.2 Electroporation

1. Electroporator (Electric Cell Fuser ECF2001, Wakenyaku, Kyoto, Japan or similar).
2. Gene Pulser Cuvette 0.4 cm (Bio-Rad Laboratories, Tokyo, Japan).
3. Cell counter (Automated Cell Counter TC20, Bio-Rad Laboratories, Tokyo, Japan).

2.3 Restriction Enzymes

1. Bam HI (100 units/μL).
2. Bgl I (20 units/μL).

3. Eco RI (200 units/μL).
4. Eco RV (100 units/μL).
5. Hind III (200 units/μL).
6. Hinf I (100 units/μL).
7. Kpn I (100 units/μL).
8. Not I (20 units/μL).
9. Pst I (200 units/μL).
10. Pvu II (100 units/μL).
11. Rsa I (100 units/μL).
12. Sma I (100 units/μL).
13. Sph I (80 units/μL).

2.4 Immuno-fluorescence

1. Cover and slide glasses.
2. Formalin/PBS: Prepare 4% (vol/vol) solution by adding 37% (wt/vol) formaldehyde solution to PBS(−). The solution can be stored at 4 °C for a few days.
3. Permeabilization solution: Prepare 0.5% (vol/vol) solution by adding 5 μL of Triton X-100 to 1 mL of PBS(−). Mix well before use. The solution can be stored at 4 °C for a few days.
4. Tris-buffered saline (10× TBS): Prepare stock solution (10×) by dissolving 24.2 g/L Tris, and 80 g/L NaCl in 475 mL of Milli-Q water. Mix well and adjust pH to 7.6 by adding 1 N HCl. Make up to 500 mL with Milli-Q water. The solution can be stored at room temperature for several months.
5. TBS containing 0.1% Tween-20 (TBS-T): Prepare 1× TBS solution by diluting 1 mL of 10× TBS to 9 mL with Milli-Q water. Add 10 μL of Tween-20 to 10 mL of 1× TBS and mix well. The solution should be prepared fresh before use.
6. Antibody solution: Dissolve 50 mg/mL skim milk in TBS-T. The solution should be prepared fresh before use.
7. DAPI solution: Prepare stock solution (10×) by dissolving 1 μg DAPI in 1 mL of Milli-Q water. The solution can be stored at −20 °C for several months.
8. Glycerol/PBS: Prepare 10% (vol/vol) solution by adding 100 μL of glycerol to 900 μL of PBS(−). The solution can be stored at 4 °C for a month.
9. CoverGrip Coverslip Sealant (Biotium, Hayward, CA, USA).
10. Fluorescent microscope.

2.5 Antibodies

1. Mouse anti-phosphoS1981 ATM monoclonal antibody (Rockland Immunochemicals, Limerick, PA, USA) (1:500, diluted in antibody solution).

2. Mouse anti-phosphoS139 histone H2AX monoclonal antibody (BioLegend, San Diego, CA, USA) (1:1000, diluted in antibody solution).
3. Rabbit anti-p53 binding protein 1 (53BP1) polyclonal antibody (BETHYL Laboratories, Montgomery, TX, USA) (1:500, diluted in antibody solution).
4. Rabbit anti-Ki-67 antibody (Abcam Japan, Tokyo, Japan) (1:250, diluted in antibody solution).
5. Alexa Fluor 488-conjugated anti-mouse IgG (Molecular Probes, Carlsbad, CA, USA) (1:1000, diluted in antibody solution).
6. Alexa Fluor 594-conjugated anti-rabbit IgG (Molecular Probes, Carlsbad, CA, USA) (1:1000, diluted in antibody solution).

3 Methods

3.1 Preparation of G0/G1-Synchronized Cells

1. Maintain NHDFs in an exponentially growing condition in a culture medium (*see* **Note 1**).
2. Wash cells with PBS(−) and incubate cells with 0.2% (wt/vol) trypsin/PBS solution for up to 5 min at room temperature. Use 1 mL of 0.2% trypsin/PBS for T25 flask.
3. Tap the T25 flask to ensure cell detachment.
4. Add 5 mL of culture medium for neutralization of the trypsin activity and make cell suspension by pipetting with a 10 mL culture pipette.
5. Collect cell suspension into a 15 mL tissue culture tube.
6. Count cell number using a cell counter.
7. Resuspend cells in culture medium at a concentration of 1×10^6/mL.
8. Plate 1×10^6 cells into T25 cell culture flasks (*see* **Note 2**).
9. Maintain culture for 7 days with changing medium every 3 days (*see* **Note 3**).

3.2 Introduction of Endonucleases

1. Set the pulse conditions before starting the experiments. Adjust pulse height and width to 400 V/cm and 1 ms, respectively (*see* **Note 4**).
2. Wash exponentially growing cells with PBS(−) and incubate cells with 0.2% (wt/vol) trypsin/PBS solution for up to 5 min at room temperature. Use 1 mL of 0.2% trypsin/PBS for T25 flask.
3. Tap the T25 flask to ensure cell detachment (*see* **Note 5**).

4. Add 5 mL of culture medium for neutralization of the trypsin activity and make cell suspension by pipetting with a 10 mL culture pipette.
5. Collect cell suspension into a 15 mL tissue culture tube.
6. Count cell number using a cell counter.
7. Resuspend cells in culture medium at a concentration of 2×10^6/mL.
8. Centrifuge 2 mL of cell suspension at $430 \times g$ for 3 min at room temperature.
9. Resuspend the cell pellet in 5 mL of PBS(−).
10. Centrifuge the cell suspension at $430 \times g$ for 3 min at room temperature.
11. Resuspend cells in PBS(−) at a concentration of 2×10^6/mL.
12. Mix 450 μL of cell suspension with 50 μL of reaction buffer, and transfer 500 L cell suspension to the Gene Pulser Cuvette (*see* **Note 6**).
13. Add restriction endonucleases into the cell suspension in the Gene Pulser Cuvette just before electroporation.
14. Place the cuvette in the electroporator.
15. Pulse two to five times (*see* Subheading 3.4).
16. Mix cell suspension with a fresh medium immediately after electroporation (*see* **Note** 7).
17. Plate cells onto sterilized 22 × 22 mm cover glasses at a density of 5×10^4 cells per cover slip.

3.3 Detection of ATM Activation and DNA Damage Response

1. Incubate cells at 37 °C in a 5% CO_2 incubator until they are ready to be fixed (*see* **Note 8**).
2. Wash cells with PBS(−) and fix in 2 mL/35-mm dish of formalin/PBS for 10 min at room temperature.
3. Remove formalin/PBS and wash cells with PBS(−) twice.
4. Add 2 mL/35-mm dish of permeabilization solution and incubate for 5 min at room temperature.
5. Wash cells with PBS(−) extensively.
6. Add primary antibodies into antibody solution and pipette 100 μL of antibody solution onto the cells (*see* **Note 9**).
7. Incubate the samples in the CO_2 incubator for 2 h at 37 °C.
8. Wash cells with PBS(−) and pipette 100 μL of Alexa Fluor 488-conjugated anti-mouse IgG or Alexa Fluor 594-conjugated anti-rabbit IgG onto the cells.
9. Incubate the samples in the CO2 incubator for 1 h at 37 °C.

10. Wash cells with PBS(−) and place the cover glasses cell-side-down into a drop of Glycerol/PBS containing 1× DAPI solution on the microscope slides.
11. Wipe excess Glycerol/PBS solution and seal the edges of the cover glass with the Coverslip Sealant.
12. Examine samples with a fluorescence microscope.

3.4 Optimization of the Experimental Conditions

1. Prepare NHDFs in PBS(−) at a concentration of 2×10^6/mL as described in Subheading 3.2.
2. Mix 450 μL of cell suspension with 50 μL of 10× reaction buffer, and transfer 500 μL cell suspension to the Gene Pulser Cuvette.
3. Add 100 U of various restriction enzymes (*see* Subheading 2.3, *see* **Note 10**) into the cell suspension in the Gene Pulser Cuvette just before electroporation (*see* **Note 11**).
4. Place the cuvette in the electroporator.
5. Pulse two to five times (Fig. 1) (*see* **Note 12**).

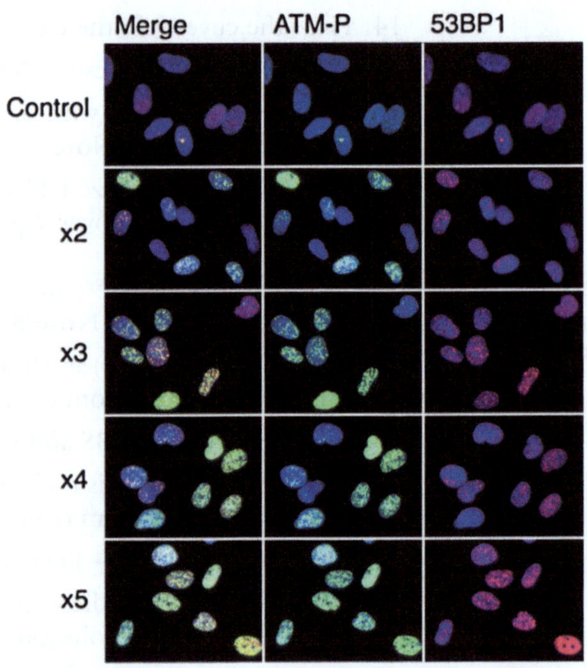

Fig. 1 Electroporation of cells with restriction endonucleases. Synchronized normal human diploid cells in G1 were collected by trypsinization and suspended in PBS buffer. Enzyme reaction buffer and *Pvu* II (100 U) were added immediately before electroporation. Pulse two to five times (×2 – ×5). Cells were incubated for 2 h before fixation. Cells were stained with anti-phosphorylated ATM antibody (ATM-P) and anti-53BP1 antibody (53BP1). Control; no treatment, ×2 – ×5; two pulses – five pulses

6. Feed a fresh medium immediately after electroporation and plate cells onto sterilized 22 × 22 mm cover glasses at a density of 5 × 10⁴ cells per cover slip.
7. Incubate cells at 37 °C in a 5% CO_2 incubator for one to 12 h (Fig. 2) (*see* **Note 13**).
8. Fix and permeabilize cells as described (*see* Subheading 3.3).
9. Incubate cells with anti-phosphorylated ATM and anti-53BP1 antibodies. Prepare the samples as described (*see* Subheading 3.3).
10. Examine samples with a fluorescence microscope (Fig. 3) (*see* **Notes 14** and **15**).

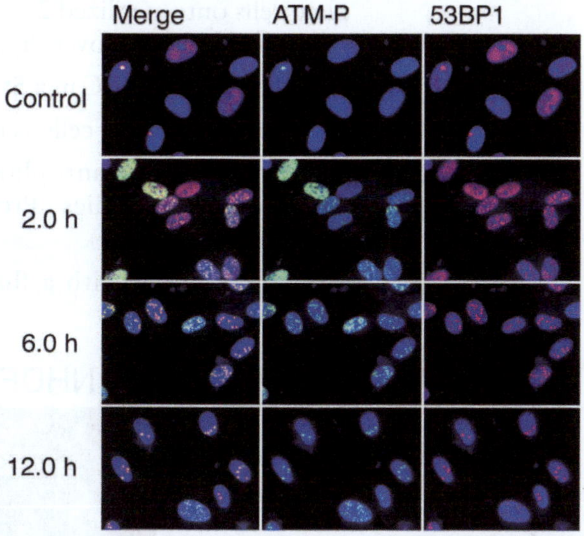

Fig. 2 Time-dependent foci formation by Pvu II. Synchronized normal human diploid cells in G1 were electroporated with *Pvu* II (100 U), and incubated for the time indicated before fixation. Cells were stained with anti-phosphorylated ATM antibody (ATM-P) and anti-53BP1 antibody (53BP1). Control; no treatment

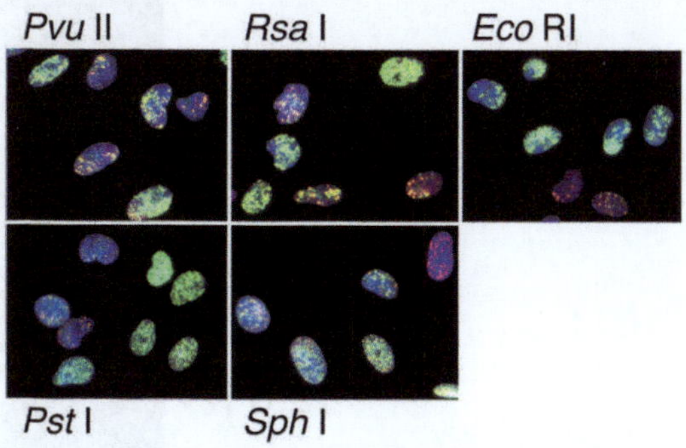

Fig. 3 Induction of foci by various restriction enzymes. Synchronized normal human diploid cells in G1 were electroporated with 100 U of various restriction enzymes, and they were incubated for 2 h before fixation. Cells were stained with anti-phosphorylated ATM antibody (*Green signals*) and anti-53BP1 antibody (*Red signals*)

3.5 Involvement of ATM Function in DNA Repair

1. Prepare NHDF and AT2KYcells in PBS(−) at a concentration of 2×10^6/mL, as described in Subheading 3.2.
2. Mix 450 μL of cell suspension with 50 μL of *Pvu* II ×1 reaction buffer.
3. Transfer 500 μL cell suspension to the Gene Pulser Cuvette.
4. Add 100 U of *Pvu* II into the cell suspension in the Gene Pulser Cuvette.
5. Place the cuvette in the electroporator.
6. Pulse five times (*see* **Note 12**).
7. Feed a fresh medium immediately after electroporation and plate cells onto sterilized 22 × 22 mm cover glasses at a density of 5×10^4 cells per cover slip.
8. Incubate cells at 37 °C in a 5% CO_2 incubator for one to 12 h.
9. Fix and permeabilize cells as described (*see* Subheading 3.3).
10. Incubate cells with anti-phosphorylated histone H2AX and anti-53BP1 antibodies. Prepare the samples as described (*see* Subheading 3.3).
11. Examine samples with a fluorescence microscope (Fig. 4) (*see* **Note 16**).

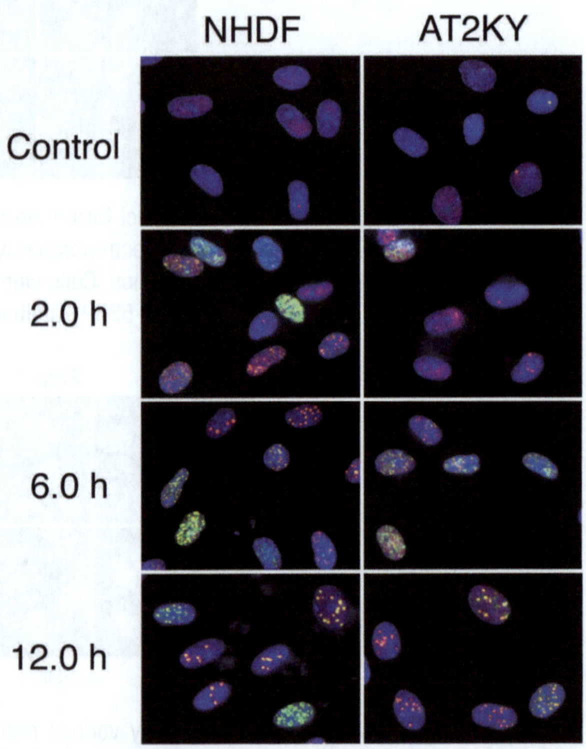

Fig. 4 Effect of ATM function on foci formation. Synchronized normal human diploid fibroblast (NHDF) and AT2KY fibroblasts (AT2KY) in G1 were electroporated with *Pvu* II (100 U), and incubated for the time indicated before fixation. Cells were stained with anti-phosphorylated histone H2AX antibody (*Green signals*) and anti-53BP1 antibody (*Red signals*). Control; no treatment

4 Notes

1. Start with lower passage NHDFs, as normal human cells significantly increase the background foci frequency with the increase in the passage number. G0/G1-synchronized cells are recommended to use, since the S-phase cells show unwanted spontaneous signals for ATM phosphorylation, which are resulted from alteration of chromatin structure during DNA replication. Also, the number of DNA damage foci is different depending on cell cycle phases. Even with the same radiation dose received, cells in G1 phase harbor twice the number of foci that the G2 cells have (*see* ref. [20]). Therefore, the usage of asynchronous cells may not be suitable for any quantitative assays.
2. Using T25 cell culture flasks is not necessary for preparing G0/G1-synchronized cells. Any type of cell culture ware can be applied.
3. The doubling time of the normal human diploid cells is approximately 24 h, so that the 7-day's cultivation is sufficient for getting the confluent cultures. However, cells with longer doubling times need to extend the culture for up to 10 days. Also, the cells used in this method are quite sensitive to cell-to-cell contact. For those relatively insensitive to contact inhibition, serum deprivation or low serum medium (3–5% FBS) may be needed.
4. Square wave pulses are used in this protocol. Optimum conditions may vary depending on the cells used. Typical conditions for the time constants and field strengths are approximately 1–20 ms and 400–900 V/cm, respectively. In general, increasing the time constants and field strengths decrease the cell recovery rate but increase the electroporation efficacy. Number of pulses is another experimental condition that needs optimization (*see* Subheading 3.4).
5. Use fresh trypsin solution to facilitate efficient cell collection. Since electroporation by itself also damages cell membrane, trypsinization should be completed as soon as possible; otherwise, cell viability will be significantly compromised.
6. The Gene Pulser Cuvettes used in this protocol have 0.4 cm width; however, any type of cuvette is applicable. Maximum volume of the cell suspension for the Gene Pulse Cuvette with 0.4 cm is 500 μL, but it is possible to use 200–500 μL cell suspension with similar electroporation efficiency.
7. Prepare 15 mL culture tube with 5 mL of culture medium before electroporation. Immediately after the pulse, take the cell suspension and transfer to the culture tube. Cell suspension should be mixed gently to avoid further cell damage.

8. DNA damage response is initiated immediately after DNA double strand breaks are induced. The initial foci will be detectable approximately 10 min after the treatment. However, it is rather difficult to prepare the samples, because the protocol needs attached cells on the cover glasses. Usually approximately two hours is required for cell attachment.
9. For 22 × 22 mm cover glasses, 100 μL of antibody solution is sufficient. It is possible to use the smaller size of cover glasses to reduce the amount of antibody solution. Also, using the plastic cover may minimize the amount of antibody solution; however, one might be cautious as plastic covers sometimes damage cells.
10. A variety of restriction enzymes were examined (Subheading 2.3). Group I includes *Pvu* II, *Rsa* I, *Bgl* I, *Eco* RV, and *Sma* I, which generate blunt ends. Group II enzymes include *Pst* I, *Sph* I, and *Kpn* I, which produce 3′-protruding ends. Group III includes *Eco* RI, *Bam* HI, *Not* I, *Hind* III, and *Hinf* I, which create 5′-protruding ends.
11. It is ideal to use smaller amount of restriction enzymes to minimize cellular damage. If high concentration restriction enzymes are unavailable, enzyme solution up to 5 μL can be applied.
12. Number of pulses needs to be optimized. For mammalian cells, two to five times are recommended. More than five times could be lethal to cells. There is a general tendency that the conditions more harmful to cells show higher electroporation efficacy. Normal human cells are relatively vulnerable to electroporation, so that the maximum pulse number will be around five. For rodent cells, increasing pulse numbers up to ten times may be useful. Cells with two pulses usually show uneven DNA damage distribution (Fig. 1), and approximately 30% of cells are phosphorylated ATM and 53BP1 foci negative. With increasing the pulse numbers, the number of cells with DNA damage is increasing. With five pulses, almost all cells show extensive DNA damage.
13. Cells incubated for 2 h after restriction enzyme treatment exhibit ATM phosphorylation and 53BP1 foci (Fig. 2). Most of the signals are diffused throughout the nuclei. The signals become countable 6 h after the treatment, indicating that cells efficiently eliminate DSBs. By 12 h after the treatment, most of the signals are disappeared, while a few foci are still detectable. Such residual foci remain for a long time after the treatment, which result in a loss of proliferative potential.
14. Induction of DNA damage was examined by introducing various restriction enzymes (100 U) (Fig. 3). Electroporation of cells with *Pvu* II, *Rsa* I, *Pst* I, *Sph* I, and *Eco* RI efficiently induces phosphorylated ATM and 53BP1 foci in more than 80%

of cells, while *Eco* RV, *Bam* HI, and *Hinf* I induce foci in approximately 50% of cells. However, little or no foci was detectable by *Bgl* I, *Sma* I, *Kpn* I, *Not* I, and *Hind* III, even if the amount of enzymes was increased.

15. Although optimized enzyme buffers are used, cells are not suspended in a Milli-Q water but in PBS(−). Therefore, the reaction conditions may not be ideal for certain types of restriction enzymes, such as *Bgl* I, *Sma* I, *Kpn* I, *Not* I, and *Hind* III.
16. Involvement of ATM function in DNA damage repair was investigated (Fig. 4). Since AT2KY cells lack ATM protein, no phosphorylated ATM signals are detectable. Therefore, phosphorylated histone H2AX was used as DNA damage marker. After the treatment with 100 U of *Pvu* II, significant abrogation of foci induction was observed compared with NHDF. After 6 h, foci signals were detectable but the number of foci was higher in AT2KY cells, demonstrating that DNA repair is partly compromised in AT2KY cells. Similar results were obtained in other restriction enzymes, except *Pst* I and *Sph* I, which create 3′-protruding ends. Thus, the current protocol reveals a role of ATM function in the end structure-dependent DNA repair.

Acknowledgments

This study was supported in part by the Global Center Of Excellence (GCOE) Program and Grant-in-Aid for Scientific Research from the Ministry of Education, Culture, Sports, Science and Technology, Japan.

References

1. Chapman JR, Taylor MR, Boulton SJ (2012) Playing the end game: DNA double-strand break repair pathway choice. Mol Cell 47:497–510
2. Bunting SF, Nussenzweig A (2013) End-joining, translocations and cancer. Nat Rev Cancer 13:443–454
3. Kakarougkas A, Jeggo PA (2014) Requirement for PBAF in transcriptional repression and repair at DNA breaks in actively transcribed regions of chromatin. Br J Radiol 87:20130685
4. Ceccaldi R, Rondinelli B, D'Andrea AD (2015) Repair pathway choices and consequences at the double-strand break. Trends Cell Biol 26:52–64
5. Polo SE, Jackson SP (2011) Dynamics of DNA damage response proteins at DNA breaks: a focus on protein modifications. Genes Dev 25:409–433
6. Goodarzi AA, Jeggo PA (2013) The repair and signaling responses to DNA double-strand breaks. Adv Genet 82:1–45
7. Lavin MF (2008) Ataxia-telangiectasia: from a rare disorder to a paradigm of cell signalling and cancer. Nat Rev Mol Cell Biol 9:759–769
8. Ciccia A, Elledge SJ (2010) The DNA damage response: making it safe to play with knives. Mol Cell 40:179–204
9. Shiloh Y, Ziv Y (2013) The ATM protein kinase: regulating the cellular response to genotoxic stress, and more. Nat Rev Mol Cell Biol 14:197–210
10. Lavin MF, Kozlov S, Gatei M, Kijas AW (2015) ATM-dependent phosphorylation of all three

members of the MRN complex: from sensor to adaptor. Biomolecules 5:2877–2902
11. Bakkenist CJ, Kastan MB (2003) DNA damage activates ATM through intermolecular autophosphorylation and dimer dissociation. Nature 421:499–506
12. Kozlov SV, Graham ME, Jakob B et al (2011) Autophosphorylation and ATM activation: additional sites add to the compleity. J Biol Chem 286:9107–9119
13. Bakkenist CJ, Kastan MB (2015) Chromatin perturbations during the DNA damage response in higher eukaryotes. DNA Repair 36:8–12
14. Paull TT (2015) Mechanisms of ATM activation. Annu Rev Biochem 84:711–738
15. Morgan SE, Kastan MB (1997) p53 and ATM: cell cycle, cell death, and cancer. Adv Cancer Res 71:1–25
16. Suzuki K, Okada H, Yamauchi M et al (2006) Qualitative and quantitative analysis of phosphorylated ATM foci induced by low-dose ionizing radiation. Radiat Res 165:499–504
17. Suzuki K, Takahashi M, Oka Y et al (2010) Recruitment of ATM-dependent pathway for the repair of a subset of DNA double-strand breaks created by restriction endonucleases. Genome Integr 1:4
18. Winegar RA, Phillips JW, Youngblom JH, Morgan WF (1989) Cell electroporation is a highly efficient method for introducing restriction endonucleases into cells. Mutat Res 225:49–53
19. Suzuki K, Kodama S, Watanabe M (1999) Recruitment of ATM protein to double strand DNA irradiated with ionizing radiation. J Biol Chem 274:25571–25575
20. Beucher A, Birraux J, Tchouanding L et al (2009) ATM and artemis promote homologous recombination of radiation-induced DNA double-strand breaks in G2. EMBO J 28:3413–3427

Chapter 8

Quantitative and Dynamic Imaging of ATM Kinase Activity by Bioluminescence Imaging

Shyam Nyati, Grant Young, Brian Dale Ross, and Alnawaz Rehemtulla

Abstract

Ataxia telangiectasia mutated (ATM) is a serine/threonine kinase critical to the cellular DNA damage response, including DNA double strand breaks (DSBs). ATM activation results in the initiation of a complex cascade of events facilitating DNA damage repair, cell cycle checkpoint control, and survival. Traditionally, protein kinases have been analyzed in vitro using biochemical methods (kinase assays using purified proteins or immunological assays) requiring a large number of cells and cell lysis. Genetically encoded biosensors based on optical molecular imaging such as fluorescence or bioluminescence have been developed to enable interrogation of kinase activities in live cells with a high signal to background. We have genetically engineered a hybrid protein whose bioluminescent activity is dependent on the ATM-mediated phosphorylation of a substrate. The engineered protein consists of the split luciferase-based protein complementation pair with a CHK2 (a substrate for ATM kinase activity) target sequence and a phosphoserine/threonine-binding domain, FHA2, derived from yeast Rad53. Phosphorylation of the serine residue within the target sequence by ATM would lead to its interaction with the phospho-serine-binding domain, thereby preventing complementation of the split luciferase pair and loss of reporter activity. Bioluminescence imaging of reporter-expressing cells in cultured plates or as mouse xenografts provides a quantitative surrogate for ATM kinase activity and therefore the cellular DNA damage response in a noninvasive, dynamic fashion.

Key words ATM, Kinase activity, In vivo, Molecular imaging, Bioluminescence, Complementation, Split-luciferase, Live cell, Reporter

1 Introduction

Protein kinases constitute one of the largest gene families, comprising ~2% of the human genome. It is estimated that approximately 30% of all cellular proteins are phosphorylated on at least one residue. Thus, protein kinases have key roles in many fundamental processes of cellular signaling in cancer as well as normal cells. Biochemical methods have been widely used to investigate whether or not a protein kinase of interest is active. Although biochemical methods are robust in vitro, they generally do not provide information about protein kinase activity in specific subcellular

Sergei V. Kozlov (ed.), *ATM Kinase: Methods and Protocols*, Methods in Molecular Biology, vol. 1599,
DOI 10.1007/978-1-4939-6955-5_8, © Springer Science+Business Media LLC 2017

Table 1
Luciferase complementation-based reporters for imaging of kinase activities

Kinase	Substrate	Peptide sequence	Phospho peptide-binding domain	Reference
AKT	FOXO4/AFX1	QSRPRSCTWPLPRPEKKK	FHA2	[9]
ATM	CHK2	LETVSTQELYSI	FHA2	[12]
EGFR	EPS15	KPANFSAYPSEEDMIE	SH2	[2]
FADD KINASES	FADD	QNRSGAMSPMSWNSDASTSEAS	FHA2	[3]
GSK3β/CKIα	β-CATENIN	SYLDSGIHSGATTTAPSLSG	FHA2	[4]
c-MET	PYK2	LSESCSIESDIYAEIPDETLR	SH2	[10]
TGFβR	SMAD2	LTQMGSPSVRCSSMS	FHA2	[6]

compartments; nor do they provide information about activity changes at the single cell level. We and others have developed optical imaging reporters to measure the kinase activity of various oncologically important kinases ([1–11], Table 1) and have utilized these reporters in subsequent studies that lead to the identification of new inhibitors and discovery of novel signaling mechanisms [12, 13].

Bioluminescence is a chemical reaction where light is emitted by a living organism. Luciferases are a large family of light-generating enzymes that catalyze the oxidation of a substrate, generically called luciferin, to yield oxyluciferin with the concomitant production of light. For in vivo bioluminescence imaging of malignancy, tumor cells or cancer-related genes are tagged with a reporter gene that encodes a light-generating enzyme, luciferase [14–16]. When this reporter is in the presence of the substrate, it emits a blue to yellow-green light with an emission spectra peaking at a wavelength between 490 nm and 620 nm [14]. An extremely sensitive cooled charged-coupled device (CCD) camera or a photomultiplier detects any low light that is emitted during the bioluminescence reaction. Due to its extreme sensitivity, broad dynamic range, and exceptionally large signal-to-noise ratio, this type of noninvasive imaging permits a real-time analysis of an ample amount of various biological events [15]. Although there are more than 30 luciferase-luciferin systems that were derived independently of each other, the most frequently used luciferase for in vivo molecular imaging is the ATP-dependent firefly (*Photinus pyralis*) luciferase [17]. The reason for this is that 30% of the light produced by firefly luciferase has an emission spectra above 600 nm, a region in which the signal attenuation by the absorbing and scattering properties of live mammalian tissue is at a minimum [15, 17]. Recently, a very bright and smaller luciferase (NanoLuc; NLuc) from deep sea shrimp (*Oplophorus gracilirostris*) has been successfully used for dual luciferase imaging in a mouse model [18].

A significant advantage of the cell-based bioluminescent kinase reporter is its adaptability for high-throughput screening. Bioluminescence generated in luciferase assays offers higher sensitivity than FRET-based systems due to amplification of the signal. In addition, luciferase is less susceptible to inference from nonspecific fluorescence of compounds. Thus, bioluminescence-based assays are highly suited for high-throughput screening. Furthermore, luciferase activity can be monitored dynamically and noninvasively, allowing bioluminescence-based cell assays to provide a unique method for identifying specific compounds that interact with the target in the correct cellular compartment and under normal cellular physiological conditions of that compartment (pH, concentrations of specific ions, etc.). Reporters wherein the firefly luciferase enzyme has been divided into two halves (N-Luc and C-Luc) were originally developed to study protein-protein interaction [19]. These split-luciferase reporters were based on either the inter-molecular or intra-molecular complementation of the luciferase fragments to generate signal in response to cellular cues.

Ataxia-Telangiectasia Mutated (ATM) is a member of the PI3-like family of serine/threonine kinases. It is a very large 370 kDa protein encoded by human chromosome 11q22-23. It plays a critical role in repair of DNA double strand breaks (DSBs), thereby maintaining genomic stability. These processes include, but are not limited to, DNA replication, DNA repair, cell cycle progression, apoptosis, and senescence. ATM exists in its inactive form as a non-covalently linked dimer where the kinase domain of one monomer is bound to the internal domain of another monomer covering the S1981 residue. In response to DSBs, the kinase domain of one monomer phosphorylates S1981 of the other interacting ATM resulting in subunit dissociation, ATM activation, and recruitment to DNA break sites [20]. Ionizing radiation-induced ATM activation results in the activation of a large number of ATM substrates [21–26] including P53, MDM2, SMC1, KAP1, BRCA1, γH2AX, and CHK2. The activated ATM triggers a sequence of events including cell cycle arrest, allowing time for the repair of the damaged DNA in sync with circadian rhythm [27]. If damaged DNA is left unrepaired it can lead to cell death, genomic instability, cancer, and/or other pathologies [28]. The 2015 award of the Nobel Prize in Chemistry for the discovery of DNA repair mechanisms highlights the importance of this pathway. Because of the important role ATM plays in cancer, therapeutics have been devised to target it [29].

In vitro kinase assays using purified substrate and kinase are routinely used to evaluate kinase activity. For traditional cell-based studies immunohistological and biochemical techniques have been utilized for evaluating the kinase activity of ATM, such as counting pATM foci, γH2AX foci, immunofluorescence, or immunoprecipitation-western blotting [26, 30, 31]. Johnson, You, and Hunter [11] described a fluorescence resonance energy transfer (FRET)-based biosensor for monitoring ATM kinase activity in live cells. Although this reporter provides direct measurement of the ATM kinase activity, it has a limitation of usability in mouse model due to tissue penetration and autofluorescence in the CFP-YFP range. In this book chapter, we provide detailed methods of use for the recently developed split firefly-based bioluminescence reporter to noninvasively, dynamically, and sensitively measure ATM kinase activity in live cells and mouse models [32].

2 Materials

2.1 Molecular Biology

1. DNA encoding open reading frame for phospho protein-binding domain (e.g., Rad53p FHA2 domain).
2. Full-length coding sequence for the *Firefly* luciferase (FLuc) or plasmids coding for the N-terminal luciferase (amino acids 1–416) and C-terminal luciferase (amino acids 398–550) fragments.
3. Expression vectors with constitutive promoters for expression in mammalian cells (e.g., pEF vector).
4. Expression vectors and packaging plasmids for generating lentiviral particles (optional).
5. High fidelity polymerases (Pfu).
6. dNTP.
7. Primers for cloning and sequencing.
8. Thermocycler.
9. Restriction endonuclease(s).
10. DNA ligase.
11. Site-directed mutagenesis kit.
12. High efficiency competent cells.
13. Bacterial growth media (LB, SOC).
14. LB agar plates.
15. Plasmid DNA extraction kits.
16. DNA gel purification kit.
17. DNA sequencing kit.

2.2 Cell Culture

1. HEK293T cells or other readily transfectable cell lines.
2. Desired cell line(s) for biologic question of interest (i.e., D54, U87).

3. Complete cell culture growth medium.
4. Antibiotic (G418) for selection of stable clones.
5. Filter paper discs and forceps for isolation of stable clones.

2.3 Cell Imaging

1. Black-walled or white-walled 96-Well clear-bottom plates for tissue culture.
2. Multichannel pipette sets for volumes from 1 to 100 μL.
3. Sterile low adherence pipette tips with barrier filter.
4. Stock solution of D-luciferin or Glo-sensor c-AMP reagent 40 mg/mL in PBS, stored in dark-colored vials at −80 °C (*Firefly* luciferase substrate).
5. Live cell bioluminescence imaging system with very high sensitivity and required software package for data generation and analysis (e.g., Xenogen IVIS, Perkin Elmer).
6. Live cell plate reader (e.g., EnVision Multilabel Reader, Perkin Elmer).
7. Liquid-handling instrument, plate-handling robot, and a cell culture incubator compatible with high-throughput instruments for high-throughput assays (optional).

2.4 Animal Imaging

1. Appropriate mouse strain for desired experimental system, such as immunocompromised mice (nude, SCID, or NSG) for human tumor xenografts.
2. Small animal shaver (optional).
3. 4 mg/mL luciferin stock in sterile PBS, store in tightly sealed dark tubes at −20 or −80 °C.
4. 28–30 gauge insulin syringe for intra-peritoneal (IP) luciferin injection in mice.
5. Bioluminescence imaging instrument with a heated platform and isoflurane anesthesia injection and controller systems.

3 Methods

3.1 Construct Firefly Luciferase Complementation-Based ATM Kinase Activity Reporter

1. Select a substrate such as CHK2 and determine the length of the substrate sequence which can be used for the construction of the reporter (*see* **Note 1**). We typically select 12–20 amino acid long substrate sequences with the target residue/s at the center of the sequence where possible (Table 1). For the construction of the ATM kinase reporter (ATMR), we selected a 12-residue sequence derived from CHK2 (Figs. 1 and 2).
2. Add a 5–7 amino acid long linker sequence at both the ends of the substrate sequence. We typically use GGSGG as the linker in our kinase reporters. Attach a phospho-peptide-binding

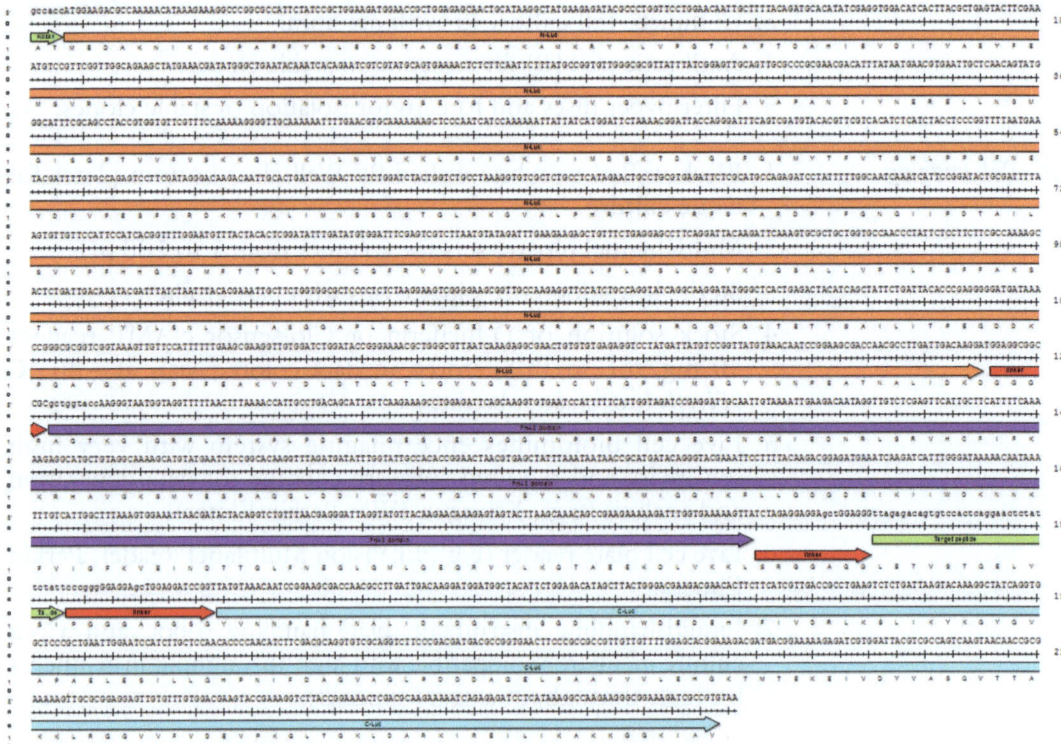

Fig. 1 The DNA coding sequence and translated amino acid sequence for all the domains of the ATM kinase activity reporter. In frame short linker sequences (linker) inserted between each functional domain provide flexibility for the intramolecular domain interaction in the chimeric reporter molecule. N-Luc denotes amino acids 1–416 of firefly luciferase and C-Luc denotes amino acids 398–550. The target peptide sequence was derived from the CHK2 coding sequence (amino acids 63–74). The Ser/Thr phospho-peptide-binding domain (FHA2) comprises amino acids 420–522 of Rad53P protein

domain such as FHA2 (residues 420–582) for Ser/Thr kinases [33], or SH2 domain for Tyr kinases (residues 374–465 of mouse shc2; [34, 35]). Use appropriate N-terminal (N-Luc) and C-terminal (C-Luc) firefly luciferase fragment pairs [19] at the flanks.

3. Generate chimeric reporter constructs using appropriate molecular biology procedures. We generally incorporate linkers, substrates, and restriction enzyme sites in primers. We typically confirm the clones by sequencing, bioluminescence, and expression by western blotting before proceeding to generate stable cell lines.
4. Generate relevant control constructs wherein the phospho-target residue/s within the substrate is mutated to neutral amino acids. We typically mutate Ser/Thr or Tyr to Ala in mutant reporters by site-directed mutagenesis (Fig. 2a).
5. Express complementation reporters in appropriate vectors for mammalian cells. Vectors should be selected with markers,

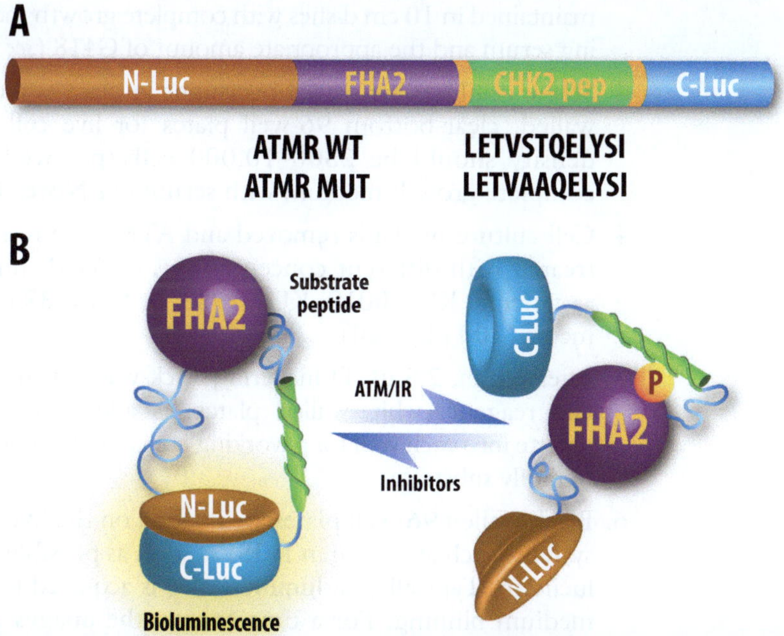

Fig. 2 The components and functional basis of the ATM kinase activity reporter. (**a**) The ATM reporter is consisted of a phospho Ser/Thr-binding domain (FHA2), substrate peptide (CHK2), and split firefly luciferase. The substrate sequence is flanked by short linker sequences at either end. The functional basis of the reporter is demonstrated in (**b**). In the presence of ATM kinase activity, the CHK2 target peptide is phosphorylated, resulting in interaction with the FHA2 domain, producing steric constraints that inhibit functional reconstitution of the luciferase. In the absence of ATM kinase activity achieved by use of small molecule inhibitors, siRNA-mediated knockdown (of ATM), or over-expression of phosphatases, the CHK2 consensus sequence is hypo-phosphorylated, allowing for luciferase enzyme reconstitution and increased bioluminescent activity

such as antibiotic resistance genes or coexpressed fluorescent proteins that are suitable for generating stable cell lines. We typically use pEF vector with Geneticin/G418 antibiotic for stable clone selection.

3.2 Cell-Based Bioluminescence Imaging of ATM Kinase Activity

1. We typically do all of our cell-based and in vivo bioluminescence assays using the reporter-expressing stable cell lines. We carefully select cell lines that represent appropriate cellular and biological context for our studies (*see* **Note 2**). Cells are transfected with the reporter plasmids and allowed to grow under the antibiotic selection media. We typically pick 12–24 single cell clones using sterilized filter paper discs and choose the best clones by measuring bioluminescence in response to specific kinase inhibitors (*see* **Notes 3** and **4**).
2. The three best reporter-expressing stable cell lines (clones) are expanded and frozen at low passages for future use. Cells are

maintained in 10 cm dishes with complete growth media containing serum and the appropriate amount of G418 (*see* **Note 5**).

3. Stable cell lines are plated overnight in black-walled or white-walled, clear-bottom 96-well plates for live cell assays. Cell density should be 2500–10,000 cells/per well in 100 μL complete growth medium with serum (*see* **Notes 6** and 7).
4. Cell culture media is removed and ATMR-expressing cells are treated with different concentrations of ATM inhibitors such as caffeine, KU-60019, KU-55933, or CGK733 in serum-free media (100 μL/well).
5. After 5 min, 2.5 μL D-luciferin (black-walled plates) or cAMP-Glo reagent (white-walled plates) is added by multichannel pipette into each well for a working concentration of 100 μg/mL of firefly substrate.
6. Black-walled 96-well plates are imaged on the live cell imaging system (such as Xenogen IVIS) as soon as possible after adding luciferin. Typically, bioluminescence is acquired for 30–60 s at medium binning. For a time-course, the images are acquired with 3–10 min delay between the reads (Fig. 3a, b).
7. The white-walled plates are read on the live cell plate reader (such as Envision) right after the addition of the substrate. Generally, each well of the plate is read for 0.01–1.0 s. For a longer time course activity measurement, a delay of 15–60 min between each read is set (Fig. 3c). For each read the robot takes the plate out from the incubator, loads it on the reader where the plate is read, and is transferred back to the incubator until the next time point (*see* **Note 8**).
8. Quantify bioluminescence acquired on the IVIS imaging system by region-of-interest (ROI) analysis using Living Image software. The bioluminescence data from the live cell plate reader is automatically saved in a quantitative form in a tab-delimited file format.
9. Since radiation activates ATM within minutes, the bioluminescence activity of ATMR can be evaluated within 15 min after irradiation.
10. All the bioluminescence measurements should be validated by western blotting in ATMR-expressing cell lines in parallel experiments.

3.3 In Vivo Imaging of ATM Kinase Activity

1. D54-ATMR cells are expanded, trypsinized, and suspended in serum-free media at 40 × 10⁶ cells/mL. 50 μL of this suspension is injected into each flank (2 × 10⁶ cells) in nude mice using a 22-gauge needle. We usually wait until the tumor reaches 60–100 mm³ size (3–4 weeks) before starting the experiments.
2. We acquire baseline bioluminescence measurements 3–6 h before starting the treatment (Fig. 4a). Each mouse is injected

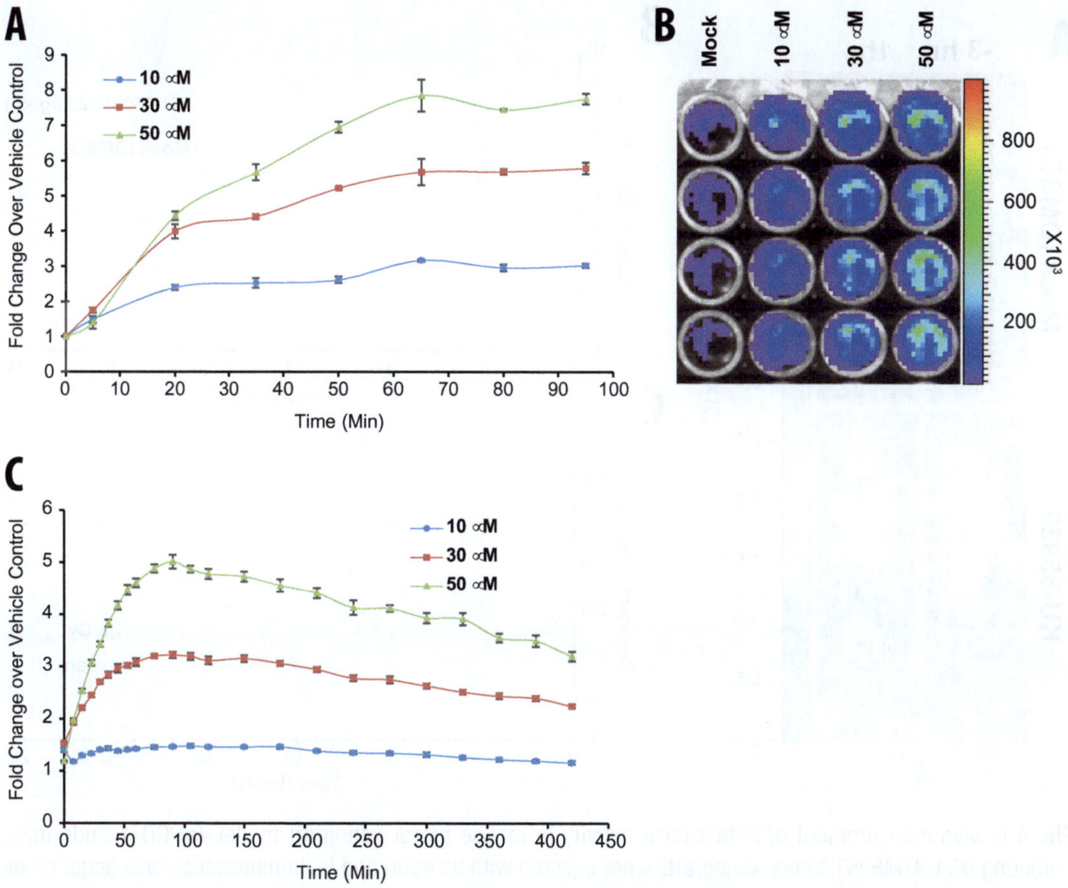

Fig. 3 Firefly complementation-based live cell assay for noninvasive monitoring of ATM kinase activity. (**a**) D54 cells stably expressing the ATM kinase activity reporter (ATMR) were plated in black-walled 96-well plates (5000 cells/well). Cells were incubated with mock (DMSO) or an increasing concentration of ATM kinase inhibitor KU-55933, which increased the signal in a dose-dependent way. (**b**) A representative image of the bioluminescence acquired in response to various concentrations of KU-55933 is shown. ROI in a grid is created and overlaid in the pseudocolored image to quantitate the photons emitted. Scale bar shows photon flux in pseudocolor with *blue* as the lowest and *red* as the highest counts. (**c**) D54-ATMR cells plated in *white*-walled 96-well plates and treated as described above and read on Envision plate reader

with 100 μL D-luciferin (400 μg/mouse), anesthetized with 1–2% isoflurane for 5 min (*see* **Note 9**).

3. Transfer mice to the bioluminescence instrument, where they are maintained under anesthesia, and acquire bioluminescence. We typically acquire data on 5 mice at once isolated by a plastic separator. Generally, a 15–30 s acquisition at medium sensitivity is sufficient. We typically acquire data for 10–20 reads with a 1–5 min delay between the reads to cover the bioluminescence peak from all the tumors in each of the mice.
4. Treat the mice with appropriate inhibitors such as KU-55933 (25 mg/kg), or activators such as radiation (5Gy) and monitor

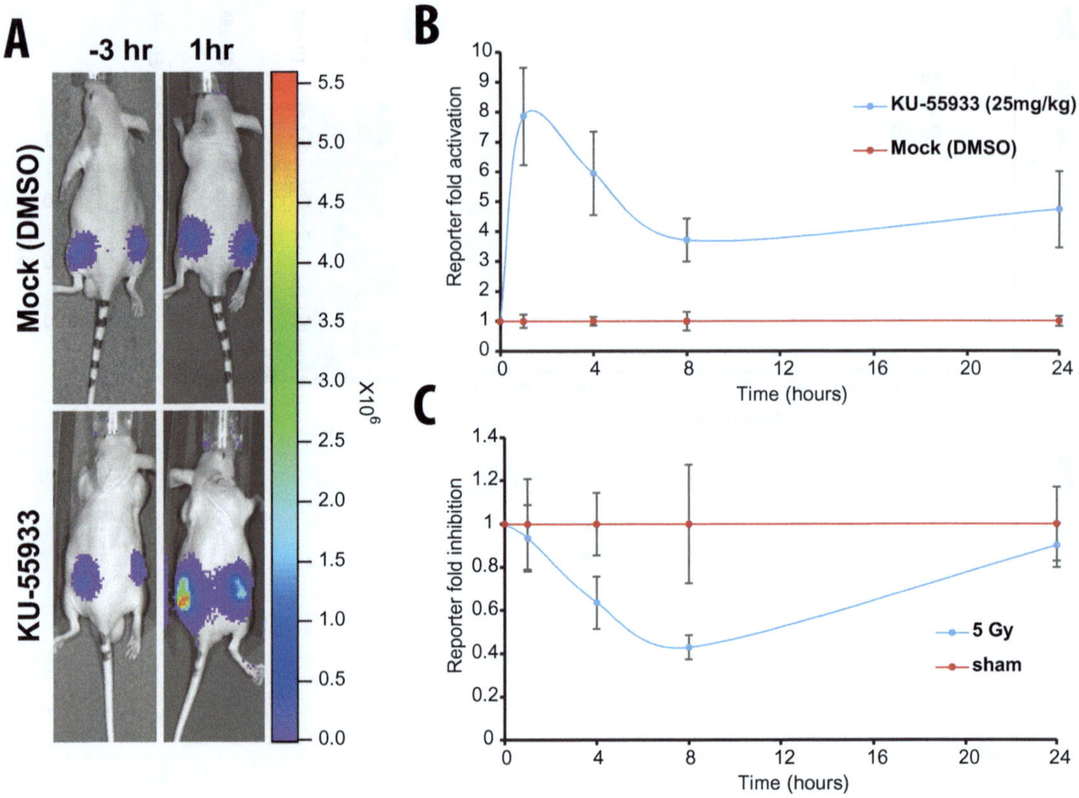

Fig. 4 In vivo measurement of ATM kinase activity in mouse tumor xenograft model. (**a**) CD-1 nude mice harboring D54-ATMR WT tumor xenografts were injected with luciferin and bioluminescence was acquired as described 3 h before treatments. (**b**) The animals were injected with KU-55933 (25 mg/kg) or vehicle control (DMSO) and bioluminescence was acquired 1, 4, 8, and 24 h posttreatment. The ATM reporter fold activation upon ATM inhibition is plotted over mock treatment. (**c**) Similarly, mouse harboring D54-ATMR WT tumor xenografts were whole body irradiated with 5Gy of radiation or sham irradiated and bioluminescence was measured for up to 24 h. About 70% decrease in the reporter activity was observed 8 h post irradiation

bioluminescence over time. Vehicle control (DMSO) or sham-irradiated mice should be used as control (Figs. 3c and 4b; *see* **Note 10**).

5. Remove mouse from imaging instrument and monitor for complete recovery from anesthesia.
6. Quantify imaging data by region-of-interest (ROI) analysis of bioluminescence produced by the tumor, using units of photon flux (Fig. 4a; *see* **Notes 11–17**).

4 Notes

1. Substrate sequence for the construction of the reporter should be decided based on literature searches. It is imperative to have good antibodies available to detect changes in the substrate

phosphorylation as this will help in validating the kinase reporter bioluminescence data by biochemical techniques.

2. The selection of a specific cell line for creating a reporter-expressing stable line should be based on the specific questions being interrogated and the intrinsic activity and detectability of the kinase and the substrate in the cell line. This should be determined by experimentation as well as literature searches. We selected the D54 glioblastoma cell line because ATM inhibition in glioblastoma may sensitize them to IR and chemotherapy [36]. D54 cells express wild-type ATM protein and respond to ATM inhibitors and irradiation as seen by western immunoblotting using antibodies against ATM and its substrate CHEK2.
3. We generally select for clones that exhibit low to moderate bioluminescence after adding luciferin. Since it is an activatable reporter that shows an increase in the light with inhibition of the ATM kinase, clones expressing the reporter in very high abundance may not yield high signal to background and thus may show a limited fold activation in response to inhibitors.
4. For picking up the clones, complete growth media with antibiotic is aspirated from the tissue culture dishes containing single cell clones. Sterilized filter paper discs pre-wetted in trypsin are transferred to clones using sterilized forceps in a laminar flow hood. After 4–5 min, the discs are lifted from the tissue culture plate and swirled around in a 24-well plate containing complete growth media with half the concentration of antibiotic used for clone selection. Generally, 12–24 single cell clones are picked. Forceps are sterilized either by heat or by dipping them in ethanol for 5–10 min. Make sure to let the ethanol evaporate before using the forceps to pick up the clones.
5. For expanding and maintaining of stable cell lines we use half the concentration of G418 than that was used for selection. The G418 concentration for stable clone selection should be empirically decided.
6. The N-Luc and C-Luc fragments used in the construction of the complementation-based ATM reporter are derived from a firefly luciferase that has been optimized to work in mammalian cells at the physiological temperature (+37 °C). Therefore, all the bioluminescence acquisitions should be performed at +37 °C.
7. Since this reporter is based on the complementation of the light-generating enzyme luciferase, it works only in live cells under physiological conditions. Methods wherein cell lysate is used to measure the bioluminescence signals are incompatible with this reporter system and will not be able to yield any detectable change in signal to background (under different treatment conditions).

8. The Envision system is built with ultra-sensitive luminescence detection technology, thus 0.01–0.1 s measurement for each well is usually sufficient. The detection time can be increased if the signal is very weak and desired signal-to-noise is not reached with lower detection times. Furthermore, the bioluminescence signal from cAMP-Glo reagent is stable for a prolonged period of time, allowing us to measure the kinase activity of ATM for longer periods.
9. In our hands we find that 400 μg luciferin/mouse gives us the best signal-to-background bioluminescence readings for split firefly luciferase-based kinase reporters. For smaller tumors or tumors generated from cell lines expressing very low levels of reporter, the amount of luciferin can be increased to 150 mg/kg body weight (i.e., 3.5–4 mg/mouse with 20–25 g average mouse weight).
10. One may need to empirically determine the optimal inhibitor concentration for the best signal-to-noise bioluminescence detection in vivo. We usually test two to three different concentrations of the drugs in mouse tumor xenograft model to find the optimal concentration that gives highest fold change over vehicle control without being toxic to animals.
11. We create separate region-of-interest (ROI) for each tumor based on its size and shape. We also make sure that this ROI does not overlap with the ROI of any other tumors. We copy-paste the same ROI for each tumor for counting total photon flux for all the time points. We may move the position of the ROI so that it covers the tumor (because the position of the same animal with the same tumor might be slightly different between different time points) but do not change the overall shape or size of any ROI. This removes the chances of including background photon counts emanating from mouse skin.
12. We generally use 4–5 mice in an experimental group for bioluminescence data acquisition on the IVIS imaging system. Based on the number of animals in a group, we choose a stage level (distance of the CCD camera and the subject) and use the same stage level for the whole experiment. Changing the distance between the CCD camera and the subject height between different time points or reads would lead to different photon counts which will be difficult to analyze.
13. We usually use medium sensitivity settings with 15 s–60 s data collection time for each read. One should acquire photon flux without any saturated pixels. All the saturated pixels (above 10^8 counts) show the same value, and therefore, cannot be accurately analyzed.
14. We typically use the maximum photon flux emitted from each tumor separately for all the calculations. Each tumor may show

the maximum emittance at different periods of time; therefore, we usually perform sequential reads with a delay between each read which allows us to collect the maximum photon flux for all the tumors.

15. We typically use the maximum photon count for each tumor for each time for the analysis. If a tumor does not respond at any time-point, we omit that reading from the analysis. We analyze and plot combined values of all the tumors from one treatment group. Photon counts from vehicle (DMSO) or sham treated animals are set as onefold and the ratio of other treatments is counted as fold change from vehicle/sham treatment.
16. Due to the sensitivity of the imaging system, one typically gets some background bioluminescence in nude and SCID mice. It is important to shave the mice if using a mouse strain with hair to reduce the background and determine the correct size, shape, and position of a ROI.
17. Bioluminescence data acquired in a mouse xenograft model should be validated by biochemical methods such as western blotting or immunohistochemistry (IHC). For the validation of the bioluminescence data for ATMR, tumor tissue should be analyzed with pATM and pCHK2 antibodies after control, KU-55933, KU-60019, or radiation treatments.

5 Conclusion

The method described herein is an adaptation of the traditional protein complementation assay for the detection of protein-protein interaction in live cells. Instead of monitoring the interaction of two proteins through the use of split reporter molecules, we have adapted the assay such that the interaction between the "bait" and the "prey" occurs in response to the activity of a specific kinase. The kinase can be a serine/threonine- or a tyrosine-kinase. The reporter has also been engineered such that increased complementation (and therefore reporter activity) occurs in response to decreased kinase activity. This approach is therefore very well suited for high-throughput screens for kinase inhibitor libraries since a positive hit would be detected as an increase in bioluminescence activity, thereby less likely to result in false positives. We have also used analogous reporters for whole genome siRNA screens. As an example, a reporter for TGF-β receptor serine/threonine kinase activity was used in a human kinome siRNA screen to yield a number of novel genes as regulators of the TGF-β receptor function [12]. Regulation of the molecular events that lead to the activation and/or inactivation of the ATM kinase activity is yet to be defined; therefore, it is anticipated that analogous whole genome siRNA screens against the ATMR will most likely yield new insights into the role of novel genes in the regulation of the cellular response to DNA damage.

Acknowledgments

We would like to acknowledge the members of the Center for Molecular Imaging at the University of Michigan for their valuable input and support for the studies. This work was supported by the National Institutes of Health grant R01CA193690 (AR), P01CA087634 (BDR, AR and SN), as well as a P30CA046592 award to the University of Michigan Cancer Center.

References

1. Bhojani MS, Nyati S, Rao HR, Rehemtulla A (2010) Molecular imaging in lung cancer metastases. In: Lung cancer metastasis. Springer, New York, pp 267–287
2. Khan AP, Contessa JN, Nyati MK, Ross BD, Rehemtulla A (2011) Molecular imaging of epidermal growth factor receptor kinase activity. Anal Biochem 417(1):57–64. doi:10.1016/j.ab.2011.05.040
3. Khan AP, Schinske KA, Nyati S, Bhojani MS, Ross BD, Rehemtulla A (2010) High-throughput molecular imaging for the identification of FADD kinase inhibitors. J Biomol Screen 15(9):1063–1070
4. Nyati S, Ranga R, Ross BD, Rehemtulla A, Bhojani MS (2010) Molecular imaging of glycogen synthase kinase-3β and casein kinase-1α kinases. Anal Biochem 405(2):246–254
5. Nyati S, Ross BD, Rehemtulla A, Bhojani MS (2010) Novel molecular imaging platform for monitoring oncological kinases. Cancer Cell Int 10:23. doi:10.1186/1475-2867-10-23
6. Nyati S, Schinske K, Ray D, Nyati M, Ross BD, Rehemtulla A (2011) Molecular imaging of TGF beta-Induced smad2/3 phosphorylation reveals a role for receptor tyrosine kinases in modulating TGF beta signaling. Clin Cancer Res 17(23):7424–7439. doi:10.1158/1078-0432.Ccr-11-1248
7. Williams TM, Nyati S, Ross BD, Rehemtulla A (2013) Molecular imaging of the ATM kinase activity. Int J Radiat Oncol Biol Phys 86(5): 969–977. doi:10.1016/j.ijrobp.2013.04.028. S0360-3016(13)00457-4 [pii]
8. Zhang L, Bhojani MS, Ross BD, Rehemtulla A (2008) Molecular imaging of protein kinases. Cell Cycle 7(3):314–317
9. Zhang L, Lee KC, Bhojani MS, Khan AP, Shilman A, Holland EC, Ross BD, Rehemtulla A (2007) Molecular imaging of Akt kinase activity. Nat Med 13(9):1114–1119
10. Zhang L, Virani S, Zhang Y, Bhojani MS, Burgess TL, Coxon A, Galban CJ, Ross BD, Rehemtulla A (2011) Molecular imaging of c-Met tyrosine kinase activity. Anal Biochem 412(1):1–8. doi:10.1016/j.ab.2011.01.028
11. Johnson SA, You Z, Hunter T (2007) Monitoring ATM kinase activity in living cells. DNA Repair 6(9):1277–1284. doi:10.1016/j.dnarep.2007.02.025
12. Nyati S, Schinske-Sebolt K, Pitchiaya S, Chekhovskiy K, Chator A, Chaudhry N, Dosch J, Van Dort ME, Varambally S, Kumar-Sinha C, Nyati MK, Ray D, Walter NG, Yu H, Ross BD, Rehemtulla A (2015) The kinase activity of the Ser/Thr kinase BUB1 promotes TGF-beta signaling. Sci Signal 8(358):ra1. doi:10.1126/scisignal.2005379
13. Schinske KA, Nyati S, Khan AP, Williams TM, Johnson TD, Ross BD, Tomás RP, Rehemtulla A (2011) A novel kinase inhibitor of FADD phosphorylation chemosensitizes through the inhibition of NF-κB. Mol Cancer Ther 10(10): 1807–1817
14. McCaffrey A, Kay MA, Contag CH (2003) Advancing molecular therapies through *in vivo* bioluminescent imaging. Mol Imaging 2(2): 75–86
15. Contag CH, Bachmann MH (2002) Advances in *in vivo* bioluminescence imaging of gene expression. Annu Rev Biomed Eng 4:235–260
16. Choy G, Choyke P, Libutti SK (2003) Current advances in molecular imaging: noninvasive *in vivo* bioluminescent and fluorescent optical

imaging in cancer research. Mol Imaging 2(4):303–312

17. Greer LF 3rd, Szalay AA (2002) Imaging of light emission from the expression of luciferases in living cells and organisms: a review. Luminescence 17(1):43–74
18. Stacer AC, Nyati S, Moudgil P, Iyengar R, Luker KE, Rehemtulla A, Luker GD (2013) NanoLuc reporter for dual luciferase imaging in living animals. Mol Imaging 12(7):1–13
19. Luker KE, Smith MC, Luker GD, Gammon ST, Piwnica-Worms H, Piwnica-Worms D (2004) Kinetics of regulated protein-protein interactions revealed with firefly luciferase complementation imaging in cells and living animals. Proc Natl Acad Sci U S A 101(33):12288–12293
20. Bakkenist CJ, Kastan MB (2003) DNA damage activates ATM through intermolecular autophosphorylation and dimer dissociation. Nature 421(6922):499–506. doi:10.1038/nature01368
21. Bensimon A, Schmidt A, Ziv Y, Elkon R, Wang SY, Chen DJ, Aebersold R, Shiloh Y (2010) ATM-dependent and -independent dynamics of the nuclear phosphoproteome after DNA damage. Sci Signal 3(151):rs3. doi:10.1126/scisignal.2001034
22. Bhatti S, Kozlov S, Farooqi AA, Naqi A, Lavin M, Khanna KK (2011) ATM protein kinase: the linchpin of cellular defenses to stress. Cell Mol Life Sci 68(18):2977–3006. doi:10.1007/s00018-011-0683-9
23. Choi S, Srivas R, Fu KY, Hood BL, Dost B, Gibson GA, Watkins SC, Van Houten B, Bandeira N, Conrads TP, Ideker T, Bakkenist CJ (2012) Quantitative proteomics reveal ATM kinase-dependent exchange in DNA damage response complexes. J Proteome Res 11(10):4983–4991. doi:10.1021/pr3005524
24. Lavin MF, Kozlov S (2007) ATM activation and DNA damage response. Cell Cycle 6(8):931–942
25. Mu JJ, Wang Y, Luo H, Leng M, Zhang J, Yang T, Besusso D, Jung SY, Qin J (2007) A proteomic analysis of ataxia telangiectasia-mutated (ATM)/ATM-Rad3-related (ATR) substrates identifies the ubiquitin-proteasome system as a regulator for DNA damage checkpoints. J Biol Chem 282(24):17330–17334. doi:10.1074/jbc.C700079200
26. Kim ST, Lim DS, Canman CE, Kastan MB (1999) Substrate specificities and identification of putative substrates of ATM kinase family members. J Biol Chem 274(53):37538–37543
27. Sancar A, Lindsey-Boltz LA, Kang TH, Reardon JT, Lee JH, Ozturk N (2010) Circadian clock control of the cellular response to DNA damage. FEBS Lett 584(12):2618–2625. doi:10.1016/j.febslet.2010.03.017
28. Stracker TH, Roig I, Knobel PA, Marjanovic M (2013) The ATM signaling network in development and disease. Front Genet 4:37. doi:10.3389/fgene.2013.00037
29. Weber AM, Ryan AJ (2015) ATM and ATR as therapeutic targets in cancer. Pharmacol Ther 149:124–138. doi:10.1016/j.pharmthera.2014.12.001
30. Shiotani B, Zou L (2009) Single-stranded DNA orchestrates an ATM-to-ATR switch at DNA breaks. Mol Cell 33(5):547–558. doi:10.1016/j.molcel.2009.01.024
31. Kozlov S, Gueven N, Keating K, Ramsay J, Lavin MF (2003) ATP activates ataxia-telangiectasia mutated (ATM) *in vitro*. Importance of autophosphorylation. J Biol Chem 278(11):9309–9317
32. Williams TM, Nyati S, Ross BD, Rehemtulla A (2013) Molecular imaging of the ATM kinase activity. Int J Radiat Oncol Biol Phys 86(5):969–977. doi:10.1016/j.ijrobp.2013.04.028
33. Durocher D, Jackson SP (2002) The FHA domain. FEBS Lett 513(1):58–66
34. Filippakopoulos P, Muller S, Knapp S (2009) SH2 domains: modulators of nonreceptor tyrosine kinase activity. Curr Opin Struct Biol 19(6):643–649. doi:10.1016/j.sbi.2009.10.001
35. Schlessinger J (1994) SH2/SH3 signaling proteins. Curr Opin Genet Dev 4(1):25–30
36. Frosina G (2009) DNA repair and resistance of gliomas to chemotherapy and radiotherapy. Mol Cancer Res 7(7):989–999. doi:10.1158/1541-7786.MCR-09-0030

Chapter 9

Zn(II)–Phos-Tag SDS-PAGE for Separation and Detection of a DNA Damage-Related Signaling Large Phosphoprotein

Eiji Kinoshita, Emiko Kinoshita-Kikuta, and Tohru Koike

Abstract

In this chapter, we provide a standard protocol for phosphate-affinity sodium dodecyl sulfate–polyacrylamide gel electrophoresis (Zn^{2+}–Phos-tag SDS-PAGE). This technique uses a dizinc(II) complex of the phosphate-binding molecule Phos-tag in conjunction with a neutral-pH gel system, Tris [tris(hydroxymethyl) aminomethane], and acetic acid (Tris–AcOH), to detect shifts in the mobility of phosphorylated ataxia telangiectasia-mutated (ATM) kinase. This protocol, which employs a 3% (w/v) polyacrylamide gel strengthened with 0.5% (w/v) agarose, permits the separation of larger phosphoproteins with molecular masses in the order of 200 kDa over a period of approximately 4 h. Subsequently, multiple phosphorylated forms of high-molecular-mass ATM kinase (350 kDa) can be clearly detected via immunoblotting as multiple upshifted migration bands on the Zn^{2+}–Phos-tag SDS-PAGE gel. The procedure described in this protocol requires a completion time of approximately 5 h from the beginning of gel preparation to the end of electrophoresis.

Key words Affinity electrophoresis, ATM kinase, Large phosphoprotein, Phos-tag, Protein phosphorylation, Zn^{2+}–Phos-tag SDS-PAGE

1 Introduction

We have been developing several techniques for phosphoproteomic studies by using a series of functional Phos-tag molecules that bind specifically to phosphate groups. Phos-tag is a binuclear metal complex that can selectively bind to a phosphate monoester in an aqueous solution at neutral pH values [1]. Among the techniques, we used a Phos-tag derivative, acrylamide-pendant Phos-tag (Fig. 1a), to demonstrate the use of phosphate-affinity sodium dodecyl sulfate–polyacrylamide gel electrophoresis (Phos-tag SDS-PAGE) to detect shifts in the mobilities of phosphoprotein species in comparison with their nonphosphorylated counterparts (Fig. 1b) [2–7]. The affinity electrophoresis, which uses a polyacrylamide-bound Mn^{2+}–Phos-tag (*see* Fig. 1a) and the Laemmli's

Sergei V. Kozlov (ed.), *ATM Kinase: Methods and Protocols*, Methods in Molecular Biology, vol. 1599,
DOI 10.1007/978-1-4939-6955-5_9,

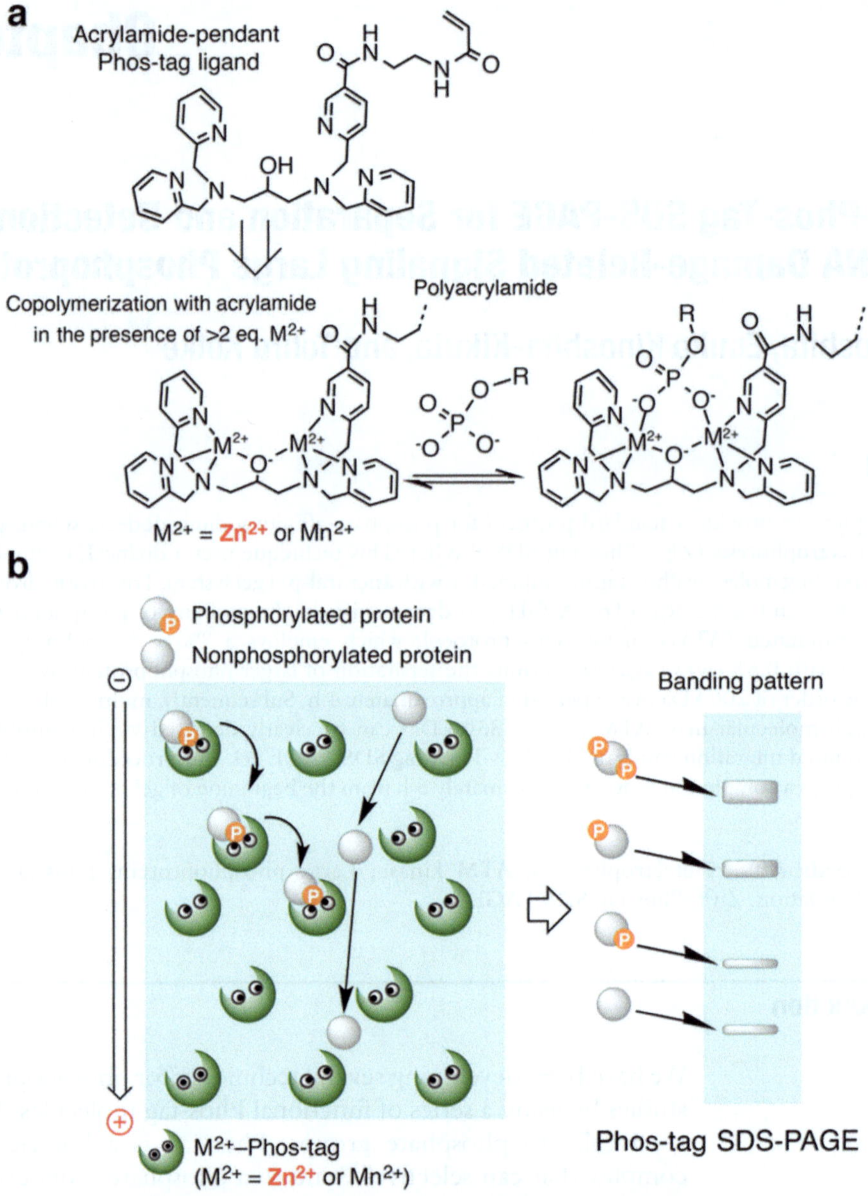

Fig. 1 Phos-tag sodium dodecyl sulfate-polyacrylamide gel electrophoresis (SDS-PAGE) for the detection of mobility shifts in phosphoprotein species. (**a**) Structure of the acrylamide-bound Phos-tag ligand and scheme for the reversible capture of a phosphomonoester dianion (R–OPO_3^{2-}) using a polyacrylamide-bound Phos-tag. (**b**) Schematic representation of the Phos-tag SDS-PAGE principle

buffer system [8], has been widely used to determine the phosphorylation states of many proteins [9–12].

Furthermore, our group has developed an improved Phos-tag SDS-PAGE technique in which a Zn^{2+}–Phos-tag complex (*see* Fig. 1a) [13–15] is used in conjunction with a neutral-pH gel system buffered with Bis-Tris [2-[bis(2-hydroxyethyl) amino]-2-(hydroxymethyl)propane-1,3-diol] and hydrochloride (Bis-Tris–HCl). Our Zn^{2+}–Phos-tag SDS-PAGE method has yielded significant improvements in the detection of phosphoprotein mobility shifts as a result of the greater affinity between phosphate groups and the dizinc(II) complex of Phos-tag under neutral pH conditions. The Phos-tag SDS-PAGE technique offers the following major advantages: (1) the phosphate-affinity procedure is almost identical to that of conventional SDS-PAGE; (2) a downstream procedure, such as gel staining, Western blotting, or mass spectrometric analysis, can be applied; (3) radioactive and chemical labels are unnecessary for kinase and phosphatase assays; (4) various phosphoprotein species can be separately detected as multiple migration bands, depending on the phosphorylation status; (5) the phosphate-binding potency is independent of the type of phosphorylated amino acid; (6) several phosphoprotein species containing the same number of phosphate groups can be separated; (7) the time-course quantitative ratio of phosphorylated to nonphosphorylated proteins can be determined; and (8) a large number of phosphorylation events producing phosphohistidine or phosphoaspartic acid intermediates can be detected in a wide variety of organisms [16–18].

The separation and detection of high-molecular-mass proteins generally require extremely porous polyacrylamide gels that contain less than 5% (w/v) polyacrylamide. We previously reported a procedure for Mn^{2+}–Phos-tag SDS-PAGE in which highly porous gels strengthened by the homogeneous addition of 0.5% (w/v) agarose were used to analyze large phosphoproteins with molecular masses in excess of 200 kDa [6, 7]. However, in the case of our Zn^{2+}–Phos-tag SDS-PAGE procedure, which uses a neutral-pH gel system buffered with Bis-Tris–HCl, we discovered a problem in the preparation of highly porous gels; specifically, all protein bands appear as smeared images because the Bis-Tris-buffered gels have inadequate sieving properties at a polyacrylamide concentration of less than 4% (w/v) [14]. Because Bis-Tris is a weakly basic amine ($pK_a = 6.5$ at 20 °C) that is partially protonated at neutral pH values, the free amine species (R_3N) are converted into an amine radical cation (R_3N^{+}) that can act as a radical quencher during acrylamide polymerization in the presence of ammonium persulfate (APS) [19]. Therefore, the abundant Bis-Tris molecules interfere with the formation of the polyacrylamide gel matrix, particularly at very low concentrations of acrylamide.

To overcome this problem, we developed an alternative procedure for the Zn^{2+}–Phos-tag SDS-PAGE analysis of large phosphoproteins that incorporates a neutral-pH gel system buffered with Tris [tris(hydroxymethyl)aminomethane] and acetic acid (Tris–AcOH) [14]. The primary amine Tris ($pK_a = 8.2$ at 20 °C) is almost fully protonated at neutral pH values, and the resulting primary ammonium species ($R–NH_3^{+}$) should reduce the inhibition of acrylamide polymerization. In fact, the use of a Tris–AcOH buffer system (pH 7.0) permits the production of a separating gel with sieving properties at polyacrylamide concentrations of less than 4% (w/v). The utility of our newly adapted Tris–AcOH system was then demonstrated through the visualization of a novel upshifted band of a high-molecular-mass protein, ataxia telangiectasia-mutated (ATM) kinase (350 kDa), on a Zn^{2+}–Phos-tag gel strengthened with 0.5% (w/v) agarose.

Here, we describe the resolving power of Zn^{2+}–Phos-tag SDS-PAGE for the separation of multiple phosphorylated forms of ATM kinase, a DNA damage signaling-related large protein.

2 Materials (*See* Note 1)

2.1 Preparation of Cell Lysate

1. Culture medium for HeLa cells: Dulbecco's modified Eagle medium (DMEM) supplemented with 10% (v/v) fetal bovine serum (FBS), 100 units/mL of penicillin, and 100 μg/mL of streptomycin.
2. Stimulation solution: 10 mM actinomycin D (Sigma-Aldrich, St. Louis, MO, USA) in dimethyl sulfoxide as a stock solution. Store at −20 °C. Dilute the stock solution to a concentration of 2.0 mM actinomycin D in distilled water immediately before using.
3. Tris-buffered saline (TBS): 10 mM Tris–HCl, pH 7.5, and 0.10 M NaCl (*see* **Note 2**).
4. Lysis buffer (1×): 65 mM Tris–HCl, pH 6.8, 1.0% (w/v) SDS, 5.0% (v/v) 2-sulfanylethanol (*see* **Note 3**), 10% (v/v) glycerol, and 0.03% (w/v) bromophenol blue (BPB). Store at −20 °C.
5. 90-mm culture dishes.
6. Benchtop centrifuge.
7. 1.5-mL microcentrifuge tubes.
8. Cell scrapers.
9. Probe sonicator.
10. Protein concentration measurement reagent (e.g., Protein Assay, Bio-Rad, Hercules, CA, USA).

2.2 Gel Casting for Electrophoresis

1. Phos-tag Acrylamide solution: 5.0 mM acrylamide-bound Phos-tag ligand (Phos-tag Acrylamide AAL-107, Wako Pure Chemical Industries, Ltd., Osaka Japan) and 3.0% (v/v) MeOH (*see* **Note 4**) in distilled water. Store at room temperature in the dark.
2. Zinc(II) chloride solution: 10 mM $ZnCl_2$ in distilled water (*see* **Note 5**). Store at room temperature.

3. Acrylamide/bis solution (30%): 30% (w/v) solution of a 29:1 mixture of acrylamide and *N,N'*-methylenebisacrylamide in distilled water (*see* **Note 6**). Store at room temperature in the dark.
4. Gel buffer (5×; *see* **Note 7**): 1.0 M Tris–AcOH buffer, pH 7.0 (*see* **Note 8**). Store at room temperature.
5. Agarose solution (*see* **Note 9**): 1.5% (w/v) SeaKem Gold agarose (Lonza Group, Basel, Switzerland) in distilled water.
6. Ammonium persulfate (APS) (*see* **Note 10**) solution and *N,N,N',N'*-tetramethylethane-1,2-diamine (TEMED) (*see* **Note 11**): 10% (w/v) APS in distilled water. Regarding TEMED, the original solution is used as received from the supplier.
7. Electrophoresis running buffer (*see* **Note 12**): 50 mM Tris, 50 mM *N*-[2-hydroxy-1,1-bis(hydroxymethyl)ethyl]glycine (tricine), and 0.10% (w/v) SDS in distilled water. Store at room temperature. Dissolve sodium bisulfite in buffer solution to a concentration of 5.0 mM immediately before use (*see* **Note 13**).
8. 50-mL centrifuge tube.
9. 100-mL flask.
10. Microwave oven.

2.3 Electroblotting

1. Blotting buffer: 25 mM Tris, 192 mM glycine, and 10% (v/v) MeOH for the wet-tank method (*see* **Note 14**). Store at room temperature.
2. Ethylenediaminetetraacetic acid disodium salt (EDTA·2Na) solution: 0.50 M EDTA–sodium hydroxide (NaOH), pH 8 (*see* **Note 15**). Store at room temperature.
3. Blotting buffer containing 1 mM EDTA: 100 mL of blotting buffer (*see* Subheading 2.3, **item 1**) containing 0.20 mL of EDTA solution (*see* Subheading 2.3, **item 2**).
4. TBS-T solution: 10 mM of Tris–HCl, pH 7.5, 0.10 M NaCl, and 0.10% (v/v) poly(oxyethylene) sorbitan monolaurate (Tween 20). Store at room temperature.
5. Polyvinylidene difluoride (PVDF) membranes (e.g., Fluorotrans W, Nippon Pall, Tokyo, Japan).
6. Blotting papers (e.g., 3MM; Whatman, Maidstone, UK).

2.4 Equipment for Electrophoresis and Blotting

1. SDS-PAGE equipment: Atto model AE-6500 mini-slab gel system (gels: 1 mm thick, 9 cm wide, and 9 cm long) (Tokyo, Japan). The setup can be readily adapted to other formats, including large gels.
2. Electroblotting equipment: Nihon Eido model NA-1511C electroblotting wet-tank unit (Tokyo, Japan). The setup can be

readily adapted to other formats but not to semidry blotting (*see* **Note 14**).

3. Power supply (e.g., Atto model AE-8750 Power Station 1000XP).

3 Methods

3.1 Preparation of a Lysate from HeLa Cells

1. Incubate HeLa cells (10^7 cells) in a 90-mm culture dish with 10 mL of culture medium overnight (16–24 h; i.e., until all cells have adhered to the culture dish) in a humidified atmosphere of 5% CO_2 and 95% air at 37 °C.
2. Treat the cells with a stimulation solution containing actinomycin D (DNA-damaging reagent) at a final concentration of 2.0 μM for 2 h.
3. Gently wash the adherent cells twice with 5 mL of TBS per wash.
4. Lyse the cells by adding 0.5 mL of lysis buffer (1×) to each culture dish.
5. Scrape off the lysed cells and transfer the lysate to a 1.5-mL microcentrifuge tube.
6. Briefly sonicate the lysates, followed by centrifugation at 14,000 × *g* and 4 °C for 10 min.
7. Transfer the supernatant to a new 1.5-mL microcentrifuge tube.
8. Measure the protein concentration using a protein assay (*see* Subheading 2.1, **item 10**).
9. Dilute the lysate with an appropriate amount of fresh lysis buffer (1×) to yield a lysate sample with a protein concentration of 2.0 mg/mL.
10. Boil the lysate samples for 5 min, and store at −20 °C if necessary.
11. Subject an aliquot (5 μL) of the resulting sample solution to Zn^{2+}–Phos-tag SDS-PAGE.

3.2 Electrophoresis

3.2.1 Preparation of the Zn^{2+}–Phos-Tag SDS-PAGE Gel

1. Ensure that the glass plates used to cast the gels are clean. It is essential to wash these plates (1 mm thick, 9 cm wide, and 9 cm long for mini-slab gels) thoroughly with a rinsable detergent and subsequently thoroughly rinse them with distilled water before casting the gels.
2. Prepare the separating gel solution that will be used for the phosphorylated ATM kinase analysis. A stacking gel is not necessary in this case. For a typical example of a Tris–AcOH-buffered Zn^{2+}–Phos-tag gel [approximately 8.5 mL, 3% (w/v)

polyacrylamide] containing 0.5% (w/v) agarose, a separating gel solution (approximately 8.5 mL) is prepared by mixing 0.85 mL of acrylamide/bis solution, 1.7 mL of gel buffer (5×), 34 μL of Phos-tag Acrylamide solution, 34 μL of zinc(II) chloride solution (two equivalents of Phos-tag), 9 μL of TEMED, and 2.997 mL of distilled water in a 50-mL centrifuge tube.

3. Prepare 1.5% (w/v) agarose solution. Suspend SeaKem Gold agarose (0.75 g) in 50 mL of distilled water in a 100 mL flask and completely melt the agarose solution by heating in a microwave oven (*see* **Note 16**).
4. Add 2.833 mL of the hot agarose solution (>90 °C) to the separating gel solution and mix gently (*see* **Note 17**).
5. Add 43 μL of APS solution and mix gently.
6. Transfer the warm solution, which should be warmer than the gelling temperature of the agarose, to the gap between the glass plates, insert a sample-well comb, and allow the acrylamide to polymerize for approximately 20 min (*see* **Note 18**).
7. Carefully remove the comb from the gel (*see* **Note 19**), and assemble the gel plate and electrophoresis apparatus (*see* **Note 20**).

3.2.2 Procedure for Electrophoresis

1. Fill the electrode chambers with the electrophoresis running buffer. Take care not to introduce bubbles to the bottom surface of the gel set (*see* **Note 21**).
2. Load the lysate samples into the wells (*see* **Note 22**).
3. Attach the leads to the power supply. Run the gels under a constant current of 15 mA/gel at room temperature until the Bromophenol Blue dye reaches the bottom of the separating gel. The time required for complete electrophoresis is approximately 4 h when a mini-slab gel is used (*see* **Note 23**).

3.3 Electroblotting

1. When the run is complete, remove the gel from the apparatus and soak it in blotting buffer containing 1 mM EDTA for 30 min (*see* **Note 24**).
2. Soak the gel in blotting buffer without EDTA for an additional 10–30 min.
3. Prepare a PVDF membrane by cutting it to the same size as the gel and soaking it for 30 s in 100% MeOH, followed by incubation in blotting buffer for more than 30 min.
4. Prepare four pieces of blotting paper by cutting them to the same size as the gel.
5. To form a "blotting sandwich" on the electroblotting screen supplied with the electroblotting equipment, assemble the gel, PVDF membrane, and blotting paper as follows. Soak the blotting sponge supplied with the electroblotting equipment

in blotting buffer and place it on the electroblotting screen. Next, soak two pieces of blotting paper in blotting buffer and place them on the sponge, followed sequentially by the gel and PVDF membrane. Avoid incorporating air between the various layers. Place two more sheets of blotting paper and one sponge soaked in blotting buffer on the membrane and close the electroblotting screen. Insert the electroblotting screen into the chamber unit of the electroblotting equipment and fill the unit with blotting buffer (*see* **Note 25**).

6. Gently add SDS solution to the blotting buffer in the chamber unit to a final concentration of 0.1% (w/v) (*see* **Note 26**) [20].
7. Attach the leads to the power supply. Run the gel at a constant voltage of 3.5 V/cm for 16 h (overnight) (*see* **Note 27**).
8. After blotting, soak the PVDF membrane in TBS-T solution before performing an immunoblotting analysis (*see* **Note 28**).

3.4 Typical Results for the Separation and Detection of Phosphorylated ATM Kinase

The Tris–AcOH buffer system used in Zn^{2+}–Phos-tag SDS-PAGE confers some advantages; for example, its use permits the casting of highly porous polyacrylamide gels containing less than 4% (w/v) polyacrylamide for the separation and analysis of large phosphoproteins (molecular masses exceeding 200 kDa). We successfully used this system to detect a novel phosphorylated form of ATM kinase (350 kDa). ATM is a DNA damage signaling-related protein, in which the Ser-367, Ser-1893, and Ser-1981 residues autophosphorylate in response to DNA damage, and subsequently initiates the kinase activity toward other molecules [21]. Recently, autophosphorylation at Ser-2996 was also reported [22]. The various phosphorylation states of ATM, therefore, form key components of the associated signaling pathway.

A set of lysates was prepared from HeLa cells after treating the cells (10^7 cells) with 0 μM (control) or 2 μM actinomycin D (DNA-damaging reagent) for 2 h. The induced and control cultures were each washed twice with a TBS solution and then lysed in 0.50 mL of a 1× sample-loading dye. The two lysates were first analyzed by Zn^{2+}–Phos-tag SDS-PAGE on a 3.0% (w/v) polyacrylamide gel strengthened with 0.5% (w/v) agarose containing 20 μM of polyacrylamide-bound Zn^{2+}–Phos-tag in conjunction with the Tris–AcOH buffer system (Fig. 2, left-hand panel). Subsequent immunoblotting with a phosphorylation-independent anti-ATM antibody yielded a single band and three additional upshifted bands from the control (left lane: −) and treated samples, respectively (right lane: +). The two lysates were subsequently analyzed using the older Mn^{2+}–Phos-tag SDS-PAGE method, which used the same concentration of agarose–polyacrylamide composite gel containing 20 μM of polyacrylamide-bound Mn^{2+}–Phos-tag in conjunction with the Laemmli buffer system (Fig. 2, right-hand panel). Only two upshifted bands, corresponding to the phosphorylated forms of ATM, were detected

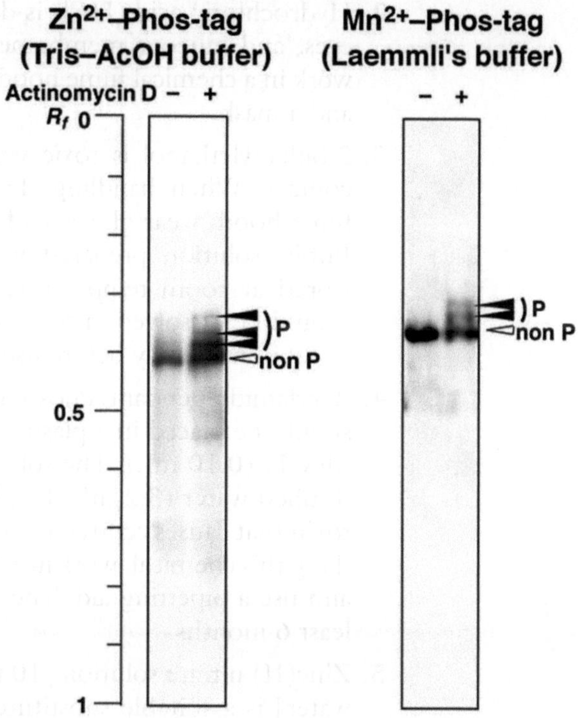

Fig. 2 Separation and detection of phosphorylated ATM in an agarose-strengthened Zn^{2+}–Phos-tag gel. The cells were treated with 0 μM (–) or 2 μM actinomycin D (+) for 2 h. The lysates were subjected to Phos-tag SDS-PAGE on a 3.0% (w/v) polyacrylamide gel strengthened with 0.5% (w/v) SeaKem Gold agarose and containing 20 μM Phos-tag, followed by immunoblotting with a phosphorylation-independent anti-ATM mouse monoclonal antibody (clone 2C1; Santa Cruz Biotechnology, Santa Cruz, CA, USA). Tris–AcOH gel buffer and Tris–Tricine running buffer were used for Zn^{2+}–Phos-tag SDS-PAGE (*left panel*), and Laemmli gel buffer and running buffer were used for Mn^{2+}–Phos-tag SDS-PAGE (*right panel*). Each lane contained 10 μg of protein. The position of the bromophenol blue dye was used to set the *Rf* value of 1.0. Reprinted with permission from ref. 14 © (2012) Wiley-VCH Verlag GmbH & Co. KGaA, Weinheim

in this case. Therefore, the Zn^{2+}–Phos-tag SDS-PAGE protocol described herein, which uses a Tris–AcOH buffer (pH 7.0), permits a more detailed detection of mobility shifts in phosphorylated species of ATM in response to DNA damage relative to the previous Mn^{2+}–Phos-tag SDS-PAGE technique.

4 Notes

1. All reagents and solvents used should be purchased at the highest commercial quality available and used without further purification. All aqueous solutions should be prepared using deionized and distilled water.

2. Hydrochloric acid (HCl) is dangerously irritating to the skin, eyes, and mucous membranes. When handing this chemical, work in a chemical fume hood and wear gloves, eye protection, and a mask.
3. 2-Sulfanylethanol is toxic via inhalation, ingestion, and skin contact. When handling this chemical, work in a chemical fume hood, wear gloves and a mask, and use a pipetting aid. Buffer solution prepared without 2-sulfanylethanol can be stored at room temperature. In this case, 2-sulfanylethanol should be dissolved in the solution to a concentration of 15% (v/v) immediately before use.
4. Acrylamide-pendant Phos-tag ligand (10 mg), which is oily, should be placed in a plastic tube and completely dissolved in MeOH (0.10 mL). The solution should then be diluted with distilled water (3.2 mL) by pipetting. MeOH is an inhalation toxin that causes central nervous system depression; when handling this chemical, work in a chemical fume hood, wear gloves, and use a pipetting aid. The Phos-tag solution is stable for at least 6 months.
5. Zinc(II) nitrate solution [10 mM $Zn(NO_3)_2 \cdot 6H_2O$ in distilled water] is a suitable substitute. Because $ZnCl_2$ and $Zn(NO_3)_2$ are deliquescent salts, the solutions should be prepared using fresh products from newly opened bottles. Aqueous solutions of $ZnCl_2$ or $Zn(NO_3)_2$ are stable for at least 6 months.
6. Because acrylamide monomer is a neurotoxin and a suspected human carcinogen and teratogen, take care to avoid exposure to this substance. When weighing powdered acrylamide, work in a chemical fume hood and wear gloves, eye protection, and a mask. Furthermore, acrylamide is unstable and can polymerize violently if heated to its melting point (84.5 °C). It is incompatible with acids, bases, oxidizing agents, reducing agents, iron and its salts, copper, aluminum, brass, and free-radical initiators.
7. This buffer is used for the separating gel.
8. Acetic acid (AcOH) is dangerously irritating to the skin, eyes, and mucous membranes. When handing this chemical, work in a chemical fume hood and wear gloves, eye protection, and a mask. Keep away from heat and flame.
9. Using Zn^{2+}–Phos-tag SDS-PAGE methodology, we have confirmed that some commercially available agarose with a gel strength exceeding 1000 g/cm^2 at 1.5% (w/v) {e.g., Agarose LO3 TAKARA [>2200 g/cm^2 at 1.5% (w/v)] purchased from Takara Bio, Agarose KANTO [900–1400 g/cm^2 at 1.5% (w/v)] purchased from Kanto Chemical, Agarose KANTO ME [1400–1700 g/cm^2 at 1.5% (w/v)] purchased from Kanto Chemical, or Agarose KANTO LE [1200–1500 g/cm^2 at 1.5% (w/v)]

purchased from Kanto Chemical} are suitable substitutes for SeaKem Gold agarose, which has a gel strength exceeding 3500 g/cm^2 at 1.5% (w/v). However, NuSieve GTG [>500 g/cm^2 at 4% (w/v)] from Lonza is not a suitable substitute for use in a Zn^{2+}–Phos-tag SDS-PAGE gel buffered with Tris–AcOH to a final concentration of 0.5% (w/v). Similarly, NuSieve 3:1 [>1400 g/cm^2 at 4% (w/v)] from Lonza does not produce sufficiently strong gels for subsequent handling during electroblotting when added to an SDS-PAGE gel to a final concentration of 0.5% (w/v).

10. Avoid skin and eye contact, ingestion, and inhalation. Prolonged exposure may result in skin burns and ulcerations. Inhalation can cause respiratory irritation. The APS solution should be prepared immediately before use.
11. TEMED should be stored in a desiccator at room temperature and should be purchased in small quantities, as its quality may degrade once the container is opened, resulting in a longer gel polymerization time. Accordingly, it should be used as soon as possible after opening. The commercially available solution should be used as received.
12. Do not adjust the pH with acid or base.
13. Sulfite ion (SO_3^{2-}) is a reducing reagent that decreases O_2 levels in the electrophoresis running buffer solution and inhibits the oxidation of reduced proteins in the gel. Buffer solution in which sodium bisulfite has been dissolved should be promptly used for electrophoresis. Do not store buffer solution containing sodium bisulfite.
14. Although the semidry method is generally the most efficient method for protein blotting in terms of time and buffer reagent consumption, it is not suitable for electroblotting from Phos-tag SDS-PAGE gels, as the protein transfer efficiency is lower with the semidry method than with the wet-tank method. Do not adjust the pH with acid or base.
15. During pH adjustment with NaOH, EDTA will dissolve in distilled water. NaOH is dangerously irritating to the skin and eyes. When handing this chemical, wear gloves and eye protection.
16. To prevent changes in the concentration of the agarose solution as a result of boiling and evaporative water loss, any volume lost after completely melting the agarose solution should be replaced with distilled water.
17. Agarose gelling may cause the separating gel solution to harden before it is poured into the gel-casting system. After the agarose is melted in the microwave oven, the solution should be mixed and poured into the casting system as quickly as possible.

If necessary, preheat the casting system and plastic pipette tips in an oven at 40–45 °C.

18. If the acrylamide does not polymerize within approximately 20 min, TEMED or APS deterioration may be possible. The use of fresh TEMED and APS is recommended.
19. If the polymerized gel shrinks or slips down between the glass plates when the clip and spacer are removed, the polyacrylamide gel may not be sufficiently strengthened by agarose. Accordingly, ensure that the final agarose concentration is 0.5% (w/v), as recommended.
20. After **step 7**, Zn^{2+}–Phos-tag gels cast in Tris–AcOH neutral buffer will be stable for at least 6 months. However, the gel, casting glass plates, and sample-well comb should be wrapped in Saran wrap to prevent the gel from desiccation. The wrapped gel can be stored at room temperature under normal laboratory illumination until required. Precast Zn^{2+}–Phos-tag SDS-PAGE gels have recently become commercially available (SuperSep Phos-tag; Wako Pure Chemical Industries, Ltd.) [23].
21. If any bubbles are observed, they should be carefully and completely removed.
22. Various contaminants (e.g., EDTA, inorganic salts, or surfactants) in the sample protein solutions can disrupt the electrophoresis bands in Zn^{2+}–Phos-tag SDS-PAGE, causing waving and/or tailing. To minimize this disruption, the samples should be desalted before loading if the concentrations of contaminants vary widely. Furthermore, to avoid distortion of the protein bands, commercially available prestained molecular-weight protein markers should not be used in Zn^{2+}–Phos-tag SDS-PAGE.
23. Agarose-containing gels frequently display partial melting during electrophoresis because of the development of heat. To avoid this problem with an agarose–polyacrylamide composite gel, a relatively lower constant voltage (15 mA/gel) should be used for a longer time (approximately 4 h).
24. The presence of Zn^{2+}–Phos-tag in the gel causes inefficient electroblotting. This can be ameliorated by treatment with EDTA, which chelates the zinc(II) ions.
25. Wet-tank equipment is strongly recommended to ensure optimal protein transfer efficiency from the Zn^{2+}–Phos-tag SDS-PAGE gel. The protein transfer efficiency from the Zn^{2+}–Phos-tag gel is much higher with the wet-tank method than with the semidry method.
26. The addition of SDS is recommended to ensure optimal protein transfer efficiency from the Zn^{2+}–Phos-tag SDS-PAGE gel.

The optimal percentage of SDS [e.g., 0.05–0.2% (w/v)] required to achieve adequate electrotransfer should be determined [20].

27. If the agarose–polyacrylamide composite gel partially sticks to the PVDF membrane and is difficult to separate from the membrane, the gel may have partially melted during electroblotting because of the generation of heat. Therefore, do not set the voltage above 3.5 V/cm when using agarose-containing gels. Higher voltages during this step are not recommended, even if an electroblotting unit with a cooling apparatus is used.
28. When crude samples, such as whole cell lysates, are applied to the gel, the resulting banding pattern is badly distorted or smeared. Possible reasons for this phenomenon are as follows: (1) contamination by small-molecule substances, such as inorganic salts or surfactants; (2) contamination by viscous genomic DNA; (3) contamination by cell debris or insoluble biomaterials; or (4) impurities, such as lipids or small molecules derived from cells, that interfere with the interaction between the Zn^{2+}–Phos-tag and phosphoprotein. To avoid problems, the following steps are recommended: (1) dialysis filtration to decrease the amounts of contaminating small-molecule substances; (2) shearing of DNA into smaller fragments via brief sonication or the addition of benzonase nuclease to the sample solution; (3) centrifugation of the sample at 14,000 × *g* for 10 min before applying the supernatant to the Phos-tag gel; and (4) sample purification with chromatography resin, such as Whatman CDR (cell debris remover; catalogue number 4025050). Spin column chromatography permits the purification of small volumes of sample solution (10–200 μL) without significant losses. In addition, if the degree of target protein migration is different in each lane and the banding pattern is distorted, the samples may contain metal-chelating agents, such as EDTA or ethylene glycol-bis(2-aminoethyl ether)-*N*,*N*,*N′*,*N′*-tetraacetic acid. To avoid this problem, add $ZnCl_2$ or $Zn(NO_3)_2$ to the sample solutions to a final concentration of 1–2 mM.

Acknowledgments

This work was supported in part by KAKENHI Grant no. 25293005 to EK, no. 25560417 to EK, no. 15K07887 to EK-K, and no. 26460036 to TK, and by research grants from the Takeda Science Foundation to EK and EK-K.

References

1. Kinoshita E, Takahashi M, Takeda H et al (2004) Recognition of phosphate monoester dianion by an alkoxide-bridged dinuclear zinc(II) complex. Dalton Trans 1189–1193. http://pubs.rsc.org/en/Content/ArticleLanding/2004/DT/b400269e#!divAbstract
2. Kinoshita E, Kinoshita-Kikuta E, Takiyama K et al (2006) Phosphate-binding tag, a new tool to visualize phosphorylated proteins. Mol Cell Proteomics 5:749–757
3. Kinoshita-Kikuta E, Aoki Y, Kinoshita E et al (2007) Label-free kinase profiling using phosphate affinity polyacrylamide gel electrophoresis. Mol Cell Proteomics 6:356–366
4. Kinoshita E, Kinoshita-Kikuta E, Matsubara M et al (2008) Separation of phosphoprotein isotypes having the same number of phosphate groups using phosphate-affinity SDS-PAGE. Proteomics 8:2994–3003
5. Kinoshita E, Kinoshita-Kikuta E, Matsubara M et al (2009) Two-dimensional phosphate-affinity gel electrophoresis for the analysis of phosphoprotein isotypes. Electrophoresis 30:550–559
6. Kinoshita E, Kinoshita-Kikuta E, Ujihara H et al (2009) Mobility shift detection of phosphorylation on large proteins using a Phos-tag SDS-PAGE gel strengthened with agarose. Proteomics 9:4098–4101
7. Kinoshita E, Kinoshita-Kikuta E, Koike T (2009) Separation and detection of large phosphoproteins using Phos-tag SDS-PAGE. Nat Protoc 4:1513–1521
8. Laemmli UK (1970) Cleavage of structural proteins during the assembly of the head of bacteriophage T4. Nature 227:680–685
9. Kinoshita E, Kinoshita-Kikuta E, Koike T (2009) Phosphate-affinity gel electrophoresis using a Phos-tag molecule for phosphoproteome study. Curr Proteomics 6:104–121
10. Yamada S, Nakamura H, Kinoshita E et al (2007) Separation of a phosphorylated histidine protein using phosphate affinity polyacrylamide gel electrophoresis. Anal Biochem 360:160–162
11. Sugiyama M, Sugiyama Y, Hatano N et al (2010) The DNA-binding activity of mouse DNA methyltransferase 1 is regulated by phosphorylation with casein kinase 1δ/ε. Biochem J 427:489–497
12. Ishiai M, Kitao H, Smogorzewska A et al (2008) FANCI phosphorylation functions as a molecular switch to turn on the Fanconi anemia pathway. Nat Struct Mol Biol 15:1138–1146
13. Kinoshita E, Kinoshita-Kikuta E (2011) Improved Phos-tag SDS-PAGE under neutral pH conditions for advanced protein phosphorylation profiling. Proteomics 11:319–323
14. Kinoshita E, Kinoshita-Kikuta E, Koike T (2012) Phos-tag SDS-PAGE systems for phosphorylation profiling of proteins with a wide range of molecular masses under neutral pH conditions. Proteomics 12:192–202
15. Kinoshita-Kikuta E, Kinoshita E, Koike T (2012) Separation and identification of four distinct serine-phosphorylation states of ovalbumin by Phos-tag affinity electrophoresis. Electrophoresis 33:849–855
16. Kinoshita E, Kinoshita-Kikuta E, Shiba A et al (2014) Profiling of protein thiophosphorylation by Phos-tag affinity electrophoresis: evaluation of adenosine 5′-O-(3-thiotriphosphate) as a phosphoryl donor in protein kinase reactions. Proteomics 14:668–679
17. Kinoshita-Kikuta E, Kinoshita E, Eguchi Y et al (2015) Functional characterization of the receiver domain for phosphorelay Ccntrol in hybrid sensor kinases. PLoS One 10: e0132598
18. Kinoshita-Kikuta E, Kinoshita E, Eguchi Y et al (2016) Validation of *cis* and *trans* modes in multistep phosphotransfer signaling of bacterial tripartite sensor kinases by using Phos-tag SDS-PAGE. PLoS One 11:e0148294
19. Sepulveda P, Binner JGP (2001) Persulfate–amine initiation systems for gelcasting of ceramic foams. Chem Mater 13:4065–4070
20. Kinoshita-Kikuta E, Kinoshita E, Matsuda A et al (2014) Tips on improving the efficiency of electrotransfer of target proteins from Phos-tag SDS-PAGE gel. Proteomics 14:2437–2442
21. Kozlov SV, Graham ME, Peng C et al (2006) Involvement of novel autophosphorylation sites in ATM activation. EMBO J 25: 3504–3514
22. Kozlov SV, Graham ME, Jakob B et al (2011) Autophosphorylation and ATM activation: additional sites add to the complexity. J Biol Chem 286:9107–9119
23. Kinoshita-Kikuta E, Kinoshita E, Koike T (2012) A laborsaving, timesaving, and more reliable strategy for separation of low-molecular-mass phosphoproteins in Phos-tag affinity electrophoresis. Int J Chem (Mumbai, India) 4(5):1–8

Chapter 10

Identification of ATM Protein Kinase Phosphorylation Sites by Mass Spectrometry

Mark E. Graham, Martin F. Lavin, and Sergei V. Kozlov

Abstract

ATM (ataxia-telangiectasia mutated) protein kinase is a key regulator of cellular responses to DNA damage and oxidative stress. DNA damage triggers complex cascade of signaling events leading to numerous posttranslational modification on multitude of proteins. Understanding the regulation of ATM kinase is therefore critical not only for understanding the human genetic disorder ataxia-telangiectasia and potential treatment strategies, but essential for deciphering physiological responses of cells to stress. These responses play an important role in carcinogenesis, neurodegeneration, and aging. We focus here on the identification of DNA damage inducible ATM phosphorylation sites to understand the importance of autophosphorylation in the mechanism of ATM kinase activation. We demonstrate the utility of using immunoprecipitated ATM in quantitative LC-MS/MS workflow with stable isotope dimethyl labeling of ATM peptides for identification of phosphorylation sites.

Key words ATM kinase, Phosphorylation sites, Dimethyl labeling, Mass spectrometry

1 Introduction

The majority of protein kinases are regulated by posttranslational modifications, with autophosphorylation being one of the most important in kinase activation mechanisms [1]. The protein ataxia-telangiectasia mutated (ATM) is a serine/threonine kinase regulated by autophosphorylation. ATM is involved in multiple aspects of cellular metabolism, such as response to genotoxic stress, proliferation, and maintaining genetic stability. ATM belongs to a family of phosphatidylinositol (PI) 3-kinase-related kinases (PIKKs), which are grouped together on the basis of similarities between their core catalytic domain. Members of this superfamily include ATM, ataxia-telangiectasia and Rad3-related (ATR), DNA-dependent protein kinase (DNA-PK), suppressor with morphogenetic effect on genitalia protein 1 (SMG1), and mammalian target of rapamycin (mTOR) [2]. All PIKK are regulated by (auto) phosphorylation. ATM is rapidly activated in the presence of DNA

Sergei V. Kozlov (ed.), *ATM Kinase: Methods and Protocols*, Methods in Molecular Biology, vol. 1599,
DOI 10.1007/978-1-4939-6955-5_10, © Springer Science+Business Media LLC 2017

double strand breaks, but the mechanism of ATM kinase activation lacks detail. Putative ATM in vitro phosphorylation sites were identified using an oriented peptide libraries approach [3, 4]. The importance of autophosphorylation in regulating ATM kinase activity was suggested by Kozlov et al. [5], followed by the identification of S1981 as a major autophosphorylation site of ATM in vivo [6]. The model for ATM activation was proposed, where rapid intermolecular autophosphorylation in response to DNA damage caused dissociation of inactive ATM dimers to form active monomers. Regulation of ATM kinase activity was by Tip60 and hMOF1-mediated acetylation was also observed [7, 8]. Protein phosphatases PP5, PP2A, and PP2C phosphatase Wip1 have been implicated in the regulation of ATM kinase activity [9–11]. Interaction of ATM with the Mre11/Rad50/Nbs1 (MRN) complex is necessary for the optimal activation of ATM kinase [12]. However, S1981 autophosphorylation is not essential for MRN-dependent activation of ATM dimers by DNA at least in an in vitro system [13]. Autophosphorylation at S1987 (the mouse site corresponding to S1981) is dispensable for mouse ATM (mATM) activation, in an *Atm* mutant model [14]. It was suggested that mATM autophosphorylation is a consequence rather than a cause for mATM activation. The role of phosphorylation in the activation of mATM remains a debated issue (e.g., [15]). More recently, a new mechanism of ATM kinase activation in response to oxidative stress was discovered [16], where cysteine oxidation (C2991 in particular) leads to formation of active dimers of ATM kinase. It is not entirely clear how different modes of ATM kinase activation are regulated and how multiple posttranslational modifications interact to achieve optimal ATM kinase signaling in various cell signaling contexts (*see* [17] for a recent review).

Initially, traditional methods of phosphopeptide analysis were used to establish the importance of autophosphorylation in the regulation of ATM kinase activity and identify ATM in vivo phosphorylation sites [5, 6]. They included in vitro and in vivo ATM radioisotope labeling with ^{32}P, immunoprecipitation, and tryptic phosphopeptide analysis by two-dimensional electrophoresis-thin layer chromatography. Detailed methodology of phosphopeptide analysis was published earlier in these series. Interested readers are referred to excellent chapters on two-dimensional phosphopeptide mapping protocols [18, 19]. While ^{32}P labeling remains useful due to the extreme sensitivity of detection of radioisotopes and the ability to quantitatively profile virtually all phosphopeptides present in a protein, these approaches were largely superseded by the development of rapid and sensitive mass spectrometry methods, which are the subject of this chapter. We extensively used the techniques described here to map the in vivo ATM phosphorylation sites [20, 21]. We combined efficient isolation of ATM by immunoprecipitation with optimized SDS-PAGE separation and a quantitative phosphoproteomics workflow. The latter workflow includes peptide derivatization, phosphopeptide enrichment, and LC-MS/MS to obtain relative amounts of identified phosphorylation sites.

2 Materials

2.1 Cell Culture and Induction of DNA Damage

1. RPMI 1640 medium with 10% fetal calf serum (FCS), 100 U/mL penicillin, and 100 U/mL streptomycin.
2. Dulbecco's modified Eagle media (DMEM) supplemented with 10% FCS, penicillin, and streptomycin.
3. Gammacell 40 Exactor research irradiator (1 Gy/min; MDS Nordion, Ottawa, Canada) or analogous radiation source (*see* **Note 1**).

2.2 Cell Lysis and Preparation of Extracts

1. Phosphate-buffered saline (PBS).
2. Universal immunoprecipitation buffer (UIP): 1% Nonidet P-40 supplemented with protease, phosphatase inhibitors, and 1 mM dithiothreitol (DTT) [22].
3. ATM kinase cell lysis buffer (0.2% Tween 20) supplemented with protease, phosphatase inhibitors, and 1 mM dithiothreitol (DTT) [23].
4. Protease Inhibitors: Complete™ tablets (Boehringer Mannheim) and 1 mM phenylmethylsulfonyl fluoride (PMSF) (*see* **Note 2**).
5. Phosphatase inhibitors: 1 mM Na_3VO_4, 1 mM NaF, 10 mM Na_2MoO_4, 20 mM β-glycerophosphate, 5 μM microcystin-LR, 5 nM okadaic acid.
6. Rotating wheel or shaking platform.
7. Refrigerated microcentrifuge.

2.3 ATM Kinase Isolation by Immunoprecipitation and SDS-PAGE Gel-Separation

1. Anti-ATM antibody suitable for immunoprecipitation (e.g., affinity-purified sheep anti-ATM antibody as described in [5]).
2. Protein quantification assay reagents (e.g., BioRad Dc protein assay kit).
3. Protein A/G-Sepharose.
4. 2× concentrated SDS-PAGE sample loading buffer.
5. Protein electrophoresis equipment (*see* **Note 3**).
6. Coomassie Brilliant Blue G-250 stain [24].
7. 20% methanol.
8. Scanning densitometer or flat-bed scanner.

2.4 In-Gel Protein Digestion

1. Use 50 mM triethylammonium bicarbonate (TEAB) pH 8.0 if peptides will be derivatized for quantification. Otherwise, 50 mM ammonium bicarbonate (AmBic) pH 8.0 can be used. No adjustment of pH is required for this concentration. AmBic is an unintended substrate of primary amine derivatization.
2. 1 M stock solution of dithiothreitol (DTT) or Tris(2-carboxyethyl)phosphine hydrochloride (TCEP).
3. 0.5 M stock solution of 2-chloroacetamide in water (prepare-fresh, light sensitive). *See* **Note 4**.
4. Trypsin or trypsin-endoproteinase Lys-C mixture (Promega). *See* **Note 5**.
5. Acetonitrile.
6. Vacuum concentrator.

2.5 Derivatization of ATM Peptides with Stable Isotopes Using Dimethylation

Caution! Reagents for dimethylation are highly toxic. Take appropriate safety precautions.

1. Sodium cyanoborohydride ($NaBH_3CN$) and sodium cyanoborodeuteride ($NaBD_3CN$).
2. Formaldehyde, using various stable isotopes; CH_2O, CD_2O, $^{13}CD_2O$.
3. Formic acid, analytical grade.
4. Ammonia solution.

2.6 ATM Phosphopeptides Enrichment Using Titanium Dioxide

1. Titanium dioxide (TiO_2) chromatography material (e.g., Titansphere TiO bulk material, GL Sciences).
2. Trifluoroacetic acid (TFA), analytical grade.
3. Acetonitrile.
4. 0.5–20 μL GELoader tips (Eppendorf) or similar pipette tips.
5. C8 Empore disk (3 M).

2.7 Quantitative LC-MS/MS Analysis of ATM Phosphopeptides

1. Formic acid, >99.99% purity.
2. Acetonitrile, analytical grade.
3. 0.5–20 μL GELoader tips (Eppendorf) or similar pipette tips for StageTip or use ZipTip (Millipore).
4. C18 Empore disk (3 M) for StageTip or use ZipTip (Millipore).
5. Ammonia solution.
6. C18 reverse phase column material.
7. Fused silica tubing, 75 μm inside diameter × 40 cm.
8. Pressure cell for column packing.
9. HPLC and mass spectrometer suitable for sensitive peptide sequencing by LC-MS/MS.

3 Methods

Successful phosphorylation site analysis is highly dependent on obtaining sufficient amount of protein. Low nanogram quantities of purified proteins or peptides are likely to be lost to surface adsorption. We recommend purifying at least 1 μg of ATM per sample. The success of the protocol is highly dependent on purifying sufficient ATM for detection and quantification above background levels. Our method can be applied to ATM from a variety of cell types, provided that sufficient quantity of cells can be grown and efficient ATM immunoprecipitation can be performed. We successfully applied the method to isolate ATM from human lymphoblastoid cell lines, primary and hTERT-transformed human fibroblasts, as wells as mouse cell lines [21].

3.1 Cell Culture, Irradiation, and Sample Collection

1. Divide growing cultures of lymphoblastoid cell lines (LCL) in half or use growing cultures of fibroblasts at 80–90% confluency. The lymphoblastoid cell lines are cultured in RPMI 1640 medium with 10% FCS, 100 U/mL penicillin, and 100 U/mL streptomycin. Fibroblast cell lines are cultured in DMEM supplemented with 10% FCS, penicillin, and streptomycin. Include ATM protein-deficient A-T cell line as a negative control to facilitate localization of the immunoprecipitated ATM band in the gel.
2. Use a separate T175 flask for each condition, e.g., Control, irradiated, or treated with neocarzinostatin or other DNA-damaging agents. Perform irradiations (6–10 Gy) and return cells to incubator for a desired period of time. ATM activation by autophosphorylation is a rapid process and DNA damage inducible up-regulation of ATM autophosphorylation can be detected within 15 min after irradiation (as soon as it is practically possible to collect cells).
3. Collect cells by centrifugation or trypsinization. Wash cells with ice-cold PBS (phosphatase inhibitors should be added to PBS). Place cell pellets in −80 °C freezer (or dry ice). Our protocols for immunoprecipitation (IP) followed by labeling of ATM by in vitro kinase assay require a freeze-thaw cycle (see below); otherwise, you can proceed directly to cell lysis (*see* **Note 6**).

3.2 Preparation of Cell Extracts

1. Our cell lysis protocol was modified from [23, 25]. The lysis protocol was optimized for the ATM kinase in vitro radioactive assays. If lysates will be used only for the purification of ATM by immunoprecipitation for MS analysis, then the UIP buffer [22] can be used. Supplement lysis buffer with protease and phosphatase inhibitors immediately before use (*see* **Note 2**). Keep lysis buffer, harvested cells, and cell extracts on ice during the procedures.

2. Lyse cell pellets by repeated pipetting in 1 mL of lysis buffer followed by incubation on a rotating wheel for 30 min in a cold room. Centrifuge lysates at the max speed for 15 min in a refrigerated centrifuge to obtain clear extracts for ATM immunoprecipitation. Transfer extracts into fresh tubes.

3.3 ATM Immunoprecipitation

1. Preclear all extracts with Protein A/G-Sepharose (resuspended according to manufacturer's instructions and washed with PBS by incubation on a rotating wheel for 1 h in a cold room).
2. Spin down in a refrigerated centrifuge, transfer precleared extracts in fresh tubes, and keep on ice.
3. Measure protein concentration by Lowry (e.g., *DC*™ Protein Assay, BioRad) or Bradford assays [26]. Use equal amounts (3–4 mg) of extracts prepared from control and irradiated cells for ATM immunoprecipitation (*see* **Note 7**).
4. Add anti-ATM antibody in a quantity sufficient to pull down majority of ATM protein. It is useful to determine optimal amount of anti-ATM for IP in a preliminary experiment monitoring depletion of ATM protein from extract by western blotting. Include pre-immune serum as a negative control for IP. As an additional negative control, perform the IP from an ATM-negative cell line. Incubate IPs overnight (*see* **Note 8**).
5. Add approx. 30–40 μL of Protein A/G-Sepharose (precoated with BSA to minimize nonspecific binding) (*see* **Note 9**). Incubate for additional 1–2 h. Collect ATM IPs by brief centrifugation at +4 °C.
6. Wash IPs with lysis buffer (1 mL) three times. Add lysis buffer containing protease and phosphatase inhibitors to the tubes, gently resuspend IPs by inverting tubes several times, and pellet by brief centrifugation. Perform additional washes if you want to label ATM with ^{32}P using in vitro kinase assay as described in [5]. Add 2× SDS loading buffer to washed IP samples. Alternatively, use 4× LDS NuPage sample buffer for the NuPAGE system (Invitrogen). Exercise caution to maintain equal volumes of reagents in all IP samples. Carefully calculate the final volume of ATM IP samples so that it matches the volume in μLs you can load in the wells of a chosen gel electrophoresis system (*see* **Note 10**).

3.4 Electrophoresis of ATM Immunoprecipitates

1. Run samples on 5% or biphasic 5%/12% SDS-PAGE (*see* **Notes 11** and **12**). The advantage of biphasic 5%/12% SDS-PAGE (AA/Bis 29:1) is the ability to visualize both ATM (5% part of the gel) and amount of anti-ATM antibody in every immunoprecipitate (12% part of the gel). This allows confirmation of the integrity of the antibodies throughout the IP procedure. Monitor the position of prestained high-molecular weight

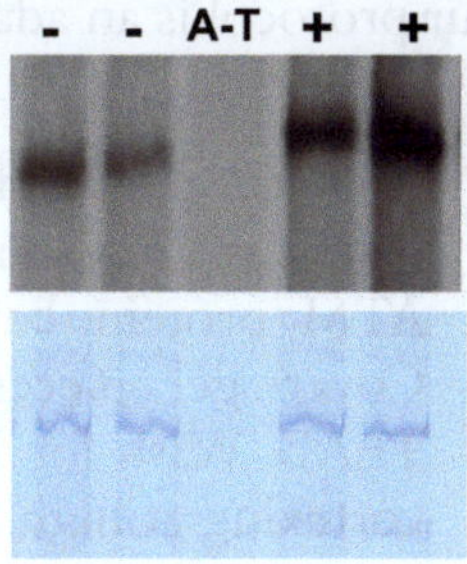

Fig. 1 ATM was immunoprecipitated from unirradiated and irradiated (10Gy) control cells, autophosphorylated by incubation with 30 μCi 32P-γATP and separated on 5% SDS-PAGE. ATM-AutoP: Autoradiogram of radioactive phosphate incorporated into ATM. *ATM* Amount of ATM protein in the autophosphorylation reaction, colloidal CBB G-250 staining. *A-T* demonstrates absence of ATM protein kinase in the immunoprecipitate from A-T cell line AT25ABR. Multiple ATM bands were cut from the gels and used for 2-D phosphopeptide mapping and identification of phosphorylation sites by mass spectrometry

markers closely during PAGE. Additional unstained MW markers may facilitate correct positioning of ATM protein in the gel.

2. After the completion of electrophoresis, remove the gel from cassette assembly, place in a clean glass or plastic tray. Do not fix the gel. Briefly wash in 20% methanol or water. Prepare fresh colloidal Coomassie Brilliant Blue G-250 stain as outlined in [24] (*see* Subheading 2). Stain the gel until ATM bands are clearly visible (*see* **Note 13**). Use negative controls (pre-immune serum IP and ATM-negative cell line IP) to assist in localizing ATM gel bands. ATM runs as approximately 360 kDa. A 250 kDa band is a nonspecific protein often visible in IPs. Briefly wash the gel in 25% methanol or water to increase contrast of gel bands (Fig. 1).

3. Scan the gel on a plastic support or a well-cleaned surface of a flatbed scanner. Save the image as record of the gel, prior to gel band excision.

4. Place the gel on a well-cleaned glass plate in a laminar flow cabinet and cut ATM bands with a fresh sterile scalpel blade (*see* **Note 14**). Handling of ^{32}P–labeled ATM bands requires special precautions and should only be performed in designated radioisotope work areas with strict adherence to principles of harm minimization and all local safety regulations. Excised ^{32}P–labeled ATM bands can be counted directly in tubes by Cerenkov counting. ATM bands can be stored at +4 °C or processed immediately, but should not be allowed to dry and adhere to tubes. Repeat immunoprecipitations until amount of ATM kinase sufficient for MS analysis is collected. The protein content from multiple gel bands may be pooled after digestion (*see* **Note 7**).

3.5 Tryptic Digest of ATM in Gel Bands

Our protocol is an adaptation of previously published methods for in-gel digestion of proteins [27, 28].

1. Cut the gel band into smaller pieces, approximately 1 cm^3, being as consistent as possible for each band. Fully destain ATM protein bands using 25% methanol. Discard liquid. Cover gel pieces with 50% acetonitrile (AcN) 50 mM TEAB. Wash for 15 min at room temperature on a rocker or vortexing at high speed.
2. Discard wash, repeat 50% AcN 50 mM TEAB wash. Gel pieces will shrink and become more opaque. Remove wash liquid. Dry the washed gel pieces briefly in a vacuum concentrator. If you dry gel pieces completely, be very careful with tubes, dry flakes have low mass and are easily blown from the tube. Proceed with reduction and alkylation of samples (*see* **Note 15**).
3. Add 10 mM DTT (or TCEP in 50 mM TEAB) and incubate at +55 °C for 10 min. Let samples cool to room temperature (~ 22 °C). Discard DTT solution and immediately add 20 mM fresh 2-chloroacetamide in 50 mM TEAB. Incubate for 30 min in the dark at room temperature.
4. Discard the solution and wash the gel pieces with 50 mM TEAB. Replace with 100% AcN. Wash for 15 min at room temperature on a rocker. Discard the solution and dry gel pieces in a vacuum concentrator as before (*see* **Note 16**).
5. Prepare trypsin solution in 50 mM TEAB. A protein-to-trypsin ratio of 20:1 is recommended. Use an estimate of ATM protein content in each band (*see* **Note 7**). Add trypsin solution to just cover gel pieces at room temperature. Gel pieces will swell and absorb trypsin. After gel pieces are fully swelled, incubate at +37 °C for at least 4 h. A second addition of the same amount of trypsin, or slightly less, for 4 h is advisable to minimize missed cleavages. Alternatively use two-step LysC/trypsin digestion as outlined in the manufacturer's protocol (Promega).
6. Remove and save the supernatant at +4 °C. Perform two extractions of tryptic peptides with 50% AcN solution. Sonicate 5 min in a water bath sonicator. Remove extracted peptides with a gel-loading tip, to carefully avoid shrunken gel pieces. Combine all supernatants. For stable isotope labeling/derivatization of primary amines, the pH is required to be basic, so no acid should be added. If there is to be no derivatization, then the pH should be adjusted to below 3 in the preparation for ZipTip/Stage Tip desalting. Reduce the volume of the peptides to <20 μL in a vacuum concentrator. Avoid completely drying the sample.

3.6 Derivatization of ATM Peptides with Stable Isotopes Using Dimethylation

This protocol describes derivatization of ATM peptide primary amines by dimethylation [21, 29]. However, other types of derivatization are possible and have various advantages and disadvantages [30, 31]. Two advantages of dimethylation are that the reagents are relatively inexpensive and the derivatization reaction is highly favorable resulting in fast and robust derivatization. One disadvantage is that it is only practical to compare a maximum of three conditions in a single LC-MS/MS run (e.g., control, irradiated, and drug treated). The dimethylation of primary amines results in the addition of +28 Da (light) to each peptide at the N-terminus and an additional +28 Da at any lysine residues. Using stable isotopes, the mass can be increased to +32 (medium) and +34/36 Da (heavy) (Fig. 2), without significantly changing the chromatographic or mass spectral detection properties of the peptides. The light, medium, and heavy peptide relative intensity in the MS spectra provide quantitative information that is proportional to the amount of that peptide in the original samples. For phosphopeptides, this will allow the determination of the relative level of phosphorylation under three different conditions. For multiple biological replicates, swap the stable isotopes so that the same isotopes are not used for the same condition each time (to avoid any label bias).

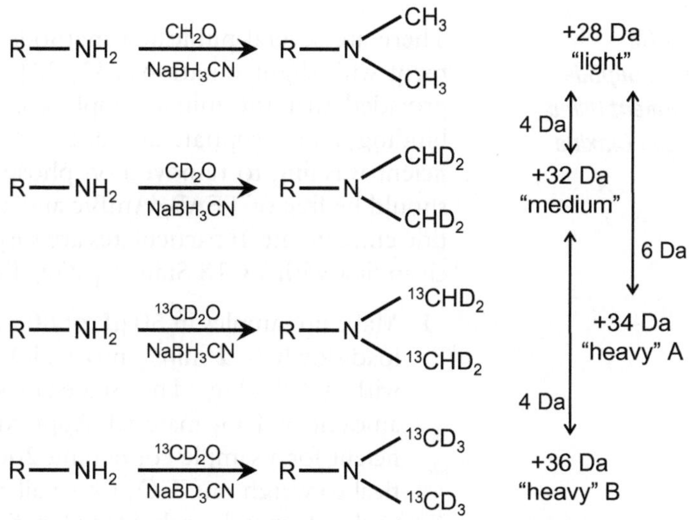

Fig. 2 Schematic of dimethyl isotope labeling reaction. The primary amines of peptides react with formaldehyde and sodium cyanoborohydride, which contain different stable isotopes to produce dimethylated peptides of different masses. The derivatized peptides retain similar properties in LC-MS/MS analysis. A minimum mass difference of 4 Da is used to allow separation of peptides within mass spectra. Two types of "heavy" label are shown, which may be used, depending on whether a duplex or triplex experiment is being performed

1. The total reaction volume for derivatization will be 50 μL. Make up the volume of ATM peptides for each condition to 35 μL. Add 5 μL of 1 M TEAB. Add 5 μL of 1% formaldehyde for the "light" condition. For a triplex experiment, add CD_2O for the "medium" condition and $^{13}CD_2O$ for the "heavy" condition. Vortex at 1000 rpm for 5 min.
2. Add 5 μL of $NaBH_3CN$ to the "light" and "medium" conditions and $NaBD_3CN$ to the "heavy" condition. Vortex at 100 rpm for 15 min. For a duplex experiment, the $NaBD_3CN$ is not required (*see* Fig. 2).
3. Add 1 μL of 25% solution of ammonium hydroxide (typically supplied as 28% ammonia in water) to quench the excess formaldehyde. Vortex for 1 min at 1000 rpm. Add 1 μL TFA to acidify the solution. At this stage, a small portion of each sample (1–5%) could be used to confirm that derivatization was successful, by sample desalting and LC-MS/MS (see below). Complete derivatization of >97% of the peptides should be readily achieved. The derivatized samples can now be mixed.
4. Reduce the volume of the peptides to <5 μL in a vacuum concentrator. Warning! The gas products from the vacuum concentrator will be toxic and must be vented according to local regulations. Avoid completely drying the sample.

3.7 ATM Phosphopeptide Enrichment Using Titanium Dioxide

There are several published methods for phosphopeptide enrichment with slight variations [32, 33]. Most procedures are robust provided that the initial sample conditions are compatible with binding, an appropriate amount of TiO_2 is used, and there is sufficient washing to remove non-phosphorylated peptides. Samples should be free of TEAB/AmBic and acidified before phosphopeptide enrichment. If particulates are suspected to be present, desalt/clean first with a C18 StageTip/ZipTip (see below).

1. Make up samples in 50 μL of 80% AcN 5% TFA 15% water and load slowly (~ 2 min) onto a TiO_2 column made as a StageTip with a C8 plug. The StageTip should have an appropriate amount of TiO_2 material. Approximately 1 mg of TiO_2 is sufficient for a sample containing 200 μg of peptide. It is impractical to weigh out TiO_2 for small amounts of peptide. Instead, pack ~1 mm length StageTip for up to 10 μg of peptide. Collect the unbound peptides, which have other uses, particularly for global normalization (see below).
2. Wash the TiO_2 with 3 × 20 μL of 50% AcN 5% TFA 45% water and collect with the unbound peptide solution.
3. Elute the phosphopeptides with 3 × 10 μL of ammonium hydroxide solution with pH > 10.5 in a 20% AcN solution. Follow with 10 μL of 50% AcN to elute any peptide bound to the C8 plug. Immediately reduce the volume of the peptides to <5 μL in a vacuum concentrator.

3.8 Quantitative LC-MS/MS Analysis of ATM Phosphopeptides

There are a number of LC-MS/MS configurations that can be used to analyze dimethylated samples and so these won't be discussed in detail. A conventional approach will be described (*see* **Note 17**). Particulates may still be present in the sample (some TiO_2 powder is very fine). Thus, samples can optionally be further purified by a C18 StageTip/ZipTip to protect the HPLC system.

1. Make up enriched phosphopeptide solution in a total of 20 μL of 0.1% TFA and load onto StageTip. A small amount of (~0.5 mm length of pack material) POROS R3 can be added on the top of the C18 to increase retention of short hydrophilic phosphopeptides. Wet the StageTip with 5 μL of 50% AcN and then re-equilibrate with 20 μL of 0.1% TFA. Load on the sample, retaining the solution of unbound material for later analysis/troubleshooting.
2. Wash with 20 μL of 0.1% TFA. Elute with 5 μL of 50% AcN. Reduce the volume of the eluate to <2 μL in a vacuum concentrator. Make up to an appropriate volume (depending on the HPLC injection method) in 0.1% TFA for LC-MS/MS.
3. Prepare a C18 column for LC-MS/MS. A robust configuration is a 75 μm inside dimeter fused silica column packed to at least 10 cm in length. Longer columns provide greater peptide separation, and focus the peptide into a narrow peak, but this requires HPLC systems designed for very high back pressures. A single protein, ATM, does not require particularly high resolving power. A commercial pressure cell can be used to pack columns of custom length with C18 material.
4. Prepare the mass spectrometer for data-independent acquisition of peptide MS/MS spectra over a typical gradient (from 0% AcN to 35% AcN over at least 30 min at 250–350 nL/min). As an alternative to data-independent acquisition, the known ATM tryptic phosphopeptides can be targeted by multiple ion monitoring [21] or the use of an inclusion list for detection of signals at specific *m/z*. Automated data processing of the samples may require a particular acquisition type. Analyze the sample by LC-MS/MS using the best available equipment (*see* **Note 17**).
5. The raw data can be processed using multiple instrument-specific and nonspecific software packages, which will automatically identify phosphopeptides based on their MS/MS spectra and then integrate the LC-MS profile of the phosphopeptides to determine their relative abundance. Fig. 3 shows part of a mass spectrum showing an ATM tryptic phosphopeptide, which has been detected with both a "light" and "heavy" labels. A commonly used and (currently) freely available software is MaxQuant [34]. Using this, or similar software, set up automated processing by selecting the appropriate light/medium/heavy mass

A

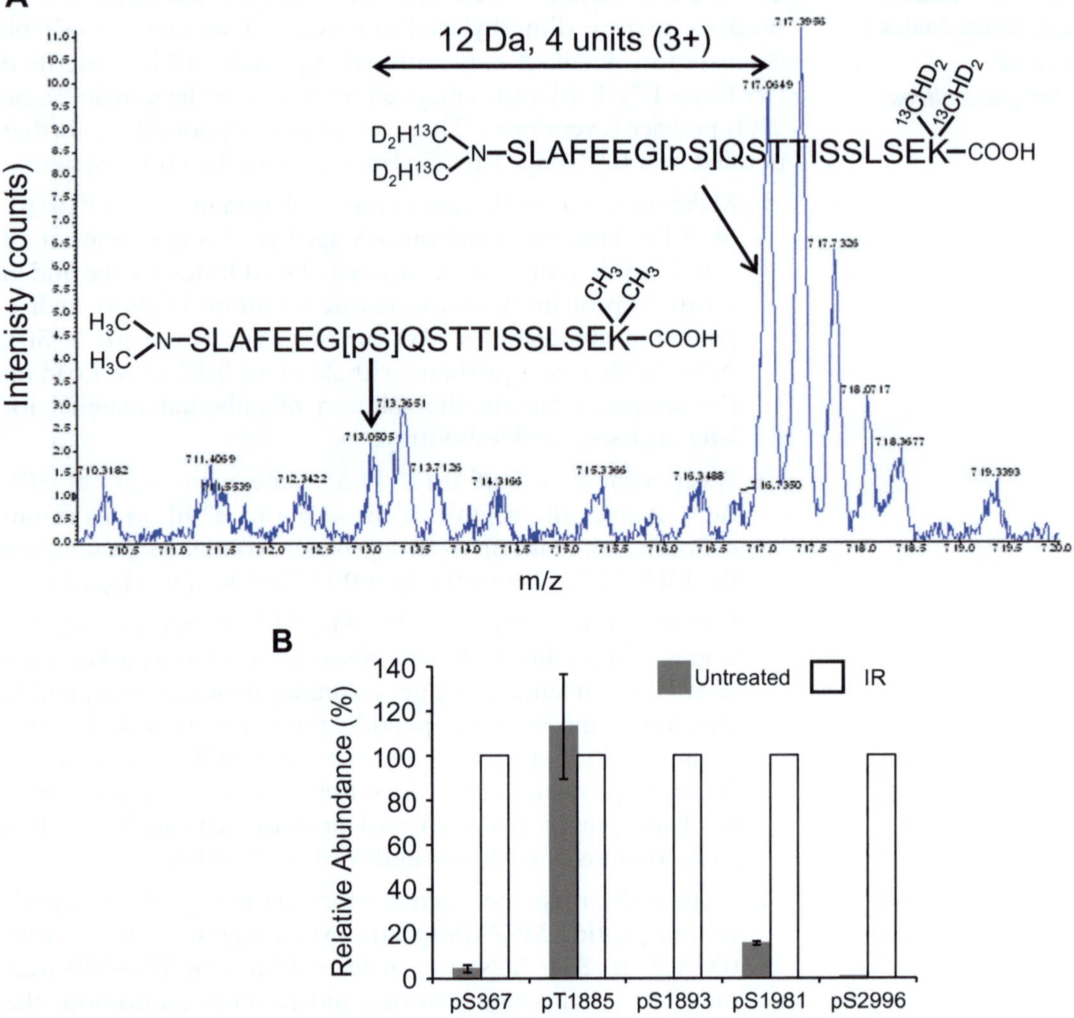

Fig. 3 Example mass spectrum and data output. (**a**) A mass spectrum in the range 710–720 *m/z* showing the triply charged hATM tryptic phosphopeptide 1974-SLAFEEG[pS]QSTTISSLSEK-1992 with a "light" and "heavy" labels. The light label was the control condition and the heavy label was the irradiated condition. The integration of a time peak in the extracted ion chromatogram of these *m/z* peaks can provide the intensity for a single experiment for this phosphopeptide ($n = 1$). (**b**) When combined with the detection of other ATM phosphopeptides and multiple independent experiments, conclusions can be made on the site-specific relative difference in abundance between different conditions such as control vs. irradiated, adapted with permission from [21]. "This research was originally published in the J Biol Chem by Sergei V. Kozlov, et al J Biol Chem. 2011 March 18; 286(11):9107–9119 © the American Society for Biochemistry and Molecular Biology"

shifts and include variable modifications of significance (e.g., deamidation, oxidation (M)), including phosphorylation. Use an appropriate database to search against, e.g., UniProtKB human proteome. The processing software should provide the sequence of the phosphopeptides with a score for its match against the database, a probability/score for phosphorylation site assignment, and the relative intensity of identified phosphopeptides in each condition. A probability of phosphorylation assignment greater than 0.75 is generally assumed to be correct without manual validation.

6. A small number of non-phosphopeptides will inevitably be present in the phosphopeptide-enriched sample and these can be useful for global normalization of the sample. It is useful to separately analyze, by LC-MS/MS, the peptides that did not bind to TiO_2 and include this raw data file in the data processing, as this will improve the normalization. Stand-alone global normalization software can also be applied to the data processing output. Normalization will likely to improve the statistical analysis by adjusting for any variation in sample preparation between conditions.
7. Examine each condition in at least three independent samples/biological replicates and apply appropriate statistics to determine any significant differences in relative phosphorylation site occupancy between different conditions (*see* **Note 18**). The quantification is linked to the phosphopeptide and the change in occupancy of the phosphorylation sites is inferred. If there are two phosphosites on a peptide then the intensity is related to the change in occupancy of both sites and is complicated by the presence of both mono- and di-phosphorylated peptides. In human ATM, the tryptic peptide 1883-STTPANLDSESEHFFR-1898 has two phosphorylation sites (phospho-T1885 and phospho-S1893) and only one is responsive to irradiation [21]. Fortunately, the two isobaric mono-phosphopeptides are separable by C18 chromatography so that the signals required for quantification do not interfere.

4 Notes

1. Neocarzinostatin or other DNA-damaging agents can be used instead.
2. It is very important to use cocktail of protease and phosphatase inhibitors in preparation of cell extracts since they are used for lengthy immunoprecipitation procedure. Extracts can be used to prepare samples for electrophoresis and western blotting of ATM substrates to check for ATM kinase activation after DNA damage.
3. Consider dedicating electrophoresis equipment exclusively for MS work to avoid contamination. It is prudent to wash electrophoresis equipment with 1% SDS, rinse with water, wash with 10% EtOH, and thoroughly rinse with water. Use powder-free gloves. The main contaminant is keratin from the laboratory

environment, people, and clothing. Use general precautions to avoid keratin contamination (e.g., keep lids closed on containers for tubes, tips, and solutions).

4. Iodoacetamide was commonly used instead of chloroacetamide. But iodoacetamide was shown to produce artifacts [35].
5. Two-stage protein digestion is common [36], using either two applications of trypsin or a combination of LysC followed by trypsin. Many commercial trypsins are available. A key consideration is the extent of chymotryptic/nonspecific proteolysis versus cost. Commercial trypsins have been compared for proteomics analysis [37].
6. Cell pellets can be safely stored in −80 °C freezer for several months.
7. It is important to estimate the amount of ATM present in immunoprecipitates, as the amount will affect mass spectrometry analysis workflow. It is possible to use protein standards to estimate amounts of protein quantities in the CBB G-250 stained bands. It was previously reported [38] that silver staining detects approximately 200–250 ng of ATM kinase in immunoprecipitates from G361 cells. It is useful to know analytical performance of available MS instruments to determine amount of ATM kinase required for the confident detection of phosphorylation sites. Note that 1 μg of ATM equals 2.85 pmol (ATM 350.644 kDa) and 3.51 μg of ATM equals 10 pmol. It is suggested that "minimum of 2–10 pmol of purified phosphorylated protein, which corresponds to 100–500 ng of a 50-kDa protein" is required for adequate sequence coverage [39].
8. The length of incubation of extracts with ATM antibodies can be shortened without detrimental effect on recovery of ATM kinase in immunoprecipitates.
9. It might be preferable to make affinity IP sorbent by crosslinking anti-ATM antibody to Protein A/G- Sepharose beads. DMP- and DSS- crosslinking protocols can be used. Crosslinking the antibodies to beads can help to reduce or eliminate large amounts of antibody present in the sample after elution of ATM kinase. While the presence of antibody does not usually drastically interfere with gel separation of ATM protein, the absence of eluted immunoglobulin bands in the immunoprecipitation samples could improve the separation quality, especially when mini-gel format is used.
10. It is convenient to use 8 control and 8 + IR ATM IP samples (plus 2 negative controls) to run them on two 10-well mini-SDS-PAGE gels. Otherwise, it is possible to run a larger number of samples on two large format gels.

11. It is important to decide on the configuration of the gel you are planning to use to purify ATM. Usually, 0.75–1 mm thick 5% SDS-PAGE or 3–8% gradient NuPage gels are suitable for tryptic digests of excised bands.
12. Un-reacted products of polymerization produce artifacts in MS/MS. It is better to polymerize SDS-PAGE gels overnight. Stacking gel can be polymerized next morning. Alkylation of cysteines with acrylamide was observed and used as an analytical tool in MS [40].
13. It is possible to use Zn^{2+} reversible stain (compatible with mass spectrometry).
14. Use only microcentrifuge tubes recommended for MS analysis. Microtubes with modified surface, siliconized tubes, and colored tubes introduce contaminants that compromise MS analysis.
15. It is possible to reduce and alkylate peptides after tryptic digest.
16. The drying step is optional. Trace AcN (< 10%) will not inhibit tryptic digestion. But care should be taken to remove the AcN consistently from each sample or there will be variable digestion efficiency [41].
17. When analyzing phosphopeptides, the HPLC should be reasonably free of metal fittings and tubing to avoid loss of phosphopeptides due to surface binding, e.g., avoid the use of metal sample loops, tubing, and metal electrospray needles. Metallic valves are difficult to avoid without high cost. Mass spectrometers with high resolving power, e.g., time-of-flight and orbitrap mass spectrometers, are preferred for quantification that uses primary amine derivatization. High resolution allows peptides of similar mass to be easily discerned and improves the accuracy of peptide and peptide fragment mass measurement. Generally, a low ion intensity threshold should be used to trigger MS/MS scans because of the low abundance of phosphopeptides. This approach has few disadvantages in a low complexity (single protein) sample while using a mass spectrometer with a fast duty cycle.
18. Each phosphorylation site identified by mass spectrometry requires validation by an alternative method. Generation of a phospho-specific antibody for each phosphorylation site is the conventional validation method (where commercial antibodies are nonexistent, which is likely). However, this is impractical for large numbers of sites. Confirmation of the phosphorylation site using an alternative quantitative phosphoproteomics workflow may be sufficient, e.g., a stable isotope dimethylation approach could be validated by a label-free approach.

Acknowledgments

M.E.G. and M.F.L. were funded by the National Health and Medical Research Council. M.E.G. was funded by the Brain Foundation and equipment grants from the Cancer Institute New South Wales, the Australian Cancer Research Foundation, and Rebecca L Cooper Medical Research Foundation.

References

1. Johnson LN, Noble ME, Owen DJ (1996) Active and inactive protein kinases: structural basis for regulation. Cell 85(2):149–158
2. Durocher D, Jackson SP (2001) DNA-PK, ATM and ATR as sensors of DNA damage: variations on a theme? Curr Opin Cell Biol 13(2):225–231
3. Kim ST, Lim DS, Canman CE, Kastan MB (1999) Substrate specificities and identification of putative substrates of ATM kinase family members. J Biol Chem 274(53): 37538–37543
4. O'Neill T, Dwyer AJ, Ziv Y, Chan DW, Lees-Miller SP, Abraham RH, Lai JH, Hill D, Shiloh Y, Cantley LC, Rathbun GA (2000) Utilization of oriented peptide libraries to identify substrate motifs selected by ATM. J Biol Chem 275(30):22719–22727. doi:10.1074/jbc.M001002200
5. Kozlov S, Gueven N, Keating K, Ramsay J, Lavin MF (2003) ATP activates ataxia-telangiectasia mutated (ATM) in vitro. Importance of autophosphorylation. J Biol Chem 278(11):9309–9317
6. Bakkenist CJ, Kastan MB (2003) DNA damage activates ATM through intermolecular autophosphorylation and dimer dissociation. Nature 421(6922):499–506. doi:10.1038/nature01368
7. Sun Y, Jiang X, Chen S, Fernandes N, Price BD (2005) A role for the Tip60 histone acetyltransferase in the acetylation and activation of ATM. Proc Natl Acad Sci U S A 102(37):13182–13187. doi:10.1073/pnas.0504211102
8. Gupta A, Sharma GG, Young CS, Agarwal M, Smith ER, Paull TT, Lucchesi JC, Khanna KK, Ludwig T, Pandita TK (2005) Involvement of human MOF in ATM function. Mol Cell Biol 25(12):5292–5305 . doi:10.1128/MCB.25.12.5292-5305.2005doi:25/12/5292 [pii]
9. Ali A, Zhang J, Bao S, Liu I, Otterness D, Dean NM, Abraham RT, Wang XF (2004) Requirement of protein phosphatase 5 in DNA-damage-induced ATM activation. Genes Dev 18(3):249–254. doi:10.1101/gad.1176004
10. Goodarzi AA, Jonnalagadda JC, Douglas P, Young D, Ye R, Moorhead GB, Lees-Miller SP, Khanna KK (2004) Autophosphorylation of ataxia-telangiectasia mutated is regulated by protein phosphatase 2A. EMBO J 23(22):4451–4461. doi:10.1038/sj.emboj.7600455
11. Shreeram S, Demidov ON, Hee WK, Yamaguchi H, Onishi N, Kek C, Timofeev ON, Dudgeon C, Fornace AJ, Anderson CW, Minami Y, Appella E, Bulavin DV (2006) Wip1 phosphatase modulates ATM-dependent signaling pathways. Mol Cell 23(5):757–764. doi:10.1016/j.molcel.2006.07.010
12. Uziel T, Lerenthal Y, Moyal L, Andegeko Y, Mittelman L, Shiloh Y (2003) Requirement of the MRN complex for ATM activation by DNA damage. EMBO J 22(20):5612–5621. doi:10.1093/emboj/cdg541
13. Lee JH, Paull TT (2005) ATM activation by DNA double-strand breaks through the Mre11-Rad50-Nbs1 complex. Science 308(5721):551–554. doi:10.1126/science.1108297
14. Pellegrini M, Celeste A, Difilippantonio S, Guo R, Wang W, Feigenbaum L, Nussenzweig A (2006) Autophosphorylation at serine 1987 is dispensable for murine Atm activation in vivo. Nature 443(7108):222–225. doi:10.1038/nature05112
15. Lavin MF, Kozlov S (2007) ATM activation and DNA damage response. Cell Cycle 6(8):931–942
16. Guo Z, Kozlov S, Lavin MF, Person MD, Paull TT (2010) ATM activation by oxidative stress. Science 330(6003):517–521. doi:10.1126/science.1192912
17. Paull TT (2015) Mechanisms of ATM activation. Annu Rev Biochem 84:711–738. doi:10.1146/annurev-biochem-060614-034335
18. Nagahara H, Latek RR, Ezhevsky SA, Dowdy SF (1999) 2-D phosphopeptide mapping. Methods Mol Biol 112:271–279

19. Boyle WJ, van der Geer P, Hunter T (1991) Phosphopeptide mapping and phosphoamino acid analysis by two-dimensional separation on thin-layer cellulose plates. Methods Enzymol 201:110–149
20. Kozlov SV, Graham ME, Peng C, Chen P, Robinson PJ, Lavin MF (2006) Involvement of novel autophosphorylation sites in ATM activation. EMBO J 25(15):3504–3514. doi:10.1038/sj.emboj.7601231
21. Kozlov SV, Graham ME, Jakob B, Tobias F, Kijas AW, Tanuji M, Chen P, Robinson PJ, Taucher-Scholz G, Suzuki K, So S, Chen D, Lavin MF (2011) Autophosphorylation and ATM activation: additional sites add to the complexity. J Biol Chem 286(11):9107–9119. doi:10.1074/jbc.M110.204065
22. Jenkins CW, Xiong Y (1996) Immunoprecipitation and immunoblotting in cell cycle studies. In: Pagano M (ed) Cell cycle-materials and methods. Springer lab manuals. Springer, New York, pp 250–264. doi:10.1007/978-3-642-57783-3
23. Sarkaria JN, Busby EC, Tibbetts RS, Roos P, Taya Y, Karnitz LM, Abraham RT (1999) Inhibition of ATM and ATR kinase activities by the radiosensitizing agent, caffeine. Cancer Res 59(17):4375–4382
24. Neuhoff V, Stamm R, Eibl H (1985) Clear background and highly sensitive protein staining with Coomassie Blue dyes in polyacrylamide gels: a systematic analysis. Electrophoresis 6(9):427–448. doi:10.1002/elps.1150060905
25. Chiang GG, Abraham RT (2004) Determination of the catalytic activities of mTOR and other members of the phosphoinositide-3-kinase-related kinase family. Methods Mol Biol 281:125–141. doi:10.1385/1-59259-811-0:125
26. Bradford MM (1976) A rapid and sensitive method for the quantitation of microgram quantities of protein utilizing the principle of protein-dye binding. Anal Biochem 72:248–254
27. Stone KL, Williams KR (1996) Enzymatic digestion of proteins in solution and in SDS polyacrylamide gels. In: Walker JM (ed) The protein protocols handbook. Humana Press, New York, pp 415–425. doi:10.1007/978-1-60327-259-9_71
28. Hellman U (2000) Sample preparation by SDS/PAGE and in-gel digestion. In: Jollès P, Jörnvall H (eds) Proteomics in functional genomics: protein structure analysis. Birkhäuser Basel, Basel, pp 43–54. doi:10.1007/978-3-0348-8458-7_3
29. Boersema PJ, Raijmakers R, Lemeer S, Mohammed S, Heck AJ (2009) Multiplex peptide stable isotope dimethyl labeling for quantitative proteomics. Nat Protoc 4(4):484–494. doi:10.1038/nprot.2009.21
30. Bantscheff M, Schirle M, Sweetman G, Rick J, Kuster B (2007) Quantitative mass spectrometry in proteomics: a critical review. Anal Bioanal Chem 389(4):1017–1031. doi:10.1007/s00216-007-1486-6
31. Bantscheff M, Lemeer S, Savitski MM, Kuster B (2012) Quantitative mass spectrometry in proteomics: critical review update from 2007 to the present. Anal Bioanal Chem 404(4):939–965. doi:10.1007/s00216-012-6203-4
32. Engholm-Keller K, Larsen MR (2016) Improving the phosphoproteome coverage for limited sample amounts using TiO2-SIMAC-HILIC (TiSH) phosphopeptide enrichment and fractionation. Methods Mol Biol 1355:161–177. doi:10.1007/978-1-4939-3049-4_11
33. Kozlov SV, Waardenberg AJ, Engholm-Keller K, Arthur JW, Graham ME, Lavin M (2016) Reactive Oxygen Species (ROS)-activated ATM-dependent phosphorylation of cytoplasmic substrates identified by large-scale phosphoproteomics screen. Mol Cell Proteomics 15(3):1032–1047. doi:10.1074/mcp.M115.055723
34. Cox J, Mann M (2008) MaxQuant enables high peptide identification rates, individualized p.p.b.-range mass accuracies and proteome-wide protein quantification. Nat Biotechnol 26(12):1367–1372. doi:10.1038/nbt.1511
35. Nielsen ML, Vermeulen M, Bonaldi T, Cox J, Moroder L, Mann M (2008) Iodoacetamide-induced artifact mimics ubiquitination in mass spectrometry. Nat Methods 5(6):459–460. doi:10.1038/nmeth0608-459
36. McDonald WH, Ohi R, Miyamoto DT, Mitchison TJ, Yates Iii JR (2002) Comparison of three directly coupled HPLC MS/MS strategies for identification of proteins from complex mixtures: single-dimension LC-MS/MS, 2-phase MudPIT, and 3-phase MudPIT. Int J Mass Spectrom 219(1):245–251. doi:http://dx.doi.org/10.1016/S1387-3806(02)00563-8
37. Bunkenborg J, Espadas G, Molina H (2013) Cutting edge proteomics: benchmarking of six commercial trypsins. J Proteome Res 12(8):3631–3641. doi:10.1021/pr4001465
38. Chan DW, Son SC, Block W, Ye R, Khanna KK, Wold MS, Douglas P, Goodarzi AA, Pelley

J, Taya Y, Lavin MF, Lees-Miller SP (2000) Purification and characterization of ATM from human placenta. A manganese-dependent, wortmannin-sensitive serine/threonine protein kinase. J Biol Chem 275(11):7803–7810

39. Dephoure N, Gould KL, Gygi SP, Kellogg DR (2013) Mapping and analysis of phosphorylation sites: a quick guide for cell biologists. Mol Biol Cell 24(5):535–542. doi:10.1091/mbc.E12-09-0677

40. Sechi S, Chait BT (1998) Modification of cysteine residues by alkylation. A tool in peptide mapping and protein identification. Anal Chem 70(24):5150–5158

41. Wall MJ, Crowell AM, Simms GA, Carey GH, Liu F, Doucette AA (2011) Implications of partial tryptic digestion in organic-aqueous solvent systems for bottom-up proteome analysis. Anal Chim Acta 703(2):194–203. doi:10.1016/j.aca.2011.07.025

Chapter 11

Studies of ATM Kinase Activity Using Engineered ATM Sensitive to ATP Analogues (ATM-AS)

Masato Enari, Yuko Matsushima-Hibiya, Makoto Miyazaki, and Ryo Otomo

Abstract

Ataxia-telangiectasia mutated (ATM) protein is a member of the phosphatidylinositol 3-phosphate kinase (PI3-K)-related protein kinase (PIKK) family and is implicated in the initiation of signaling pathways following DNA double strand breaks (DSBs) elicited by exposure to ionizing irradiation (IR) or radiomimetic compounds. Loss of function of the ATM gene product results in the human genetic disorder ataxia-telangiectasia (A-T) characterized by neurodegeneration, immunodeficiency, genomic instability, and cancer predisposition. In response to DSBs, ATM is activated and phosphorylates Ser/Thr-Gln (S/T-Q) sequences on numerous proteins participating in DNA-damage responses. Among these proteins, phosphorylation of the tumor suppressor p53 at Ser15 is known as a target for ATM, which leads to the dissociation of MDM2, an E3 ubiquitin ligase, from p53 to prevent MDM2-dependent p53 degradation. Ser46 on p53 is phosphorylated in response to DSBs and contributes to the preferential transactivation of pro-apoptotic genes, such as p53AIP1, Noxa, and PUMA, to prevent tumor formation. Our group have shown that not only ATM preferentially phosphorylates S/T-Q sequences, but also Ser46, which is a non-canonical site with an S-P sequence for ATM. Ser46 on p53 is directly phosphorylated by ATM in a p53 conformation-dependent manner using the ATP analogue-accepting ATM mutant (ATM-AS) system. This protocol summarizes an approach to identify direct numerous targets for ATM kinase and is used to elucidate ATM signaling pathways in the DNA damage responses.

Key words Ataxia–telangiectasia, ATM, ATP analogues, DNA damage response, p53, PI3-kinase family, Protein phosphorylation

1 Introduction

Posttranslational modifications (PTMs) are crucial to transduce signal into cells. Among PTMs, protein phosphorylation is a major and controls substrate enzymatic activity and protein-protein interactions [1–3]. The Ser/Thr protein kinase ataxia-telangiectasia mutated (ATM) is discovered as a responsible gene for the human genomic instability syndrome ataxia-telangiectasia (A-T) [3–5]. The major hallmarks of A-T are neurodegeneration, immunodeficiency, genomic instability, and cancer predisposition, elicited by a severe

Sergei V. Kozlov (ed.), *ATM Kinase: Methods and Protocols*, Methods in Molecular Biology, vol. 1599,
DOI 10.1007/978-1-4939-6955-5_11, © Springer Science+Business Media LLC 2017

defect of a system repairing the double strand break (DSB) [6, 7]. ATM belongs to a family of phosphatidylinositol 3-phosphate kinase (PI3-K)-related protein kinases (PIKKs) including ataxia- and Rad3-related (ATR), DNA-dependent protein kinase catalytic subunit (DNA-PKcs), mammalian target of rapamycin (mTOR), suppressor of morphogenesis in genitalia-1 (SMG-1), and transformation/transcription domain-associated protein (TRRAP), which preferably phosphorylate Ser or Thr residues followed by Gln (S/T-Q motif) [8–10]. ATM activates in response to genotoxic stresses such as DNA-damaging agents and ionizing radiation, and plays an important role in repairing the damaged DNA [11–13].

The tumor suppressor p53 is frequently mutated or deleted in various human cancers and activated in response to various stresses, including DNA damage, to prevent tumorigenesis [14–17]. Upon DNA damage, p53 activates numerous target genes involved in DNA repair, cell-cycle arrest, and apoptosis [15, 18–20]. In growing cells, the protein level of p53 is very low due to ubiquitination of p53 by an E3 ubiquitin ligase MDM2, followed by its proteasome degradation [21–24]. In response to repairable DNA damage, ATM and ATR activate and phosphorylate p53 at Ser15 and their respective downstream mediators Chk2 at Thr68 and Chk1 at Ser345 [11, 25, 26]. Activated Chk2 and Chk1 phosphorylate p53 at Ser20. Phosphorylations of p53 at Ser15 and Ser20 lead to the dissociation of p53 with MDM2. As a result, p53 stabilizes in nuclei and activates p53-target genes involved in DNA repair and cell-cycle arrest such as p21^{waf1} to repair injured DNA [27, 28]. In the case of severe nonrepairable DNA damage, further phosphorylation of p53 at Ser46 links to the activation of p53-target genes involved in apoptosis such as p53AIP1, PUMA, and Noxa to remove cells with too much injured DNA [18, 28–30].

Several kinases including p38 mitogen-activated protein kinase (MAPK), protein kinase C δ (PKCδ), homeodomain-interacting protein kinase 2 (HIPK2), and dual-specificity tyrosine phosphorylation-regulated kinase 2 (DYRK2) responsible for Ser46 phosphorylation on p53 have so far been identified and we also found that ATM directly phosphorylates p53 at Ser46 in response to severe DNA damage using the ATP analogue-accepting ATM mutant (ATM-AS) system [13, 31–35]. This system utilizes the substitution of a conserved bulky residue in the ATP-binding pocket of kinase for small amino acids [13, 36–38]. Such mutant with an enlarged ATP-binding pocket, but not wild-type kinase, is theoretically able to bind ATP analogues with large side chain. This strategy has succeeded in finding out bona fide substrates of various kinases including the Src family kinase [36], stress-activated protein kinase/Jun N-terminal kinase (SAPK/JNK) [37], cyclin-dependent kinase 1 (cdk 1) [38], and ATM [13]. Based on the

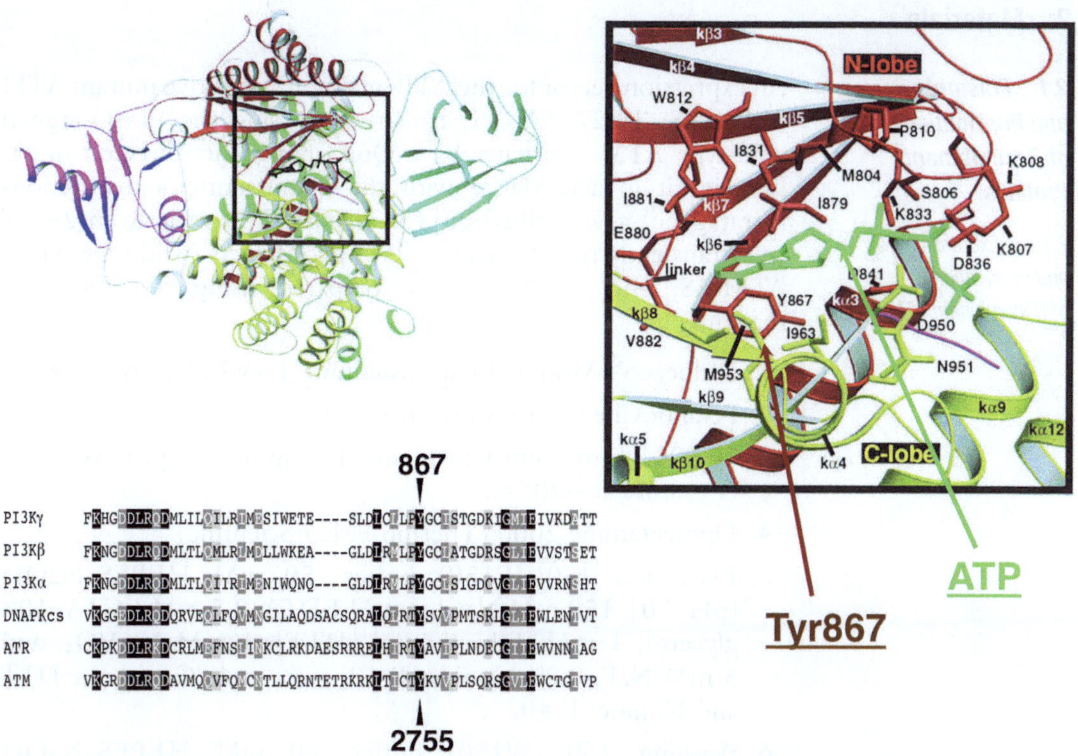

Fig. 1 The ATP-binding site of PI3Kγ and alignment of amino acid sequences around ATP-binding pocket in PI3-Kinase family proteins

X-ray crystallographic structure of ATP-bound PI3-Kγ belonging to the PI3-K family, the Tyr867 residue on PI3-Kγ contacts the N6 position of the adenine ring of ATP [39] (Fig. 1). The sequence alignment of ATP-binding pockets among members of the PI3-K family showed that this Tyr residue is conserved and corresponds to Tyr2755 in ATM (Fig. 1). Indeed, when this Tyr2755 residue in ATM is replaced with alanine, ATM becomes sensitive to ATP analogues (ATM-AS). Tyr2755 is deep in the ATP-binding pocket and far from the substrate-binding site in order not to alter the substrate specificity of the kinase. Using this method, we elucidated that ATM directly phosphorylates p53 at Ser46; nevertheless, the sequence around Ser46 in p53 is not a typical S/T-Q motif [13]. For this assay, natural ATP or various N6-substituted ATP analogues were used as phosphate donors, and ATM kinase activity was determined using a phospho-Ser15-specific antibody. Here, we would like to show the detailed protocols of the method for identification of bona fide substrates of PI3-Kinases family.

2 Materials

2.1 Plasmids and Purification of Recombinant Proteins

2.1.1 Expression and Purification of Recombinant ATM

An expression vector for the ATP-analogues-sensitive mutant ATM (ATM-AS; Tyr2755Ala) is constructed using the FLAG-tagged wild-type ATM construct by a QuikChange site-directed mutagenesis kit in accordance with the manufacturer's instructions (Stratagene) as described in [13]. Total cell extracts from 293 T cells transiently transfected with ATM constructs (wild-type ATM, ATM-AS, and ATM-KD) are prepared using the following reagents.

1. Dulbecco's Modified Eagle Medium (D-MEM), store at 4 °C.
2. Fetal bovine serum, store at −20 °C.
3. 10,000 U/ml Penicillin G and 10 mg/ml streptomycin sulfate, store at −20 °C.
4. Lipofectamine 2000 (Thermo Fisher Scientific).
5. Extraction 150 (E150) buffer: 50 mM HEPES–NaOH [pH 7.0], 150 mM NaCl, 1 mM EDTA, 2.5 mM EGTA, 10% glycerol, 1 mM dithiothreitol [DTT], 1 mM Na_3VO_4, and 5 mM NaF, 0.3% Nonidet P-40, store at 4 °C without DTT and Nonidet P-40.
6. Washing 150 (W150) buffer: 50 mM HEPES–NaOH [pH 7.0], 150 mM NaCl, 1 mM EDTA, 2.5 mM EGTA, 10% glycerol, 1 mM dithiothreitol [DTT], 1 mM Na_3VO_4, and 5 mM NaF, 0.03% Nonidet P-40, store at 4 °C without DTT and Nonidet P-40.
7. Washing 500 (W500) buffer: 50 mM HEPES–NaOH [pH 7.0], 500 mM NaCl, 1 mM EDTA, 2.5 mM EGTA, 10% glycerol, 1 mM dithiothreitol [DTT], 1 mM Na_3VO_4, and 5 mM NaF, 0.03% Nonidet P-40, store at 4 °C without DTT and Nonidet P-40.
8. Protease inhibitor cocktail (100×): 1 mg/ml pepstatin A, 1 mg/ml antipain, 1 mg/ml chymostatin, 1 mg/ml leupeptin, 1 mg/ml E-64, and 1 mg/ml phenylmethylsulfonyl fluoride (PMSF), store at −20 °C.
9. Anti-FLAG M2 agarose, store at 4 °C.
10. 3× FLAG peptides (Sigma-Aldrich), store at −20 °C.

2.1.2 Preparation and Purification of Recombinant p53

GST-p53 fusion proteins are expressed in *Escherichia coli* BL21 cells from pGEX-4 T-1 or pGEX-6P-1 vectors as the manufacturer's instructions (GE Healthcare).

1. Bacterial stock (pGEX6P1-p53wt in DH5alpha).
2. Luria-Bertani (LB) medium: 1% Bactotrypton, 0.5% Yeast extract, 0.5% NaCl, store at 4 °C.

3. 50 mg/ml sodium ampicillin, store at −20 °C.
4. 1 M isopropyl-b-d-thiogalactopyranoside (IPTG); store at −20 °C.
5. Phosphate-buffered saline (PBS): 137 mM NaCl, 2.7 mM KCl, 8.1 mM Na_2HPO_4, 1.5 mM KH_2PO4.
6. PBS-T: PBS, 0.05% Tween.
7. 1 mg/ml phenylmethylsulfonyl fluoride (PMSF), store at −20 °C.
8. Sonicator eg BIORUPTOR (COSMO BIO) Ultrasonic Cell Disrupter.
9. Glutathione-immobilized Sepharose (GE Healthcare), store at 4 °C.
10. Elution buffer: 50 mM Tris–HCl, 10 mM reduced glutathione (pH 8.0).
11. Protein assay kit.

2.1.3 Sodium Dodecyl Sulfate Polyacrylamide Gel Electrophoresis (SDS-PAGE)

Acrylamide concentration depends on the molecular weight of proteins (ATM and p53) to be resolved. The protocol for casting gels and for Sample buffer is described in Tables 1 and 2, respectively.

1. Water bath for heating samples.
2. Running buffer: 25 mM Tris base, 192 mM glycine, 0.1% SDS (w/v).
3. Equipment for SDS-PAGE.

2.1.4 Western Blotting

1. PVDF membrane (Millipore).
2. Blocking buffer: 5% skim milk (w/v) in TBS-T [pH 7.4].
3. Western Blotting Substrate Plus (Thermo Fisher Scientific).

Table 1
Preparation of the stacking gel and separating gels

Stacking gel		**Separating gel**						
Stock solution	4%	20.0%	17.5%	15.0%	12.5%	10.0%	7.5%	5.0%
30% Acrylamide (30:0.8)	0.67 ml	6.7 ml	5.8 ml	5.0 ml	4.2 ml	3.3 ml	2.5 ml	1.7 ml
1.5 M Tris–HCl, pH 8.8	–	2.5 ml	2.5 ml	2.5 ml	2.5 ml	2.5 ml	2.5 ml	2.5 ml
0.5 M Tris–HCl, pH 6.8	1.25 ml	–	–	–	–	–	–	–
10% SDS	0.05 ml	0.1 ml	0.1 ml	0.1 ml	0.1 ml	0.1 ml	0.1 ml	0.1 ml
10% APS	0.05 ml	0.1 ml	0.1 ml	0.1 ml	0.1 ml	0.1 ml	0.1 ml	0.1 ml
TEMED	0.005 ml	0.02 ml	0.02 ml	0.02 ml	0.02 ml	0.02 ml	0.02 ml	0.02 ml
Distilled H_2O	2.98 ml	0.58 ml	1.48 ml	2.28 ml	3.08 ml	3.98 ml	4.78 ml	5.58 ml

Table 2
Preparation of 2× Sample buffer

Stock solution (ml)	Volume	Final concentration (1×)
10% (w/v) SDS	2.0 ml	2% (w/v)
1.5 M Tris–HCl, pH 6.8	1.667 ml	250 mM
0.5 M EDTA, pH 8.0	20 μl	1 mM
Glycerol	2.0 g	20% (w/v)
2-Mercaptoethanol	0.5 ml	5% (v/v)
1% (w/v) Bromophenol blue	0.1 ml	0.01% (w/v)
Total	5 ml	

These buffers are divided into 50 μl aliquots and stored at −20 °C

4. Transfer buffer: 12.5 mM Tris base, 192 mM glycine, 20% (v/v) methanol [pH 8.3].
5. Tris-buffered saline (TBS): 25 mM Tris–HCl [pH 7.4], 150 mM NaCl.
6. TBS-T: TBS, 0.05% Tween 20.
7. Specific primary antibodies to proteins of interest: Anti-ATM antibody (2C1) (GeneTex, Inc.).
8. Anti-p53 antibody (9282) (Cell Signaling Technology).
9. Anti-phospho-Ser15-of-p53 antibody (9284) (Cell Signaling Technology).
10. Anti-phospho-Ser1981-of-ATM antibody (10H11. E12) (Rockland Inc.).
11. Anti-FLAG (M2) antibody (Sigma-Aldrich).
12. Anti-phospho-Ser46-of-p53 antibody (Laboratory of Monoclonal Antibody Co., Inc.).
13. Secondary antibodies conjugated with horseradish peroxidase (HRP).
14. Can Get Signal® immunostain Immunoreaction Enhancer Solution 1 (Toyobo).
15. Apparatus for electro-transfer (Bio-Rad).
16. X-Ray film, Hyperfilm ECL (GE Healthcare).
17. X-Ray film processor (e.g., Fujifilm film processor).

2.2 In Vitro Kinase Assay

1. ATP.
2. ATP analogues: N6-benzyl ATP [N6-Bn-ATP].
3. N6-phenylethyl ATP [N6-PhEt-ATP].
4. N6-1-methylbutyl ATP [N6-1-MeBu-ATP].

5. N6-2-methylbutyl ATP [N6-2-MeBu-ATP].
6. Assay buffer: 25 mM Tris–HCl [pH 7.5], 5 mM β-glycerophosphate, 2 mM DTT, 0.1 mM Na_3VO_4, and 10 mM $MgCl_2$.
7. Incubator.

2.3 Confocal Immunofluorescent Microscopy

1. 4-well chamber slide (Lab-Tek II Chamber Slide; Thermo Fisher Scientific).
2. 4% Paraformaldehyde (PFA): EM grade paraformaldehyde in PBS.
3. Permeabilization buffer: 0.1%Triton X-100 in PBS.
4. Blocking solution: 3% Bovine serum albumin, 0.1% normal goat serum in PBS.
5. Can Get Signal® immunostain Immunoreaction Enhancer Solution A (Toyobo).
6. Specific primary antibodies to proteins of interest: anti-p53 (Ab-6; Calbiochem).
7. Anti-phospho-Ser1981 of ATM (10H11. E12; Rockland Inc.).
8. Anti-phospho-Ser15 of p53 (a homemade antibody).
9. Anti-phospho-Ser46 of p53 (a homemade antibody).
10. Anti-phospho-Thr68 of Chk2 (2661; Cell Signaling Technology).
11. Anti-phospho-Ser957 of SMC1 (NB100-205; Novus Biologicals).
12. Anti-γ-H2AX (JBW103; Upstate Biotechnology Inc.).
13. Anti-PML antibody (PG-M3; Santa Cruz Biotechnology).
14. Fluorescent-Labeled Secondary Antibody Conjugates (Thermo Fisher Scientific).
15. Vectashield mounting solution with DAPI (4′,6-diamidino-2-phenylindole) (Vector Laboratories Inc.).
16. Cover glasses.
17. Confocal fluorescent microscope using a Zeiss LSM5 Exciter system equipped with Zen software (Carl Zeiss Inc.).

3 Methods

3.1 Purification of Recombinant Proteins

3.1.1 Purification of Recombinant ATM

1. Culture Human embryonic kidney (HEK) 293 T cells in Dulbecco's Modified Eagle Medium (D-MEM) supplemented with 10% fetal bovine serum (FBS), 100 U/ml penicillin G, and 100 μg/ml streptomycin sulfate.
2. Seed cells (1.5×10^7) in 15 cm dish and cultured for 24 h at 37 °C in a 5% CO_2 incubator.

3. Transfect plasmids (20 μg) into cells using Lipofectamine 2000 in accordance with the manufacturer's instructions (Thermo Fisher Scientific).
4. At 6 h after transfection, replace cell culture media to fresh one.
5. Incubate further for 42 h at 30 °C in a 5% CO_2 incubator.
6. Harvest cells using a cell scraper.
7. Centrifuge cell suspension at 600 × *g* for 5 min.
8. Lyse collected cells for 30 min on ice in 2 ml of E150 extraction buffer supplemented with 1× protease inhibitor cocktail, followed by centrifugation at 20,400 × *g* for 15 min at 4 °C.
9. Collect supernatant.
10. Add 100 μl (50% slurry) of anti-FLAG antibody-conjugated agarose in the supernatant.
11. Rotate for 2 h at 4 °C.
12. Wash beads twice by W150 buffer, twice by W500 buffer, and twice by W150 buffer without detergent [W150(−)].
13. Elute FLAG-tagged ATM from agarose beads using W150(−) buffer with 0.15 mg/ml of 3 × FLAG peptides.
14. Check protein preparations on SDS-PAGE gels after colloidal Coomassie staining.

3.1.2 Purification of GST-Fusion Proteins

GST fusion proteins are expressed in *Escherichia coli* DH5alpha cells using pGEX-4 T-1 or pGEX-6P-1 vectors as per the manufacturer's instructions (GE Healthcare).

1. Inoculate bacterial stock (an expression vector for GST or GST-p53 in DH5alpha) into 1 ml of LB medium.
2. Preculture at 37 °C overnight.
3. Add 9 ml of LB medium (tenfold dilution).
4. Incubate at 30 °C for 1 h.
5. Add IPTG at a final concentration of 0.1 mM.
6. Incubate at 30 °C for 4 h.
7. Harvest bacteria at 6000 × *g* for 15 min.
8. Resuspend 1 ml of ice-cold 1× PBS containing protease inhibitors and 1 mM DTT.
9. Store at − 80 °C.
10. Sonicate on ice (output = 4-5, duty = 50%, sonication, 30 s, rest 1 min, 6 times) until lysates were clear.
11. Centrifuge at 20,400 × *g* for 15 min.
12. Collect supernatant and add 100 μl of Glutathione–Sepharose beads (50% slurry).
13. Rotate at 4 °C for 2 h.

14. Wash beads four times with 1 ml of ice-cold 1× PBS.
15. Elute with 100 μl of 50 mM Tris–HCl (pH 8.0) containing 10 mM glutathione (rotation for >30 min at 4 °C).
16. Dialyze eluates twice with 200 ml of 1× PBS at 4 °C.
17. Check protein preparations on SDS-PAGE gels after colloidal Coomassie staining.

3.2 *In Vitro* Kinase Assay

3.2.1 Preparation of Samples

1. Mix GST-fusion proteins with purified recombinant ATM kinase in Assay buffer supplemented with 10 mM $MnCl_2$ and 0.1 mM ATP or ATP analogues.
2. Incubate at 30 °C for 60 min.
3. Stop the reaction by the addition of 1× SDS sample buffer.
4. Boil the samples.
5. Analyze kinase activity by Immunoblotting with phospho-specific antibodies.

3.2.2 Immunoblotting

1. Separate the samples prepared above on SDS-PAGE gels at 30 mA for 1 h.
2. Blot onto Immobilon-P transfer membrane.
3. Block the membranes with blocking solution containing 5% nonfat dry milk in PBS-T for 1 h at room temperature.
4. Incubate the membranes with primary antibodies (*see* Subheading 2.1.4) diluted in Can Get Signal Solution 1 overnight at 4 °C.
5. Wash the membranes with PBS-T three times for 10 min.
6. Incubate for 1 h with secondary antibodies in PBS-T containing 1% nonfat dry milk at room temperature.
7. Soak the membranes with enhanced chemiluminescence (ECL) Western Blotting Substrate Plus solution as per the manufacturer"s instructions (Thermo Fisher Scientific).
8. Expose to Hyperfilm ECL (GE Healthcare) to visualize bands of interest (Fig. 2).

3.3 Confocal Immunofluorescent Microscopy

1. Seed cells (3×10^4) in 4-well chamber slides.
2. Incubate overnight at 37 °C.
3. Treat these cells with DNA-damaging agents or expose to ionizing radiation.
4. Incubate for 0.5–4 h at 37 °C.
5. Fix cells with 3.7% formaldehyde in PBS for 10 min at room temperature.
6. Permeabilize cells with 0.5% Triton X-100 in PBS for 10 min at room temperature.
7. Block cells with PBS containing 3% BSA and 0.1% goat serum for 1 h at room temperature.

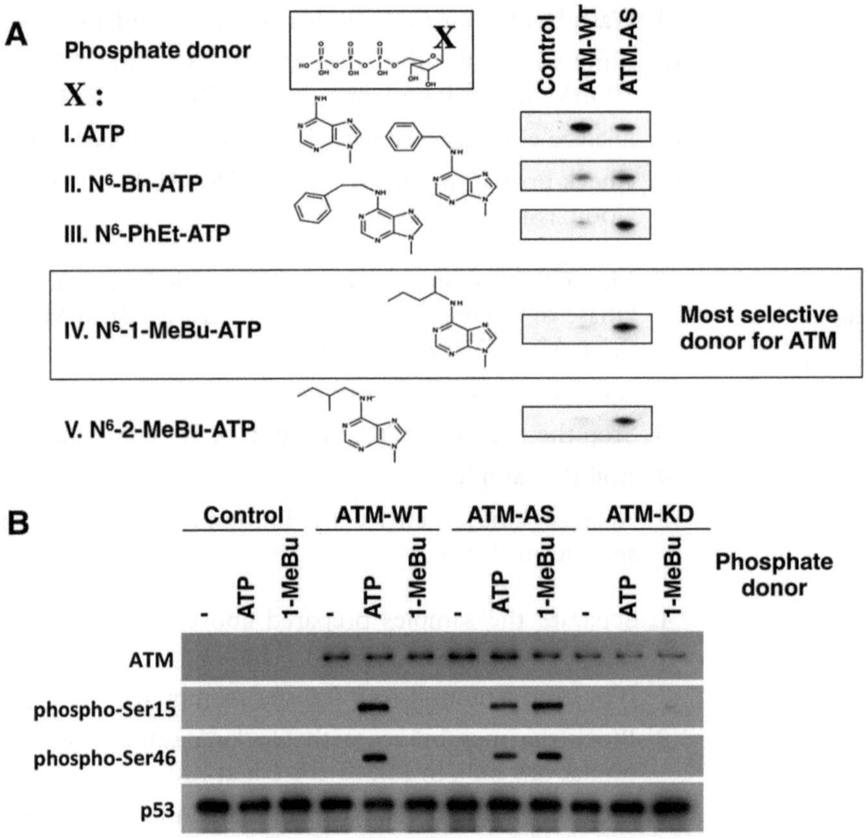

Fig. 2 ATM directly phosphorylates p53 at Ser46 as well as Ser15. (**a**) Construction of ATM sensitive to ATP analogues (ATM-AS). Recombinant ATM was prepared from 293 T cells expressing FLAG-tagged ATM using anti-FLAG antibody and comfirmed by immunoblotting with anti-ATM antibody. Kinase activity of ATM-WT or ATM-AS was measured by in vitro kinase assays using a GST-p53 substrate. As a phosphate donors, natural ATP or various N6-substituted ATP analogues (*row I*, ATP; *row II*, N6-benzyl ATP [N6-Bn-ATP]; *row III*, N6-phenylethyl ATP [N6-PhEt-ATP]; *row IV*, N6-1-methylbutyl ATP [N6-1-MeBu-ATP]; *row V*, N6-2-methylbutyl ATP [N6-2-MeBu-ATP]) were shown in this figure. α-p-Ser15 of p53 antibody was used for the determination of the kinase activities. (**b**) In vitro kinase assays of ATM-WT, ATM-AS, and ATM-KD. Purified ATM proteins from 293 T cells transfected with empty vector (control), ATM-WT, ATM-AS, or ATM-KD were mixed with GST-p53 and immunoblotted with indicated antibodies. ATP or N6-1-MeBu-ATP (1-MeBu) was used as a phosphate donor for the kinase reaction. "–" indicates a sample without a phosphate donor in the reaction mixture. Modified with permission from [13] Kodama M et al. (2010) Mol Cell Biol 30(7):1620–1633 Copyright © 2010, American Society for Microbiology

8. Incubate cells with primary antibodies (*see* Subheading 2.3) overnight at 4 °C. All primary antibodies are diluted in Can Get Signal Immunostain Solution A (Toyobo).
9. Wash cells three times with PBS for 10 min.
10. Incubate cells with secondary antibodies (Alexa Fluor 488 goat anti-rabbit IgG or Alexa Fluor 594 goat anti-mouse IgG) for 1 h at room temperature.

11. Mount cells using Vectashield with DAPI.
12. Observe cells under a confocal immunofluorescent microscope using a Zeiss LSM5 Exciter system equipped with Zen software.

Acknowledgments

The authors are supported by the following grants: Project for the Development of Innovative Research on Cancer Therapeutics (P-DIRECT) from Ministry of Education, Culture, Sports, Science, and Technology—Japan, Applied Research for Innovative Treatment of Cancer from Ministry of Health, Labour and Welfare—Japan (M.E.).

References

1. Larance M, Lamond AI (2015) Multidimensional proteomics for cell biology. Nat Rev Mol Cell Biol 16(5):269–280
2. MacLaine NJ, Hupp TR (2011) How phosphorylation controls p53. Cell Cycle 10(6): 916–921
3. Shiloh Y, Ziv Y (2013) The ATM protein kinase: regulating the cellular response to genotoxic stress, and more. Nat Rev Mol Cell Biol 14(4):197–210
4. Savitsky K, Bar-Shira A, Gilad S, Rotman G, Ziv Y, Vanagaite L et al (1995) A single ataxia telangiectasia gene with a product similar to PI-3 kinase. Science 268(5218):1749–1753
5. Shiloh Y (2014) ATM: expanding roles as a chief guardian of genome stability. Exp Cell Res 329(1):154–161
6. Biton S, Barzilai A, Shiloh Y (2008) The neurological phenotype of ataxia-telangiectasia: solving a persistent puzzle. DNA Repair (Amst) 7(7):1028–1038
7. Shiloh Y, Kastan MB (2001) ATM: genome stability, neuronal development, and cancer cross paths. Adv Cancer Res 83:209–254
8. De Cicco M, Rahim MS, Dames SA (2015) Regulation of the target of rapamycin and other phosphatidylinositol 3-kinase-related kinases by membrane targeting. Membranes (Basel, Switz.) 5(4):553–575
9. Lovejoy CA, Cortez D (2009) Common mechanisms of PIKK regulation. DNA Repair (Amst) 8(9):1004–1008
10. Abraham RT (2004) PI 3-kinase related kinases: 'big' players in stress-induced signaling pathways. DNA Repair (Amst) 3(8-9):883–887
11. Matt S, Hofmann TG (2016) The DNA damage-induced cell death response: a roadmap to kill cancer cells. Cell Mol Life Sci 73(15): 2829–2850
12. Guleria A, Chandna S (2016) ATM kinase: much more than a DNA damage responsive protein. DNA Repair (Amst) 39:1–20
13. Kodama M, Otsubo C, Hirota T, Yokota J, Enari M, Taya Y (2010) Requirement of ATM for Rapid p53 phosphorylation at Ser46 without Ser/Thr-Gln sequences. Mol Cell Biol 30(7):1620–1633
14. Vogelstein B, Lane D, Levine AJ (2000) Surfing the p53 network. Nature 408(6810): 307–310
15. Levine AJ, Oren M (2009) The first 30 years of p53: growing ever more complex. Nat Rev Cancer 9(10):749–758
16. Hollstein M, Sidransky D, Vogelstein B, Harris CC (1991) p53 mutations in human cancers. Science 253(5015):49–53
17. Hollstein MC, Metcalf RA, Welsh JA, Montesano R, Harris CC (1990) Frequent mutation of the p53 gene in human esophageal cancer. Proc Natl Acad Sci U S A 87(24): 9958–9961
18. Vousden KH, Prives C (2009) Blinded by the light: the growing complexity of p53. Cell 137(3):413–431
19. Laptenko O, Prives C (2006) Transcriptional regulation by p53: one protein, many possibilities. Cell Death Differ 13(6):951–961
20. Riley T, Sontag E, Chen P, Levine A (2008) Transcriptional control of human p53-regulated genes. Nat Rev Mol Cell Biol 9(5):402–412
21. Kubbutat MH, Jones SN, Vousden KH (1997) Regulation of p53 stability by Mdm2. Nature 387(6630):299–303

22. Haupt Y, Maya R, Kazaz A, Oren M (1997) Mdm2 promotes the rapid degradation of p53. Nature 387(6630):296–299
23. Hock AK, Vousden KH (2014) The role of ubiquitin modification in the regulation of p53. Biochim Biophys Acta 1843(1):137–149
24. Wade M, Li YC, Wahl GM (2013) MDM2, MDMX and p53 in oncogenesis and cancer therapy. Nat Rev Cancer 13(2):83–96
25. Efeyan A, Serrano M (2007) p53: guardian of the genome and policeman of the oncogenes. Cell Cycle 6(9):1006–1010
26. Caspari T (2000) How to activate p53. Curr Biol 10(8):R315–R317
27. Beckerman R, Prives C (2010) Transcriptional regulation by p53. Cold Spring Harb Perspect Biol 2(8):a000935
28. Ohata H, Miyazaki M, Otomo R, Matsushima-Hibiya Y, Otsubo C, Nagase T et al (2013) NuMA is required for the selective induction of p53 target genes. Mol Cell Biol 33(12): 2447–2457
29. Oda K, Arakawa H, Tanaka T, Matsuda K, Tanikawa C, Mori T et al (2000) p53AIP1, a potential mediator of p53-dependent apoptosis, and its regulation by Ser-46-phosphorylated p53. Cell 102(6):849–862
30. Enari M, Ohmori K, Kitabayashi I, Taya Y (2006) Requirement of clathrin heavy chain for p53-mediated transcription. Genes Dev 20(9): 1087–1099
31. Bulavin DV, Saito S, Hollander MC, Sakaguchi K, Anderson CW, Appella E et al (1999) Phosphorylation of human p53 by p38 kinase coordinates N-terminal phosphorylation and apoptosis in response to UV radiation. EMBO J 18(23):6845–6854
32. D'Orazi G, Cecchinelli B, Bruno T, Manni I, Higashimoto Y, Saito S et al (2002) Homeodomain-interacting protein kinase-2 phosphorylates p53 at Ser 46 and mediates apoptosis. Nat Cell Biol 4(1):11–19
33. Hofmann TG, Moller A, Sirma H, Zentgraf H, Taya Y, Droge W et al (2002) Regulation of p53 activity by its interaction with homeodomain-interacting protein kinase-2. Nat Cell Biol 4(1):1–10
34. Taira N, Nihira K, Yamaguchi T, Miki Y, Yoshida K (2007) DYRK2 is targeted to the nucleus and controls p53 via Ser46 phosphorylation in the apoptotic response to DNA damage. Mol Cell 25(5):725–738
35. Yoshida K, Liu H, Miki Y (2006) Protein kinase C delta regulates Ser46 phosphorylation of p53 tumor suppressor in the apoptotic response to DNA damage. J Biol Chem 281(9): 5734–5740
36. Shah K, Liu Y, Deirmengian C, Shokat KM (1997) Engineering unnatural nucleotide specificity for Rous sarcoma virus tyrosine kinase to uniquely label its direct substrates. Proc Natl Acad Sci U S A 94(8):3565–3570
37. Habelhah H, Shah K, Huang L, Burlingame AL, Shokat KM, Ronai Z (2001) Identification of new JNK substrate using ATP pocket mutant JNK and a corresponding ATP analogue. J Biol Chem 276(21): 18090–18095
38. Ubersax JA, Woodbury EL, Quang PN, Paraz M, Blethrow JD, Shah K et al (2003) Targets of the cyclin-dependent kinase Cdk1. Nature 425(6960):859–864
39. Walker EH, Pacold ME, Perisic O, Stephens L, Hawkins PT, Wymann MP et al (2000) Structural determinants of phosphoinositide 3-kinase inhibition by wortmannin, LY294002, quercetin, myricetin, and staurosporine. Mol Cell 6(4):909–919

Chapter 12

Functional Characterization of ATM Kinase Using Acetylation-Specific Antibodies

Yingli Sun and Fengxia Du

Abstract

The activation of ATM is critical in the DNA double strand breaks repair pathway. Acetylation of ATM by Tip60 histone acetyltransferase (HAT) plays a key role in the activation of ATM kinase activity in response to DNA damage. ATM forms a stable complex with Tip60 through the FATC domain of ATM. Tip60 acetylates lysine3016 of ATM, and this acetylation induces the activation of ATM. Several techniques are included in the study of ATM acetylation by Tip60, such as in vitro kinase assay, systematic mutagenesis, western blots. Here, we describe how to study the acetylation of ATM using acetylation-specific antibodies.

Key words ATM acetylation, ATM activation, Western blot, Tip60, Acetylation

1 Introduction

The ATM protein kinase is essential for cells to survive genotoxic events and is activated by DNA double strand breaks. Several factors contribute to the activation of ATM, including MRN complex [1–4], p18 tumor suppressor protein [5], chromatin remodeling [6–9], and Tip60 [10, 11]. Acetylation of lysine 3016 in the FATC domain of ATM by Tip60 is a key factor for the activation of ATM.

Tip60's acetyltransferase activity is rapidly activated by ionizing radiation, leading to the acetylation and activation of the ATM kinase [10]. Tip60 and ATM form a complex in which Tip60 associates with the highly conserved FATC domain at the extreme C-terminus of ATM [10–12]. This interaction promotes the acetylation of lysine 3016 of ATM by Tip60 [10–12]. Tip60-dependent acetylation of ATM is a key step in the activation of ATM kinase activity. Other researchers confirmed a role for Tip60 and acetylation of ATM in regulating ATM activity [13–15]. The ATM acetylation site is located in the PIKK Regulatory Domain (PRD) of ATM, which is wedged between the C-terminal FATC domain and the kinase domain of ATM. The ATM acetylation site is highly

Sergei V. Kozlov (ed.), *ATM Kinase: Methods and Protocols*, Methods in Molecular Biology, vol. 1599, DOI 10.1007/978-1-4939-6955-5_12, © Springer Science+Business Media LLC 2017

conserved among higher eukaryotes [12], suggesting that acetylation is an evolutionarily conserved event in ATM regulation.

Here, we mainly discuss the study of ATM acetylation functioning in ATM activation. We choose western blot to detect the acetylation of ATM. The difficulty of this detection is due to the isolation of ATM as a high molecular weight protein. Here, we use a specific lysis buffer to efficiently release nuclear proteins. Then we use the ATM immunoprecipitation to obtain more and purer ATM protein. This step makes the subsequent immune blotting with acetylation-specific antibodies easier and better.

2 Materials

1. HeLa cells (American Type Culture Collection, VA).
2. Dulbecco's Modified Eagle Medium (DMEM).
3. Bleomycin.
4. Polyclonal Rabbit anti-ATM antibody (PC116) (EMD Biosciences, CA).
5. Acetylated lysine antibody (AcLys; Upstate Biotech, NY).
6. ATM antibodies 5C2 or 2C1 (Genetex).
7. Anti-phosphoserine 1981 ATM antibody (Rockland Immunochemicals Inc.).
8. Goat Anti-Rabbit IgG H&L (HRP) secondary antibody (ab6721, Abcam).
9. PBS: 8 g NaCl, 0.2 g KCl, 3.58 g $Na_2HPO_3 \cdot 12H_2O$, 0.31 g $NaH_2PO_3 \cdot 2H_2O$, add deionized water to 1 L.
10. ATM lysis buffer: 20 mM HEPES–NaOH(pH 7.4), 150 mM NaCl, 0.2% Tween 20, 1.5 mM $MgCl_2$, 1 mM EGTA, 2 mM dithiothreitol, 50 mM NaF, 500 μM Na_3VO_4, 1 mM phenylmethylsulfonyl fluoride, 0.1 μg/ml aprotinin, 0.1 μg/ml leupeptin.
11. High salt wash buffer: 100 mM Tris–HCl (pH 7.5), 600 mM NaCl, 1 mM PMSF.
12. Basic wash buffer: 10 mM HEPES–NaOH (pH 7.5), 10 mM $MgCl_2$, 50 mM NaCl, 1 mM PMSF.
13. 2× SDS loading buffer: 0.1 M Tris–HCl (pH 6.8), 4% (w/v) SDS, 20% (v/v) glycerol, 2% (v/v) beta-mercaptoethanol (β-ME), and 0.2%(w/v) Bromophenol Blue in deionized water.
14. Protein A-Sepharose beads (GE Healthcare).
15. 70% Ethanol in deionized water.
16. 1.5 mL microcentrifuge tubes.

17. 6% SDS-Polyacrylamide gels.
18. Tris–glycine–SDS running buffer (10×): 30.2 g Tris base, 188 g glycine, and 10 g SDS in 1 L deionized water.
19. Prestained protein molecular weight markers.
20. SDS-PAGE Gel system (e.g., Bio-Rad Laboratories).
21. Tris–glycine transfer buffer: 30 g Tris base, 144 g glycine, 200 mL ethanol, add deionized water to 1 L.
22. PVDF membrane (Millipore).
23. PBS-T: 1× PBS containing 0.1% Tween-20.
24. PBS-T with 0.5%Tween.
25. Blocking solution: 5% nonfat milk, add 2.5 g nonfat milk powder to 50 ml of PBS-T buffer.
26. Enhanced chemiluminescent (ECL) reagents (e.g., Thermo Fisher).
27. High sensitivity X-ray film for imaging chemiluminescent Western blots (e.g., Thermo Fisher).
28. Restore™ Western Blot Stripping Buffer (Catalog#21059 Thermo Scientific).

3 Methods

3.1 Purification of ATM Kinase Using ATM Antibody Immunoprecipitation

1. Prepare cold AT lysis buffer before the experiment (*see* **Note 1**).
2. Treat cells with bleomycin or other DNA-damaging agents to activate ATM kinase. Incubate for a desired time.
3. Take out the cells from the incubator and pour the medium out.
4. Wash cells on ice with cold PBS.
5. Scrape cells off in 1 ml of cold PBS, transfer to 1.5 ml tube.
6. Spin at 800 rpm in benchtop microcentrifuge for 4 min, remove PBS and freeze sample at −80 °C for at least 30 min (*see* **Note 2**).
7. Lyse 50–100 μl cell pellet in 1.5 ml of cold AT lysis buffer (*see* **Note 3**). Resuspend pellet with a blue tip (*see* **Note 4**), transfer to a 1.5 ml microcentrifuge tube, and place on a rotator in a cold room for 1 h.
8. Spin at 15,000 × *g* for 10 min at 4 °C and transfer supernatant to fresh tubes.
9. Add 1 μg of ATM antibody PC116, and incubate for 4–16 h on the rotator at 4 °C (*see* **Note 5**).
10. Cut off the end of a yellow tip, resuspend Protein A-Sepharose beads (*see* **Note 6**), and add 40 μl of Sepharose A beads to extracts. Incubate for 60–80 min on a rotator.

11. Spin down samples for10 s in microcentrifuge at max speed and remove supernatant.
12. Wash immunoprecipitated ATM with 3 × 1 ml of AT lysis buffer, followed by 1 ml of High salt wash and 1 ml of Basic wash buffer.
13. Carefully remove remaining buffer with a needle.
14. Add equal volume of 2× SDS buffer (35 μl), mix gently, heat to 96 °C for 3.5 min. Spin samples to get beads back to the bottom of the tube.

3.2 Western Blot of Acetyl-ATM Using Acetyllysine-Specific Antibody

1. Pour a 6% SDS-PAA minigel (*see* **Note** 7). Load 26 μl of immunoprecipitated ATM sample, run until dye comes off the botto of the gel.
2. For transfer, use transfer buffer without methanol (*see* **Note 8**).
3. Wet the PVDF membrane and assemble the transfer sandwich.
4. Transfer the gel in the cold room (*see* **Note 9**).
5. Block the membrane with 5% milk in PBST (0.5%Tween) for 30 min.
6. Wash the membrane three times in PBST.
7. Incubate for 2 h with anti-acetyl lysine antibody (AcLys; Upstate Biotech, NY). Remember to save antibody for reuse when you finish.
8. Wash membrane three times in PBST.
9. Add 10 μl of secondary antibody to 10 ml PBST and incubate membrane for 45 min.
10. Wash three times with PBST.
11. Develop the blot with the Enhanced chemiluminescent (ECL) reagents.
12. Expose membrane to high sensitivity X-ray film. The time of exposure should be as long as possible since the signal of acetylated ATM protein may be very weak.
13. Strip the membrane with Restore™ Western Blot Stripping Buffer and re-probe with antibodies against total ATM and pS1981 ATM protein (Fig. 1).

4 Notes

1. The AT buffer, high salt buffer and base buffer should not be preserved too long. Use the freshly prepared buffer each time.
2. Frozen cell pellets can be kept at −80 °C for up to 6 months.
3. The AT lysis buffer should be no less than 1 ml and the lysis time should be no less than 1 h so as to lyze the cells adequately.

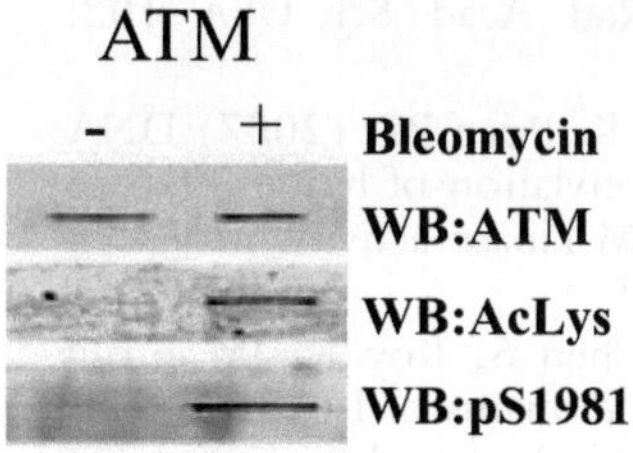

Fig. 1 Western blot of ATM, acetylation, and phosphorylation of ATM using antibodies to ATM, AcetylLys, and pS1981 ATM. ATM is acetylated and phosphorylated when cells are exposed to bleomycin. Reprinted with permission from [10] Sun Y et al. (2005) Proc Natl Acad Sci U S A 102(37):13,182–187 *Copyright (2005) National Academy of Sciences, U.S.A*

4. Use a blue tip to resuspend the cell pellet so that the chromatin proteins will not be destroyed.
5. The time for antibody incubation should be no less than 4 h for good immunoprecipitation.
6. Cut off the end of the tip when pipetting the ProteinA-Sepharose beads to avoid damaging the beads.
7. Since the ATM is a large molecular weight protein, it is better to choose the 6% gel.
8. It is better to use the ethanol instead of methanol in the transfer buffer.
9. Low voltage (80 V) and long time (120 min) will be better for the transfer.

References

1. Lee JH, Paull TT (2004) Direct activation of the ATM protein kinase by the Mre11/Rad50/Nbs1 complex. Science 304:93–96
2. Lee JH, Paull TT (2005) ATM activation by DNA double-strand breaks through the Mre11-Rad50-Nbs1 complex. Science 308:551–554
3. Uziel T, Lerenthal Y, Moyal L, Andegeko Y, Mittelman L, Shiloh Y (2003) Requirement of the MRN complex for ATM activation by DNA damage. EMBO J 22:5612–5621
4. Carson CT, Schwartz RA, Stracker TH, Lilley CE, Lee DV, Weitzman MD (2003) The Mre11 complex is required for ATM activation and the G2/M checkpoint. EMBO J 22:6610–6620
5. Park BJ, Kang JW, Lee SW, Choi SJ, Shin YK, Ahn YH, Choi YH, Choi D, Lee KS, Kim S (2005) The haploinsufficient tumor suppressor p18 upregulates p53 via interactions with ATM/ATR. Cell 120:209–221
6. Kusch T, Florens L, Macdonald WH, Swanson SK, Glaser RL, Yates JR, Abmayr SM, Washburn MP, Workman JL (2004) Acetylation by Tip60 is required for selective histone variant exchange at DNA lesions. Science 306:2084–2087
7. Carrozza MJ, Utley RT, Workman JL, Cote J (2003) The diverse functions of histone acetyltransferase complexes. Trends Genet 19:321–329
8. Bird AW, Yu DY, Pray-Grant MG, Qiu Q, Harmon KE, Megee PC, Grant PA, Smith MM, Christman MF (2002) Acetylation of histone H4 by Esa1 is required for DNA double-strand break repair. Nature 419:411–415
9. Bakkenist CJ, Kastan MB (2003) DNA damage activates ATM through intermolecular autophosphorylation and dimer dissociation. Nature 421:499–506
10. Sun Y, Jiang X, Chen S, Fernandes N, Price BD (2005) A role for the Tip60 histone acetyltransferase in the acetylation and activation

of ATM. Proc Natl Acad Sci USA 102: 13182–13187

11. Sun Y, Xu Y, Roy K, Price BD (2007) DNA damage-induced acetylation of lysine 3016 of ATM activates ATM kinase activity. Mol Cell Biol 27:8502–8509
12. Jiang X, Sun Y, Chen S, Roy K, Price BD (2006) The FATC domains of PIKK proteins are functionally equivalent and participate in the Tip60-dependent activation of DNA-PKcs and ATM. J Biol Chem 281:15741–15746
13. Kim YC, Gerlitz G, Furusawa T, Catez F, Nussenzweig A, Oh KS et al (2009) Activation of ATM depends on chromatin interactions occurring before induction of DNA damage. Nat Cell Biol 11:92–96
14. Li H, Balajee AS, Su T, Cen B, Hei TK, Weinstein IB (2008) The HINT1 tumor suppressor regulates both gamma-H2AX and ATM in response to DNA damage. J Cell Biol 183:253–265
15. Eymin B, Claverie P, Salon C, Leduc C, Col E, Brambilla E et al (2006) p14ARF activates a Tip60-dependent and p53-independent ATM/ATR/CHK pathway in response to genotoxic stress. Mol Cell Biol 26:4339–4350

Chapter 13

Identification of ATM-Interacting Proteins by Co-immunoprecipitation and Glutathione-S-Transferase (GST) Pull-Down Assays

Amanda L. Bain, Janelle L. Harris, and Kum Kum Khanna

Abstract

The ATM kinase is a master regulator of the DNA damage response, and can interact with more than 700 proteins in response to DNA damage. These interactions play a critical role in fine-tuning the response of ATM to multiple cellular stressors, and can play both a positive or negative role in regulating its activity. Here, we detail using protein-protein interaction methods, including co-immunoprecipitation and Glutathione-S-transferase (GST) fusion protein pull-down assays to understand the molecular interactions of ATM. These assays give valuable functional insights into the role of ATM, as they are easy to establish within the laboratory, are not overly laborious, and are easily reproducible.

Key words ATM, DNA damage response, Immunoprecipitation, GST pull-down, Protein interaction

1 Introduction

1.1 Protein-Protein Interaction Methods: Co-immunoprecipitation and Glutathione-S-Transferase (GST) Fusion Protein Pull-down Assays

During the DNA damage response, a highly specific and orchestrated network of protein-protein interactions and posttranslational modifications mediates a signaling cascade eventuating in the repair of genetic lesions [1]. One of the best tools for investigating the function of a protein comes from characterization of its molecular interactions. Co-immmunoprecipitation and Glutathione-S-Transferase (GST) fusion protein pull-down assays are two commonly utilized techniques for probing the molecular interactions of a protein of interest. Both of these techniques rely on the same basic principle of isolating protein complexes using affinity beads or resin, allowing their purification. Interacting proteins present in these complexes can then be assessed via multiple methodologies, such as western-blotting for a putative interaction partner, or mass-spectrometry screening to identify a host of interaction partners. In co-immunoprecipitation, an antibody specific to the endogenous

Sergei V. Kozlov (ed.), *ATM Kinase: Methods and Protocols*, Methods in Molecular Biology, vol. 1599,
DOI 10.1007/978-1-4939-6955-5_13, © Springer Science+Business Media LLC 2017

antigen of interest (or to an epitope-tagged fusion protein of interest which has been exogenously expressed) is used to enrich for complexes associated with the antigen from cells or tissue. These complexes are then "pulled-down" via affinity binding of the antibody and associated protein complexes with Protein A/G agarose or resin (*see* **Note 1**). In contrast, pull-down assays use purified, GST-fused recombinant protein that is first bound with high affinity to Glutathione agarose or resin, followed by "pull-down" of interacting proteins via interaction of bead-protein complexes with the cell/tissue lysate. Both techniques have advantages and disadvantages—while immunoprecipitation studies of endogenous interacting proteins retain their native conformation, GST fusion protein pull-down assays rely on purified proteins, which may not fully mimic a protein's native conformation or posttranslational modification mediating the interaction. However, one major caveat of co-immunoprecipitations is that they will pull down associated protein complexes with a protein of interest, and identified interacting proteins may not be the result of a direct interaction (but via a third-party intermediate). Moreover, an antibody with high specificity for the antigen of interest is required to minimize potential false-positive interactions, and potentially large volumes of antibody can be required, which is costly. In contrast, as GST pull-downs use recombinant protein they are more cost-effective, and larger quantities of "bait" can be utilized for more sensitive detection of interactions. GST pull-down assays are also amenable to more specific investigations of protein-protein interactions, where Bead:GST-protein complexes can be incubated with an exogenously expressed, purified putative interaction partner, without contamination from additional protein complex components. Moreover, GST-pull-down assays can be readily adapted to "map" the region of interest mediating the protein-protein interaction, via recombinant expression and purification of truncated protein fragments, or proteins where the residue mediating the interaction has been mutated to encode an amino acid non-permissive of the interaction.

The Ataxia-Telangiectasia Mutated (ATM) kinase is a master regulator of the DNA damage response, and belongs to a family of phosphoinositide kinase-like-kinases (PIKKs), including DNA-protein kinase catalytic subunit (DNA-PKcs), Ataxia-Telangiectasia and Rad-3 Related (ATR), Mammalian Target of Rapamycin (mTOR), and Transformation/Transcription Domain-Associated Protein (TRRAP) [2]. ATM canonically responds to the presence of DNA double-strand breaks (DSBs), which can be caused by both endogenous stresses (e.g., replication stress), or exogenous agents (e.g., ionizing radiation, X-rays, and certain chemotherapeutic agents). Moreover, ATM can be activated noncanonically in the absence of DSBs, in response to oxidative damage [3], as well as chromatin alterations [4]. Therefore, careful consideration of the conditions in which ATM would normally activate is required when performing assays to assess its function. In response to DNA damage, ATM interacts with and phosphorylates a plethora of proteins on the canonical serine- or threonine-, followed by glutamine residues [3]. In fact, it has been estimated that this phosphorylation network spans more than 700 different protein targets, spanning multiple cellular pathways [5]. Moreover, a number of proteins, which can positively or negatively regulate the activation and activity of ATM have been described [4].

Studies of ATM interactions have historically been difficult. Due to its large size (approximately 350 kDa), ATM is difficult to express in cell culture, and so a highly specific antibody against the endogenous protein must be used. Its large size also hinders recombinant protein expression in bacteria, and many studies utilize recombinantly expressed, overlapping fragments of ATM to probe its molecular interactions. Here, we describe in detail utilizing these two methods, immunoprecipitation and GST fusion protein pull-down assays to investigate the molecular functions of ATM.

2 Materials

2.1 Immunoprecipitation of ATM

2.1.1 Materials

1. 100 or 150 mm tissue culture grade plates.
2. 75 or 150 cm² tissue culture flasks.
3. Plastic cell scraper or rubber policeman.
4. Styrofoam esky.
5. 1.5 mL Eppendorf Tubes.
6. Insulin needles (approximately 21 G).
7. Kimwipes.

2.1.2 Equipment

1. Tissue culture incubator and biolaminar flow hood suitable for cell culture.
2. Source of DNA damage: ionizing radiation source or X-ray generator; alternatively, a radiomimetic such as bleomycin can be used (*see* **Note 2**).
3. Refrigerated centrifuge (suitable for 15 mL falcon tubes) and microcentrifuge (for eppendorf sized tubes).
4. Sonicator.
5. Spectrophotometer.
6. Rotary spinner or roller (setup in 4 °C cold room).
7. Heating block.
8. SDS-PAGE apparatus and power supply.
9. Transfer apparatus.
10. X-ray film.
11. Film developer or Chemiluminescent imaging system.

2.1.3 Reagents

1. Cell line of interest (we recommend Hek293T cells as they proliferate rapidly, and large amounts of protein can be isolated).
2. Complete cell culture medium (RPMI 1640 supplemented with 10% fetal bovine serum).
3. Radiomimetic Drug or DNA damaging agent (if not using ionizing or X-ray source of radiation, *see* Subheading 2.1.2 above).
4. Phosphate-Buffered Saline (PBS), ice cold.
5. Universal Immunoprecipitation Buffer: 50 mM Tri–HCl pH 7.4, 150 mM NaCl, 2 mM EGTA, 2 mM EDTA, 25 mM β-glycerophosphate, 0.2% (v/v) TritonX-100, 0.3% NP-40 (Igepal), protease and phosphatase inhibitors cocktail (add phosphatase inhibitors cocktail and protease inhibitors cocktail fresh just prior to use).
6. Complete™ EDTA-free Protease inhibitor cocktail (Roche) or a home-prepared cocktail of leupeptin, pepstatin, aprotinin, and freshly added PMSF.
7. PhosSTOP™ Phosphatase inhibitor cocktail (Roche) or a home-prepared cocktail of NaF, Na_3VO_4, and okadaic acid.
8. Bradford protein assay reagent.
9. Specific antibody against ATM (we use an in-house generated sheep antibody [6]).
10. IgG from same species as primary antibody.
11. Antibody raised against putative interactor of interest.
12. *Protein A or G Sepharose* (*see* **Note 1**).
13. Benzonase Nuclease (*see* **Note 3**).
14. 2X Laemmli sample buffer: 10% (w/v) Sodium Dodecyl Sulfate (SDS), 20% glycerol, 120 mM Tris–HCl (pH 6.8), 0.02% Bromophenol blue (w/v).
15. TBS-T: Tris-Buffered Saline with 0.1% Tween-20 added.
16. 5% Skim milk powder in TBS-T.
17. Secondary Antibodies: Horseradish peroxidase (HRP)-conjugated IgG (anti-sheep).
18. Enhanced chemiluminescent (ECL) substrate.
19. X-ray or high-sensitivity chemiluminescent film. Alternatively, a chemiluminescent reader documentation system.

2.2 GST-ATM Fusion Protein Pull-Down Assays

2.2.1 Materials

1. 1.5 mL eppendorf tubes.
2. 15 mL falcon tubes.
3. 50 mL falcon tubes.
4. 30 mL reusable centrifuge tubes.

5. 400–600 mL reusable centrifuge tubes.
6. Insulin needles.
7. Heating block.
8. Western blotting and transfer rigs.
9. 10% SDS-PAGE gel.
10. large plastic weighing trays (to fit SDS-PAGE gel).
11. kimwipes.

2.2.2 Equipment

1. Oven and shaking incubator set at 37 °C.
2. Refrigerated centrifuges (with various rotors suitable for volumes from 1 to 500 mL).
3. Sonicator.
4. Rotary spinner or roller (setup in cold room).
5. Heating block.
6. SDS-PAGE apparatus and power supply.
7. Transfer apparatus.
8. X-ray film.
9. Film developer or Chemiluminescent imaging system.

2.2.3 Reagents

1. pGEX plasmids containing cDNA fragments spanning the length of ATM gene (*see* [7] as well as GST-only expressing negative control plasmid).
2. Bacterial competent cells suitable for recombinant protein expression (e.g., BL21).
3. 2YT broth: 16 g tryptone, 10 g yeast extract, 5 g NaCl per liter, made to 1 L in milli-Q water, and sterilized by autoclaving.
4. Isopropyl β-D-1-thiogalactopyranoside (IPTG) dissolved in milli-Q H_2O to a concentration of 1 M.
5. STE buffer: 10 mM Tris–HCl (pH 8.0), 1 mM EDTA, 100 mM NaCl. Supplement with 1 mg/mL lysozyme just before use.
6. Lysozyme stocks at 10 mg/mL dissolved in 10 mM Tris–HCl (pH 8.0).
7. Protease inhibitors cocktail.
8. Dithiothreitol (DTT), 1 M stock.
9. 10% sarcosyl (w/v) in STE buffer.
10. 10% TritonX-100 (v/v) in STE buffer.
11. Glutathione- Sepharose 50:50 slurry, prewashed in PBS prior to use.
12. NETN wash buffer: 10 mM Tris-HCl (pH 8.0), 1 mM EDTA, 100 mM NaCl, 0.5% NP-40, 1 mM PMSF.

13. NETN/0.5 M NaCl: NETN as above but with NaCl concentration altered to 500 mM.
14. Coomassie Blue solution: 45% (v/v) methanol, 10% (v/v) glacial acetic acid, 45% (v/v) water, 3 g/L Coomassie Brilliant Blue R250.
15. Destain Solution: 50% (v/v) methanol in water with 10% (v/v) glacial acetic acid.
16. Anti-GST antibody.
17. Antibody raised against putative interactor of interest.

2.3 Identification of ATM-Interacting Proteins by Mass Spectrometry

2.3.1 Materials

1. Protective wear: Long cuff gloves, hairnets, and facemask (*see* **Note 4**).
2. Filter Tips.
3. 1.5 mL tubes, presoaked in 50% acetonitrile 1% TFA in MilliQ water for at least 30 minutes, and stored in a sealed container (*see* **Note 5**).
4. sealable, clean container to fit eppendorf tubes.
5. Scalpel blades.
6. Western blotting and transfer rigs.
7. Commercial precast gels (*see* **Note 6**).
8. C-18 Zip Tips (*see* **Note 7**) For optional Tryptic clean-up step.

2.3.2 Equipment

1. Laminar Flow hood.
2. Vacuum centrifuge (Speedy-vac).
3. lyophilizer.

2.3.3 Reagents

1. 50% acetonitrile –1% TFA in MilliQ water.
2. Colloidal Coomassie blue solution: 0.08% Coomassie Brilliant Blue G-250 in 1.6% ortho-phosphoric acid, 8% ammonium sulfate, and 20% methanol) or the Biorad Silver Stain kit as per the manufacturer's instructions (catalogue #1610449).
3. Ammonium Bicarbonate Solution: 0.2 M NH_4HCO_3 (pH 8.0) stock, 500 mL (*see* **Note 8**).
4. Silver Destain solution (if your gel pieces are silver stained): 2 g/L potassium ferricyanide ($K_3Fe(CN)_6$) in 0.2 g/L sodium thiosulfate (0.2 g/L $Na_2S_2O_3$). Depending on the number of samples, prepare 5–10 mL (*see* **Note 9**).
5. Coomassie Destain solution: 200 mM NH_4HCO_3 in 50% acetonitrile (ACN).
6. 20 mM DTT in 25 mM NH_4HCO_3.
7. 50 mM iodoacetamide (IAA) in 25 mM NH_4HCO_3. Prepare 5–10 mL depending on the number of samples (*see* **Note 10**).

8. Sequencing Grade Trypsin, reconstituted to 1 μg/μL in the supplied 1 mM HCl (*see* **Note 11**).

For optional tryptic clean-up step:

1. Rehydration solution: 0.1% TFA in 50% ACN.
2. Wash solution: 0.1% TFA.
3. Elution solution: 0.1% TFA in 60% ACN.

3 Methods

3.1 Co-immunoprecipitation

3.1.1 Treatment and Harvesting of Cell Lysates

Day 1:

1. Plate cell line of interest into two large tissue culture plates to be approximately 60–80% confluent on the following day.

Day 2:

1. Remove one of the plates from the incubator and treat with 6 Gy of ionizing radiation (or alternative DNA damaging agent to activate ATM), and carefully mark the treated plate. Replace the flask back to the incubator at +37 °C for 1 h.
2. Remove both untreated and irradiated plates from the incubator, place on ice, and carefully aspirate or pour off cell culture media. Wash the plate once with 10 mL of ice-cold PBS, gently rocking the plate to ensure all cells are washed. Aspirate PBS off.
3. Pipette 500 μL of ice-cold UIP buffer with protease and phosphatase inhibitors onto the plate, and scrape cells using a plastic cell scraper or rubber policeman.
4. Carefully pipette cell suspension into 1.5 mL eppendorf tubes and place on ice for 20 min.
5. Sonicate cell lysate on ice at output of 50% until clear (we use a Branson Sonifier with a Microtip and power set at ~50% for approximately 20 s).
6. Spin cell lysates at full speed for 20 min at 4 °C.
7. Remove supernatant to fresh 1.5 mL eppendorf tube on ice and quantitate protein concentration by Bradford assay
8. For each sample, take 2 mg of protein lysate and divide each into 2 × 1.5 mL eppendorf tubes: one tube will have antibody added, the other tube will have IgG added (negative control). Bring all sample volumes up to 1 mL with cold UIP buffer containing protease and phosphatase inhibitors.

3.1.2 Preclearing of Lysate (Removing Proteins Which Bind Nonspecifically to Beads)

9. Using a yellow pipette tip with the end snipped off with scissors, carefully pipette 20 μL of protein A or G agarose. Wipe the outside of the yellow tip with a kimwipe to remove excess beads, and gently pipette into each cell lysate. Pipette a few times to ensure

all of the beads have entered the lysate solution. Make sure the lid of each tube is tightly closed (wrap in parafilm if necessary) and rotate tubes in the cold room for 1 h to preclear lysate (we usually place eppendorf tubes inside a 50 mL falcon tube to rotate).

10. Spin tubes at 1500 × *g* for 5 min, and remove supernatant to a fresh 1.5 mL eppendorf tube.

3.1.3 Antibody Binding

11. Add 10 U/reaction Benzonase to cell lysate to disrupt protein:DNA and protein:RNA interactions (*see* **Note 3**).
12. Add 1–2 μg of antibody or equivalent amount of IgG to eppendorf tubes for each condition and wrap tubes in parafilm.
13. Rotate tubes in cold room for 2 h to overnight to bind antibody to epitope.

3.1.4 Binding of Antibody: Antigen Complexes to Beads

14. Add appropriate volume of Protein A or G agarose beads (prewashed in PBS, 50:50 slurry) to tubes (*see* **Note 1**).
15. Wrap tubes in parafilm and incubate overnight at 4 °C to bind antibody: antigen complexes to beads.

Day 3:

3.1.5 Washing

1. Centrifuge tubes at 1500 × *g* at 4 °C to pellet beads.
2. Using a separate insulin syringe for each sample (*see* **Note 12**), carefully remove as much supernatant as possible without disrupting the bead pellet.
3. Add 500 μL of cold, complete UIP buffer to Eppendorf tubes, invert several times to mix and pellet beads at 1500 × *g* for 2 min, and discard supernatant. Repeat step thrice.
4. For the final wash, add 500 μL of cold, complete UIP buffer to eppendorf tubes and rotate in the cold room for 10 min. Remove supernatant and discard (*see* **Note 13**).
5. Add 20 μL of Laemmli buffer to each tube.
6. Place tubes on a heating block set at 95 °C and boil for 5 min to detach protein complexes from beads.
7. Remove supernatant and place in a fresh 1.5 mL eppendorf tube. Complexes can now be stored at −80 °C for future use or run directly for western blot.

3.1.6 Western Blot

8. Prepare two polyacrylamide gels for loading samples on. One gel will be to assess for equal ATM pull-down between samples (we recommend a 6% gel) and the other to probe with an antibody raised against the putative protein interactor (choose gel percentage appropriately). Load the gels as follows:

   Lane 1: *Protein Ladder.*

   Lane 2: *3% of whole-cell extract input (untreated).*

   Lane 3: *3% of whole-cell extract input (treated).*

Lane 4: IgG control untreated.

Lane 5: IgG control treated.

Lane 6: ATM IP untreated.

Lane 7: ATM IP treated.

9. Transfer SDS-PAGE gels to PVDF or nitrocellulose membrane using standard procedures.
10. Block membranes in 5% Skim milk powder/TBST for 1 h at room temperature (25 °C).
11. Probe each gel overnight as follows: On the first gel, probe with antibody raised against putative interaction partner of interest. On the second gel, probe with anti-ATM antibody (this is to ensure equal input and to ensure the primary immunoprecipitation has worked).

Day 4:

1. Wash 3× TBS-T and incubate in horse-radish peroxidase (HRP)-conjugated secondary antibody against the host species of the primary antibody according to manufacturers' instructions for 1–2 h at room temperature.
2. Develop using chemiluminescent substrate according to manufacturers' instructions.

3.1.7 Assessing Results

What to Expect: First, assess the gel probed with the ATM antibody has ATM present in input lanes, as well as equivalent amounts of ATM pulled down in control and experimental IP lanes and ATM should not be detected in the IgG negative control lanes. Second, check the gel probed with antibody directed against protein of interest to assess for presence in input lane. Assess IP lanes to determine if the protein is present. If a *bona-fide* interaction has occurred, a band for the putative interactor will be present in the experimental IP lane(s) but not in the IgG control lanes. Also assess to determine if the interaction changes in the presence of DNA damage (treated versus untreated lanes) (*see* Fig. 1 for example).

3.2 GST-ATM Fusion Protein Pull-Down Assay

Pull-down assays can be performed with recombinant purified protein of interest or with cell lysate. Here, we describe a method for assessing protein-protein interactions via utilization of cell lysate. As ATM commonly interacts with many proteins following DNA damage, we recommend preparing cell lysates both without, and with the presence of DNA damage. This can be induced with numerous agents including radiomimetics and ionizing radiation treatment (described here). We recommend using the human embryonic kidney cell line HEK293T for protein purification, as it grows rapidly, and large amounts of protein can be isolated from these cells. We recommend having cells at approximately 70% confluency prior to treatment with DNA damaging agents.

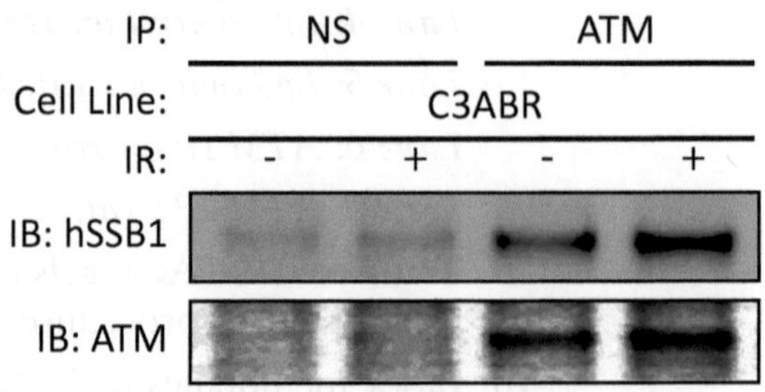

Fig. 1 Example of ATM immunoprecipitation. C2ABR LCL cells were treated with (+) or without (−) Ionizing Radiation (IR) and ATM protein immunoprecipitated (IP) with an anti-ATM antibody, or nonspecific serum (NS) before probing for interacting protein hSSB1 by immunoblot (IB). Note the absence of ATM pulled down in the NS condition as expected, and that the interaction between ATM and hSSB1 is strengthened in the presence of IR. Figure reprinted with permission from [9]

Day 1:

3.2.1 Bacterial Induction

1. Inoculate single colonies from transformation plates of ATM fragments cloned into pGEX vector and pGEX vector (GST alone) (or from glycerol stocks) into 5 mL of 2YT broth with appropriate selection antibiotic (ampicillin) and grow overnight at 37 °C with shaking.

Day 2:

3.2.2 Inducing Recombinant Protein Expression

1. Add starter cultures to 500 mL of 2YT broth containing selection antibiotic and grow at 37 °C with shaking until the OD600 reaches 0.6–0.8.
2. Add IPTG to a final concentration of 0.1 mM (1:1000 dilution from a 1 M stock solution). Shake at 37 °C for 4–6 h to induce protein expression (*see* **Note 14**).
3. Centrifuge cultures at 20,000 × *g* for 15 min at 4 °C.
4. Wash bacterial pellet in PBS and transfer contents to 30 mL tubes. Spin at 20,000 × *g* for 15 min.
5. Freeze pellets overnight or until needed at −70 °C.

Day 3:

3.2.3 Sarcosyl Lysis of Bacterial Pellets

1. Thoroughly resuspend pellets in STE buffer with 1 mg/mL lysozyme (approximately 6 mL per 100 mL of bacterial culture volume) and leave on ice for 30 min to lyse.
2. Add Sarcosyl to a final concentration of 1.5% (from 10% Sarcosyl in STE stock). Add DTT to a final concentration of 5 mM, PMSF to a final concentration of 1 mM, and protease

inhibitor cocktail (according to manufacturer's recommended concentration).

3. Vortex for 5 s to mix.
4. Sonicate lysates using Branson sonifier on output limit of "50% microtip limit" and 50% cycle for 2–3 min on ice, or until lysates are clear.
5. Spin lysates at 20,000 × *g* for 20 min at 4 °C.
6. If the supernatant remains viscous repeat the sonication and spin.
7. Transfer supernatant to a fresh tube. Add Triton X-100 to a final concentration of 2% (from 10% Triton X-100 in STE buffer) and vortex briefly to mix.
8. Lysates can be frozen if desired at this stage.

3.2.4 Binding of GST Fusion Protein to Beads

9. Add prewashed Glutathione-Sepharose beads (250–500 μL beads per 100 mL of culture volume, 50% bead/buffer slurry) to each lysate and rotate tubes for 3 h at 4 °C.
10. Spin at 500 × *g* for 5 min at 4 °C and remove supernatant.
11. Resuspend beads in 10 mL of NETN buffer and transfer to 10 mL tube.
12. Rotate tubes for 5 min to wash beads, and spin tubes at 500 × *g* for 5 min at 4 °C. Remove supernatant.
13. Resuspend beads in NETN with 0.5 M NaCl, and rotate for 10 min.
14. Spin at 500 × *g* for 5 min at 4 °C and remove supernatant.
15. Resuspend beads in a volume of NETN (with Complete™ protease inhibitors) equal to the bead volume. Store at 4 °C.

Day 4:

3.2.5 Analysis of GST-Fusion Proteins

1. Mix 5–10 μL of GST fusion protein bound to beads with 2× Laemmli buffer.
2. Run samples on 10% SDS-PAGE alongside BSA standards of known concentrations (1, 2, 5, and 10 μg of BSA) as well as a molecular weight marker.
3. Stain gel with Coomassie Blue for 3–4 h with gentle agitation at room temperature and destain with destain solution for 4–24 h with gentle agitation, changing destain solution a few times as necessary.
4. Assess quality of and concentration of purified GST-ATM fusion proteins. A dominant band should be observed at the predicted size of the GST-fusion fragment. Estimate fusion protein concentration by comparison with BSA standard band intensities (*see* **Note 15**) (Fig. 2).

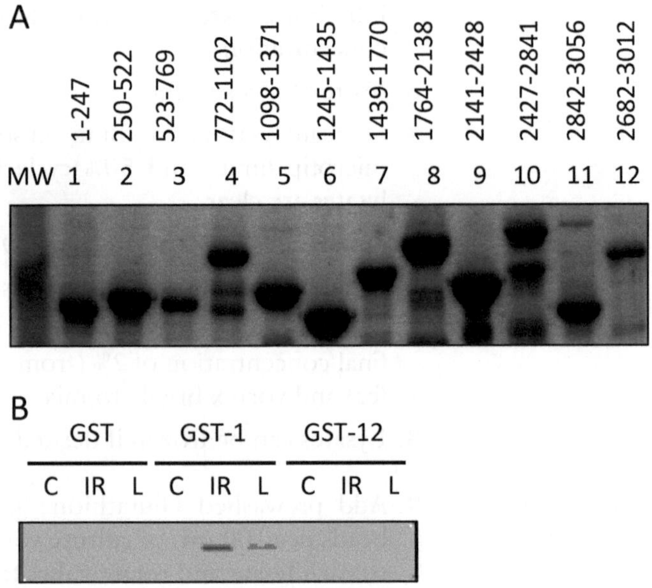

Fig. 2 Example of GST-ATM fusion construct pull-down assay. (**a**) GST-ATM fusion constructs (1–12) were bacterially expressed, purified, resolved by SDS-PAGE, and stained with Coomassie blue. (**b**) Whole cell extracts from control (**c**) irradiated (IR) or ALLN (L)-treated cells were mixed with glutathione beads containing GST, GST-ATM construct 1, or GST-ATM construct 12 and probed by western-blot with an antibody against p53. Figure reprinted with permission from [7]

Day 5:

3.2.6 Preparation of Cell Lysate for Pull-Down Assay

1. Irradiate (or leave untreated) 293T cells at approximately 70% confluency with 5–6 Gy of ionizing radiation.
2. 1–2 h following irradiation, aspirate medium from plate(s) and wash twice in cold PBS.
3. Collect cells by scraping with a plastic cell scraper or rubber policeman in cold PBS and collect cells on a 15 mL tube on ice.
4. Rinse plastic scraper or policeman with extra cold PBS to collect any residual cells.
5. Centrifuge collected cells at 3000 × *g* in a swinging bucket rotor at 4 °C.
6. Add 1 mL of ice-cold UIP buffer containing protease inhibitors cocktail per 10^7 cells, mix by trituration, and leave on ice for 20–30 min to lyse.
7. *Optional*: sonicate samples using Branson Sonifier set at 50% output for 15 × 1 s pulses.
8. Spin cell lysates at 15,000 × *g* at 4 °C, remove supernatant, and keep on ice. Estimate protein concentration by Bradford assay.
9. Aliquot 0.5–1 mg of protein lysate into 13× tubes (one tube for each ATM fragment plus GST only control tube).

10. Add 10–15 μg of GST-bound ATM protein fragments (calculated from polyacrylamide gel concentration estimation in Subheading 3.2.5) or GST alone (negative control) to cell lysates.
11. Tumble bead-lysate mixtures at 4 °C for 2–4 h (this step can be performed overnight).
12. Collect GST-ATM beads with interacting protein complexes by centrifugation at 4000 × *g* for 5 min.
13. Add 500 μL of cold, complete UIP buffer to eppendorf tubes, invert several times to mix and pellet beads at 1500 × *g* for 2 min, and discard supernatant. Repeat step thrice (*see* **Note 13**).
14. Add bead equivalent volumes of 2× Laemmli buffer to the samples and boil for 5 min to release protein complexes from beads.

3.2.7 SDS-PAGE and Western Blot for Interactor Identification

1. Load two identical gels with the following: Load pull-down samples on an appropriate percentage SDS-PAGE gel for protein of interest (if unknown, we recommend using a 10% or 4–15% gradient gel) and resolve. Also include lanes to load 1–3% of the input (this is to ensure the putative protein being tested for interaction is expressed in the cell lysate input).
2. Transfer SDS-PAGE gels to PVDF or nitrocellulose membrane using standard procedures.
3. Block membranes in 5% Skim milk powder/TBST for 1 h at room temperature (25 °C).
4. Probe each gel overnight as follows: On the first gel, probe with antibody raised against putative interaction partner of interest. On the second gel, probe with an anti-GST antibody (this is to ensure equal input).
5. Wash 3× TBS-T and incubate in horseradish peroxidase (HRP)-conjugated secondary antibody against the host species of the primary antibody according to manufacturers' instructions.
6. Develop using chemiluminescent substrate according to manufacturers' instructions.

3.2.8 Assessing Results

Putative protein interactor of interest: A band should be present for the protein in the input lanes. If a band of the same size is observed in any of the GST-ATM fusion protein pull-down lanes, this indicates an interaction between the protein of interest and the corresponding GST-ATM region of interest. It is of interest to cross-reference any important functional domains within ATM which map to the region (e.g., kinase domain), as well as to determine if the interaction differs in the IR-treated cell lysate (indicates a DNA-damage-induced interaction) (*see* Fig. 2).

3.3 Identification of ATM-Interacting Proteins by Mass Spectrometry

In gel tryptic digestion is a method to obtain peptides from samples resolved on 1D or 2D acrylamide gels. Given recent improvements in sensitivity of mass spectrometers, generally if a band can be seen on a Coomassie stained gel there is probably sufficient material for detection by mass spec. However, it should be considered that high molecular weight proteins like ATM may be present in your samples at quantities visible on a Coomassie stain but still have poor mass spec detection. While these proteins can produce darkly staining bands, their molar mass is very high and so the number of moles in such bands may be below the detection limit of some instruments.

3.3.1 Running and Staining Gels

1. Validate your IPs. We recommend splitting IP samples into two reactions. Ten percentage of your final IP sample will be analyzed by western blot to validate successful immunoprecipitation. Remember to include pre-IP lysates samples on this western as a positive control for the western. If ATM is detected in this validation western blot, then resolve the remaining 90% of the samples by SDS-PAGE and staining.

3.3.2 Cutting Gel Slices

1. Place the gel on a clean piece of glass or plastic. The long gel plate is ideal.
2. Carefully excise regions of interest using a scalpel blade, and place the gel pieces into clean pre-rinsed 1.5 mL tubes. Use a fresh blade for each gel piece. Take care not to lean too close to the gel and drag or drop hair onto it.
3. Cut gel pieces into can be placed in 5% acetic acid to keep them wet and stored at 4 °C (*see* **Note 16**).

3.3.3 Destaining

Coomassie-Stained Gel Pieces

1. Remove any liquid in the tubes, being careful to not dislodge the gel slices.
2. Destain by soaking pieces in 200 μL of 200 mM NH_4HCO_3 in 50% acetonitrile (ACN) for 20 min at room temperature. Large gel pieces can be cut into two to three smaller pieces by crushing them a little with a P200 tip. This will speed up their downstream processing. The gel pieces should lose their blue color and shrink.
3. Remove and discard the liquid. If pieces are still blue, repeat this process until they are clear.

Silver Stained Gel Pieces

1. Destain by incubation with fresh 2 g/L $K_3Fe(CN)_6$ dissolved in 0.2 g/L sodium thiosulfate.
2. Incubate each slice with 200 μL of this destained solution for 15 min. Repeat this if bands are still visible.
3. Wash destained slices in MilliQ water (200 μL per wash, three to five washes of 20 min each). After this process the slices should be completely clear.

3.3.4 Drying Samples

1. For either stain type, once pieces are clear, remove the residual liquid.
2. Puncture a small hole in the lid of each tube using a beveled needle and place in a vacuum centrifuge for 30–60 min (*see* **Note 17**).

3.3.5 Reduction and Alkylation

1. Prepare 20 mM DTT in 25 mM NH_4HCO_3. Add 100uL to each gel piece and incubate at 65 °C for 1 h. Discard the supernatant.
2. Prepare 50 mM IAA in 25 mM NH_4HCO_3. Add 100 μL to each gel piece and incubate at 37 °C for 40 min in the dark (for example in a standard incubator). Remove this supernatant and discard.
3. Wash the slices in 25 mM NH_4HCO_3 to remove residual DTT and IAA (3 × 200 μL washes, each with a 20 min incubation). Once again discard the liquid and dry slices in a vacuum centrifuge as before.

3.3.6 Tryptic Digestion

1. Prepare a fresh solution of 40 mM NH_4HCO_3 in 10% ACN.
2. Dilute the master stock trypsin in the 40 mM NH_4HCO_3/10% ACN to a final concentration of 20 ng/μL.
3. Add 20 μL of this diluted trypsin to each gel slice. Larger slices may need up to 50 μL. Allow the slices to re-hydrate at 4 °C for 1 h. A properly re-hydrated gel piece should have a few μL of liquid left over. If a slice looks dry after re-hydration, add a few μL more of the diluted trypsin. If there is more than a couple of μL excess liquid, remove it and discard.
4. Add an additional 50 μL of 40 mM NH_4HCO_3/10% ACN to each slice and incubate at 37 °C overnight.

3.3.7 Peptide Extraction

1. Remove the liquid from the overnight digestions and collect it into a fresh tube.
2. For each gel slice add 50 μL of 0.1% TFA and incubate for 45 minutes at 37 °C with occasional mixing. Pool this extraction with the previous supernatant.
3. Repeat this TFA wash another two times, each time pooling the supernatant. Be careful not to transfer any gel into your new sample tubes.
4. Tryptic peptides can be stored at −20 °C as aqueous samples or freeze dried until further cleanup or analysis.
OPTIONAL:

3.3.8 Peptide Cleanup with Zip-Tips

Peptides prepared from gel slices can have carry over contaminants that affect the quality of the mass spectrum. These can be reduced by an additional reverse phase chromatography step. While this step is not necessary for high abundance proteins, it does improve the quality of MALDI data and can be the critical difference between getting an identification or not for low abundance peptides or those that don't "fly" well. This step is of no benefit if your downstream sample analysis is based on LC-MS, as in this case chromatography is already part of the final sample analysis.

Due to the large volumes obtained by multiple peptide extractions, your peptides will need to be concentrated by freeze drying overnight and then reconstitution in a smaller volume. Resuspend dried peptides in 10–20 μL of 0.1% TFA in 5% ACN.

Important: Pipette slowly to avoid introducing air into the packed resin throughout this procedure.

1. Transfer out a 20 μL aliquot of Elution solution into clean pre-labeled tubes corresponding to each sample. Also label separate empty tubes for each sample.
2. Put the zip tip onto a 10 or 20 μL pipette set to 10 μL. Pushing the pipette to a dead-stop, draw through Rehydration solution, then push most of this liquid back out into a clean waste tube. It is fine to leave 1–2 μL of liquid still in the tip, be careful to not pump air through the resin. Repeat.
3. Wash the tip in Wash solution three times. Each time discard the flow through.
4. Slowly aspirate your sample up and down through the resin bed. Leaving the tip in your sample tube, pipette up and down ten times to ensure good contact between the resin and the molecules in your sample. Finally, expel most of the sample liquid back into the original sample tube.
5. Draw up Wash solution. Expel most of the liquid into a waste container. Repeat this another two times.
6. Go to the pre-labeled tubes containing aliquots of Elution solution prepared earlier. Touch the Zip Tip to the elution solution droplet and draw up a small ~3–5 μL amount of this solution. Slowly expel this into the empty pre-labeled tube. Again draw up a fresh 3–5 μL of Elution solution and expel it into the collection tube. Do this a total of three times.
7. You now have "cleaned up" peptides. A small sample can be mixed directly with MALDI matrix and spotted, or they can be vacuum dried and resuspended in a smaller volume to spot the whole sample.

4 Notes

1. Protein A or G agarose can be purchased from a number of suppliers. Proteins A and G have different affinities depending

on immunoglobulin type of the antibody. A comparison of protein A vs protein G binding can be found at: https://www.neb.com/tools-and-resources/selection-charts/affinity-of-protein-ag-for-igg-types-from-different-species. Please refer to the manufacturers' instructions to ascertain the binding affinity of the beads.

2. Many interactions of proteins with ATM kinase occur following DNA damage. We recommend treating samples with and without a suitable DNA damaging agent to induce the DNA damage response and test if the interaction between the two proteins changes in these conditions. It is also important to carefully choose the timing at which the interaction is tested, as the DNA Damage Response is a dynamic process. Generally, we utilize ionizing radiation (6 Gy, 1 h), but other sources may be utilized.

3. Working with DNA-binding proteins adds an extra level of complexity to immunoprecipitation studies. This is due to DNA or RNA contamination in the pull-down, which brings with it a number of DNA/RNA binding proteins that may not specifically interact with the protein of interest that is being tested. To circumvent this issue, we strongly recommend the addition of a reagent to break DNA-protein bonds, such as Benzonase Nuclease. Alternatively, Ethidium Bromide [8], or DNAase can be used.

4. It is critical to work in an area that is free of dust. We use clean horizontal or vertical laminar flow hoods. Users should wear long cuff gloves and be especially careful to not touch their gloves to ungloved skin or hair. Users with long hair should tie it back and wear a hair net for the whole procedure. Gloves should be changed regularly and after any instance where the user may have touched ungloved skin or a contaminated surface (for example door handles).

5. Many brands of tubes can leak contaminants into samples that then affect downstream experiments. We therefore presoak our eppendorf tubes. To do this, presoak tubes in 50% acetonitrile 1% TFA in MilliQ water for at least 30 min. Aspirate the liquid and discard as solvent waste, and leave the open tubes to air-dry in a laminar-flow hood. These can be prepared in advance and stored in a sealed container until needed. Do not use siliconized tubes.

6. Use precast commercial gels, or if using hand cast gels ensure that they are prepared the day before so that the acrylamide polymerization reaction has time to complete. This will allow improved resolution and reduce unwanted chemical modification of proteins by unpolymerized gel reagents. Resolve as normal and then stain.

7. We use C-18 Zip Tips from Millipore. Each mini-chromatography cartridge is the size of a P10 tip and contains a small amount of chromatography resin. They are supplied dry, and need to be wetted and equilibrated before sample binding and washing, and finally elution.
8. Ammonium bicarbonate (NH_4HCO_3) solutions can be stored for reuse for not more than 1 month. This can be diluted to generate lower concentration ammonium bicarbonate solutions used throughout the procedure. Prepare 500 mL of this solution. All other chemicals should be prepared fresh for every use.
9. Silver stain and silver destain solutions are toxic so do not dispose of them down the sink. Collect unused and used solutions in a waste container for disposal as toxic chemical waste.
10. IAA is light sensitive. Prepare immediately before use and do not expose to light.
11. Sequencing grade trypsin is supplied freeze dried, and most suppliers provide a resuspension solution (1 mM HCl) to reconstitute it. We use Promega trypsin and reconstitute a master stock to 1 μg/μL in the supplied 1 mM HCl. This solution can be aliquoted and stored at −80 °C for later use. Do not freeze and thaw aliquots multiple times.
12. It can be difficult to thoroughly remove washing buffer from the beads without dislodging, or pipetting the beads themselves. To circumvent this issue, we recommend using insulin syringes to remove supernatant during bead washes. As the gauge of the needle is finer than the bead diameter, this allows thorough removal of any remaining liquid in the sample.
13. We recommend use of a gentle buffer when initially testing interactions by immunoprecipitation. However, if significant nonspecific binding is an issue after initial testing, a more stringent buffer can be used for final washes, by increasing the salt concentration to 500 mM.
14. It may be necessary to titrate the amount of IPTG necessary for optimal induction.
15. In our experience, it is necessary to load a much smaller amount of the GST alone purification (approximately 10%) in order to accurately estimate its concentration. Take this into consideration when estimating protein concentrations.
16. Samples can be stored for up to a month before further processing.
17. Given that mass spectrometry is very sensitive to contamination, we recommend that users DO NOT dry tubes in a vacuum centrifuge with completely open lids. Instead, puncture the closed lids with a beveled needle. This will allow your sam-

ples to vacuum dry and reduce the chance of contamination from communal lab equipment. Pieces should take 30–60 min to dry and should appear like small crisp flakes.

References

1. Khanna KK, Jackson SP (2001) DNA double-strand breaks: signaling, repair and the cancer connection. Nat Genet 27(3):247–254. doi:10.1038/85798
2. Bhatti S, Kozlov S, Farooqi AA, Naqi A, Lavin M, Khanna KK (2011) ATM protein kinase: the linchpin of cellular defenses to stress. Cell Mol Life Sci 68(18):2977–3006. doi:10.1007/s00018-011-0683-9
3. Guo Z, Kozlov S, Lavin MF, Person MD, Paull TT (2010) ATM activation by oxidative stress. Science 330(6003):517–521. doi:10.1126/science.1192912
4. Bakkenist CJ, Kastan MB (2003) DNA damage activates ATM through intermolecular autophosphorylation and dimer dissociation. Nature 421(6922):499–506. doi:10.1038/nature01368
5. Matsuoka S, Ballif BA, Smogorzewska A, McDonald ER III, Hurov KE, Luo J, Bakalarski CE, Zhao Z, Solimini N, Lerenthal Y, Shiloh Y, Gygi SP, Elledge SJ (2007) ATM and ATR substrate analysis reveals extensive protein networks responsive to DNA damage. Science 316(5828):1160–1166. doi:10.1126/science.1140321
6. Kozlov S, Gueven N, Keating K, Ramsay J, Lavin MF (2003) ATP activates ataxia-telangiectasia mutated (ATM) in vitro. Importance of autophosphorylation. J Biol Chem 278(11):9309–9317
7. Khanna KK, Keating KE, Kozlov S, Scott S, Gatei M, Hobson K, Taya Y, Gabrielli B, Chan D, Lees-Miller SP, Lavin MF (1998) ATM associates with and phosphorylates p53: mapping the region of interaction. Nat Genet 20(4):398–400. doi:10.1038/3882
8. Lai JS, Herr W (1992) Ethidium bromide provides a simple tool for identifying genuine DNA-independent protein associations. Proc Natl Acad Sci U S A 89(15):6958–6962
9. Richard DJ, Bolderson E, Cubeddu L, Wadsworth RI, Savage K, Sharma GG, Nicolette ML, Tsvetanov S, McIlwraith MJ, Pandita RK, Takeda S, Hay RT, Gautier J, West SC, Paull TT, Pandita TK, White MF, Khanna KK (2008) Single-stranded DNA-binding protein hSSB1 is critical for genomic stability. Nature 453(7195):677–681. doi:10.1038/nature06883

Chapter 14

ATM Activation and H2AX Phosphorylation Induced by Genotoxic Agents Assessed by Flow- and Laser Scanning Cytometry

Hong Zhao, H. Dorota Halicka, Jorge Garcia, Jiangwei Li, and Zbigniew Darzynkiewicz

Abstract

Activation of *Ataxia Telangiectasia Mediated* protein kinase (ATM) by its phosphorylation on serine 1981 and phosphorylation of histone H2AX on serine 139 (γH2AX) are the key events reporting DNA damage, primarily formation of DNA double strand breaks. These events are detected immunocytochemically in individual cells using phospho-specific Abs. The protocols are presented that describe the methodology of immunofluorescent labeling of cells in conjunction with specific staining of cellular DNA. Flow- and imaging-cytometry, the latter exemplified as laser scanning cytometry, is used to quantify intensity of cellular fluorescence reporting activation of ATM and induction of γH2AX with respect to cellular DNA content, which in turn reports the cell cycle phase. Different protocols are presented for analysis of cells either grown in suspension or attached to surface of culture vessels. Examples of ATM activation and H2AX phosphorylation in response to DNA damage in leukemic HL-60 cells by DNA topoisomerase I inhibitor topotecan, and in lung carcinoma A549 cells by hydrogen peroxide, are presented.

Key words DNA damage, DNA double strand breaks, Imaging cytometry, Topotecan, H_2O_2, Phospho-specific antibodies, DNA content, Cell cycle

1 Introduction

DNA damage triggers a cascade of molecular changes termed the DNA damage response (DDR). These events comprise a variety of posttranslational modifications of proteins that activate the DNA damage-related checkpoints having the four critical goals: (1) to enable accessibility of DNA repair machinery to the damage site; (2) activate and engage this machinery; (3) halt cell progression through the cell cycle thereby preventing transfer of damaged DNA to cell progeny, and (4) mobilize mechanisms preparing cell to submit itself either to apoptosis or to replicative senescence, as the means to eradicate cells with unrepairable DNA damage [1–5].

Sergei V. Kozlov (ed.), *ATM Kinase: Methods and Protocols*, Methods in Molecular Biology, vol. 1599,
DOI 10.1007/978-1-4939-6955-5_14, © Springer Science+Business Media LLC 2017

The DDR is regulated by three phosphatidyl inositol 3' kinase-related kinases (PIKKs): Ataxia telangiectasia mutated (ATM), ATM and Rad3-related (ATR), and DNA-dependent protein kinase (DNA-PK). In many instances, these PIKKs have redundant activities and can back up each other in terms of phosphorylation of the same proteins involved in DNA repair, cell cycle progression, or induction of apoptosis as well as cellular senescence.

Among the PIKKs activated in response to DNA damage, the most extensively studied was ATM, which is the key component of the signal transduction pathways mobilized by the induction of double-strand DNA breaks (DSBs) [6–9]. Activation of ATM is provided through its autophosphorylation on Ser1981 located within the ATM dimer and is preceded by its prior acetylation that is mediated by the Tip60 histone acetyltransferase. Phosphorylation of ATM results in dissociation of the dimer onto ATM monomers having kinase enzymatic activity [6–9]. The catalytic domain of ATM is outside the Ser1981 site and becomes accessible to the kinase substrates when ATM is in its monomeric conformation. The MRN complex consisting of Mre11, Rad50, and Nbs1 proteins plays a key role in the process of ATM activation by detecting DSBs, recruiting ATM to the damage site, and targeting ATM to the respective substrates to carry on their phosphorylation [6–9].

Histone H2AX, one of the variants of histone H2A, plays a critical role in surveillance of genome integrity [2–4, 10]. In response to DNA damage, particularly when the damage involves induction of DSBs, H2AX is phosphorylated on Ser139 and thus modified H2AX is termed γH2AX. The phosphorylation can be mediated by ATM-, ATR-, and/or DNA-PKcs and takes place on H2AX in nucleosomes along a mega-base DNA domain on both flanks of DSBs. Phosphorylation of H2AX as the key substrate of ATM which is activated by a variety of mutagens is widely studied as a marker of DNA damage, mainly in regards to the formation of DSBs [4–9].

Flow cytometry and imaging-assisted cytometry, the latter represented by laser scanning cytometry (LSC) [11] are the key instrumentation methodologies used to assess DDR in individual cells. The major advantage offered by flow or imaging cytometry is the possibility of multiparameter analysis which makes it possible to associate particular DDR events with the cell cycle phase, induction of apoptosis or cellular senescence, or other functional states of the cell [12–16]. The intensity of DDR measured by cytometry has been also attempted to connect with the actual DNA damage that is cell-lethal [17] or with the clinical response to the DNA damaging anticancer treatments [18–20]. The LSC make it possible to record images of the studied cells and therefore correlate the images and cellular morphometry with actual cytometric measurement of fluorescence intensity from each individual cell [11].

The method presented in this chapter demonstrates activation of ATM by virtue of its phosphorylation on Ser1981 and the presence of γH2AX in individual cells after treatment with the DNA topoisomerase I inhibitor topotecan (Tpt) [21], or with the exogenous oxidant H_2O_2 [22]. Both Tpt and H_2O_2 are known to damage DNA inducing DSBs preferentially in DNA replicating (S-phase) cells [21, 22] and thereby phosphorylate ATM and H2AX in these cells. The choice of these DNA damage inducers allows one to distinguish, within the same sample, between the cells having phosphorylated ATM (ATM-S1981^P) or H2AX (γH2AX), i.e., the S phase cells, and the predominantly phosphorylation-negative G_1 and G_2M cells [21, 22]. This can be accomplished when the cellular DNA is additionally stained and thus based on differences in DNA content the G_1, S, and G_2M cells can be identified on the bivariate (DNA content versus ATMP or versus γH2AX immunofluorescence) distribution scattergrams (Figs. 1 and 2; *see* **Note 1**).

In this protocol, following incubation with camptothecin (CPT) or H_2O_2, the cells are briefly fixed in methanol-free formaldehyde and then transferred into 70% ethanol in which they can be

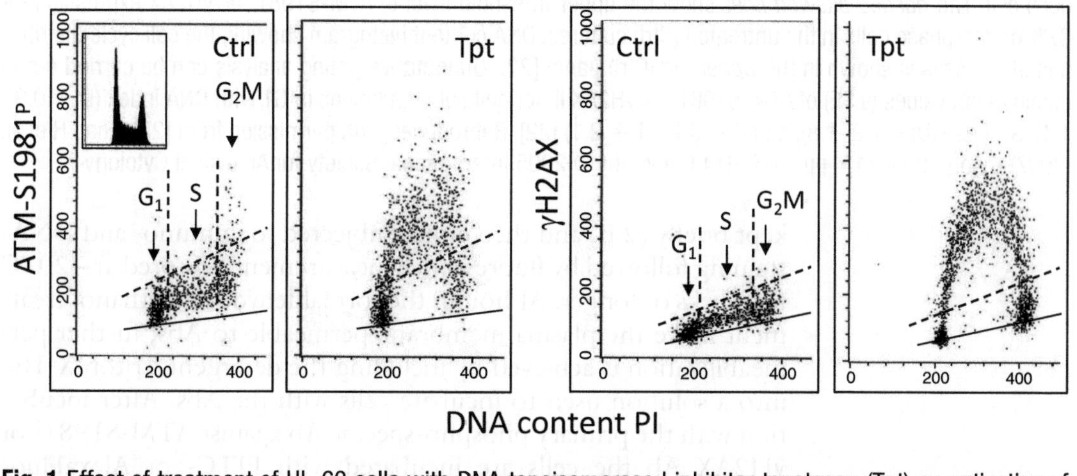

Fig. 1 Effect of treatment of HL-60 cells with DNA topoisomerase I inhibitor topotecan (Tpt) on activation of ATM and phosphorylation of H2AX. Bivariate distributions (scatterplots) showing expression of cellular DNA versus ATM-S1981^P or versus γH2AX in cells treated in suspension culture with 0.15 μM Tpt for 60 min. Expression of ATM-S1981^P or γH2AX was detected using indirect immunofluorescence with the secondary Ab tagged with FITC. Cellular DNA was stained with PI. Cellular fluorescence was measured by flow cytometry (FACScan; Becton Dickinson, San Jose, CA). The *solid skewed lines* represent the upper threshold of fluorescence of cells stained with irrelevant IgG instead of ATM-S1981^P or γH2AX as primary Ab (negative control). DNA content frequency *histogram* representing the cell cycle distribution of the cells is shown in the *upper left* (Ctrl) panel. Reproduced with permission from [21]: Kurose A et al. (2005), Cytometry A, 68A, pp. 1–9. Copyright © 2005 International Society for Analytical Cytology

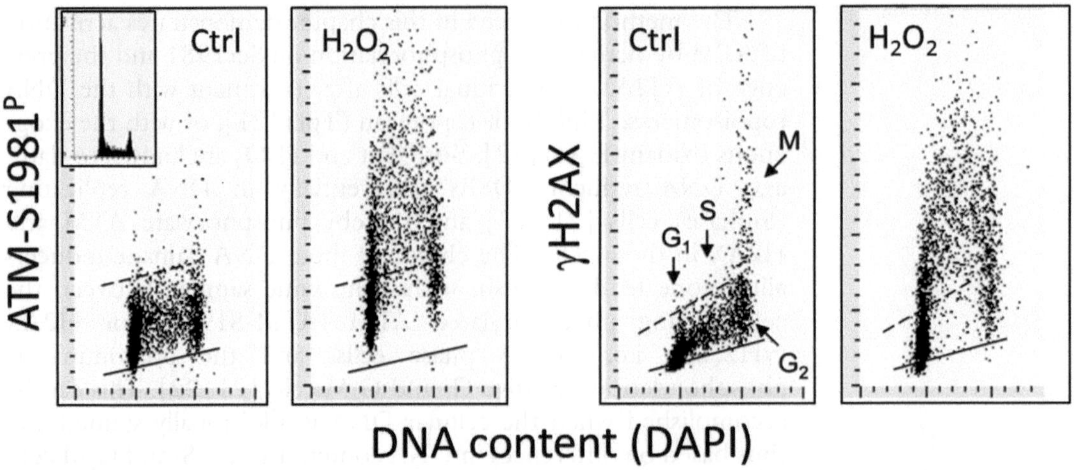

Fig. 2 Effect of H_2O_2 on activation of ATM and phosphorylation of H2AX in A549 cells. Bivariate distributions representing expression of cellular DNA content versus ATM-S1981^P and γH2AX in cells untreated (Ctrl) or treated with 200 μM of H_2O_2 for 1 h. The cells were growing attached on slides and intensity of their fluorescence was measured using a LSC (iCys®), utilizing standard filter settings; fluorescence was excited with 488 nm argon ion and violet diode lasers, respectively. Cellular DNA was stained with DAPI, expression of ATM-S1981^P and γH2AX was detected using indirect immunofluorescence with the secondary Ab tagged with AlexaFluor 488. The intensities of maximal pixel and integrated fluorescence were measured and recorded for each cell. The *dashed skewed lines* show the upper threshold level of ATM-S1981^P or γH2AX expression for 97% of interphase cells in the untreated (Ctrl) cultures. DNA content histogram showing the cell cycle distribution of the cells is shown in the *upper left* (Ctrl) panel [22]. Quantitative gating analysis can be carried out to obtain mean values (±SE) of ATM-S1981^P or γH2AX IF for cell subpopulations in G1 with DNA index (DI = 0.9–1.1), S-phase (DI = 1.2–1.8), and G_2M (DI = 1.9–2.1) [22]. Reproduced with permission from [22]: Zhao H et al. (2007) Cytometry A 71A, pp. 905–914 Copyright © 2005 International Society for Analytical Cytology

kept briefly (2 h) and then either subjected to immuno- and DNA-staining followed by fluorescence measurement or stored at −20 °C for weeks or longer. Although the formaldehyde and ethanol treatment make the plasma membrane permeable to Abs, further permeabilization is achieved by including the detergent Triton X-100 into a solution used to incubate cells with the Abs. After incubation with the primary phospho-specific Ab against ATM-S1981^P or γH2AX Ab the cells are incubated with FITC- or AlexaFluor 488-tagged secondary Ab. Alternatively, they may be directly labeled with the primary phospho-specific Ab conjugated with FITC or AlexaFluor 488 Ab (*see* **Note 2**). Cellular DNA is counterstained with propidium iodide (PI) in the presence of RNase A to remove RNA, which otherwise can also be stained with PI, similar to DNA. Intensity of cellular green (FITC or Alexa Fluor 488) and red (PI) fluorescence is measured by flow-cytometry- or laser scanning cytometry. Cellular DNA alternatively may be stained with 4,6-diamidino-2-phenylindole (DAPI) that is a more specific DNA fluorochrome than the PI is and therefore does not require the presence of RNase (*see* **Note 3**).

It should be noted that ATM and H2AX undergo constitutive endogenous phosphorylation in healthy cells, untreated by genotoxic agents. The level of the constitutive phosphorylation of these proteins is more pronounced in S and G_2M than in G_1 cells, and in large part it reports oxidative DNA damage caused by the metabolically generated oxidants in mitochondria [23]. Furthermore, extensive DNA fragmentation that takes place during apoptosis leads to the formation of a very large number of DSBs which triggers extensive H2AX phosphorylation [13, 16]. Strategies are presented in this chapter to differentiate between the DDR induced by CPT or H_2O_2 (or potentially other exogenous genotoxic agents) and the constitutive damage occurring in untreated cells (*see* **Note 4**), or the apoptosis-associated (AA) DSBs (*see* **Note 5**).

2 Materials

2.1 Reagents and Glassware

1. Phosphate-buffered saline (PBS), pH 7.4.
2. Formaldehyde (methanol-free, "ultrapure").
3. 70% Ethanol.
4. Triton X-100.
5. Bovine serum albumin (BSA).
6. Propidium iodide (PI) or 4'6'-diamidino-2-phenylindole (dichloride) (DAPI).
7. DNase-free RNase A.
8. 12 × 75 mm polypropylene tubes.
9. Centrifuge and rotor capable of 300 × *g*.
10. Camptothecin.
11. Hydrogen peroxide (H_2O_2).
12. Polypropylene 5 mL tubes.
13. To be used in conjunction with analysis by LSC—Microscope slides or single- or multi-chambered CultureSlides (LAB-TEK Brand, ThermoFisher Scientific; also BD Biosciences).
14. To be used in conjunction with the analysis by LSC—Coplin jars.
15. To be used in conjunction with LSC—Parafilm "M" (American National Can, Greenwich, CT).
16. To be used in conjunction with the analysis by LSC—glycerol or the antifade cell mounting medium.

2.2 Instrumentation

1. Fluorescence microscope, or multiparameter Laser Scanning Cytometer (LSC, iCys® model), available from Thorlabs Inc. (Newton, NJ, USA). Flow cytometers of different types,

offered by several manufacturers, can be used to measure cell fluorescence following staining according to the procedures described below. The manufacturers of the most common flow cytometers are Beckman/Coulter Corporation (Miami, FL), BD Biosciences (Franklin Lakes, NJ; formerly Becton Dickinson Immunocytometry Systems; San Jose, CA), Dako North America (formerly DAKO/Cytomation (Fort Collins, CO), ACEA Bioscences (San Diego, CA) and PARTEC GmbH (Zurich, Switzerland).

2. The software to deconvolute the DNA content frequency histograms, to analyze the cell cycle distributions, is available from Phoenix Flow Systems San Diego, CA) or Verity Software House (Topham, MA).

2.3 Reagents to Be Prepared

1. Methanol-free formaldehyde fixative: Prepare 1% (v/v) solution of methanol-free formaldehyde in PBS. This solution may be stored at 4 °C for up to 2 weeks.
2. PBS-BSA solution: Dissolve BSA in PBS to obtain 1% (w/w) solution.
3. BSA-T-PBS: Dissolve BSA in PBS to obtain 1% (w/v) BSA solution. Add Triton X-100 to obtain 0.1% (v/v) of concentration. This solution may be stored at 4 °C for up to 2 weeks.
4. PI stock solution: Dissolve PI in distilled water to obtain 1 mg/mL solution. This solution can be stored at 4 °C in the dark (e.g., in the tube wrapped in aluminum foil) for several months.
5. PI staining solution: Dissolve RNase A (DNase-free) in PBS to obtain 10 μg/mL solution. Add an appropriate aliquot of PI stock solution to obtain its 5 μg/mL final concentration. Store the PI staining solution in the dark. This solution may be stored at 4 °C for up to 2 weeks.
6. DAPI staining solution: Dissolve 4,6-diamidino-2-phenylindole (DAPI) in PBS to obtain its 1 μg/mL final concentration.
7. Antibodies: Phospho-specific Abs, mono-, as well as polyclonal, unconjugated and FITC- or AlexaFluor 488-conjugated, applicable to cytometry, that can be used to detect ATM-S1981^{P} or γH2AX are commercially available from different sources [e.g., from BD Biosciences (Franklin Lakes, NJ), BioLegend (San Diego, CA), Cell Signaling/Santa Cruz Biotechnology (Danvers, MA), or Molecular Probes/Thermo Fisher Scientific (Eugene, OR) (*see* **Note 6**).

3 Methods

3.1 Analysis of Cells by Flow Cytometry

3.1.1 Cells Growing in Suspension

1. Prepare the cells for analysis by treating them in cultures when they are at exponential phase of growth (2–4 × 10^5 cells/mL) with the agents to be studied with respect to induction of ATM activation or/and H2AX phosphorylation (*see* **Note 7**).
2. Transfer the cells still suspended in culture medium into 5 mL centrifuge tubes and centrifuge at 300 × *g* for 4 min at room temperature. Suspend cell pellet (~10^6 cells) in 0.5 mL of PBS.
3. With a Pasteur pipette transfer this cell suspension into polypropylene tube (*see* **Note 8**) containing 4.5 mL of ice-cold 1% methanol-free formaldehyde solution in PBS. Keep on ice for 15 min to fix cells.
4. Centrifuge at 300 × *g* for 4 min at room temperature and suspend cell pellet in 4.5 mL of PBS. Centrifuge again as above and suspend cell pellet in 0.5 mL of PBS. With a Pasteur pipette, transfer the suspension to a tube containing 4.5 mL of ice-cold 70% ethanol. The cells should be maintained in 70% ethanol at −20 °C for at least 2 h, but may be stored under these conditions for several weeks.
5. Centrifuge at 200 × *g* for 4 min at room temperature, remove the ethanol, and suspend cell pellet in 2 mL of BSA-T-PBS solution.
6. Centrifuge at 300 × *g* for 4 min at room temperature and suspend the cells again in 2 mL of BSA-T-PBS. Keep at room temperature for 5 min.
7. Centrifuge at 300 × *g* for 4 min at room temperature and suspend the cells in 100 μL of BSA-T-PBS containing 1 μg of the primary ATM-S1981^P or γH2AX Ab (*see* **Notes 2** and **6**).
8. Cap the tubes to prevent drying and incubate them overnight at 4 °C (*see* **Note 9**).
9. Add 2 mL of BSA-T-PBS and centrifuge at 300 × *g* for 4 min at room temperature.
10. Centrifuge at 300 × *g* for 4 min at room temperature and suspend the cells in 2 mL of BSA-T-PBS.
11. Centrifuge at 300 × *g* for 4 min at room temperature and suspend the cells pellet in 100 μL of BSA-T-PBS containing the appropriate (anti-mouse or anti-rabbit, depending on the source of the primary Ab) FITC- or AlexaFluor 488-tagged secondary Ab (*see* **Note 6**).
12. Incubate for 1 h on a gentle shaker, or occasionally gently shake, at room temperature. Add 5 mL of BSA-T-PBS and after 2 min centrifuge at 300 × *g* for 4 min at room temperature.

13. Suspend cells pellet in 1 mL of the PI staining solution. Incubate at room temperature for 30 min in the dark (*see* **Note 3**).
14. Set up and adjust the flow cytometer for excitation with light at blue wavelength (488 nm laser line or if using UV light, BG-12 excitation filter).
15. Measure intensity of green (530 ± 20 nm) and red (>600 nm) fluorescence of the cells whose DNA was stained with PI, by flow cytometry. Record and analyze the data by computer (Figs. 1 and 2) (*see* **Note 3**).

3.1.2 Cells Growing Attached to Cell Flasks or Petri Dishes

1. Treat cells in cultures when they are at less than 50% confluency with agents to be studied with respect to possible induction of ATM activation or/and H2AX phosphorylation. Incubate with such agents for appropriate period(s) of time.
2. Subject the cells to trypsin or EDTA treatment to dislodge them from culture vessels (*see* **Note 10**).
3. Transfer ~10^6 cells suspended in 0.5 mL of solution still containing trypsin or EDTA. Add 4.5 mL of PBS-BSA solution.
4. Centrifuge at 300 × *g* for 4 min at room temperature. Suspend cell pellet in 5 mL of PBS-BSA solution and centrifuge again at 300 × *g* for 4 min at room temperature.
5. Suspend cells (~10^6) in 0.5 mL of PBS.
6. Follow **steps 3–15** of Subheading 3.1.1.

3.2 Analysis of Cells by LSC

3.2.1 Cells Growing Attached to Culture Vessels

1. Prepare cells attached to microscope slides by growing them in culture vessels having microscope slides at the bottom of the chamber (ChamberSlides, ThermoFisher Scientific). Maintain such cultures to obtain cell densities at about 50% of the slide surface area (*see* **Note 11**).
2. Treat the cells with agents to be investigated with respect to possible activation of ATM or/and phosphorylation of H2AX for appropriate time periods.
3. Discard the medium and wash each chamber with PBS once.
4. Add ice-cold 1% formaldehyde into each chamber and fix for 15 min on ice.
5. Remove formaldehyde, wash each chamber with PBS once (*see* **Note 3**).
6. Remove the chamber sides (walls) as described with the ChamberSlide kit, and immerse a slide into 75% ethanol in a Coplin jar. Keep the cells in ethanol, at 4 °C for 4 h. The cells can be then subjected to incubations with phospho-specific Abs (as below) or stored in 70% ethanol in Coplin jars at −20 °C for up to 4 weeks.
7. Wash the slide twice with PBS, for 5 min, each time.

8. Incubate with 0.1% TritonX-100 in PBS for 15 min at room temperature.
9. Incubate with PBS-BSA for 30 min at room temperature.
10. Incubate with 100 μL PBS-BSA containing 1 μg of the primary ATM or γH2AX Ab in each chamber area overnight at 4 °C. Rigorously prevent drying by covering the site containing cells and antibody solution with a coverslip made of polyethylene foil, and keeping the slides in the moist chamber.
11. Wash slide by PBS for 5 min once.
12. Incubate with 100 μL of PBS-BSA containing 1 μg AlexaFlour 488 or 647 secondary Ab in each chamber area (100–200 μL to each slide area) for 45 min at room temperature preventing drying.
13. Wash slide by PBS for 5 min once.
14. Incubate with 100 μL of PI or DAPI staining solution for 20 min at room temperature (*see* **Note 3**).
15. Wash a slide by PBS and put one drop of mounting (antifade) medium on each chamber (or cytospin area), then cover it with the cover glass.
16. Measure cell fluorescence by LSC.
17. Cellular red (PI) (or blue, DAPI) and green (AlexaFluor 488) fluorescence emission is measured simultaneously in the same cells using LSC utilizing standard filter settings; blue (DAPI) fluorescence with violet diode ion laser, the green (ATM-S1981^P or γH2AX) as well the red (PI) is excited with 488 nm argon ion laser. The intensities of integrated fluorescence and maximal pixel are measured and recorded for each cell.

4 Notes

1. On the bivariate distributions (scatterplots) subpopulations of cells in G_1, versus S versus G_2/M are distinguished based on differences in their DNA content (intensity of PI or DAPI fluorescence; *see* Figs. 1 and 2). To assess the mean extent of expression of ATM-S1981^P or γH2AX for cells at a particular phase of the cycle, the mean values of ATM-S1981^P or γH2AX immunofluorescence (IF) are calculated separately for G_1, S, and G_2/M cell populations distinguished based on differences in DNA content, by the computer-interactive "gating" analysis. The gating analysis should be carried out to obtain mean values of γH2AX or ATM-S1981^P IF for G_1 (DNA Index; DI = 0.9–1.1), S (DI = 1.2–1.8), and G_2M (DI = 1.9–2.1) cell subpopulations, respectively. The mean values of the control cells, untreated with potential DNA damaging agents, are then

subtracted for each cell cycle phase cell. It should be noted that the expression of ATM-S1981^{P} or γH2AX of the untreated control cells represents their constitutive level, resulting to a large degree from DNA damage induced by endogenous oxidants [23].

2. If the primary Ab conjugated with fluorochrome (e.g., with FITC or AlexaFluor 488) is being used, replace the unlabeled primary Ab in **step 7**, Subheading 3.1.1, skip **steps 11** and **12** and move to **step 13**. It should be noted that it is often more advantageous to use the indirect immunofluorescence because of the better choice of the secondary Abs that have a higher intensity of fluorochrome labeling. However, using the primary fluorochrome-tagged Ab obviously shortens the labeling procedure.
3. As an alternative to PI cellular DNA can be stained with DAPI. An advantage of DAPI is that there is no need to treat cells with RNase and resolution of cell cycle phases may be more distinct compared to PI. However, excitation of DAPI is in UV, or near UV (405-nm) wavelength and not all flow cytometers have such a laser. In **step 13** (Subheading 3.1.1) instead with PI the cells may be stained with DAPI at concentration 1 μg/mL in PBS, at room temperature, for 10 min.
4. A low level of ATM-S1981^{P} or γH2AX IF seen in the cells that have not been treated with inducers of DSBs represents the "background," constitutive H2AX phosphorylation and ATM activation, most of which reflects oxidative DNA damage by metabolically generated reactive oxygen species (ROS) [23]. The level of constitutive γH2AX phosphorylation and ATM activation is generally higher in S- and G_2M- than in G_1-phase cells and varies between different cell lines. To quantify ATM-S1981^{P} or γH2AX IF induced by external factors that damage DNA (e.g., topoisomerase inhibitor or H_2O_2), the constitutive component of ATM-S1981^{P} or γH2AX IF has to be subtracted. Toward this end the means of ATM-S1981^{P} or γH2AX IF of G_1, S, and G_2/M-phase of the untreated cells are subtracted from the respective means of the G_1, S, and G_2/M subpopulations of the drug-treated cells, respectively (*see* Figs. 1 and 2). After the subtraction, *the extent of increase in intensity* of ATM-S1981^{P} (ΔATM-S1981^{P} IF) or γH2AX (ΔγH2AX IF) over the untreated sample represents the *treatment-induced phosphorylation* of these proteins. The irrelevant isotype control can be used to estimate the nonspecific Ab binding component, although its use may be unnecessary when one is interested only in the assessment of the drug-induced increase (*Δ*) in IF.
5. DNA undergoes extensive fragmentation during apoptosis [21, 24] which leads to the appearance of a multitude of DSBs

in apoptotic cells, triggering ATM activation and phosphorylation H2AX [21]. It is often desirable, therefore, to distinguish between primary DSBs induced by DNA damaging agents (e.g., topo inhibitors) versus DSBs generated during apoptosis. The following attributes of ATM-S1981^P and γH2AX IF allow one to distinguish the cells with the radiation-, drug-, or carcinogen-induced ATM or H2AX phosphorylation from the cells that have phosphorylation of this histone triggered by apoptosis-associated (AA) DNA fragmentation. (a) The γH2AX IF induced by external DNA damaging agents is seen rather early during the treatment (10 min to 2 h), whereas AA γH2AX IF is seen later (>3 h) [21, 24]; (b) The intensity AA γH2AX IF is generally an order of magnitude higher than that of DNA damage-induced γH2AX IF, unless the cells are at late stage of apoptosis [21]; (c) The induction of AA-induced ATM activation and H2AX phosphorylation is prevented by cell treatment with the caspase inhibitor z-VAD-FMK, which precludes activation of endonuclease responsible for DNA fragmentation. In its presence, thus, the AA-γH2AX IF is suppressed [23]; (d) The AA H2AX phosphorylation occurs concurrently with activation of caspase-3 in the same cells. Multiparameter analysis (active caspase-3 versus γH2AX IF), thus, provides the direct approach to distinguish cells in which DSBs were caused by inducers of DNA damage (active caspase-3 is undetectable) from the cells that have H2AX phosphorylation and ATM activation additionally triggered in response to apoptotic DNA fragmentation (active caspase-3 is present). The presence of activated caspase-3 is detected immunocytochemically with the Ab targeting the activated (cleaved) caspase-3 epitope. These strategies are discussed in more detail elsewhere [16].

6. Quality of the primary and of secondary antibody is of utmost importance. Select the primary Ab that is compatible with immunocytochemical, applicable to cytometry, detection of the phosphoproteins. The ability to detect γH2AX or ATM-S1981^P is often lost during improper transport or storage conditions of the Ab. We occasionally observed that the Abs provided by the vendor were defective. Also of importance is use of the Abs at optimal concentration. It is recommended that with the first use of every new batch of the primary or secondary Ab, test them at serial dilution (e.g., within the range between 0.2 and 2.0 μg/100 μL) to find their optimal titer for the detection of ATM-S1981^P or γH2AX. The optimal titer is recognized as giving a maximal signal-to-noise ratio, i.e., the maximal ratio of the mean IF intensity of the drug-treated to the untreated cells. The titer recommended by the vendor not always is the optimal one.

7. Analysis of cells that are at exponential phase of growth is important to reveal the cell cycle specificity, particularly vis-à-vis DNA replication (S phase), induction of ATM-S1981^{P} or γH2AX by the investigated genotoxic agent.
8. If the sample initially contains a small number of cells they may be lost during repeated centrifugations. To minimize cell loss, polypropylene or siliconized glass tubes are recommended. Since transferring cells from one tube to another causes electrostatic attachment of a large fraction of cells to the surface of each new tube, all the steps of the procedure (including fixation) preferably should be done in the same tube. Addition of 1% (w/v) BSA to rinsing solutions also decreases cell loss. When the sample contains very few cells, carrier cells (e.g., chick erythrocytes) may be included; they may be recognized during analysis based on differences in DNA content (intensity of PI or DAPI fluorescence). Furthermore, to minimize cell loss for samples with paucity of cells tubes of smaller volume (e.g. 1 mL or 2 mL) can be used. Proportions of the respective components of the solutions used for cell fixation, storage and rinsing should be then proportionally adjusted.
9. Alternatively, incubate with the primary ATM or γH2AX Ab for 1 h at 22–24 °C. The overnight incubation at 4 °C, however, appears to yield somewhat higher intensity and signal-to-noise ratio of ATM-S1981^{P} or γH2AX IF compared to 1 h–incubation.
10. The cells have to be thoroughly dispersed, isolated from each other, with no cell clumps/aggregates that can clog nozzle of flow cytometer. Repeated channeling through orifice of Pasteur pipette often is able to disperse cell aggregates. Passage (filtering) of cells through 54 μm mesh is frequently required to remove cell aggregates. When extensive cell death occurs in cultures, the DNA is released from dead cells and it works as a glue to "paste" (aggregate) large number of cells. To disperse cells in such a case the cells collected from the culture, prior to fixation, should be incubated with 50 μg/mL of DNase, in PBS for 20 min, then rinsed with PBS twice and fixed.
11. Cells that normally grow in suspension can be attached to glass slides by electrostatic forces. This is due to the fact that sialic acid residues that cover the cell surface have a net negative charge in contrast to the glass surface that is positively charged. Incubation of cells on microscope slides in the absence of any serum or serum proteins (which otherwise neutralize the charge), thus, leads to their attachment. Toward this end the cells taken from culture should be thoroughly rinsed in PBS to remove serum proteins from the cell culture media and then suspended in PBS at a concentration ~ 2×10^{5} cells/mL. Alternatively, to enforce cells to adhere, the surface of a

slide should be rinsed in a 0.1% w/v solution of polylysine and then thoroughly dried. An aliquot (50–100 μL) of this suspension should be deposited within a shallow well prepared with a pen that deposits a hydrophobic barrier ("Isolator," Shandon Scientific, Pittsburgh, PA), nail polish, or melted paraffin on the horizontally placed microscope slide. To prevent drying, a small piece (~2 × 2 cm) of a thin polyethylene foil or Parafilm may be placed atop of the cell suspension drop. A short (15–20 min) incubation of such cell suspension at room temperature in a closed box containing wet tissue or filter paper that provides 100% humidity is adequate to ensure that most cells will firmly attach to the slide surface. Such preparations can then be fixed in formaldehyde without a significant loss of cells from the slide. However, as in the case of cell growth on glass, mitotic and late apoptotic cells have a tendency to detach even after initial attachment. It should be stressed that the microscope slide to which cells are going to be attached electrostatically should be extra clean. Fingerprints leave oils on the slide that interferes with cell attachment. To remove possible contamination of glass surface that may interfere with cell attachment, soak microscope slides in a household detergent then rinse in water and 100% ethanol respectively. Slides should be allowed to air dry and used the same day they were cleaned. Alternatively, the cells can be deposited on microscopic slides by cytocentrifugation. The Cytospin 3 (or Cytospin 4; Shandon Scientific) is the cell preparation system using cytocentrifugal force to attach cells to microscope slides.

Acknowledgment

Supported by NIH R01 28 705 and The Robert A. Welke Cancer Research Foundation.

References

1. Bakkenist CJ, Kastan MB (2004) Initiating cellular stress responses. Cell 118:9–17
2. Bonner WM, Redon CE, Dickey JS, Nakamura AJ, Sedelnikova OA, Solier S, Pommier Y (2008) γH2AX and cancer. Nat Rev Cancer 8:957–966
3. Kastan MB (2008) DNA damage responses: mechanisms and roles in human disease. G.H.A. cloves memorial award lecture. Mol Cancer Res 6:517–524
4. Nakamura AJ, Rao VA, Pommier Y, Bonner WM (2010) The complexity of phosphorylated H2AX foci formation and DNA repair assembly at DNA double-strand breaks. Cell Cycle 9:389–398
5. Lord CJ, Ashworth A (2012) The DNA damage response and cancer therapy. Nature 481:287–294
6. Helt CE, Cliby WA, Keng PC, Bambara RA, O'Reilly MA (2005) Ataxia telangiectasia mutated (ATM) and ATM and Rad3-related protein exhibit selective target specificities in response to different forms of DNA damage. J Biol Chem 280:1186–1192
7. Shiloh Y (2014) ATM: expanding roles as a chief guardian of genome stability. Exp Cell Res 329:154–161
8. Lavin MF, Kozlov S, Gatei M, Kijas AW (2015) ATM-dependent phosphorylation of all three

members of the MRN complex: from sensor to adaptor. Biomolecules 5:2877–2902
9. Paull TT (2015) Mechanisms of ATM activation. Annu Rev Biochem 84:711–738
10. Siddiqui MS, François M, Fenech MF, Leifert WR (2015) Persistent γH2AX: a promising molecular marker of DNA damage and aging. Mutat Res Rev Mutat Res 766:1–19
11. Pozarowski P, Holden E, Darzynkiewicz Z (2013) Laser scanning cytometry: Principles and applications. an update. Methods Mol Biol 913:187–212
12. Olive PL, Durand RE, Banath JP, Johnston PJ (2001) Analysis of DNA damage in individual cells. Methods Cell Biol 64:235–249
13. Huang X, Okafuji M, Traganos F, Luther E, Holden E, Darzynkiewicz Z (2004) Assessment of histone H2AX phosphorylation induced by DNA topoisomerase I and II inhibitors topotecan and mitoxantrone and by DNA crosslinking agent cisplatin. Cytometry A 58A:99–110
14. Tanaka T, Huang X, Halicka HD, Zhao H, Traganos F, Albino AP, Dai W, Darzynkiewicz Z (2007) Cytometry of ATM activation and histone H2AX phosphorylation to estimate extent of DNA damage induced by exogenous agents. Cytometry A 71A:648–661
15. Zhao H, Traganos F, Darzynkiewicz Z (2010) Kinetics of the UV-induced DNA damage response in relation to cell cycle phase. Correlation with DNA replication. Cytometry A 77A:285–293
16. Darzynkiewicz Z, Zhao H, Halicka HD, Rybak P, Dobrucki J, Wlodkowic D (2012) DNA damage signaling assessed in individual cells in relation to the cell cycle phase and induction of apoptosis. Crit Rev Clin Lab Sci 49:199–217
17. Olive PL (2011) Retention of γH2AX foci as an indication of lethal DNA damage. Radiother Oncol 101:18–23
18. Halicka HD, Ozkaynak MF, Levendoglu-Tugal O, Sandoval C, Seiter K, Kajstura M et al (2009) DNA damage response as a biomarker in treatment of leukemias. Cell Cycle 8:1720–1724
19. Li P, Du CR, Xu WC, Shi ZL, Zhang Q et al (2013) Correlation of dynamic changes in γ-H2AX expression in peripheral blood lymphocytes from head and neck cancer patients with radiation-induced oral mucositis. Radiat Oncol 8:155
20. Vici P, Di Benedetto A, Ercolani C, Pizzuti L, Di Lauro L, Sergi D et al (2015) Predictive significance of DNA damage and repair biomarkers in triple-negative breast cancer patients treated with neoadjuvant chemotherapy: an exploratory analysis. Oncotarget 6:42773–42780
21. Kurose A, Tanaka T, Huang X, Halicka HD, Traganos F, Dai W, Darzynkiewicz Z (2005) Assessment of ATM phosphorylation on Ser-1981 induced by DNA topoisomerase I and II inhibitors in relation to Ser-139-histone H2AX phosphorylation, cell cycle phase and apoptosis. Cytometry A 2005(68A):1–9
22. Zhao H, Tanaka T, Halicka HD, Traganos F, Zarebski M, Dobrucki J, Darzynkiewicz Z (2007) Cytometric assessment of DNA damage by exogenous and endogenous oxidants reports the aging-related processes. Cytometry A 71A:905–914
23. Tanaka T, Halicka HD, Huang X, Traganos F, Darzynkiewicz Z (2006) Constitutive histone H2AX phosphorylation and ATM activation, the reporters of DNA damage by endogenous oxidants. Cell Cycle 5:1940–1945
24. Gorczyca W, Gong J, Darzynkiewicz Z (1993) Detection of DNA strand breaks in individual apoptotic cells by the in situ terminal deoxynucleotidyl transferase and nick translation assays. Cancer Res 53:1945–1951

Chapter 15

Peptide Immunoaffinity Enrichment with Targeted Mass Spectrometry: Application to Quantification of ATM Kinase Phospho-Signaling

Jeffrey R. Whiteaker, Lei Zhao, Regine M. Schoenherr, Jacob J. Kennedy, Richard G. Ivey, and Amanda G. Paulovich

Abstract

Peptide immunoaffinity enrichment coupled with targeted mass spectrometry is a quantitative approach for the robust and reproducible quantification of peptide analytes. The approach is capable of multiplexed quantification of peptides, including posttranslational modifications such as phosphorylation. Anti-peptide antibodies are used to enrich analytes and heavy stable isotope-labeled standards. The enriched peptides are directly measured by multiple reaction monitoring (MRM), a well-characterized quantitative mass spectrometry-based method. Quantification is performed by measuring the analyte (light) peptide response relative to the heavy standard, which is spiked at a known concentration. Here, we describe the methodology for multiplexed measurement of phosphorylated peptides on the ATM kinase and their nonmodified peptide analogs in cellular lysates. The method provides quantitative measurements of phospho-signaling and can be extended to a number of other phosphopeptides and sample types.

Key words Targeted proteomics, DNA damage response, Multiplex, Quantitation, LC/MS

1 Introduction

Targeted mass spectrometry techniques applied to proteomics have seen widespread growth in recent years [1, 2]. The advantages of targeted approaches, like multiple reaction monitoring (MRM; also known as selected reaction monitoring, SRM), include high specificity, precision, interlaboratory reproducibility, and the ability to multiplex measurements of peptides [3, 4]. Analyzing proteotypic peptides (i.e., peptides unique to a given protein of interest and analyzable by the mass spectrometer), released upon proteolysis of biospecimens, allows for quantification of proteins in a wide array of sample types. However, for many peptide analytes of interest, an enrichment step must be performed in order for MRM assays to have sufficient sensitivity to quantify endogenous levels of analyte in a complex biological matrix. This is particularly

Sergei V. Kozlov (ed.), *ATM Kinase: Methods and Protocols*, Methods in Molecular Biology, vol. 1599,
DOI 10.1007/978-1-4939-6955-5_15, © Springer Science+Business Media LLC 2017

true for posttranslational modifications, including phosphorylation. Coupling a peptide immunoaffinity enrichment step with MRM, resulting in an immuno-MRM assay, enables sufficient sensitivity for many low-level analytes [5–9]. The benefits of immuno-MRM include excellent specificity, a wide dynamic range, ease of sample handling, transferability of methods across laboratories, and the capability of multiplexing many analytes together.

We previously demonstrated the application of a multiplexed immuno-MRM assay targeting phospho-signaling in response to DNA damage [10]. The assay measured the response of 69 analytes (phosphorylated and nonmodified peptides) to demonstrate: (1) multiplexed quantitative analysis of signaling events, (2) simultaneous analysis of modified and nonmodified isoforms of peptides, and (3) applicability to a variety of sample types and conditions. Here, we describe in detail the method for applying the multiplexed assay, with focus on measurement of two phosphopeptides derived from the ATM kinase. The two phosphopeptides targeted by the assay contain Ser 367 and Ser 2996 phosphosites. The method describes reagent preparation (e.g., cross-linking antibodies on magnetic beads) in addition to sample preparation (e.g., cell lysis, trypsin digestion, immunoaffinity enrichment, and analysis by targeted MRM mass spectrometry). While the method is focused on the analysis of the ATM kinase, it is generally applicable to any multiplexed immuno-MRM assay applied to cell lysates.

2 Materials

2.1 Cell Culture and Induction of DNA Damage

1. HeLa cell line (adherent cells and suspension cell cultures can also be used) (*see* **Note 1**).
2. Irradiator.

2.2 Cross-Linking Solutions and Buffers

1. 1× PBS: 137 mM NaCl, 2.7 mM KCl, 11.9 mM Phosphate. Dilute from 10× PBS using HPLC-grade water. Store at room temperature for 3 months.
2. Cross-linking solution: 20 mM dimethyl pimelimidate (DMP) in 200 mM triethanolamine (TEA), pH 8.5. To make 200 mM TEA, dissolve 10.6 mL of neat triethanolamine into 400 mL HPLC-grade water; adjust pH to 8.5 with 5 M HCl. Store in the dark in a well-ventilated cabinet. Dissolve 1.04 g of DMP in 200 mL of 200 mM TEA. Make fresh and store at room temperature.
3. Cross-linking quench solution: 150 mM monoethanolamine, pH 9. Dissolve 3.62 mL of neat monoethanolamine in 400 mL HPLC-grade water; adjust pH to 9.0 with 5 M HCl. Store in the dark in a well-ventilated cabinet.

4. Cross-linked antibodies wash solution: 1× PBS with 0.03% CHAPS. First, prepare 5% (w/v) CHAPS (3-((3-cholamidopropyl) dimethylammonio)-1-propanesulfonate) by dissolving 1.0 mg CHAPS powder in 20 mL of LCMS-grade water. Then, prepare 1 × PBS/0.03% CHAPS by diluting 100 mL of 10 × PBS and 6.0 mL of 5% CHAPS to 1 L using LCMS-grade water. Store at room temperature for 3 months.
5. Cross-linked antibodies storage solution: 1 × PBS, 0.03% CHAPS, 0.1% sodium azide (NaN_3). First, prepare 10% (w/v) NaN_3 by dissolving 10 g of NaN_3 in 100 mL of HPLC-grade water. Then, prepare 1× PBS, 0.03% CHAPS, 0.1% NaN_3 solution by diluting 50 mL of 10× PBS, 3.0 mL of 5% CHAPS, and 5.0 mL of 10% NaN_3 to 500 mL using HPLC-grade water. Store at room temperature.
6. Bicinchoninic acid protein assay (BCA assay).

2.3 Cell Lysis Solutions and Buffers

1. 1× DPBS:138 mM NaCl, 2.7 mM KCl, 8 mM Na_2HPO_4-$2H_2O$, 1.5 mM KH_2PO_4.
2. Trypsin Neutralization Solution (TNS) (5% FBS in DPBS).
3. 7.5 M Urea: Add 6 mL HPLC-grade water to a 15 mL Falcon conical tube. Add 4.50 g urea and mix until the urea is in solution. Bring final volume to 10 mL with HPLC-grade water. Must be made fresh daily. Store at room temperature.
4. 5× Lysis buffer stock solution: combine 12.5 mL 1 M Tris–HCl (pH 8.0), 1.0 mL 0.5 M EDTA, and 1.0 mL 0.5 M EGTA. Bring to 100 mL with HPLC-grade water. Filter with 0.22 μm filter. Store at room temperature for 6 months.
5. Lysis buffer: Add 4 Parts 7.5 M Urea, 1 Part 5× Lysis buffer stock solution, 1% phosphatase cocktail 2 (Sigma), 1% phosphatase cocktail 3 (Sigma), and 1% Protease Inhibitor (Sigma) to a final volume of 100 mL. Mix well. Must be made fresh daily, store on ice.
6. 0.2 M Tris–HCl, pH 8.0. Add 4 parts HPLC-grade water and 1 part 1 M Tris–HCl, pH 8.0.

2.4 Digestion Solutions and Buffers

1. Reducing agent: 0.5 M TCEP (tris(2-carboxyethyl)phosphine) bond breaker solution.
2. Alkylating agent: 0.5 M Iodoacetamide (IAM). To one 56 mg vial of iodoacetamide (Sigma, St. Louis, MO), add 605 μL of 0.2 M Tris-HCL, pH 8.0 (made by diluting from 1.0 M Trizma hydrochloride buffer solution, pH 8.0 (Sigma, St. Louis, MO) using HPLC-grade water). Mix until dissolved. Prepare immediately before use and keep out of light.
3. Trypsin solution: Sequencing-grade trypsin (Promega, Madison, WI) at 0.5 μg/μL in 50 mM acetic acid solution.

4. Digestion quenching solution: 20% Formic acid. Mix 4 parts HPLC-grade water and 1 part neat formic acid.

2.5 Immunoaffinity Enrichment Solutions and Buffers

1. 1× PBS with 0.01% CHAPS: Prepare a 5% CHAPS stock solution by dissolving 5 mg CHAPS in 100 mL LCMS-grade water. Add 100 mL of 10× PBS and 2 mL of 5% CHAPS to LCMS-grade water, for a total of 1 L. Store at room temperature for 3 months.
2. 0.1× PBS with 0.01%CHAPS. Add 10 mL of 10× PBS and 2 mL of 5% CHAPS to LCMS-grade water, for a total of 1 L. Store at room temperature for 3 months.
3. Peptide elution solution: 3% v/v acetonitrile (ACN), 5% v/v acetic acid (AcOH), 50 mM citrate. First, prepare a 200 mM citric acid solution by dissolving 8.4 g of citric acid monohydrate in 200 mL of LCMS-grade water. Then, mix 3.0 mL LCMS-grade ACN, 5.0 mL AcOH (ACS reagent grade, ≥99.7%), and 25.0 mL of 200 mM citric acid in LC-MS-grade water, for a total of 100 mL.
4. Peptide storage solution: 30% acetonitrile (LCMS-grade), 0.1% formic acid in LCMS-grade water.

2.6 Liquid Chromatography/Mass Spectrometry Solutions and Buffers

1. Mobile phase A: 0.1% formic acid. Mix 1.0 mL formic acid in 1 L of LCMS-grade water. Store at room temperature for 3 months. Sonicate before use and degas regularly.
2. Mobile phase B: 0.1% formic acid in 90% acetonitrile in water. Mix 100 mL LCMS-grade water and 1.0 mL formic acid with LCMS-grade acetonitrile, for a total of 1 L. Store at room temperature for 3 months. Sonicate before use and degas regularly.

2.7 Equipment and Supplies

1. Magnet: A suitable magnet is required for separation. Formats vary depending on if tubes or plates are used. The protocol described requires tube-based (Antibody Cross-linking) and plate-based (Peptide Immunoaffinity Enrichment) magnets.
2. Refrigerated micro-centrifuge.
3. Centrifuge for use with 96-well plates.
4. Coulter Counter or hemocytometer.
5. Sonic dismembrator (e.g., model 550, ThermoFisher Scientific).
6. Lyophilizer.
7. Positive Pressure-96 Processor Manifold (Waters, Milford, MA).
8. Autosampler and nanoLC system.
9. Column heater for capillary column (Phoenix S&T, Chester, PA).

10. Hybrid triple quadrupole-ion trap mass spectrometer (model 6500 QTRAP, Sciex, Framingham, MA).
11. (Optional) King-Fisher magnetic bead handling platform (ThermoFisher Scientific) equipped with a 96-well deep well magnet head and a 96-well PCR magnet head, with associated plastic tip combs.
12. Tube shaker/rotator.
13. Oasis HLB 96-plate, 30 mg Sorbent per Well, 30 μm Particle Size (Waters) or equivalent solid phase extraction system.
14. 2 mL polypropylene 96-well deep well plates.
15. 200 μL 96-well plates.
16. PCR plates.
17. Square Matrix CapMats for 2 mL blocks (ThermoFisher Scientific).
18. Chemically resistant sealing foil.
19. ThermalSeal® PCR Sealing Films (Genesee Scientific, San Diego, CA).
20. 96-well Silicone Sealing Mat (Axygen Scientific, Union City, CA).
21. Protein-G magnetic beads, suspended as supplied by vendor (*see* **Note 2**).

2.8 Specialized Reagents, Standards, and Controls

1. Antibody master mixes (*see* **Note 3**).
   (a) Antibody stock solutions: Purified monoclonal antibody stocks from the vendor should be in 1× PBS-0.1% sodium azide. Store at 4 °C (for long-term storage, use −20 or −80 °C).
   (b) Antibody working master mix: Combine equivalent amounts of antibodies cross-linked onto magnetic beads, with a final concentration of 1 μg of each antibody per 50 μL. Store in cross-linked antibodies storage solution (*see* Subheading 2.2, **item 5**) sodium azide at 4 °C.
2. Heavy Stable Isotope-Labeled Synthetic Peptide Solutions and Mixtures
   (a) Heavy peptide stock solutions: Heavy peptides may be obtained from synthetic peptide vendors. Prepare stock solutions of individual heavy peptides at 100–500 μM (pmol/μL) in peptide storage solution (*see* Subheading 2.5, **item 4**) in sealed screw-top microcentrifuge tubes. Store aliquots at −80 °C for 6–12 months (*see* **Note 4**).
   (b) Heavy peptide working solution (SIS mix): Generate a master mix of the standard heavy peptides to yield an equal

molar mix at 2 pmol/μL in 3% acetonitrile, 0.1% formic acid solution. Store aliquots at −80 °C for 3 months.

3. Quality control standards.
   (a) MS QC System suitability standard: Prepare a mixture of standard peptides spanning the retention time range of the analytes (such as an iRT standard (Pierce or Biognosys) or mixture of digested proteins). Dilute to a final concentration of 2.5 fmol/μL in 0.1% formic acid immediately prior to analysis.
   (b) Process QC standard: A background matrix consisting of HeLa cells harvested 2 h post-exposure to 10 Gy ionizing radiation (*see* **Note 5**). Prepare lysates as described below (Subheading 3.2). Aliquots are stored in liquid nitrogen and are used as process controls and background matrix for characterization experiments.
   (c) Response curve standards: Prepare a series of samples with varying concentrations of heavy peptides by serially diluting the heavy peptide working solution in previously digested aliquots of Process QC standard. Prepare 7 or 8 concentration points ranging from 0.1 to 1000 fmol in 500 μg of background digested lysate, suspended in 1× PBS with 0.01% CHAPS (*see* Subheading 2.5.1). Add equal amount (100 fmol) of light peptide to each sample. A blank (no heavy peptide added) and double blank (no light or heavy peptide added) are also prepared. Prepare sufficient samples to run each concentration point in triplicate. The same pipettor tip is used for the serial dilution.

3 Methods

3.1 Antibody Cross-Linking

1. Antibodies are separately cross-linked to magnetic Protein G beads depending on the loading capacity of the magnetic beads used. For the beads described in this protocol, a ratio of 5 μg antibody to 1 μL beads is applied. This method describes cross-linking of 500 μg of an antibody to 100 μL beads, which provides enough material for approximately 500 enrichment experiments (1 μg of each antibody per capture).
2. Measure the concentration of antibody solution by bicinchoninic acid assay (BCA assay).
3. Wash 100 μL of the magnetic beads twice using 500 μL aliquots of 1× PBS (pH 7) in microcentrifuge tubes and a magnet. Remove 1× PBS and add 500 μg of the antibody to the washed beads. If the antibody concentration is very low, 5 or 15 mL tubes should be used.

4. Incubate the mixture overnight at 4 °C with tumbling.
5. Using a magnet, pull the beads to the side of the tube and remove the supernatant. Measure the protein concentration in the supernatant by BCA (*see* **Note 6**).
6. Add 900 μL of cross-linking solution (*see* Subheading 2.2, **item 2**). Mix at room temperature for 30 min (*see* **Notes 7** and **8**).
7. Remove the supernatant and quench the coupling reaction by adding 900 μL of cross-linking quench solution (*see* Subheading 2.2, **item 3**). Mix at room temperature for 30 min.
8. Remove the quench solution and wash the antibody-coupled beads three times for 5 min each time, using 900 μL aliquots of antibody wash solution 1 (*see* Subheading 2.2, **item 4**) (*see* **Note 9**).
9. After the final wash, resuspend the antibody-coupled beads in 1000 μL antibody storage solution, at a final antibody concentration of approximately 0.5 μg/μL (disregarding the volume of the beads), and store at 4 °C in 1.5 mL screw-cap tubes until use (*see* **Note 10**).

3.2 Cell Lysis

1. To prepare for cell lysis: turn on the refrigerated micro-centrifuge and cool to 4 °C, turn on benchtop centrifuge and cool to 4 °C, turn on Coulter Counter and prime aperture, thaw phosphatase and protease inhibitors, label and precool 50 mL tubes, label and precool micro-centrifuge tubes, label and precool Cryo-vials, and make fresh Lysis buffer.
2. Remove media and rinse cells with half volume DPBS. Harvest cells by trypsinization and incubate cells at room temperature until the cells lift from the culture surface (as seen under the microscope).
3. Add Trypsin Neutralization Solution (TNS) and collect cells in a precooled 50 mL centrifuge tube. Rinse plates with an additional aliquot of TNS to ensure all cells are collected. If needed, use a cell lifter to remove all cells from the culture surface.
4. Centrifuge cells at 400 × *g* for 8 min at 4 °C. Remove the supernatant (i.e., TNS).
5. Resuspend and pool cells (if necessary) from the same cell line in ice-cold DPBS (note the volume), remove 150 μL for cell counting by Coulter Counter, and check viability.
6. Add ice-cold DPBS to 50 mL and centrifuge cells at 400 × *g* for 8 min at 4 °C, while counting cells with Coulter Counter of hemocytometer. Remove supernatant.
7. Add ice-cold Lysis buffer to a final concentration of 5×10^7 cells/mL.

8. Disperse cell pellet by dragging the tube along a microfuge tube rack. Do not pipette.
9. Sonicate cells 2 × 10 s (Sonic Dismembrator knob set to 1). Wipe down probe with water and 70% ethanol between samples. Place lysate on ice for ~30 s between sonications.
10. Transfer lysate by pipette to a micro-centrifuge tube, vortex for 15 s. Place the tube in a precooled micro-centrifuge and spin 20,000 × *g* for 10 min at 4 °C.
11. Transfer supernatant to a 1.0 mL cryo-vial and store lysates in liquid nitrogen (*see* **Note 11**).

3.3 Trypsin Digestion

1. Adjust the cell lysate concentration to 5 mg/mL with Lysis buffer.
2. Transfer 100 μL (500 μg) of cell lysate to a 2 mL polypropylene 96-deep-well plate. Digest duplicate aliquots of Process QC standard along with each batch of samples.
3. Add 6 μL of reducing agent (0.5 M TCEP).
4. Incubate with mixing at 700 rpm on a heater/shaker block for 30 min at 37 °C.
5. Add 14 μL of alkylating agent (0.5 M IAM). Incubate at room temperature for 30 min in the dark.
6. Add 880 μL of 0.2 M Tris, pH 8.0 to each sample to decrease the urea concentration to ~0.6 M.
7. Add 20 μL trypsin solution to each digest to achieve a 1:50 enzyme-to-substrate ratio.
8. Incubate with mixing at 700 rpm on a heater/shaker for 2 hr. at 37 °C.
9. Add 10 μL trypsin solution to each digest (a 1:100 enzyme-to-substrate ratio).
10. Incubate with mixing at 700 rpm on a heater/shaker for ~16 h (overnight) at 37 °C.
11. Add 50 μL of digestion quenching solution (20% formic acid) to each digest to quench the digestion (should have a final formic acid concentration of ~1%).
12. Add 10 μL of SIS mix. Before spiking the heavy peptide master mix into the digested samples, dilute the master mix with 3% acetonitrile/0.1% formic acid to a 10 fmol/μL concentration, to be able to spike 10 μL of the master mix into each sample (100 fmol of each heavy peptide per sample).
13. Desalt the digested peptides on an Oasis HLB 96-well plate, or an equivalent solid phase extraction medium.
    (a) Condition the wells with 4 × 0.4 mL of 0.1% formic acid in 50% acetonitrile.
    (b) Equilibrate the wells using 4 × 0.4 mL of 0.1% formic acid in water.

(c) Apply the digests to the wells.

(d) Wash with 4 × 0.4 mL of 0.1% formic acid in water.

(e) Elute peptides with 3 × 0.4 mL of 0.1% formic acid in 50% acetonitrile. Collect eluates in a 2 mL polypropylene 96-deep-well plate.

14. Lyophilize the digests and store frozen (−70 °C) as dried powder until the next step.

3.4 Peptide Immunoaffinity Enrichment

1. Resuspend the digested lysate in 200 μL of 1 × PBS, 0.01% CHAPS.
2. Adjust the pH to between 7.5 and 8.0 with 1 M Tris–HCl, pH 8.0. Check pH by pipetting 1 μL onto a pH 5–10 test strip. Verify that pH is between 7.5 and 8.0 before continuing to addition of antibody-beads.
3. Transfer the resuspended digests to 200 μL 96-well plates (the plates actually hold up to 400 μL of liquid).
4. Add the Antibody working master mix (*see* Subheading 2.8, **item 1b**) to each sample well to deliver an equivalent of 1 μg of each antibody (*see* **Note 12**).
5. Cover the plate with a Square Matrix CapMat and seal it by pressure to ensure that no liquid can leak from any of the wells (*see* **Note 13**).
6. Incubate samples overnight at 4 °C on a rotator/shaker (*see* **Note 14**).
7. Remove samples from the rotator and centrifuge at 800 × *g* for 30 s to remove liquid that may be on the mat surface. Then carefully remove the mat.
8. Wash the beads and elute the peptides (*see* **Note 15**).

   (a) Wash the beads twice with 200 μL of 1 × PBS-0.01% CHAPS and once with 200 μL of 0.1 × PBS-0.01% CHAPS by mixing beads for 1.5 min in each wash solution and using a magnet to separate the beads from the supernatant in between washes.

   (b) The peptides are eluted in PCR plates by mixing the beads for 5 min in 26 μL of Peptide Elution Solution (*see* Subheading 2.5, **item 3**). Separate the beads from the eluate using a magnet (*see* **Note 16**).

9. Cover the plate containing the eluates with ThermalSeal® PCR Sealing Film and spin the plate at 800 × *g* for 30 s.
10. Place the elution plate on a plate magnet for 5 min.
11. Transfer and split the eluates to two clean PCR plates by carefully drawing up 2 × 12 μL of the supernatant using a multichannel pipet, without touching the bottom or sides of the

wells, and transferring into corresponding wells of two clean PCR plates. Cover one plate with chemically resistant sealing foil, spin down the plate briefly, and store it at −80 °C.

12. If analyzing samples immediately, proceed to the next step using the second plate. Otherwise, cover the plate with chemically resistant sealing foil, spin it down briefly, and store at −80 °C.

3.5 Analysis of Samples by LC-MRM

1. When ready for analysis, thaw the samples (if necessary) and spin the plate at 800 × *g* for 30 s.
2. Add MS QC system suitability standards and blank injections (Peptide Elution Solution, *see* Subheading 2.5, **item 3**) to empty wells, and response curve standards to regions of the plate not containing samples.
3. Cover the plate with a 96-well Silicone Sealing Mat, spin it down briefly, and place it on the autosampler (*see* **Note 17**).
4. Set up the HPLC instrument configuration according to Table 1.
   (a) Degas solvents for mobile phases A and B. Purge pumps. Calibrate flow rates.
   (b) Autosampler method: Pick up 2 μL of mobile phase A (0.1% formic acid in water). Pick up 11 μL of sample. Briefly rinse the outside of the needle by inserting in a separate vial of 0.1% formic acid in water. Pick up 2 μL of mobile phase A. Switch autosampler valve to inject.
   (c) Perform the HPLC Gradient method according to Table 1.
5. Set up the MS instrument configuration and method according to Table 2 (*see* **Note 18**).
   (a) Set up a Skyline [11] document containing the targeted ATM peptides.
   (b) An average of five light and five heavy transitions were monitored per peptide. The optimized transitions, collision energies (CEs), and declustering potentials (DPs) are reported in Table 2.
6. Set up an analysis batch file and submit samples for analysis (*see* **Note 19**).

3.6 Data Analysis

1. Typical results for peptides captured from the Process QC sample are shown in Fig. 1. Peak profiles are similar and retention times are identical for the light and heavy peptides. The relative areas of transitions for each set of peptides should be within 20% when comparing the light and heavy peptides Fig. 1
2. Import the raw data into the Skyline document containing the targeted peptides.

Table 1
HPLC and autosampler parameters

HPLC parameters			
Autosampler loop volume	10 μL		
Autosampler needle volume	2.6 μL		
Autosampler tray temperature	4–8 °C		
Trap column	5 × 0.3 mm, PepMap 100 C18, 5 μm particle, 100 Å (Dionex, ThermoFisher Scientific)		
Analytical column	100 × 0.075 mm, ReproSil-Pur C18-AQ, 3 μm particle (Dr. Maisch GmbH)		
Column temperature	45 °C		
Gradient	**Time**	**Mobile phase B (%)**	**Flow rate**
Loading	0	2	10
	1.5	2	10
Elution	1.5	2	0.3
	3	2	0.3
	18	40	0.3
	19	90	0.3
	20	90	0.3
	21	2	0.3
	30	2	0.3

3. Check the integration and retention time alignment of the endogenous peptides with the heavy peptide standards and relative peak areas of transitions. Annotate samples according to sample type and treatment (if applicable).
4. Export the results from Skyline to a .csv file. Analysis can be performed in any quantitative analysis package (e.g., R platform) or with flat files (e.g., Excel files). Endogenous light peptide concentrations in fmol/mg are determined by multiplying the light-to-heavy peak area ratios by the known, spiked-in heavy peptide concentration (*see* **Note 20**).

Table 2
Mass spectrometer settings and parameters

Mass spec parameters						
Ion mode	Positive ion					
IonSpray voltage	1900–2200 V					
CAD gas	Medium					
Interface heater temperature	150 °C					
Q1/Q3 resolution	Unit resolution					
Settling time	0 ms					
Pause between masses	3 ms					
Declustering potential	100 V					
MRM detection window	120 s					
Target scan time	0.5 s					
Compound-dependent parameters						
Peptide	**Light Q1**	**Light Q3**	**Heavy Q1**	**Heavy Q3**	**CE (V)**	**Fragment ion**
ATM.SLEISQSYTTTQR++ (1)	757.38	1071.51	762.38	1081.51	36.8	y^{9+}
ATM.SLEISQSYTTTQR++ (2)	757.38	856.42	762.38	866.42	33.8	y^{7+}
ATM.SLEISQSYTTTQR++ (3)	757.38	984.47	762.38	994.48	34.8	y^{8+}
ATM.SLEISQSYTTTQR++ (4)	757.38	1184.59	762.38	1194.60	32.8	y^{10+}
ATM.SLEISQSYTTTQR++ (5)	757.38	769.38	762.38	779.39	34.8	y^{6+}
ATM.SLEIpSQSYTTTQR++ (1)	797.36	856.42	802.37	866.42	41.9	y^{7+}
ATM.SLEIpSQSYTTTQR++ (2)	797.36	583.79	802.37	588.80	36.9	y^{10}–98^{++}
ATM.SLEIpSQSYTTTQR++ (3)	797.36	606.32	802.37	616.33	39.9	y^{5+}
ATM.SLEIpSQSYTTTQR++ (4)	797.36	769.38	802.37	779.39	42.9	y^{6+}
ATM.SLEIpSQSYTTTQR++ (5)	797.36	1151.47	802.37	1161.48	34.9	y^{9+}
ATM.NLSDIDQSFNK++ (1)	640.81	1053.49	644.82	1061.50	27.7	y^{9+}
ATM.NLSDIDQSFNK++ (2)	640.81	966.45	644.82	974.47	28.7	y^{8+}
ATM.NLSDIDQSFNK++ (3)	640.81	738.34	644.82	746.36	26.7	y^{6+}
ATM.NLSDIDQSFNK++ (4)	640.81	851.43	644.82	859.44	28.7	y^{7+}
ATM.NLSDIDQSFNK++ (5)	640.81	495.26	644.82	503.27	33.7	y^{4+}
ATM.NLpSDIDQSFNK++ (1)	680.79	1133.45	684.80	1141.47	26.8	y^{9+}
ATM.NLpSDIDQSFNK++ (2)	680.79	738.34	684.80	746.36	32.8	y^{6+}
ATM.NLpSDIDQSFNK++ (3)	680.79	1035.47	684.80	1043.49	33.8	y^{9}–98^{+}
ATM.NLpSDIDQSFNK++ (4)	680.79	966.45	684.80	974.47	32.8	y^{8+}
ATM.NLpSDIDQSFNK++ (5)	680.79	851.43	684.80	859.44	33.8	y^{7+}

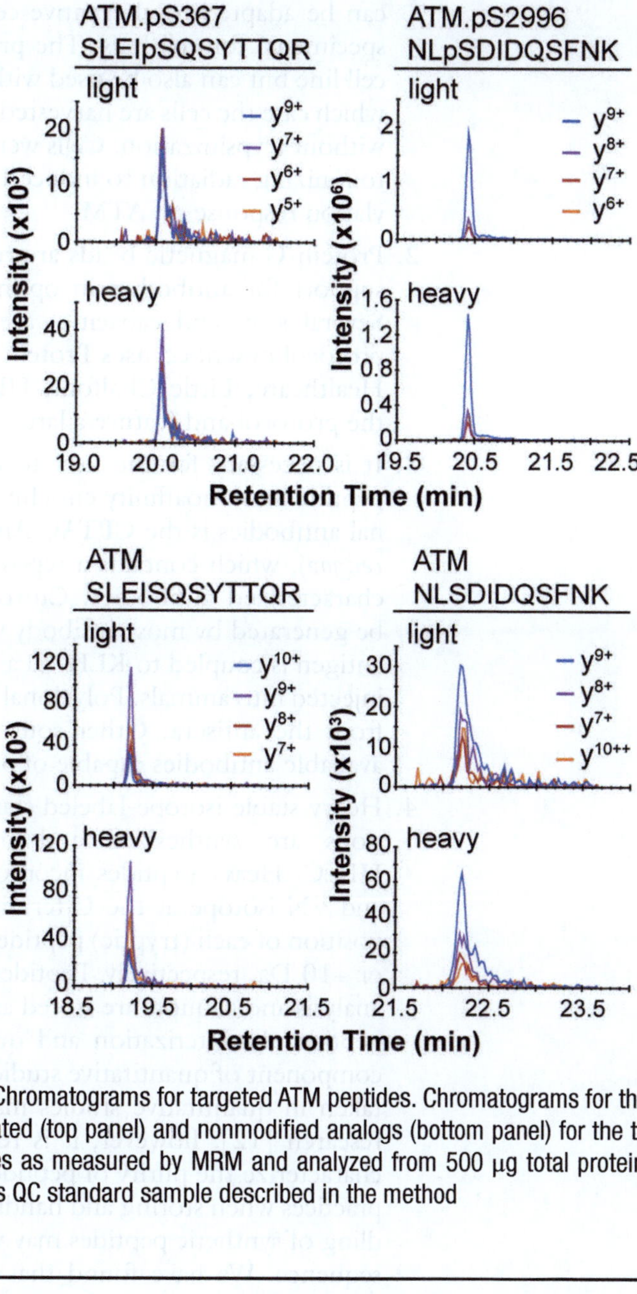

Fig. 1 Chromatograms for targeted ATM peptides. Chromatograms for the phosphorylated (top panel) and nonmodified analogs (bottom panel) for the targeted peptides as measured by MRM and analyzed from 500 μg total protein of the Process QC standard sample described in the method

4 Notes

1. Whole cell lysates prepared from cultured immortalized cell lines, primary human cells, and tissue specimens have been evaluated. Detection of endogenouos analyte depends on expression levels and treatment conditions in individual samples. The method has been tested using sufficient sample material to provide 500 μg of total protein as input material. The method describes application to cultured HeLa cell lines but

can be adapted to alternative cell lines or inputs (e.g., tissue specimens) for cell lysis. The procedure assumes an adherent cell line but can also be used with cells grown in suspension, in which case the cells are harvested directly from growth medium without trypsinization. Cells were harvested 2 h post-exposure to ionizing radiation to induce DNA damage and a phosphorylation response on ATM.

2. Protein-G magnetic beads are recommended to provide solid support for antibodies in optimum orientation for binding. Several sizes and capacities are commercially available. The protocol described uses Protein G Mag Sepharose beads (GE Healthcare, Little Chalfont, UK) which have been tested in the protocol and feature a large binding capacity.

3. It is necessary for the user to obtain suitable antibodies for peptide immunoaffinity enrichment. One source for monoclonal antibodies is the CPTAC Antibody Portal (*antibodies.cancer.gov*), which contains a repository of publicly available and characterized antibodies. Custom-made polyclonals can also be generated by most antibody vendors. Typically, the peptide antigen is coupled to KLH via a C-terminal cysteine linker and injected into animals. Polyclonal antibodies are affinity-purified from the antisera. Other sources may include commercially available antibodies capable of peptide immunoprecipitation.

4. Heavy stable isotope-labeled standards and matched light versions are synthesized and purified to >95% purity by HPLC. Heavy peptides incorporate a fully atom labeled ^{13}C and ^{15}N isotope at the C-terminal lysine (K) or arginine (R) position of each (tryptic) peptide, resulting in a mass shift of +8 or +10 Da, respectively. Peptides are quantified by amino acid analysis and aliquots are stored as described until use. Synthetic peptide characterization and quality control is an important component of quantitative studies. The degree to which care is taken in quantitative studies may depend on the goal of the research [12]; however, it is recommended that laboratories characterize the purity of peptide standards and adhere to best practices when storing and handling peptides. Storage and handling of synthetic peptides may vary depending on the peptide sequence. We have found that general recommendations for peptide standards as outlined in Hoofnagle et al. [13] are suitable for the ATM peptides measured in this method.

5. A suitable background matrix that closely replicates the nature of the intended sample with confirmed expression of the target can be substituted for the HeLa cell lysate.

6. The approximate coupling efficiency can be determined by comparing the amount of antibody in the supernatant pre- and post-coupling to magnetic beads. Coupling efficiencies are typically in the range of 60–80% using the procedure outlined.

7. Bead handling steps can be performed manually or using an automated bead handling system.
8. Due to the toxicity of DMP, triethanolamine, and monoethanolamine, it is advisable to perform steps using these reagents in a fume hood or well-ventilated area.
9. The method describes coupling of purified monoclonal antibodies. When coupling purified polyclonal antibodies, a slight modification is made to the procedure to remove any peptide that may be bound to the antibody. Wash the beads twice with 3% acetonitrile—5% acetic acid and then wash with 1 × PBS—0.03% CHAPS.
10. Check the crosslinking step by subjecting an aliquot of antibody cross-linked beads to elution with 3% acetonitrile—5% acetic acid. Run the eluate using a 4–20% gradient SDS-PAGE gel system and silver stain. The lanes should be void of bands corresponding to antibody.
11. If storing aliquots of a lysate, first transfer the lysate to a fresh micro-centrifuge tube to ensure homogeneous mixing of the lysate before aliquoting. In addition, an aliquot is used to determine protein concentration by BCA to inform the digestion step and the amount of trypsin to add. Lysates can be stored in liquid nitrogen for up to 1 month or more. For extended storage, it is recommended to lyophilize the lysate and store the dried powder frozen at −80 °C or in a liquid nitrogen dewar.
12. Ensuring the antibody-coupled beads are evenly distributed and well mixed is important to ensure consistent application of equivalent amount of antibody and ultimately the reproducibility of the method. Vortexing, tumbling, agitating, or vigorously pipetting up-and-down are preferred methods for mixing beads.
13. Performance can vary depending on the type of beads used in the assay. Some beads do not interact with adhesive film and other types of beads can adhere to adhesive sealing film.
14. This is a critical step to the reproducibility of the method. We have found that tumbling the beads end-over-end is the most effective way to achieve good mixing. This is accomplished by sealing the plate and fixing it to a tube rotator using lab tape, rubber bands, or Velcro strips. Mixing is conducted for 16 h at 4 °C.
15. Like previous bead-handling steps, the wash and elution steps can be performed by an automated magnetic particle processor (e.g., KingFisher). If using a KingFisher (ThermoFisher Scientific), we recommend using the PCR magnetic head and a PCR deep well tip comb.
16. Following elution, the initial samples (i.e., analyte-depleted protein digests) can be saved for potential sequential capture

experiments [14] and the antibody-coupled magnetic beads can be saved for potential recycling [15]. Transfer or resuspend used beads in Cross-linked antibodies storage solution (*see* Subheading 3.4, **item 8b**). Seal the sample incubation plate and the collected antibody-bead plate with adhesive foil and store at −80 and 4 °C, respectively. The utility of sequential capture or recycling experiments should be validated prior to use.

17. It is advisable to use a plate magnet on the autosampler tray to keep any residual beads out of the HPLC system.
18. The following settings and configuration are reported for a Sciex 6500 QTRAP. Adjust settings to corresponding values for alternative triple-quadrupole systems. Table 2 provides the top five most intense transitions for the peptides determined on the 6500 QTRAP. Transitions may vary slightly depending on the instrument used and interferences arising from different sample matrices. Collision energy values were determined using Skyline optimization of the heavy peptides.
19. Inject the MS QC system suitability standard with an unscheduled MRM method a minimum of once per day. Injection of samples is dependent on satisfactory response of the MS QC system suitability standard compared to historical values. Inject the Process QC standard a minimum of twice per sample batch (i.e., twice per day), once before and once after samples, and monitor the results. Carryover can be determined by running a blank injection following the Process QC standard and monitoring the signal in the light and heavy peptide channels. Carryover for the ATM peptides is typically low (i.e., less than 1%). If carryover is a concern, inject wash runs using a mixture of methanol:acetonitrile:water (50:40:10) between samples.
20. If desired, annotate peaks and replicates for downstream analysis. SampleGroup, Treatment, Concentration, IS Spike, and Replicate annotations are useful for response curve and sample analysis. Peak annotations can be incorporated for split peaks, peaks not found, or unstable spray. Response curves can be analyzed directly in Skyline or by using add-ons such as QuaSAR [16] for analysis of response curves and identifying potential interferences. Additional tools for analysis directly in Skyline are continually being developed. Check with https://skyline.gs.washington.edu/ for updates.

Acknowledgments

This work was funded by the Clinical Proteomic Tumor Analysis Consortium (CPTAC) of the US National Cancer Institute (U24CA160034).

References

1. Picotti P, Bodenmiller B, Aebersold R (2013) Proteomics meets the scientific method. Nat Methods 10:24–27. doi:10.1038/nmeth.2291
2. Gillette MA, Carr SA (2013) Quantitative analysis of peptides and proteins in biomedicine by targeted mass spectrometry. Nat Methods 10:28–34. doi:10.1038/nmeth.2309
3. Addona TA, Abbatiello SE, Schilling B et al (2009) Multi-site assessment of the precision and reproducibility of multiple reaction monitoring-based measurements of proteins in plasma. Nat Biotechnol 27:633–641. doi:10.1038/nbt.1546
4. Kennedy JJ, Abbatiello SE, Kim K et al (2014) Demonstrating the feasibility of large-scale development of standardized assays to quantify human proteins. Nat Methods 11:149–155. doi:10.1038/nmeth.2763
5. Whiteaker JR, Paulovich AG (2011) Peptide immunoaffinity enrichment coupled with mass spectrometry for peptide and protein quantification. Clin Lab Med 31:385–396. doi:10.1016/j.cll.2011.07.004
6. Madian AG, Rochelle NS, Regnier FE (2013) Mass-linked immuno-selective assays in targeted proteomics. Anal Chem 85:737–748. doi:10.1021/ac302071k
7. Ackermann BL, Berna MJ (2007) Coupling immunoaffinity techniques with MS for quantitative analysis of low-abundance protein biomarkers. Expert Rev Proteomics 4:175–186. doi:10.1586/14789450.4.2.175
8. Boström T, Takanen JO, Hober S (2015) Antibodies as means for selective mass spectrometry. J Chromatogr B Analyt Technol Biomed Life Sci. doi:10.1016/j.jchromb.2015.10.042
9. Anderson NL, Anderson NG, Haines LR et al (2004) Mass spectrometric quantitation of peptides and proteins using stable isotope standards and capture by anti-peptide antibodies (SISCAPA). J Proteome Res 3:235–244
10. Whiteaker JR, Zhao L, Yan P et al (2015) Peptide immunoaffinity enrichment and targeted mass spectrometry enables multiplex, quantitative pharmacodynamic studies of phospho-signaling. Mol Cell Proteomics 14:2261–2273. doi:10.1074/mcp.O115.050351
11. MacLean B, Tomazela DM, Shulman N et al (2010) Skyline: an open source document editor for creating and analyzing targeted proteomics experiments. Bioinformatics (Oxford) 26:966–968. doi:10.1093/bioinformatics/btq054
12. Carr SA, Abbatiello SE, Ackermann BL et al (2014) Targeted peptide measurements in biology and medicine: best practices for mass spectrometry-based assay development using a fit-for-purpose approach. Mol Cell Proteomics 13:907–917. doi:10.1074/mcp.M113.036095
13. Hoofnagle AN, Whiteaker JR, Carr SA et al (2016) Recommendations for the generation, quantification, storage, and handling of peptides used for mass spectrometry-based assays. Clin Chem 62:48–69. doi:10.1373/clinchem.2015.250563
14. Whiteaker JR, Zhao L, Lin C et al (2012) Sequential multiplexed analyte quantification using peptide immunoaffinity enrichment coupled to mass spectrometry. Mol Cell Proteomics 11:M111.015347. doi:10.1074/mcp.M111.015347
15. Zhao L, Whiteaker JR, Voytovich UJ et al (2015) Antibody-coupled magnetic beads can be reused in immuno-MRM assays to reduce cost and extend antibody supply. J Proteome Res 14:4425–4431. doi:10.1021/acs.jproteome.5b00290
16. Mani DR, Abbatiello SE, Carr SA (2012) Statistical characterization of multiple-reaction monitoring mass spectrometry (MRM-MS) assays for quantitative proteomics. BMC Bioinformatics 13(Suppl 16):9. doi:10.1186/1471-2105-13-S16-S9

Chapter 16

Mass Spectrometry-Based Proteomics for Quantifying DNA Damage-Induced Phosphorylation

Marina E. Borisova, Sebastian A. Wagner, and Petra Beli

Abstract

Protein phosphorylation plays central regulatory roles in DNA damage repair and signaling. Protein kinases of the phosphatidylinositol 3-kinase-related kinase family ATM, ATR, and DNA-PKcs mediate phosphorylation of hundreds of substrates after DNA damage and thereby orchestrate the cellular response to DNA damage. Protein phosphorylation can be studied using antibodies that specifically recognize phosphorylated protein species; however, this approach is limited by existing antibodies and does not permit unbiased discovery of phosphorylation sites or analyzing phosphorylation sites in a high-throughput manner. Mass spectrometry (MS)-based proteomics has emerged as a powerful method for identification of phosphorylation sites on individual proteins and proteome-wide. To identify phosphorylation sites, proteins are digested into peptides and phosphopeptides are enriched using titanium dioxide (TiO_2)-based chromatography followed by the identification by liquid chromatography-tandem mass spectrometry (LC-MS/MS). Quantitative proteomics approaches, such as stable isotope labeling with amino acids in cell culture (SILAC), enable relative quantification of phosphopeptide abundance in different conditions. Here, we describe a streamlined protocol for enrichment of phosphopeptides using TiO_2-based chromatography, and outline the application of quantitative phosphoproteomics for the identification of DNA damage-induced phosphorylation and substrates of kinases functioning after DNA damage.

Key words ATM, ATR, DNA damage signaling, Mass spectrometry-based proteomics, Phosphoproteomics, Phosphorylation, Protein kinase, SILAC, TiO_2

1 Introduction

Protein phosphorylation plays central regulatory roles in DNA damage repair and signaling. DNA damage activates protein kinases of the phosphatidylinositol 3-kinase-related kinase (PIKK) protein family: ATM (Ataxia Telangiectasia Mutated), ATR (Ataxia Telangiectasia and Rad3 related), and DNA-PKcs (DNA-dependent protein kinase, catalytic subunit). Activation of these kinases initiates a signaling cascade that involves phosphorylation of hundreds of substrate proteins and orchestrates the cellular response to DNA damage [1, 2]. The substrates of ATM and ATR are involved in

Sergei V. Kozlov (ed.), *ATM Kinase: Methods and Protocols*, Methods in Molecular Biology, vol. 1599,
DOI 10.1007/978-1-4939-6955-5_16, © Springer Science+Business Media LLC 2017

diverse cellular processes including DNA repair, chromatin organization, and RNA metabolism [3–9]. In addition to ATM, ATR, and DNA-PKcs, a number of other kinases are activated in response to DNA damage and contribute to the maintenance of genome and chromatin integrity [10].

Protein phosphorylation can be investigated using phosphorylation-specific antibodies that specifically recognize modified species of proteins. Although this approach is suitable for the targeted analysis of a limited set of phosphorylation sites for which antibodies are available, it does not enable de novo discovery of phosphorylation sites or high throughput studies of phosphorylation-signaling induced by a particular stimuli or perturbation. Mass spectrometry (MS)-based proteomics has become an indispensable tool for analyzing protein phosphorylation [11]. In phosphoproteomics workflows, proteins are extracted from cells or tissues and digested into peptides using specific proteases. Phosphopeptides are subsequently enriched and identified using high-pressure liquid chromatography coupled to tandem mass spectrometry (LC-MS/MS) [12].

A major obstacle for proteome-wide analysis of phosphopeptides by LC-MS/MS is the sub-stoichiometric occurrence of post-translationally modified protein species and the inability of current mass spectrometers to sequence all peptides resulting from digestion of the cellular proteome. Therefore, the enrichment of phosphopeptides prior to analysis by LC-MS/MS is a critical step in phosphoproteomics workflows [12, 13]. The negative charge of the phosphoryl group allows specific enrichment of phosphopeptides using positively charged affinity matrices. The most commonly employed methods for enrichment of phosphopeptides include immobilized metal ion affinity chromatography (IMAC) and titanium dioxide (TiO_2) chromatography [12, 14, 15]. In IMAC, the negatively charged phosphoryl groups bind to positively charged metal ions such as Fe^{3+} or Ga^{3+} that are trapped in a chelating matrix. In TiO_2 chromatography, the negatively charged phosphoryl group forms a bridging bidentate surface complex with the positively charged titanium atoms. Phosphopeptides are subsequently eluted from the TiO_2 matrix using a high pH buffer containing ammonia [16]. Recent reports have demonstrated that high concentrations of the ion-pairing reagent TFA can greatly enhance the specificity of the binding of phosphopeptides to TiO_2 [17–19].

Typically, fractionation methods including strong cation exchange (SCX) chromatography, strong anion exchange (SAX) chromatography, hydrophilic interaction chromatography (HILIC), or high pH reversed phase chromatography are employed prior to enrichment of phosphopeptides to decrease the complexity of the peptide mixture [20–26]. However, some of these strategies require dedicated equipment such as automated liquid chromatography systems and might hamper their adoption.

Quantitative proteomics approaches, such as stable isotope labeling with amino acids in cell culture (SILAC), enable relative quantification of phosphopeptides in different conditions [27]. Therefore, metabolic labeling of cellular proteins using SILAC can be employed for quantification of phosphorylation site abundance after cellular perturbations such as DNA damage or kinase inhibition. For instance, MS-based proteomics has been employed to analyze phosphorylation signaling induced after different types of DNA damage, including ionizing radiation and topoisomerase II inhibition, and for mapping of substrates of DNA damage-activated kinases ATM and ATR [3–7, 9].

Here, we report a streamlined phosphoproteomics protocol based on the enrichment of phosphopeptides using TiO_2 matrix and post-enrichment fractionation by micro tip-based SCX chromatography. We outline the application of this protocol for quantifying phosphorylation signaling after DNA damage and for identification of substrates of DNA damage-activated kinases.

2 Materials

All solutions should be prepared with Milli-Q water of 18.2 MΩ·cm resistivity at 25 °C.

2.1 Common Solutions and Equipment

1. Acetic acid.
2. Acetonitrile (ACN).
3. Chloroacetamide (CAA).
4. Dithiothreitol (DTT).
5. Formic acid.
6. NuPAGE LDS Sample Buffer (4×) (Thermo Fisher Scientific).
7. Methanol.
8. Trifluoroacetic acid (TFA).
9. NanoDrop 2000 spectrophotometer.
10. Sonicator with a micro tip.
11. Thermomixer.
12. Vacuum concentrator.

2.2 Stable Isotope Labeling with Amino Acids in Cell Culture (SILAC)

1. Dulbecco's phosphate-buffered saline (DPBS).
2. Cell culture media without arginine and lysine.
3. Amino acids: L-arginine (Arg0), L-lysine (Lys0), L-arginine-U-^{13}C6 99% (Arg6), L-lysine-4,4,5,5,-D4 96–98% (Lys4), L-arginine-U-^{13}C6-^{15} N4 99% (Arg10), L-lysine-U-^{13}C6-^{15} N2 99% (Lys8).

4. Fetal bovine serum (FBS) dialyzed by ultrafiltration with a 10,000 Da molecular weight cutoff.
5. Antibiotics for cell culture (penicillin, streptomycin).
6. L-glutamine.
7. Sodium pyruvate.

2.3 Cell Treatment and Lysis

1. Modified RIPA buffer: 50 mM Tris–HCl, pH 7.5, 150 mM NaCl, 1 mM EDTA, 1% IGEPAL CA-630, 0.1% sodium deoxycholate.
2. Phosphatase inhibitor stock solutions (100×): 500 mM β-glycerophosphate, 500 mM sodium fluoride, 100 mM sodium orthovanadate.
3. Protease inhibitors cocktail.
4. 5 M sodium chloride (NaCl).
5. Quick Start Bradford 1× Dye Reagent.

2.4 Protein Precipitation and Digestion

1. Acetone.
2. Denaturation buffer: 8 M urea in 10 mM HEPES–NaOH, pH 8.0.
3. Lysyl endopeptidase (LysC): 0.5 μg/μl in 50 mM ammonium bicarbonate, pH 8.0.
4. Sequencing grade Trypsin: 0.5 μg/μl in 50 mM acetic acid.

2.5 Purification of Peptides

1. Sep-Pak C_{18} Classic Cartridge (360 mg sorbent per cartridge, 55–105 μm particle size) (Waters).
2. 10 ml syringe.

2.6 Titanium Dioxide (TiO_2)-Based Enrichment of Phosphopeptides

1. Titansphere TiO bulk material (TiO_2) 10 μm (GL Sciences Inc).
2. 60% TFA.
3. Binding buffer: 50% ACN, 6% TFA.
4. Wash buffer: 50% ACN, 0.1% TFA.
5. C_8 47 mm Extraction Disk (3 M, Empore).
6. Elution buffer 1: 5% ammonia (NH_4OH).
7. Elution buffer 2: 10% NH_4OH, 25% ACN.

2.7 Micro Tip-Based Strong Cation Exchange Chromatography (Micro-SCX)

1. Cation 47 mm Extraction Disk (3 M, Empore).
2. SCX wash buffer: 40% ACN, 0.1% TFA.
3. SCX elution buffers: 40 mM acetic acid, 40 mM boric acid, 40 mM phosphoric acid. Adjust pH to 3.0, 3.5, 4.5, 6.0, and 8.5 with sodium hydroxide. Add 40% ACN before use.

2.8 Desalting and Concentration of Peptides

1. C_{18} 47 mm Extraction Disk (3 M, Empore).
2. Buffer A: 0.1% formic acid.
3. Buffer B: 80% ACN, 0.1% formic acid.
4. C_{18} elution buffer: 50% ACN, 0.1% formic acid.
5. Buffer A*: 5% ACN, 0.1% TFA.
6. 96-well plate.

2.9 Analysis of Peptides by LC-MS/MS

1. Reprosil-Pur Basic C18, 1.9 μm (Dr. Maisch HPLC GmbH).
2. 30 cm fused silica emitter, 360 μm outer diameter, 75 μm inner diameter, 8 μm laser-pulled tip (PicoTip) (New Objective, Inc.).
3. Pressure injection cell.
4. EASY-nLC 1000 liquid chromatograph (Thermo Scientific).
5. Hybrid Quadrupole-Orbitrap mass spectrometer: Q Exactive Plus (Themo Scientific).

3 Methods

3.1 Stable Isotope Labeling with Amino Acids in Cell Culture (SILAC)

1. Collect cells growing in culture and dispense equal amounts of cells to three tubes, pellet by centrifugation, and wash twice with DPBS to remove residual media.
2. Resuspend cells in light, medium, or heavy SILAC media (*see* **Note 1**).
3. Grow cells in SILAC media for 5–10 doublings (*see* **Note 2**).

3.2 Cell Treatment and Lysis

1. Treat cells with kinase inhibitor followed by a DNA damage-inducing agent according to your experimental design. A possible experimental design for the identification of DNA damage- and kinase-dependent phosphorylation sites is outlined in Fig. 1 (*see* **Note 3**).
2. Wash adherent cells twice with DPBS. In case of suspension cells, collect cells by centrifugation and wash twice with DPBS. The following steps should be performed on ice.
3. Add ice-cold modified RIPA buffer supplemented with phosphatase and protease inhibitors to the cells (~500 μl lysis buffer per 1 × 10^7 cells).
4. Collect cells using a cell scraper in a 15 ml tube (adherent cells).
5. Add 1:10 volume of 5 M NaCl to cells collected in lysis buffer. Sonicate cell lysates to shear released DNA until the lysate loses viscosity (*see* **Note 4**).
6. Pellet cell debris by high-speed centrifugation (16,000 × *g*) at 4 °C for 15 min (*see* **Note 5**).
7. Measure protein concentration and combine equal amounts of protein from each SILAC condition in a 50 ml tube (*see* **Note 6**).

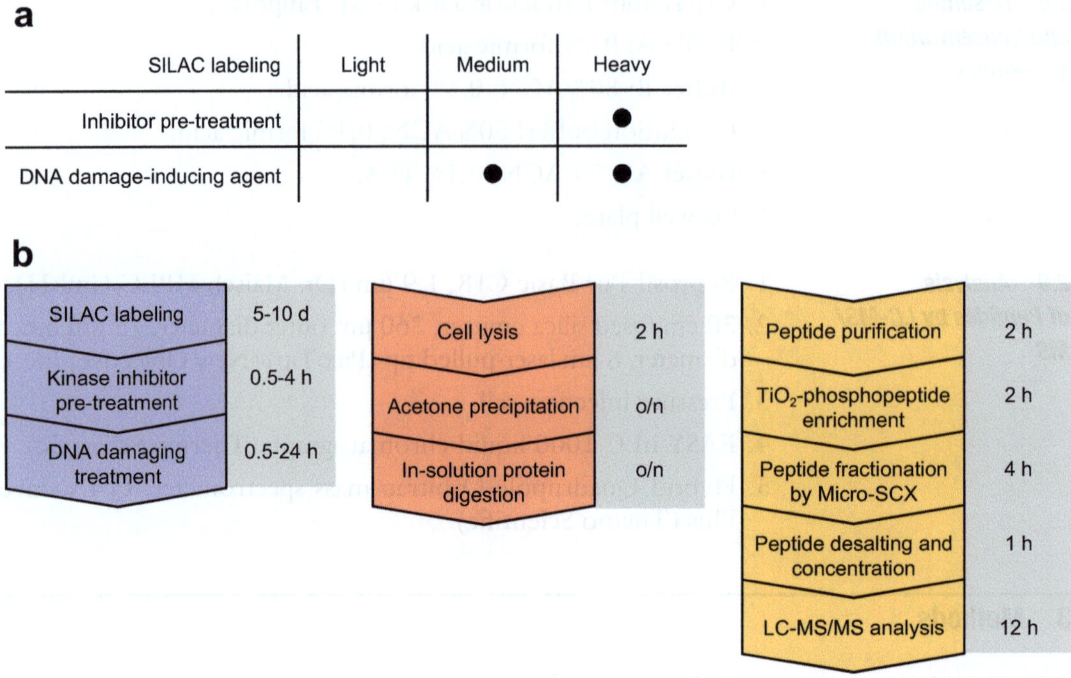

Fig. 1 Phosphoproteomics for quantifying DNA damage- and kinase-dependent phosphorylation sites. (**a**) Stable isotope labeling with amino acids in cell culture (SILAC) is employed to determine the relative abundance of phosphopeptides in different conditions. Light-labeled cells are mock-treated, whereas medium- and heavy-labeled cells are treated with a DNA damage-inducing agent. In heavy-labeled cells, the kinase activity is inhibited by knockdown/knockout or chemical inhibition. (**b**) Outline of the described phosphoproteomics protocol. The estimated time required for each step is indicated

3.3 Protein Precipitation and Digestion

1. Add 4 volumes of ice-cold acetone to precipitate the proteins. Incubate at −20 °C for 4 h to overnight (*see* **Note 7**).
2. Pellet the acetone-precipitated proteins by centrifugation at 1000 × *g* for 5 min.
3. Dissolve protein pellet in Denaturation buffer (*see* **Note 8**).
4. Add DTT to a final concentration of 1 mM and incubate at room temperature for 1 h with gentle shaking.
5. Add CAA to a final concentration of 5.5 mM and incubate at room temperature for 1 h in the dark with gentle shaking (*see* **Note 9**).
6. Add 1 μg of LysC per 150 μg of protein and digest for 3 h to overnight at room temperature (*see* **Note 10**).
7. Dilute sample with four volumes of water (*see* **Note 11**).
8. Add 1 μg of Trypsin per 150 μg of protein and digest overnight at room temperature.
9. Add TFA to a final concentration of 0.5% to acidify the sample and stop the enzymatic digestion.

10. Allow precipitates to form by incubating the sample at 4 °C for a minimum of 1 h.
11. Centrifuge acidified peptides for 10 min at 4000 × *g* and transfer supernatant to a fresh tube.

3.4 Purification of Peptides

1. Attach a C_{18} cartridge to a 10 ml syringe and wash once with 5 ml of 100% ACN.
2. Wash the C_{18} cartridge three times with 5 ml 0.1% TFA in water.
3. Load the clarified peptide solution onto the C_{18} cartridge.
4. Wash the C_{18} cartridge three times with 10 ml 0.1% TFA in water (*see* **Note 12**).
5. Elute the peptides with 4.5 ml 50% ACN, 0.1% TFA in water.

3.5 Titanium Dioxide (TiO_2)-Based Enrichment of Phosphopeptides

1. Quantify peptide amount by NanoDrop spectrophotometer (A280).
2. Incubate 10 mg of TiO_2 spheres with Binding buffer for 5 min on a rotation wheel.
3. Add 0.5 ml of 60% TFA to the peptide solution.
4. Incubate peptides with TiO_2 spheres for 1 h on a rotation wheel (*see* **Note 13**).
5. Wash peptide-loaded spheres twice with 1 ml Binding buffer (centrifuge at 1000 × *g* in between washes).
6. Wash peptide-loaded spheres twice with 1 ml Wash buffer.
7. Use a 17-gauge Hamilton syringe to cut out one disk from a C_8 47 mm extraction disk and place into a 200 μl pipette tip.
8. Transfer peptide-loaded spheres to the C_8 filter tip placed in a StageTip adapter [28] and dry the spheres by centrifugation at 400–800 × *g*.
9. Elute phosphorylated peptides with 100 μl Elution buffer 1 by centrifugation at 400 × *g*.
10. Elute phosphorylated peptides with 100 μl Elution buffer 2 by centrifugation at 400 × *g* to the same tube.
11. Vacuum concentrate the eluted peptides at 45 °C for 15 min to remove NH_4OH.
12. Adjust pH to ~2 using TFA.

3.6 Micro Tip-Based Strong Cation Exchange Chromatography (Micro-SCX)

The micro tip-based strong cation exchange chromatography protocol is based on the following publications [28–30] (*see* **Note 14**).

1. Use a 17-gauge Hamilton syringe to cut out six disks from a Cation 47 mm extraction disk and place into a 200 μl pipette tip.
2. Wash SCX tip once with 50 μl methanol by centrifugation at 400–800 × *g* (*see* **Note 15**).

3. Wash SCX tip once with 50 µl SCX elution buffer with pH 8.5.
4. Wash SCX tip once with 50 µl SCX wash buffer.
5. Load the sample by centrifugation at 400–600 × *g*.
6. Elute the peptides with 100 µl SCX elution buffer in five pH steps (3.0, 3.5, 4.5, 6.0, 8.5) by centrifugation at 500–800 × *g*. Collect eluates in separate tubes.
7. Vacuum concentrate the eluted peptides at 45 °C for 15 min to remove ACN.
8. Adjust pH to ~2 using TFA.

3.7 Desalting and Concentration of Peptides

1. Use a 17-gauge Hamilton syringe to cut out two disks from a C_{18} 47 mm extraction disk and place into a 200 µl pipette tip [28].
2. Wash C_{18} tips once with 25 µl methanol by centrifugation at 400–800 × *g*.
3. Wash C_{18} tips once with 25 µl Buffer B.
4. Wash C_{18} tips twice with 25 µl Buffer A.
5. Load acidified samples onto the C_{18} tips.
6. Wash C_{18} tips once with 50 µl Buffer A (*see* **Note 16**).
7. Elute the peptides directly into a 96-well plate by passing 50 µl C_{18} elution buffer through the C_{18} tips using a syringe.
8. Vacuum concentrate peptides at 45 °C for 25 min to reduce the sample volume to 4.5–5 µl.
9. Add 0.5–1 µl Buffer A* to the sample.

3.8 Analysis of Peptides by LC-MS/MS

1. Pack a nanospray column (15 cm length, 75 µl inner diameter) with C_{18} reversed phase chromatography material (1.9 µm bead size) using a pressure injection cell [31].
2. Use a nano-flow UHPLC system to load the peptide sample onto the C_{18} column.
3. Elute the peptides with a linear gradient from 5 to 50% ACN in 2 h.
4. The mass spectrometer is operated in data-dependent mode automatically switching between MS and MS^2 acquisitions [32, 33].
5. MS spectra (m/z 300–1650) are acquired in the Orbitrap mass analyzer with a resolution of 70,000 at $m/z = 200$ after accumulation of ions to a target value of 3e6 estimated based on predictive automatic gain control from the previous full scan.
6. The ten most intense ions are isolated using the quadrupole mass filter (maximum injection time 120 ms, isolation window 2.6 m/z, AGC target 1e5) and subsequently fragmented in the higher-energy C-trap dissociation (HCD) cell [34].
7. MS^2 spectra are acquired in the Orbitrap mass analyzer with a resolution of 35,000 at $m/z = 200$.

3.9 Data Analysis

1. Analyze raw MS data using the MaxQuant software [35, 36].
2. Search MS^2 spectra against a database containing protein sequences obtained from the UniProtKB using the Andromeda search engine [37]. Spectra are searched with a parent ion mass tolerance of 6 ppm, fragment ion mass tolerance of 20 ppm, strict trypsin specificity and allowing up to three miscleavages. Cysteine carbamidomethylation is searched as fixed modification, whereas protein N-terminal acetylation, methionine oxidation and phosphorylation of serine, threonine, and tyrosine as variable modifications.
3. Filter reverse hits and potential contaminants from the MaxQuant output table containing all identified phosphorylation sites.
4. In case multiple biological replicates have been performed, identify significantly regulated phosphorylation sites using appropriate statistical approaches [38, 39]. An example scatter plot of normalized SILAC ratios quantified in a phosphoproteomics experiment designed to identify DNA damage- and kinase-dependent phosphorylation sites is shown in Fig. 2.

4 Notes

1. For the preparation of SILAC media supplement lysine and arginine-free media with dialyzed FBS and antibiotics. L-glutamine and sodium pyruvate can be added if required. Add amino acids to a final concentration of 28 mg/L and 48.7 mg/L of arginine and lysine, respectively. This concentration corresponds to one third of the concentration of arginine and lysine in DMEM media and is well tolerated by commonly used cell lines such as HEK293T, HeLa, and U-2 OS. For other cell lines the optimal conditions have to be determined experimentally [40].
2. The time required for complete labeling of cellular proteins with isotope-containing amino acids depends on the cell line. Testing the incorporation of isotope-containing amino acids by LC-MS/MS is recommended before proceeding with the experiment [41].
3. Biological replicate experiments should be performed to be able to determine significantly regulated phosphorylation sites.
4. High salt concentrations are used during cell lysis to facilitate the extraction of chromatin-associated proteins.
5. The knockdown or inhibition of the kinase as well as the induction of DNA damage should be tested by western blotting.
6. The recommended starting amount is 7–10 mg of protein (~3 mg of protein per SILAC condition in case of a triple

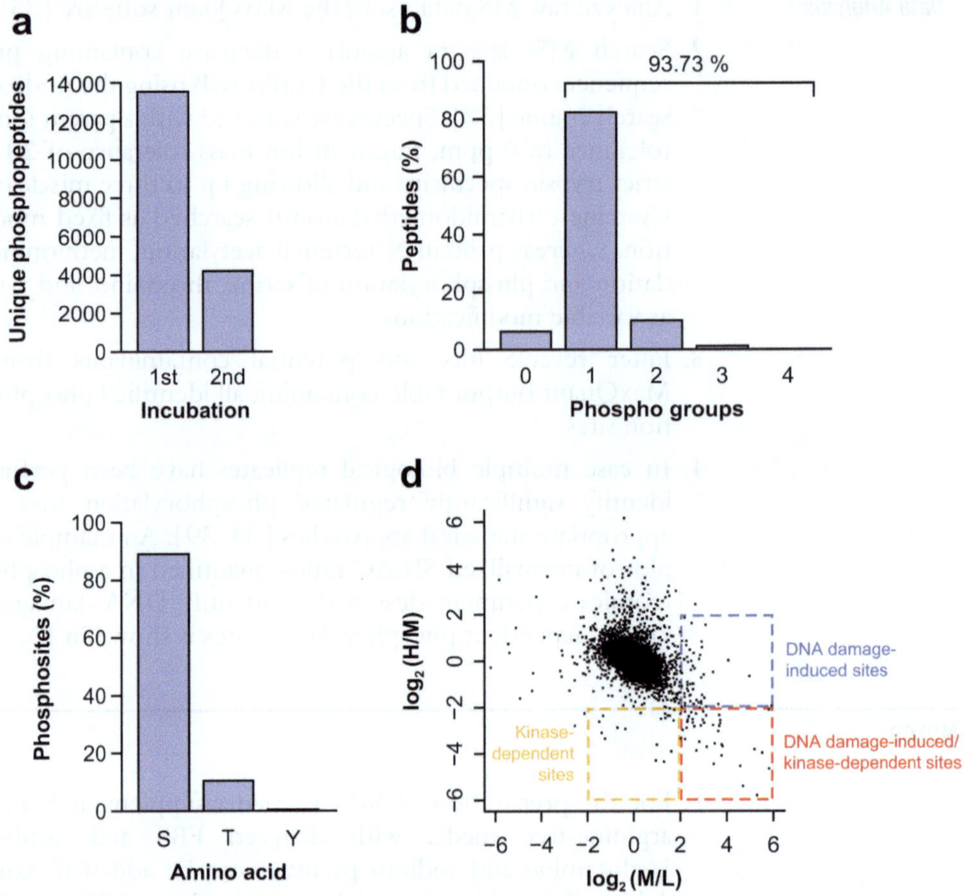

Fig. 2 Overview of the results obtained from a typical quantitative phosphoproteomics experiment performed as described in this protocol [9]. (**a**) The bar graph shows the number of unique phosphopeptides identified using the described protocol and two rounds of enrichment with TiO_2 spheres. (**b**) Fraction of unmodified, singly, doubly, triply, and quadruply phosphorylated peptides. (**c**) Fraction of phosphorylation sites identified on serine, threonine, and tyrosine residues. (**d**) The normalized SILAC ratios M/L and H/M of all quantified phosphopeptides are presented in a scatter plot

SILAC experiment). An aliquot (50–100 μg) of the combined protein extract can be used for proteome analysis.

7. Acetone-precipitated proteins can be stored for several months at −20 °C.

8. Use 2–4 ml of Denaturation buffer for dissolving acetone-precipitated proteins. Dissolve proteins by shaking vigorously (1000 rpm). Protein denaturation is required for the subsequent proteolysis.

9. Alkylation of cysteines with CAA is required to prevent unspecific side-reactions.

10. LysC is an endoproteinase that cleaves proteins C-terminally of lysine residues. Pre-digestion with LysC is recommended to inhibit protease activity at lower concentrations of urea.
11. High concentrations of urea inhibit the activity of Trypsin.
12. Peptide-loaded C_{18} cartridges can be stored at 4 °C for several months.
13. Multiple incubations of the peptides with TiO_2 spheres can be performed to increase the depth of the analysis (Fig. 2a).
14. The Micro-SCX fractionation can be omitted. This will decrease the number of identified phosphopeptides.
15. Avoid drying out of the SCX tip.
16. Peptide-loaded C_{18} tips can be stored for several months at 4 °C.

Acknowledgments

The research in PB's group is supported by the German Research Foundation (Emmy Noether Program, BE 5342/1-1 and SFB 1177 on Selective Autophagy) and the Marie Curie Career Integration Grant from the European Commision (grant agreement number: 630763). The research in SAW's group is supported by the LOEWE program Ubiquitin Networks (Ub-Net) of the State of Hesse (Germany).

References

1. Polo SE, Jackson SP (2011) Dynamics of DNA damage response proteins at DNA breaks: a focus on protein modifications. Genes Dev 25:409–433. doi:10.1101/gad.2021311
2. Ciccia A, Elledge SJ (2010) The DNA damage response: making it safe to play with knives. Mol Cell 40:179–204. doi:10.1016/j.molcel.2010.09.019
3. Beli P, Lukashchuk N, Wagner SA et al (2012) Proteomic investigations reveal a role for RNA processing factor THRAP3 in the DNA damage response. Mol Cell 46:212–225. doi:10.1016/j.molcel.2012.01.026
4. Matsuoka S, Ballif BA, Smogorzewska A et al (2007) ATM and ATR substrate analysis reveals extensive protein networks responsive to DNA damage. Science 316:1160–1166. doi:10.1126/science.1140321
5. Bennetzen MV, Larsen DH, Bunkenborg J et al (2010) Site-specific phosphorylation dynamics of the nuclear proteome during the DNA damage response. Mol Cell Proteomics 9:1314–1323. doi:10.1074/mcp.M900616-MCP200
6. Bensimon A, Schmidt A, Ziv Y et al (2010) ATM-dependent and -independent dynamics of the nuclear phosphoproteome after DNA damage. Sci Signal 3:rs3. doi:10.1126/scisignal.2001034
7. Stokes MP, Rush J, MacNeill J et al (2007) Profiling of UV-induced ATM/ATR signaling pathways. Proc Natl Acad Sci 104:19855–19860. doi:10.1073/pnas.0707579104
8. Blasius M, Forment JV, Thakkar N et al (2011) A phospho-proteomic screen identifies substrates of the checkpoint kinase Chk1. Genome Biol 12:R78. doi:10.1186/gb-2011-12-8-r78
9. Wagner SA, Oehler H, Voigt A et al (2015) ATR inhibition rewires cellular signaling networks induced by replication stress. Proteomics. doi:10.1002/pmic.201500172

10. Bensimon A, Aebersold R, Shiloh Y (2011) Beyond ATM: the protein kinase landscape of the DNA damage response. FEBS Lett 585:1625–1639. doi:10.1016/j.febslet.2011.05.013
11. Larance M, Lamond AI (2015) Multidimensional proteomics for cell biology. Nat Rev Mol Cell Biol 16:269–280. doi:10.1038/nrm3970
12. Macek B, Mann M, Olsen JV (2009) Global and site-specific quantitative phosphoproteomics: principles and applications. Annu Rev Pharmacol Toxicol 49:199–221. doi:10.1146/annurev.pharmtox.011008.145606
13. Thingholm TE, Jensen ON, Larsen MR (2009) Analytical strategies for phosphoproteomics. Proteomics 9:1451–1468. doi:10.1002/pmic.200800454
14. Nita-Lazar A, Saito-Benz H, White FM (2008) Quantitative phosphoproteomics by mass spectrometry: past, present, and future. Proteomics 8:4433–4443. doi:10.1002/pmic.200800231
15. Rigbolt KTG, Blagoev B (2012) Quantitative phosphoproteomics to characterize signaling networks. Semin Cell Dev Biol 23:863–871. doi:10.1016/j.semcdb.2012.05.006
16. Engholm-Keller K, Larsen MR (2011) Titanium dioxide as chemo-affinity chromatographic sorbent of biomolecular compounds—applications in acidic modification-specific proteomics. J Proteomics 75:317–328. doi:10.1016/j.jprot.2011.07.024
17. Zhou H, Low TY, Hennrich ML et al (2011) Enhancing the identification of phosphopeptides from putative basophilic kinase substrates using Ti (IV) based IMAC enrichment. Mol Cell Proteomics 10:M110.006452. doi:10.1074/mcp.M110.006452
18. Zhou H, Ye M, Dong J et al (2008) Specific phosphopeptide enrichment with immobilized titanium ion affinity chromatography adsorbent for phosphoproteome analysis. J Proteome Res 7:3957–3967. doi:10.1021/pr800223m
19. Imanishi SY, Kochin V, Ferraris SE et al (2007) Reference-facilitated phosphoproteomics: fast and reliable phosphopeptide validation by microLC-ESI-Q-TOF MS/MS. Mol Cell Proteomics 6:1380–1391. doi:10.1074/mcp.M600480-MCP200
20. McNulty DE, Annan RS (2008) Hydrophilic interaction chromatography reduces the complexity of the phosphoproteome and improves global phosphopeptide isolation and detection. Mol Cell Proteomics 7:971–980. doi:10.1074/mcp.M700543-MCP200
21. Beausoleil SA, Jedrychowski M, Schwartz D et al (2004) Large-scale characterization of HeLa cell nuclear phosphoproteins. Proc Natl Acad Sci U S A 101:12130–12135. doi:10.1073/pnas.0404720101
22. Olsen JV, Blagoev B, Gnad F et al (2006) Global, in vivo, and site-specific phosphorylation dynamics in signaling networks. Cell 127:635–648. doi:10.1016/j.cell.2006.09.026
23. Gauci S, Helbig AO, Slijper M et al (2009) Lys-N and trypsin cover complementary parts of the phosphoproteome in a refined SCX-based approach. Anal Chem 81:4493–4501. doi:10.1021/ac9004309
24. Batth TS, Francavilla C, Olsen JV (2014) Off-line high-pH reversed-phase fractionation for in-depth phosphoproteomics. J Proteome Res 13:6176–6186. doi:10.1021/pr500893m
25. Wiśniewski JR, Nagaraj N, Zougman A et al (2010) Brain phosphoproteome obtained by a FASP-based method reveals plasma membrane protein topology. J Proteome Res 9:3280–3289. doi:10.1021/pr1002214
26. Nühse TS, Stensballe A, Jensen ON, Peck SC (2003) Large-scale analysis of in vivo phosphorylated membrane proteins by immobilized metal ion affinity chromatography and mass spectrometry. Mol Cell Proteomics 2:1234–1243. doi:10.1074/mcp.T300006-MCP200
27. Ong S-E, Mann M (2005) Mass spectrometry-based proteomics turns quantitative. Nat Chem Biol 1:252–262. doi:10.1038/nchembio736
28. Rappsilber J, Mann M, Ishihama Y (2007) Protocol for micro-purification, enrichment, pre-fractionation and storage of peptides for proteomics using StageTips. Nat Protoc 2:1896–1906. doi:10.1038/nprot.2007.261
29. Weinert BT, Scholz C, Wagner SA et al (2013) Lysine succinylation is a frequently occurring modification in prokaryotes and eukaryotes and extensively overlaps with acetylation. Cell Rep 4:842–851. doi:10.1016/j.celrep.2013.07.024
30. Wiśniewski JR, Zougman A, Mann M (2009) Combination of FASP and StageTip-based fractionation allows in-depth analysis of the hippocampal membrane proteome. J Proteome Res 8:5674–5678. doi:10.1021/pr900748n
31. Ishihama Y, Rappsilber J, Andersen JS, Mann M (2002) Microcolumns with self-assembled particle frits for proteomics. J Chromatogr A 979:233–239
32. Michalski A, Damoc E, Hauschild JP et al (2011) Mass spectrometry-based proteomics using Q Exactive, a high-performance benchtop quadrupole Orbitrap mass spectrometer. Mol Cell Proteomics 10(M111):011015. doi:10.1074/mcp.M111.011015
33. Kelstrup CD, Young C, Lavallee R et al (2012) Optimized fast and sensitive acquisition methods for shotgun proteomics on a quadrupole orbitrap mass spectrometer. J Proteome Res 11:3487–3497. doi:10.1021/pr3000249
34. Olsen JV, Macek B, Lange O et al (2007) Higher-energy C-trap dissociation for peptide

modification analysis. Nat Methods 4:709–712. doi:10.1038/nmeth1060
35. Cox J, Mann M (2008) MaxQuant enables high peptide identification rates, individualized p.p.b.-range mass accuracies and proteome-wide protein quantification. Nat Biotechnol 26:1367–1372. doi:10.1038/nbt.1511
36. Tyanova S, Mann M, Cox J (2014) MaxQuant for in-depth analysis of large SILAC datasets. Methods Mol Biol 1188:351–364. doi:10.1007/978-1-4939-1142-4_24
37. Cox J, Neuhauser N, Michalski A et al (2011) Andromeda: a peptide search engine integrated into the MaxQuant environment. J Proteome Res 10:1794–1805. doi:10.1021/pr101065j
38. Smyth GK (2004) Linear models and empirical bayes methods for assessing differential expression in microarray experiments. Stat Appl Genet Mol Biol 3:Article3. doi:10.2202/1544-6115.1027
39. Tusher VG, Tibshirani R, Chu G (2001) Significance analysis of microarrays applied to the ionizing radiation response. Proc Natl Acad Sci U S A 98:5116–5121. doi:10.1073/pnas.091062498
40. Bendall SC, Hughes C, Stewart MH et al (2008) Prevention of amino acid conversion in SILAC experiments with embryonic stem cells. Mol Cell Proteomics 7:1587–1597. doi:10.1074/mcp.M800113-MCP200
41. Cox J, Matic I, Hilger M et al (2009) A practical guide to the MaxQuant computational platform for SILAC-based quantitative proteomics. Nat Protoc 4:698–705. doi:10.1038/nprot.2009.36

Chapter 17

Statistical Analysis of ATM-Dependent Signaling in Quantitative Mass Spectrometry Phosphoproteomics

Ashley J. Waardenberg

Abstract

Ataxia-telangiectasia mutated (ATM) is a serine/threonine protein kinase, which when perturbed is associated with modified protein signaling that ultimately leads to a range of neurological and DNA repair defects. Recent advances in phospho-proteomics coupled with high-resolution mass-spectrometry provide new opportunities to dissect signaling pathways that ATM utilize under a number of conditions. This chapter begins by providing a brief overview of ATM function, its various regulatory roles and then leads into a workflow focused on the use of the statistical programming language R, together with code, for the identification of ATM-dependent substrates in the cytoplasm. This chapter cannot cover statistical properties in depth nor the range of possible methods in great detail, but instead aims to equip researchers with a set of tools to perform analysis between two conditions through examples with R functions.

Key words Phosphoproteomics, Bioinformatics, Statistics, Ranked product analysis, Phosphorylation

1 Introduction

Ataxia-telangiectasia mutated (ATM) is a serine/threonine protein kinase that is best known for its association to the autosomal-recessive disease ataxia-telangiectasia (A-T), the progressive loss of cerebellar function [1, 2]. Associated symptoms include ionizing radiation sensitivity, immunodeficiency, and a particular susceptibility to malignancies [1]. A-T patients begin illustrating symptoms before 5 years of age and death is usually by 20 years of age. ATM belongs to a family of phosphatidylinositol-3-kinase-related protein kinases (PIKKs) involved in regulating the DNA damage response (including response to DNA double strand breaks), DNA repair, cell cycle regulation (coordination of checkpoint activation), as well as oxidative stress [3].

ATM is among 518 documented human protein kinases [4] that are responsible for the phosphorylation of proteins through the addition of phosphates to substrate proteins. Phosphorylation is a key cellular signaling mechanism. Phosphorylation provides

Sergei V. Kozlov (ed.), *ATM Kinase: Methods and Protocols*, Methods in Molecular Biology, vol. 1599,
DOI 10.1007/978-1-4939-6955-5_17,

just one mode of transmitting extracellular and intracellular signals to the nucleus for orchestration of cellular responses—signal transduction. At least 50% of all proteins, if not most, are phosphorylated under some condition [4] and aberrant kinase activity is associated with disease and cancers [5]. Kinases recognize sequence-specific substrate sequences, or motifs-specific linear combinations of amino acids. ATM has a preference for the S/T(Q) motif [6], targeting substrate peptides containing a Serine (S) or a Threonine (T), next to a Glutamine (Q). Identifying all the substrate sequences that ATM acts on is of interest for the deconvolution of signal transduction pathways that are ATM activity dependent and potential identification of druggable targets for preventative medicines. The large number of potential substrates that are condition specific leads to the requirement of statistical and bioinformatics methodologies for identification of robust sites and to relate these changes to the specific role of ATM.

High-throughput phosphoproteomics is an unbiased approach concerned with the detection and quantification of phosphorylated proteins by mass-spectrometry. Although this chapter focuses on phosphorylation, other forms of protein modifications, including methylation, acetylation, glycosylation, and ubiquitylation, are also involved in signal transduction and amenable to high-throughput proteomics [7]. To obtain data of the phosphoproteome, a series of steps are typically followed—protein isolation, tryptic digestion, labeling (in the case of isotope dimethyl labeling or other methods), phospho-peptide enrichment, followed by liquid chromatography separation, and identification/measurement using mass-spectrometry ready for data analysis. A number of methodological combinations for phosphopeptide targeting, enrichment, and data acquisition exist, which are outside the scope of this chapter.

To elucidate the pathways and role of ATM in the DNA damage response, a number of studies have coupled induction of DNA damage, via chemical or irradiation exposure, with high-throughput phosphoproteomics [8–13]. Bensimon et al. [8] identified 753 phosphosites responding to neocarzinostatin treatment; Beli et al. [9] identified 1470 phosphosites responding to etoposide or irradiation; Bastos de Oliveria et al. [10] identified 232 phosphopeptides altered in response to methyl methanesulfonate or hydroxyurea that were Tel1/Mec dependent (the yeast version of mammalian ATM/ATR); Kozlov et al. [11] identified 62 phosphopeptides responding to hydrogen peroxide, of which 43 were lost in ATM-null cells; Mazouzi et al. [12] identified ~50 phosphosites that were ATM or ATM interactor (ATMIN) dependent in response to aphidicolin; and Bennetzen et al. [13] identified 594 phosphosites from 209 proteins responding to ionizing radiation. Each of these studies varied in terms of treatment, cell line, and the statistical method/stringencies for the identification of phosphopeptides, primarily due to the inclusion or lack of inclusion of replicates in their experimental design and consequently there was a large variation in the number of phosphorylation changes reported. Currently, there are no gold standards for the analysis of phosphoproteomics data and typically methodologies developed for gene expression analyses are utilized for proteomics data analysis (for a recent overview related to label-free mass-spectrometry, *see* ref. 14).

Having obtained sets of phosphopeptides associated with conditions tested, typical downstream bioinformatics approaches include the discovery of significant amino acid patterns surrounding phosphorylation sites, such as via motif-x [15], inference of effector kinases based on amino acid patterns recognized by kinases, e.g., using GPS "group based prediction system" [16], gene ontology enrichment for the discovery or validation of expected biological processes/pathways, e.g., DAVID [17], and network-based methods that integrate a range of public protein-protein interaction or association data, e.g., STRING [18]. Not surprisingly, for these studies, gene ontology terms/pathways identified included "chromosome organization," "DNA replication," "double strand break repair," and "DNA repair," consistent with the experimental design and proteins anticipated to be involved in these processes. More powerful, however, is the unbiased identification of phosphopeptides and their potentially novel roles in responding to DNA damage. For example, Mazouzi et al. reported that 27.3% of sites detected had not been previously reported, Beli et al. linked phosphorylation of RNA processing factors PPM1G and THRAP3 to the DNA damage response, and Kozlov et al. identified a number of novel ATM substrates responding to DNA damage in the cytoplasm. Consistently, these studies highlighted a smaller proportion of S/T(Q) containing phosphopeptides (ATM consensus motif) than expected, inferring and predicting (e.g., via GPS) additional kinases likely to be involved in the DNA damage response. Thus, these studies have provided new opportunities to explore the role of previously unidentified proteins and/or phosphosites in responding to DNA damage.

Previously, we demonstrated an analysis strategy for the identification of ATM-dependent phosphorylation sites from phosphoproteomics data derived from the cytoplasm of A-T-derived human fibroblasts exposed to hydrogen peroxide (i.e., oxidative stress) [11]. Here, we focus on this dataset to demonstrate an example of phospho-proteomics analysis. This chapter aims to provides a series of steps, explanations, and code for performing analysis of the phosphoproteome from A-T patient-derived human fibroblast cell lines that were exposed to H_2O_2 (inducing oxidative damage) and comparison to normal human fibroblasts (NF). Processing steps include ***exploratory analysis*** (method for viewing/visualizing the properties of a data set), ***normalization***, ***imputation*** of missing values, ***correction*** of nonbiological sources of variation and statistical comparison using the nonparametric *rank product analysis* method. This chapter is not intended to be comprehensive, each section could easily expand into a chapter, but to provide a toolkit for getting started in downstream analyses of phosphoproteomics analysis.

2 Materials

2.1 Computer Requirements

To perform statistical analyses of phosphoproteomics data, access to any personal computer will suffice. For the purposes of this chapter, it is assumed the user is based on a Windows PC, but the workflows and code are adaptable to any working environment supported by R (e.g., Linux). Specific requirements include:

1. Installation of R. R/Bioconductor is a free statistical programming software [19] available for download here: https://cran.r-project.org/.
2. Bioconductor packages ("impute" [20], "pcaMethods" [21], "plyr" [22], "preprocessCore," "RankProd" [23], "sva" [24], and "RCurl").
3. Internet access.

3 Methods

Here, we are concerned with the down-stream analysis of phosphoproteomics data following (1) identification (mapping of peptides to proteins and phosphosite identification) and (2) initial measurement of identified phosphorylated peptides. This process can be broken into five steps, (1) acquisition of data, (2) import, formatting, and filtering, (3) exploratory analysis, (4) normalization and correction, and (5) statistical testing.

3.1 Acquisition of Data

The methods and examples below utilize data obtained from the recent phospho-proteomics screen of ATM-dependent phosphorylation that was responsive to H_2O_2 (oxidative damage) in the cytoplasm [11]. Complete raw data, although not discussed here, are also publically available from ProteomeXchange (PXD002850), http://proteomecentral.proteomexchange.org/cgi/GetDataset?ID=PXD002850. Following sample enrichment, sites were identified and quantified according to matching of spectra to human protein database derived from UniProtKB [25] and identification of phosphorylated sites utilizing MaxQuant [26]. *See* original paper for full description [11]. MaxQuant outputs a table consisting of the modified peptide sequence as well as a range of scores and values for the peptides for each experimental condition/replicate (a guide to the use of MaxQuant is available here: [27]). The data used for this protocol are available as supplemental data to the

Table 1
Columns from supplementary Table 1 to keep during formatting

Column name	Excel column	Description
Positions within proteins	B	Position of phosphorylation sites
Leading proteins	C	Protein identifier
Gene names	F	Official Gene Symbol encoding protein
Sequence window	J	Centered Amino Acid sequence
Phospho (STY) Probabilities	Q	Probability for each phosphosite
Ratio M/L normalized NFFRep1 (log2)	X	Normalized log2 data for normal fibroblast exposed to IR vs. control
Ratio H/L normalized NFFRep2 (log2)	Y	Normalized log2 data for normal fibroblast exposed to IR vs. control
Ratio H/L normalized NFFRep3 (log2)	Z	Normalized log2 data for normal fibroblast exposed to IR vs. control
Ratio H/L normalized NFFRep4 (log2)	AA	Normalized log2 data for normal fibroblast exposed to IR vs. control
Ratio H/L normalized A-TRep1 (log2)	AG	Normalized log2 data for A-T patient-derived fibroblast exposed to IR vs. control
Ratio H/L normalized A-TRep2 (log2)	AH	Normalized log2 data for A-T patient-derived fibroblast exposed to IR vs. control
Ratio H/L normalized A-TRep3 (log2)	AI	Normalized log2 data for A-T patient-derived fibroblast exposed to IR vs. control

original publication: http://www.mcponline.org/content/15/3/1032/suppl/DC1 (Supplemental Table1).

Begin by downloading Supplemental Table1 (a Microsoft Excel Spreadsheet) from http://www.mcponline.org/content/15/3/1032/suppl/DC1. This spreadsheet contains five tabs. The tab named "MaxQuant_output" contains the data we will be concerned with. This table can be formatted either within Microsoft Excel or once imported into R. Format the table to keep the following columns (Table 1) and save this worksheet as CSV (Comma delimited) (*.csv) to a working directory, for example: "C:/phosphodata/Supplemental_Table1.csv". Alternatively, the formatted table can be downloaded from "https://github.com/awaardenberg/Kozlov_MCP_2016" or imported directly into R using the RCurl method described in Subheading 3.2.1.

3.2 Import, Formatting, and Filtering

Having installed R, open a new R session. The RGui will initiate with a console where you type commands after ">". To learn more about a function type a question mark "?" prior to the function, for example "?read.csv" will open details about the function for importing tables in csv format. Ensure that all the packages are installed and loaded, either using the Packages > Install package(s) drop down menu of the gui or installing on the command line. Each of the boxes below contain R code that can be pasted into the R GUI console. For example, to install and load the "plyr" package paste the following commands:

```
source("http://bioconductor.org/biocLite.R")
biocLite("plyr")
library(plyr)
```

Continue to install the remaining packages: "impute," "pcaMethods," "preprocessCore," "RankProd," "sva," and "RCurl."

3.2.1 Import and Format Data

To import the dataset:

```
#import data formatted as per 3.1:
phospho.data <- read.csv("C:/phosphodata/
Supplemental_Table1.csv", header=T)
#or read directly from the online GitHub
(first install RCurl package)
myfile <- getURL("https://raw.githubusercon-
tent.com/awaardenberg/Kozlov_2016/master/Sup-
plemental_Table1.csv", ssl.verifyhost=FALSE,
ssl.verifypeer=FALSE)
phospho.data <- read.csv(textConnection
(myfile), header=T)
```

It is always a good practice to check that the dimensions of the imported table are as expected and/or after table manipulations:

```
dim(phospho.data)
```

This should report 9,612 rows and 12 columns (assuming only the columns in Table 1 were imported). Prior to further analysis, a unique identifier will be created as the row names for each peptide, by merging the information from three columns using the "paste" command:

```
rownames(phospho.data) <- paste(phospho.
data$Gene.names, phospho.data$Leading.pro-
teins, phospho.data$Positions.within.pro-
teins, sep=":")
```

The dataset is now ready for further analysis.

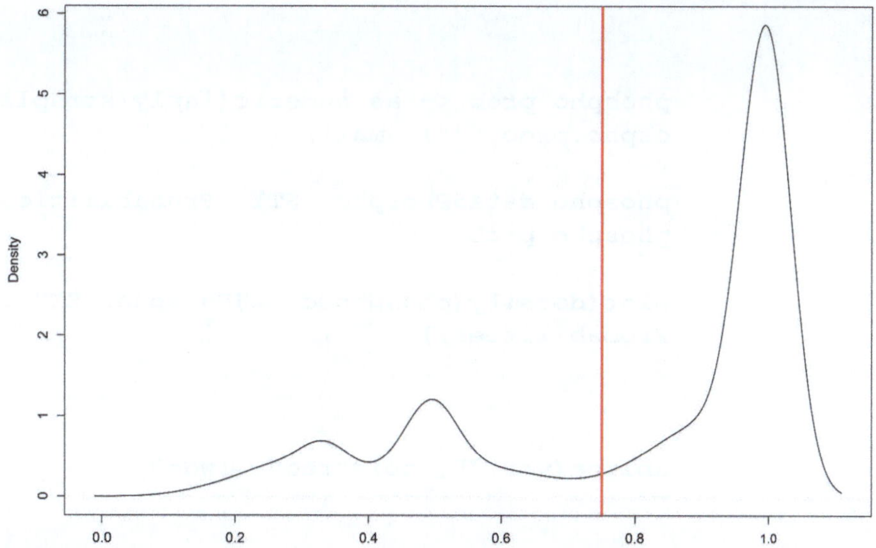

Fig. 1 Density distribution of phospho-peptide site probabilities. *Red line* indicates the threshold set at 0.75

3.2.2 Filter Data

Phospho-site probability. It is often a standard practice to set a threshold for probability of phosphosite assignment, (usually 0.75) to select phosphopeptides with a high confidence of site assignment. Here, the probability ranges from 0 to 1, closer to 1 being a more confident probability. This data is usually provided in the "Phospho (STY) Sites.csv" from the MaxQuant output. To check if this threshold is reasonable for this dataset, we will first plot the density distribution of site probabilities by extracting the maximum probability for the sites identified for each phosphopeptide. This data is available in the "Phospho.STY. Probabilities" column, but first requires some formatting. A vector of the maximum site probability for each phosphopeptide (*phospho.prob*) is created, and then replaces the existing probabilities column and the results are plotted as a density distribution (Fig. 1):

```
phospho.prob <- phospho.data$Phospho..STY..
Probabilities
#strip alphabetical characters:
phospho.prob <- gsub("([[:alpha:]])", "",
phospho.prob)
#if more than one site was identified, here
noted by adjacent brackets ")(", join by ";"
phospho.prob <- gsub(")(", ";", phospho.
prob, fixed=T)
#remove all remaining brackets:
phospho.prob <- gsub("\\)|\\(", "", phospho.
prob)
```

```
#determine the maximum probability (utilis-
ing a plyr function "laply"):
phospho.prob <- as.numeric(laply(strsplit(ph
ospho.prob,";"), max))
#replace entry in data frame:
phospho.data$Phospho..STY..Probabilities <-
phospho.prob
#plot results
plot(density(phospho.data$Phospho..STY..
Probabilities))
#to add a line to the plot, at the "v" ver-
tical point, 0.75, "col" colour, red, and
"lwd" line width of 3:
abline(v=0.75, col="red", lwd=3)
```

This threshold is reasonable and captures the majority of high probability site assignments (Fig. 1) and we therefore continue by keeping only phospho-peptides with a site probability of at least 0.75:

```
phospho.data <- phospho.data[phospho.
data$Phospho..STY..Probabilities >= 0.75,]
```

Replicate representation. Due to the number of processing steps that can influence the capturing of a peptide by the mass-spectrometer (e.g., variable elution during chromatography, fragmentation and stochastic sampling in complex mixtures), a peptide may not always be captured and quantified in all replicates, irrespective of whether it was actually present in the sample. It is important to be mindful that genetic perturbation may result in a set of peptides being absent and/or poorly represented. Nonetheless, for a direct comparison of data between two conditions (or a two-class comparison: here A-T versus NF), peptides are assumed to be present in both populations, but at varying concentrations. Peptides not present in at least two of the samples for the NF (four replicates) and A-T samples (three replicates) versus controls are therefore removed by applying the following threshold:

```
#here NA's (missing values) are counted for
NF and AT, using the rowSums function.
phospho.data.two.class <- phospho.
data[(rowSums(is.na(phospho.data[6:9])) <=
```

```
2 & rowSums(is.na(phospho.data[10:12])) <=
1),]
#to view which columns are being selected,
use: colnames(phospho.data). Here 6:9 are NF
and 10:12 at AT.
```

3.2.3 Exploratory Analysis

Having removed lower quality phosphosite assignments and phosphopeptides with poor representation, an *exploratory analysis* of the data is conducted. The main purposes of exploratory analyses are to determine (1) if the conditions tested can be separated based on the measurements obtained, (2) if normalization or rescaling of the data is required, and (3) if correction for unwanted effects is required. When comparing values obtained from different replicates and conditions, it can be important to ensure that they are on the same scale. For the purposes of a two-class comparison using the Rank Product method (more below), this method can be sensitive to normalization/scaling of data [23].

Distribution. To check the distribution of data and to assess whether global normalization might be required, plotting a boxplot is a simple method to determine centering and spread of the data. Plotting the density distributions of each sample is an alternative to the boxplot, and pair-wise scatter plots and MA (mean versus average) plots can be utilized to assess linearity over the range of values between pairs of samples (*see* ref. 28 for more details and normalization considerations). The latter plots are left to the reader to explore.

```
boxplot(phospho.data.two.class[6:12],
names=c("NF-1", "NF-2", "NF-3", "NF-4", "AT-
1", "AT-2", "AT-3"))
abline(h=0, col="red", lwd=3)
```

As can be observed in Fig. 2, most samples are reasonably centered on zero, except for AT-3 and the spread (width of the boxes) of the data varies between replicates. This variation between samples indicates a requirement for global normalization.

3.2.4 Normalization

A number of normalization approaches exist that are well beyond the scope of this chapter. The fundamental principle of normalization is to rescale data in order for their values to be comparable. Here, we utilize *quantile normalization*, which rescales all data to exactly the same distribution [29]. For these purposes we assume that values are missing at random. A new data frame is created (*phospho.data.two.class2*) containing the normalized values for continuing analysis. This also enables an analysis of the effects of normalization.

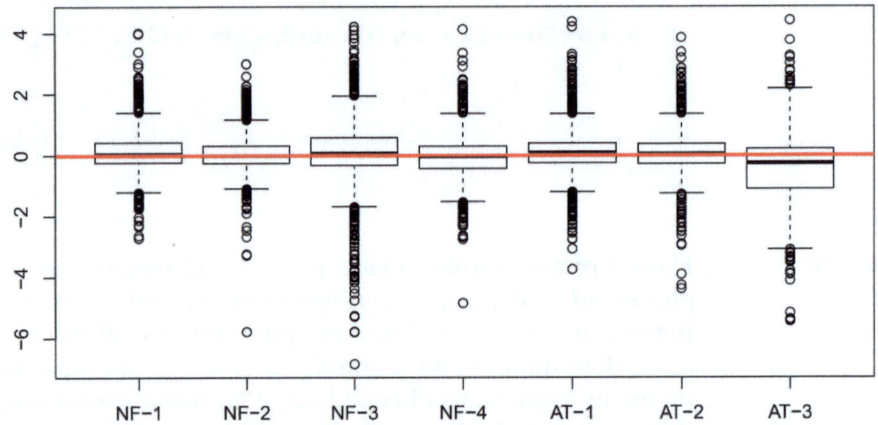

Fig. 2 Boxplot—exploratory analysis. *Y*-axis is the log-ratio between NF + H_2O_2 and control or A-T + H_2O_2 and control

```
phospho.data.two.class2 <- normalize.
quantiles(as.matrix(phospho.data.two.
class[6:12]))
rownames(phospho.data.two.class2) <-
rownames(phospho.data.two.class)
colnames(phospho.data.two.class2) <-
colnames(phospho.data.two.class[6:12])
#plot the results
boxplot(phospho.data.two.class2,
names=c("NF-1", "NF-2", "NF-3", "NF-4", "AT-
1", "AT-2", "AT-3"))
abline(h=0, col="red", lwd=3)
```

As can now be seen (Fig. 3), each replicate has exactly the same distribution. Again, density distribution plots, pairwise plots, and MA plots can also be utilized before and after normalization for visual inspection of normalization effects.

Principle Component Analysis. Principle Component Analysis (or PCA) is a data transformation method for reducing complex datasets to a set of *components* that are uncorrelated and in decreasing order of variance for explaining separation of variables [30]. Nontechnically, PCA is a very useful tool for determining if there is an underlying pattern that separates the data (such as NF from A-T), without explicitly separating the data based on what was known. A number of utilities for PCA exist, but here we use it as a visual aid to determine if the data obtained naturally separates A-T and NF. The number of components in PCA is equal to or less than the number of variables; here, we define this to be the number of samples. It is always useful to plot the first three components

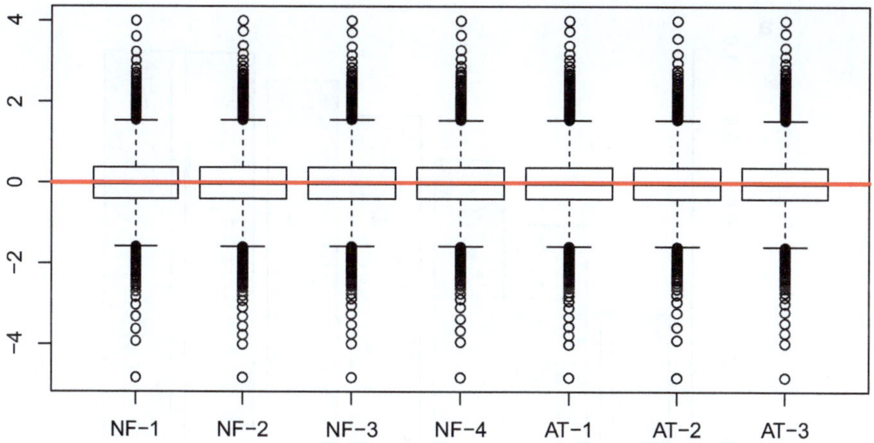

Fig. 3 Boxplot—after quantile normalization. *Y*-axis is the log-ratio between NF + H_2O_2 and control or A-T + H_2O_2 and control

against each other which typically, but not always, explain the largest proportion of variance and therefore contain the majority of information that separates identified variables. This can be confirmed by plotting the cumulative variance captures for components (below).

```
#by default this will use Singular Value De-
composition when the data is complete - test
with complete and non-complete to analyse
the effects.
pc <- pca(t(phospho.data.two.class2),
center=TRUE, nPcs=ncol(phospho.data.two.
class2))
pc.scores <- as.data.frame(scores(pc))
#plot the cumulative variance (how many com-
ponents are required):
barplot(R2cum(pc), ylim=c(0, 1))
#plot the first two components, pc.scores[,1]
and pc.scores[,2]:
plot(pc.scores[,1], pc.scores[,2], cex = 2,
col = c(rep("green", 4), rep("red", 3)),
pch=19)
text(pc.scores[,1], pc.scores[,2], c("NF-1",
"NF-2", "NF-3", "NF-4", "AT-1", "AT-2", "AT-
3"), cex=0.5, pos=3, col="black")
abline(h = 0, v = 0, col = "gray", lty = 2)
#plot the first and third, as well as second
and third components against each by substi-
tuting the 1 and 2, with 1 and 3, or 2 and 3.
```

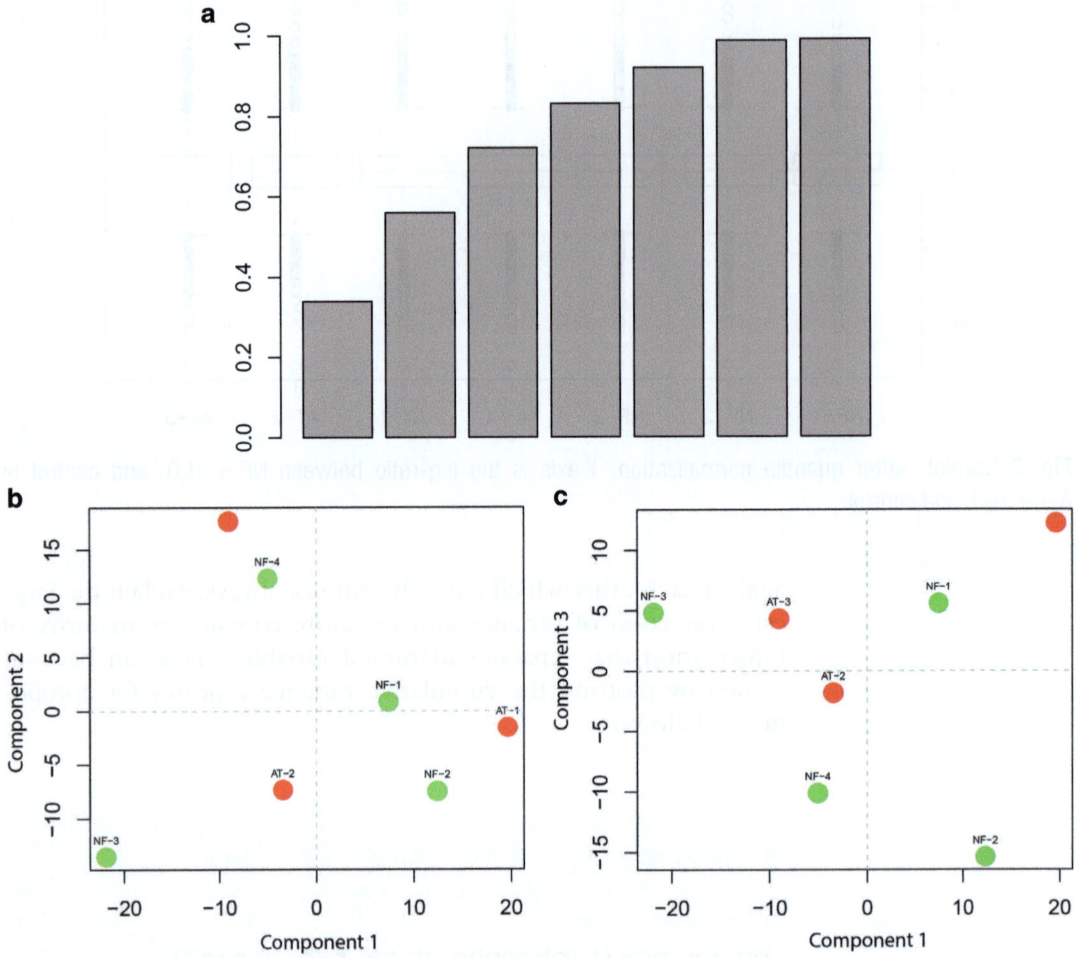

Fig. 4 (**a**) Accumulated variance. (**b**) PCA of first and second components. (**c**) PCA of first and third components

As can be observed (Fig. 4), the data does not separate the NFF and A-T samples well. To determine if this separation can be improved, we now supervise the analysis by correcting for potential nonbiological sources of variance.

Missing value imputation and correction of nonbiological variance. The samples collected were done so at different times and run on independent mass-spectrometry machines. One of the normal human fibroblast samples was run independently of the other (NF-1). It is possible that a large level of technical variation is overcoming the biological signal. To determine if the data can be corrected, we apply imputation [20] (a requirement of the correction method—to fill in missing values) and then correct the data by fitting a null and supervised model to the data using surrogate variable analysis (or SVA) [24].

```
#impute missing values:
phospho.data.two.class.impute <- impute.
knn(as.matrix(phospho.data.two.class2))$data
#set-up a matrix describing the layout of
the samples - supervising
mod.matrix <- data.frame("sample"=as.factor(
c("NF","NF","NF","NF","AT","AT","AT")))
rownames(mod.matrix) <- colnames(phospho.
data.two.class2)
mod <- model.matrix(~as.factor(sample),
data=mod.matrix)
# set-up a null model - assume that all sam-
ples are no different
mod0 <- model.matrix(~1, data=mod.matrix)
#fit the SVA iteratively reweighted model
(irw), using a model of samples of interest
and null.
svaobj <- sva(as.matrix(phospho.data.two.
class.impute), mod, mod0, method="irw")
#reweight the data and obtain the new values
v.2 <- fsva(phospho.data.two.class.impute,
mod=mod, sv=svaobj, newdat=phospho.data.two.
class.impute, method="exact")
```

When performing PCA and plotting the corrected data (we leave the user to reuse the above code to do so as an exercise) (Fig. 5), there is still an effect of NF-1 separating from the other NF samples (when comparing components 1 and 2), but the combination of components 1 and 3 now separates NF and A-T samples, indicating the surrogate variable analysis improved separation of NF and A-T samples.

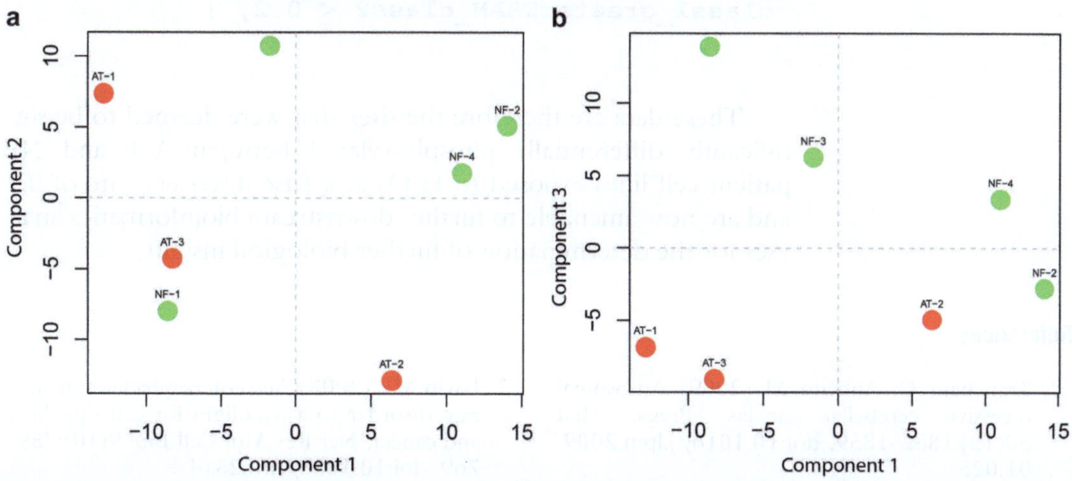

Fig. 5 (**a**) PCA of first and second components. (**b**) PCA of first and third components

3.2.5 Statistical Testing: Nonparametric Rank Product Analysis

Having reviewed the data, corrected for nonbiological sources of variation which now demonstrate separation of variables of interest (A-T and NF), we conclude by performing the rank product analysis [23], which is a nonparametric analysis, for the determination of significantly differentially phosphorylated phosphopeptides between the NF and A-T conditions. These sites will indicate the A-T-dependent changes. Rank product analysis computes the differences in the ranks between conditions. As previously stated, normalization can however affect the rank product analysis. This is because prior to generating differences of ranks between two conditions, all combinations of ratios are first considered to form a new list. *P*-values are computed directly for the ranks and here 1000 permutations are utilized to determine the proportion of false positives (pfp); here, we name the false discovery rate (FDR), by shuffling the data to determine how many times the ranked product is observed by chance [23].

```
set.seed(1234)
rp <- RP(v.2$new, cl=c(0,0,0,0,1,1,1), num.
perm=1000)
#extract information of interest and form
new data frame:
rp.merge <- data.frame(rp$AveFC, rp$pfp,
rp$pval)
#tidy column names:
colnames(rp.merge) <- c("AveFC", "FDR_
class1_lessTHAN_class2", "FDR_class1_grea-
terTHAN_class2", "pVAL_class1_lessTHAN_
class2", "pVAL_class1_greaterTHAN_class2")
#select sites with a FDR of < 0.2
rp.merge.final <- rp.merge[rp.merge$FDR_
class1_lessTHAN_class2 < 0.2 | rp.merge$FDR_
class1_greaterTHAN_class2 < 0.2, ]
```

These data are therefore the sites that were deemed to be significantly differentially phosphorylated between A-T and NF patient cell lines exposed to H_2O_2 at a false discovery rate of 0.2 and are now amenable to further downstream bioinformatics analyses for the determination of further biological insight.

References

1. Tranchant C, Anheim M (2009) Autosomal recessive cerebellar ataxias. Presse Med 38(12):1852–1859. doi:10.1016/j.lpm.2009.01.025
2. Lavin MF (2008) Ataxia-telangiectasia: from a rare disorder to a paradigm for cell signalling and cancer. Nat Rev Mol Cell Biol 9(10):759–769. doi:10.1038/nrm2514

3. Paull TT (2015) Mechanisms of ATM activation. Annu Rev Biochem 84:711–738. doi:10.1146/annurev-biochem-060614-034335
4. Gross S, Rahal R, Stransky N, Lengauer C, Hoeflich KP (2015) Targeting cancer with kinase inhibitors. J Clin Invest 125(5):1780–1789. doi:10.1172/JCI76094
5. Manning G, Whyte DB, Martinez R, Hunter T, Sudarsanam S (2002) The protein kinase complement of the human genome. Science 298(5600):1912–1934. doi:10.1126/science.1075762
6. Kim ST, Lim DS, Canman CE, Kastan MB (1999) Substrate specificities and identification of putative substrates of ATM kinase family members. J Biol Chem 274(53):37538–37543
7. Choudhary C, Mann M (2010) Decoding signalling networks by mass spectrometry-based proteomics. Nat Rev Mol Cell Biol 11(6):427–439. doi:10.1038/nrm2900
8. Bensimon A, Schmidt A, Ziv Y, Elkon R, Wang SY, Chen DJ, Aebersold R, Shiloh Y (2010) ATM-dependent and -independent dynamics of the nuclear phosphoproteome after DNA damage. Sci Signal 3(151):rs3. doi:10.1126/scisignal.2001034
9. Beli P, Lukashchuk N, Wagner SA, Weinert BT, Olsen JV, Baskcomb L, Mann M, Jackson SP, Choudhary C (2012) Proteomic investigations reveal a role for RNA processing factor THRAP3 in the DNA damage response. Mol Cell 46(2):212–225. doi:10.1016/j.molcel.2012.01.026
10. Bastos de Oliveira FM, Kim D, Cussiol JR, Das J, Jeong MC, Doerfler L, Schmidt KH, Yu H, Smolka MB (2015) Phosphoproteomics reveals distinct modes of Mec1/ATR signaling during DNA replication. Mol Cell 57(6):1124–1132. doi:10.1016/j.molcel.2015.01.043
11. Kozlov SV, Waardenberg AJ, Engholm-Keller K, Arthur JW, Graham ME, Lavin M (2016) Reactive oxygen species (ROS)-activated ATM-dependent phosphorylation of cytoplasmic substratesidentifiedbylarge-scalephosphoproteomics screen. Mol Cell Proteomics 15(3):1032–1047. doi:10.1074/mcp.M115.055723
12. Mazouzi A, Stukalov A, Muller AC, Chen D, Wiedner M, Prochazkova J, Chiang SC, Schuster M, Breitwieser FP, Pichlmair A, El-Khamisy SF, Bock C, Kralovics R, Colinge J, Bennett KL, Loizou JI (2016) A comprehensive analysis of the dynamic response to aphidicolin-mediated replication stress uncovers targets for ATM and ATMIN. Cell Rep. doi:10.1016/j.celrep.2016.03.077
13. Bennetzen MV, Larsen DH, Bunkenborg J, Bartek J, Lukas J, Andersen JS (2010) Site-specific phosphorylation dynamics of the nuclear proteome during the DNA damage response. Mol Cell Proteomics 9(6):1314–1323. doi:10.1074/mcp.M900616-MCP200
14. Pursiheimo A, Vehmas AP, Afzal S, Suomi T, Chand T, Strauss L, Poutanen M, Rokka A, Corthals GL, Elo LL (2015) Optimization of statistical methods impact on quantitative proteomics data. J Proteome Res 14(10):4118–4126. doi:10.1021/acs.jproteome.5b00183
15. Schwartz D, Gygi SP (2005) An iterative statistical approach to the identification of protein phosphorylation motifs from large-scale data sets. Nat Biotechnol 23(11):1391–1398. doi:10.1038/nbt1146
16. Xue Y, Zhou F, Zhu M, Ahmed K, Chen G, Yao X (2005) GPS: a comprehensive www server for phosphorylation sites prediction. Nucleic Acids Res 33(Web Server issue):W184–W187. doi:10.1093/nar/gki393
17. Dennis G Jr, Sherman BT, Hosack DA, Yang J, Gao W, Lane HC, Lempicki RA (2003) DAVID: database for annotation, visualization, and integrated discovery. Genome Biol 4(5):P3
18. Snel B, Lehmann G, Bork P, Huynen MA (2000) STRING: a web-server to retrieve and display the repeatedly occurring neighbourhood of a gene. Nucleic Acids Res 28(18):3442–3444
19. Gentleman RC, Carey VJ, Bates DM, Bolstad B, Dettling M, Dudoit S, Ellis B, Gautier L, Ge Y, Gentry J, Hornik K, Hothorn T, Huber W, Iacus S, Irizarry R, Leisch F, Li C, Maechler M, Rossini AJ, Sawitzki G, Smith C, Smyth G, Tierney L, Yang JY, Zhang J (2004) Bioconductor: open software development for computational biology and bioinformatics. Genome Biol 5(10):R80. doi:10.1186/gb-2004-5-10-r80
20. Troyanskaya O, Cantor M, Sherlock G, Brown P, Hastie T, Tibshirani R, Botstein D, Altman RB (2001) Missing value estimation methods for DNA microarrays. Bioinformatics 17(6):520–525
21. Stacklies W, Redestig H, Scholz M, Walther D, Selbig J (2007) pcaMethods—a bioconductor package providing PCA methods for incomplete data. Bioinformatics 23(9):1164–1167. doi:10.1093/bioinformatics/btm069
22. Wickham H (2011) The split-apply-combine strategy for data analysis. J Stat Softw 40(1):29. doi:10.18637/jss.v040.i01
23. Breitling R, Armengaud P, Amtmann A, Herzyk P (2004) Rank products: a simple, yet powerful, new method to detect differentially regulated genes in replicated microarray experiments. FEBS Lett 573(1–3):83–92. doi:10.1016/j.febslet.2004.07.055

24. Leek JT, Storey JD (2007) Capturing heterogeneity in gene expression studies by surrogate variable analysis. PLoS Genet 3(9):1724–1735. doi:10.1371/journal.pgen.0030161
25. Boutet E, Lieberherr D, Tognolli M, Schneider M, Bairoch A (2007) UniProtKB/Swiss-Prot. Methods Mol Biol 406:89–112
26. Cox J, Mann M (2008) MaxQuant enables high peptide identification rates, individualized p.p.b.-range mass accuracies and proteome-wide protein quantification. Nat Biotechnol 26(12):1367–1372. doi:10.1038/nbt.1511
27. Cox J, Matic I, Hilger M, Nagaraj N, Selbach M, Olsen JV, Mann M (2009) A practical guide to the MaxQuant computational platform for SILAC-based quantitative proteomics. Nat Protoc 4(5):698–705. doi:10.1038/nprot.2009.36
28. Dudoit S, Yang YH, Callow MJ, Speed TP (2002) Statistical methods for identifying differentially expressed genes in replicated cDNA microarray experiments. Stat Sin 12: 111–139
29. Bolstad BM, Irizarry RA, Astrand M, Speed TP (2003) A comparison of normalization methods for high density oligonucleotide array data based on variance and bias. Bioinformatics 19(2):185–193
30. Abdi H, Williams LJ (2010) Principal component analysis. Wiley Interdiscip Rev Comput Stats 2(4):433–459. doi:10.1002/wics.101

Chapter 18

ChIP Technique to Study Protein Dynamics at Defined DNA Double Strand Breaks

Jie Wen and Patrick Concannon

Abstract

Information about the timing of appearance and composition of protein aggregates, termed foci, that arise in eukaryotic cells at sites of DNA double strand breaks (DSBs) has been mainly obtained through immunostaining and thus is limited by the resolution of light microscopy and the availability of appropriate antibodies. In this chapter, we describe a system using direct protein transduction of homing endonuclease, I-*Ppo*I, into human cells to generate site-specific DSBs, allowing for detection of target proteins using chromatin immunoprecipitation (ChIP), enabling molecular probing of the cellular response to a DNA DSB. Following the introduction of I-*Ppo*I and generation of DSBs, genomic DNA and protein are cross-linked and analyzed by ChIP to determine the spatial distribution and temporal accumulation of specific proteins at the site of breaks. Direct transduction of I-*Ppo*I protein results in rapid accumulation and turnover of I-*Ppo*I in live cells, facilitating comparisons across multiple cell lines. This system allows the direct detection of protein and chromatin dynamics at the site of the break, as well as timing and extent of DNA DSB repair in human cells.

Key words I-PpoI, DNA double strand breaks (DSBs), DNA repair, Chromatin immunoprecipitation

1 Introduction

DNA double strand breaks (DSBs) are among the most dangerous of genomic lesions, and can result from both normal metabolic activities and exposure to environmental factors such as ionizing radiation or radiomimetic chemicals. DSBs cause structural damage to DNA molecules which may alter or eliminate the transcription of genes that the affected DNA encodes, impair DNA replication, or affect chromosomal integrity.

Given the risk posed by DSBs, organisms have evolved a range of elegant strategies to respond to such lesions. DNA damage responses require coordinated and timely sensing of damage, signaling to relevant cellular processes of the presence of lesions and activation of appropriate effector functions to arrest cell cycle progression, initiate DNA repair, or trigger apoptosis. Initial DNA

Sergei V. Kozlov (ed.), *ATM Kinase: Methods and Protocols*, Methods in Molecular Biology, vol. 1599,
DOI 10.1007/978-1-4939-6955-5_18, © Springer Science+Business Media LLC 2017

damage signal amplification leads to the recruitment of hundreds to thousands of molecules at or around the site of DNA breaks [1, 2]. These sites can be visualized by immunostaining for specific protein components. However, environmental DSB-inducing agents, such as ionizing radiation, produce randomly distributed DNA damage that, while observable in individual cells by microscopy, does not provide for biochemical analysis of the structure of repair foci. Directed targeting of lesions can be achieved through the use of UV lasers to cut chromosomes but is still limited by the resolution of microscopy as a readout [3, 4]. Thus, while these methods have provided a wealth of information on the kinetics of damage-responsive protein recruitment to DSB lesions, they remain limited by the inability to specify the exact locations of individual proteins relative to DSBs in active DNA damage response foci.

Alternative methods for the induction of DNA DSBs at specifically targeted sites in the genome allow for improved resolution of the spatio-temporal features of DNA damage responses, but may not accurately reproduce the range of damage induced by environmental mutagens such as ionizing radiation. Such alternative approaches have traditionally used the expression of infrequently cutting restriction endonucleases such as I-*Sce*I homing endonuclease from budding yeast [5]. Introduction of a target sequence for I-*Sce*I into human cells has been used in a number of studies of DSB response and repair upon expression of I-*Sce*I [6, 7]. Other homing endonucleases such as I-*Ppo*I and I-*Asi*SI, which have varying numbers of endogenous sites in human cells, have similarly been used [8–10]. A limitation of these endonuclease-based approaches is that they typically require genomic integration of an expression construct producing the endonuclease whose expression must then be rigorously regulated to avoid redundant cycles of cutting and repair.

In this chapter, we describe a method for protein transduction of bacterially produced and purified I-*Ppo*I to create DNA DSBs at defined endogenous sites in the human genome. Since the approach does not involve integration of either target sequence nor a gene encoding the endonuclease into human cells, this system can be easily applied to isogenic cell lines bearing mutations in genes encoding selected DNA damage response proteins. This I-*Ppo*I system allows tracking and comparison of protein recruitment and dissociation at or around identified I-*Ppo*I sites by ChIP assay as well as monitoring of the repair of DSBs via quantitative real-time PCR [11].

The serine-threonine kinase ATM plays a key role in the sensing and signaling of DNA damage, localizing quickly to the sites of DNA DSBs and phosphorylating many of the proteins that aggregate into repair foci [12]. Among these substrates is the protein NBN [13–16], which, like ATM, was originally identified because of its mutation in a rare, recessive disorder characterized by radiation hypersensitivity [17, 18]. NBN associates with the nuclease MRE11 and the ATPase RAD50 forming the MRE11/RAD50/NBN (MRN) complex [18]. The MRN complex localizes to DSBs within minutes following their formation, and facilitates the recruitment and retention of ATM at these sites via interaction between ATM and the C-terminal end of NBN [19, 20]. ATM phosphorylates many of the DNA damage response proteins that accumulate at the site of a DSB including all three components of the MRN complex [21]. ATM phosphorylates NBN on as many as three serine residues, serine 278, 343 and 397, in response to DNA damage.

In the current chapter, we provide a detailed protocol for the use of protein transduction to introduce I-*Ppo*I into human cells and to assay its effects on DNA repair and focus formation by quantitative PCR and ChIP assay, respectively. We illustrate the utility of the approach in the context of ATM function, applying the assay to multiple cell lines with mutations in NBN that affect its expression or phosphorylation state. The cell lines utilized include ILB1 cells, which fail to express full-length NBN protein due to a homozygous frame shift mutation in the NBN gene, ILB1-[NBN] cells, in which NBN expression has been restored by transfection with a retroviral vector expressing wild-type NBN, ILB1-[NBN-S>A] cells expressing full-length NBN with alanine substitutions at the ATM phosphorylation sites, and ILB1-[NBN-S>E] cells expressing full-length NBN with glutamic acid substitutions at ATM phosphorylation sites. The ability to reproducibly induce DNA DSBs at specific sites in these isogenic cell lines using the described I-*Ppo*I transduction system allows for a comparison of the effects of blocking or mimicking ATM phosphorylation of NBN and its time-dependent effects on retention of ATM and NBN at the sites of DNA DSBs.

2 Materials

2.1 Reagents

1. I-*Ppo*I open-reading frame (synthesized by GenScript USA Inc.).
2. *Not*I.
3. *Xho*I.
4. pET45b (+) vector (Novagen, EMD Chemicals Inc.).
5. BL21 (DE3) competent cells.
6. SOC medium for *E. coli*.
7. Isopropyl-β-D-thiogalactopyranoside (IPTG).
8. 0.2 M zinc acetate.
9. 0.5 M MOPS–NaOH buffer (pH 7.5).
10. 0.5 M EDTA, pH 8.0.

11. Leupeptin.
12. Chymostatin.
13. Phenylmethanesulfonylfluoride (PMSF).
14. Ni-NTA beads.
15. Protein assay reagents (Bio-Rad).
16. NuPAGENovex (3–8% Tris–Acetate or 4–12% Bis-Tris) gels (Invitrogen).
17. Immobilon-P transfer membranes.
18. ECL Western blotting detection reagents.
19. Dulbecco's Modified Eagle's Medium (DMEM).
20. Fetal Bovine Serum (FBS).
21. Penicillin-Streptomycin.
22. Nonessential amino acids (NEAA).
23. Sodium pyruvate.
24. 0.25% Trypsin-EDTA.
25. Pep-1 (synthesized by AnaSpect, Fremont, CA, USA).
26. Chariot™ Protein Delivery Reagent (Active Motif).
27. 37% (wt/vol) formaldehyde solution.
28. Herring sperm DNA.
29. Protease inhibitor cocktails.
30. Phosphatase inhibitor cocktails.
31. Anti-ATM antibody (GeneTex Inc.).
32. Anti-ATM pSer1981 antibody (Rockland Immunochemicals Inc.).
33. Anti-NBN antibody (Novus Biologicals Inc.).
34. Anti-NBN pSer343 antibody (Upstate Cell Signaling Solutions).
35. Anti-γH2AX antibody (Millipore).
36. Anti-His-tag antibody (GE Healthcare).
37. Protein G-agarose beads.
38. QIAquick® PCR Purification Kit (Qiagen).
39. SYBR Green Master Mix.

2.2 Buffers

1. PBS lysis buffer: 1 mL of 0.5 M MOPS–NaOH, 50 μL of 0.5 M EDTA, 200 μL of Leupeptin, 200 μL of Chymostatin, 200 μL of PMSF, 25 mg of Lysozyme.
2. Wash buffer (protein purification): PBS with 25 mM imidazole, pH 7.4.
3. Elution buffer (protein purification): PBS with 250 mM imidazole, pH 7.4.

4. ChIP sonication buffer: 1% Triton X-100, 0.1% Deoxycholate, 50 mM Tris–HCl pH 8.1, 150 mM NaCl, 5 mM EDTA.
5. TE buffer: 10 mM Tris–HCl pH 7.5, 1 mM EDTA.
6. Elution buffer (ChIP assay): 0.1% SDS, 0.1 M $NaHCO_3$, 0.01 mg/mL Herring sperm DNA.

2.3 Major Instruments

- French Press Cell Disruptor (SLM Instruments, Rochester, NY, USA).
- Branson Sonifier 250 (Branson Ultrasonics, Danbury, CT, USA).
- Real-time PCR machine.
- BiomekFX^{P} Lab Automation Workstation (Beckman Coulter, BiomeckFX^{P}).

3 Methods

Steps overview. I-*Ppo*I protein is expressed and purified from *E. coli* and transiently transduced into target cells using cell-penetrating peptides (CPP). Cells are then crosslinked with formaldehyde, prepared for chromatin fragmentation and immunoprecipitated with antibodies against proteins suspected to be present at the damage site. Following immunoprecipitation, the crosslinked protein/DNA complexes are eluted, crosslinking is reversed, and DNA is purified. Quantitative real-time PCR analysis of the purified DNA allows the spatial and temporal analysis of the recruitment and dissociation of targeted damage responsive proteins to the DSBs immediately after DNA damage. Alternatively, by using specific oligonucleotide primers flanking the I-*Ppo*I-mediated DSBs site, creation and repair of the DSBs can be quantitated by PCR.

3.1 I-PpoI Protein Purification

1. I-*Ppo*I open-reading frame is chemically synthesized and cloned into pUC57 vector between *Not*I and *Xho*I sites.
2. I-*Ppo*I insert is excised with *Not*I and *Xho*I and ligated into a compatible vector, pET45b(+). This results in the addition of a 6-Histidine epitope tag to the N-terminus of the I-*Ppo*I open-reading frame, and brings its expression under the control of a T7 RNA polymerase promoter.

Day 1:

3. The product from the ligation reaction is transformed into BL21 (DE3) competent cells using one vial of competent cells (50 μL) for each transformation. Add 5–10 μL (10 pg to 100 ng) of a ligation reaction per transformation. Mix very gently and rotate the tube horizontally.

4. Incubate the tube on ice for 30 min and heat shock the cells for 30 s at 42 °C. Place the tube immediately on ice for at least 2 min. Add 250 μL of warmed SOC medium to each tube and incubate for 1 h at 37 °C with vigorous shaking.
5. After incubation, plate cell suspension on a pre-warmed LB-agar plate containing the appropriate antibiotic. Incubate plate inverted overnight at 37 °C.

Day 2:

6. Pick a single colony from plate using sterile tips, place in 5 mL of LB media containing appropriate antibiotic, incubate overnight at 37 °C, and shake at 250 rpm.

Day 3:

7. 450 μL of the non-induced overnight culture is added to ~450 mL of LB media (1:1000 dilution) containing appropriate antibiotic and 2.25 mL of 0.2 M zinc acetate (1 mM final concentration). Incubate at 37 °C, shake at 250 rpm for several hours till O.D. of 0.4 is reached (*see* **Note 1**).
8. Add IPTG to a final concentration of 1 mM and zinc acetate to 2 mM (another 2.5 mL of 0.2 M zinc acetate is added to the culture) and continue shaking at 37 °C for 8–10 h (*see* **Note 1**).
9. Pellet the bacteria at 6000 × *g*, 4 °C for 15 min and discard supernatant. Suspend cell pellet with 45 mL of cold PBS, then collect cells by centrifugation at 6000 × *g*, 4 °C for 15 min, discard the supernatant.
10. Prepare 20 mL cold PBS buffer containing: 1 mL of 0.5 M MOPS–NaOH, 50 μL of 0.5 M EDTA, 200 μL of Leupeptin, 200 μL of Chymostatin, 200 μL of PMSF, 25 mg of Lysozyme.
11. Resuspend cell pellet with the supplemented cold PBS buffer described above and transfer cell suspension into two 15 mL conical tubes. Freeze on dry ice and store at −80 °C overnight or longer (*see* **Note 2**).

Day 4:

12. Thaw cell pellet on ice and add 0.2 mL of 0.1 M $MgCl_2$ to cell suspension (*see* **Note 3**).
13. Cell suspension is passed through French Press Cell Disruptor twice. Cell debris is removed by centrifugation at 13,000 × *g* at 4 °C for 30 min. Supernatant is collected into a 50 mL conical tube and kept on ice.
14. Wash 400 μL Ni-NTA beads (800 μL if 50% slurry) once with cold PBS and keep on ice. Add Ni-NTA beads to cell lysate and mix gently by shaking at 4 °C for 30 min (*see* **Note 4**).

15. Centrifuge at 700 × *g*, 4 °C for 10 min, discard the supernatant.
16. Wash three times with 30 mL of wash buffer, shake 5 min at 4 °C, centrifuge at 700 × *g*, 4 °C for 10 min.
17. Check the protein levels of the washes using Bio-Rad protein assay reagent (100 μL of Bio-Rad reagent +10 μL Wash 1/Wash 2/Wash 3), Wash 1 should be dark blue and Wash 3 should be very light blue.
18. After the third wash, add 1 mL of wash buffer, transfer beads to the elution column and drain out the wash buffer.
19. Elute I-*Ppo*I protein with at least twelve 100 μL fractions of elution buffer, each elution fraction is collected in an individual 1.5 mL tube.
20. Use Bio-Rad protein assay reagent to check protein concentration in the collected fractions, combine tubes if necessary.
21. Confirm the purity of the I-*Ppo*I on SDS-PAGE and aliquot into 1.5 mL tubes. Store at −80 °C for later use.

3.2 Generation of DSBs, I-PpoI Protein Transduction, and ChIP Assay

Target cells are maintained in regular growth medium, such as Dulbecco's Modified Eagle's Medium supplemented with 10% fetal bovine serum (FBS), 1% pen/strep, 1% nonessential amino acids (NEAA), and 1% sodium pyruvate. For all experiments, cells are maintained at 37 °C in 5% CO_2/95% air. I-*Ppo*I can be introduced into target cells using the cell-penetrating peptide (CPP), Pep-1, or the protein delivery reagent, Chariot, following manufacturer's instructions. Diluted I-*Ppo*I is added to the diluted Pep-1/Chariot and incubated at room temperature to allow the protein/CPP complex to form. Before transfection, cells are washed once with PBS and overlaid with proper volume of I-*Ppo*I/CPP complex. Serum-free medium is added to the cell culture, and cells are incubated at 37 °C, complete medium is added at 1 h after transfection for prolonged incubation.

Day 1:

1. Cells are plated in 15-cm dishes and incubated at 37 °C in a humidified atmosphere containing 5% CO_2 and allowed to grow to about 80–90% confluence. The appropriate cell density needs to be validated and optimized for each specific cell line but is typically between 3×10^6 and 1×10^7 cells per plate.

Day 2: Generation of DSBs

1. Preparation of diluted I-*Ppo*I protein: 1 μg of I-*Ppo*I is diluted into 200 μL of cold PBS and kept on ice.
2. Preparation of diluted Chariot reagent or CPP: in a separate tube, dilute 50 μL of 2 μg/μL Chariot/CPP stock solution into 200 μL of sterile H_2O for one 15-cm dish. For multiple

transfections, do not use a master mix that exceeds the volume required for two transfections per tube (≤100 μL of 2 μg/μL Chariot/CPP stock solution into 400 μL of sterile H_2O), as this may cause aggregation.

3. Preparation of Chariot/CPP-I-*Ppo*I complex: Add the diluted I-*Ppo*I from **step 1** to the diluted Chariot/CPP from **step 2** and incubate at room temperature for 30 min to allow the Chariot/CPP-I-*Ppo*I complex to form. (*see* **Note 5**)
4. Transfection of cells: aspirate growth medium from cells to be transfected and wash cells once with room temperature PBS. For each 15-cm dish, overlay cells with ~500 μL of Chariot/CPP-I-*Ppo*I complex and then add 7.5 mL of serum-free medium to the dish (final transfection volume for 15-cm dish is 8 mL). Gently rock the plate to ensure even delivery of the transfection reagent.
5. Incubate cells at 37 °C in a humidified atmosphere containing 5% CO_2 for 1 h.
6. Continue to incubate at 37 °C in a humidified atmosphere containing 5% CO_2 for 30 min up to 4 h. If cells are sensitive to serum starvation, complete growth medium can be added to the cells. Do not remove the Chariot/CPP-I-*Ppo*I complex (Fig. 1).

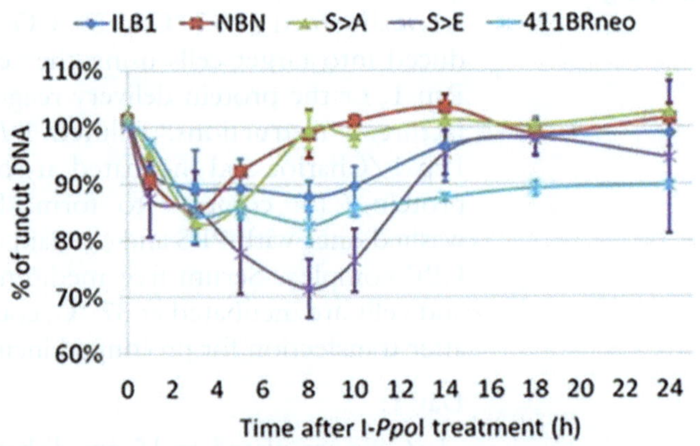

Fig. 1 Transduced I-*Ppo*I induces site-specific DNA cleavage in human cells. Yield of PCR product at a single I-*Ppo*I site in the DAB1 gene on chromosome 1 relative to a reference sequence lacking an I-*Ppo*I site on the same chromosome was compared in five isogenic cell lines, including ILBI cells expressing no full-length NBN; ILBI-[NBN] cells expressing full-length wild-type NBN; ILBI-[S>A] cells expressing full-length NBN carrying serine to alanine mutation at three ATM phosphorylation sites; ILBI-[S>E] cells, expressing full-length NBN carrying serine to glutamic acid mutations at three ATM phosphorylation sites; as well as 411BRneo cells, which lack detectable ligase IV activity due to a homozygous missense mutation at a highly conserved residue (R278H). The data represent the mean ± s.d. from independent triplicate experiments each including two technical replicates. Reprinted with permission from [11]

7. Activation of DNA damage response should be detectable in as early as 1 h after transfection by western blotting with ATM phospho-S1981 antibody (Fig. 2a *lower panel*); by immunofluorescent staining with γH2AX (Fig. 2c) or by ChIP assay (Fig. 3), as described below.

Preparation of samples for ChIP analysis:

8. After transfection, at specific time points, bound proteins are crosslinked with DNA by directly adding 27 μL of formaldehyde (37% stock, wt/vol) per 1 mL of tissue culture medium, to a final concentration of 1%. Incubate cells for 10 min at 37 °C.
9. Quench crosslinking by adding glycine to a final concentration of 0.125 M. Incubate cells at room temperature for 5 min with continuous agitation.
10. Wash cells twice with ice-cold PBS and then harvest cells with ice-cold PBS (1 mL per 15-cm dish) containing protease and phosphatase inhibitor cocktails (1 inhibitor cocktail tablet per 20 mL PBS) and collect in fresh tubes. This should be carried out on ice.
11. Centrifuge cell lysate at 1000 × *g* for 5 min at 4 °C.
12. Remove supernatant and resuspend cell pellet in 1 mL of ChIP sonication buffer containing protease and phosphatase inhibitor cocktails (one inhibitor cocktail tablet per 20 mL ChIP sonication buffer). Incubate on ice for 10 min.
13. Sonicate cell suspensions using Branson Sonifier 250. It is suggested to do 25 × 5 s sonications at constant output level 3.0 which will shear the chromatin to 700–1000 bp fragments. The effectiveness of shearing can be confirmed by sizing a DNA aliquot on 1% agarose gel.
14. After sonication, centrifuge cell suspensions at 13,000 rpm (~2000 × *g*) for 15 min at 4 °C in a microcentrifuge. Carefully collect supernatant (fragmented chromatin) into a new clean tube and label as whole cell extracts (WCEs). WCE can be stored at −80 °C for several weeks or carried out with the following steps directly.
15. Dilute WCEs to ChIP sonication buffer, 0.2 mL of WCEs (corresponding to ~3×10^6 cells) to 0.8 mL of ChIP sonication buffer containing protease and phosphatase inhibitor cocktails. Incubate with either no antibody or 1–3 μg of antibodies targeting ATM, NBN, or other proteins suspected to be present at the damage site. Another 0.2 mL of the WCE is saved for later standard curve analysis.
16. Immunoprecipitation reaction is carried out overnight at 4 °C with constant agitation.
17. Calculate the amount of beads needed for the ChIP assay according to the number of samples in the experiment. Prepare protein G-agarose beads based on the ratio of 45 μL of pro-

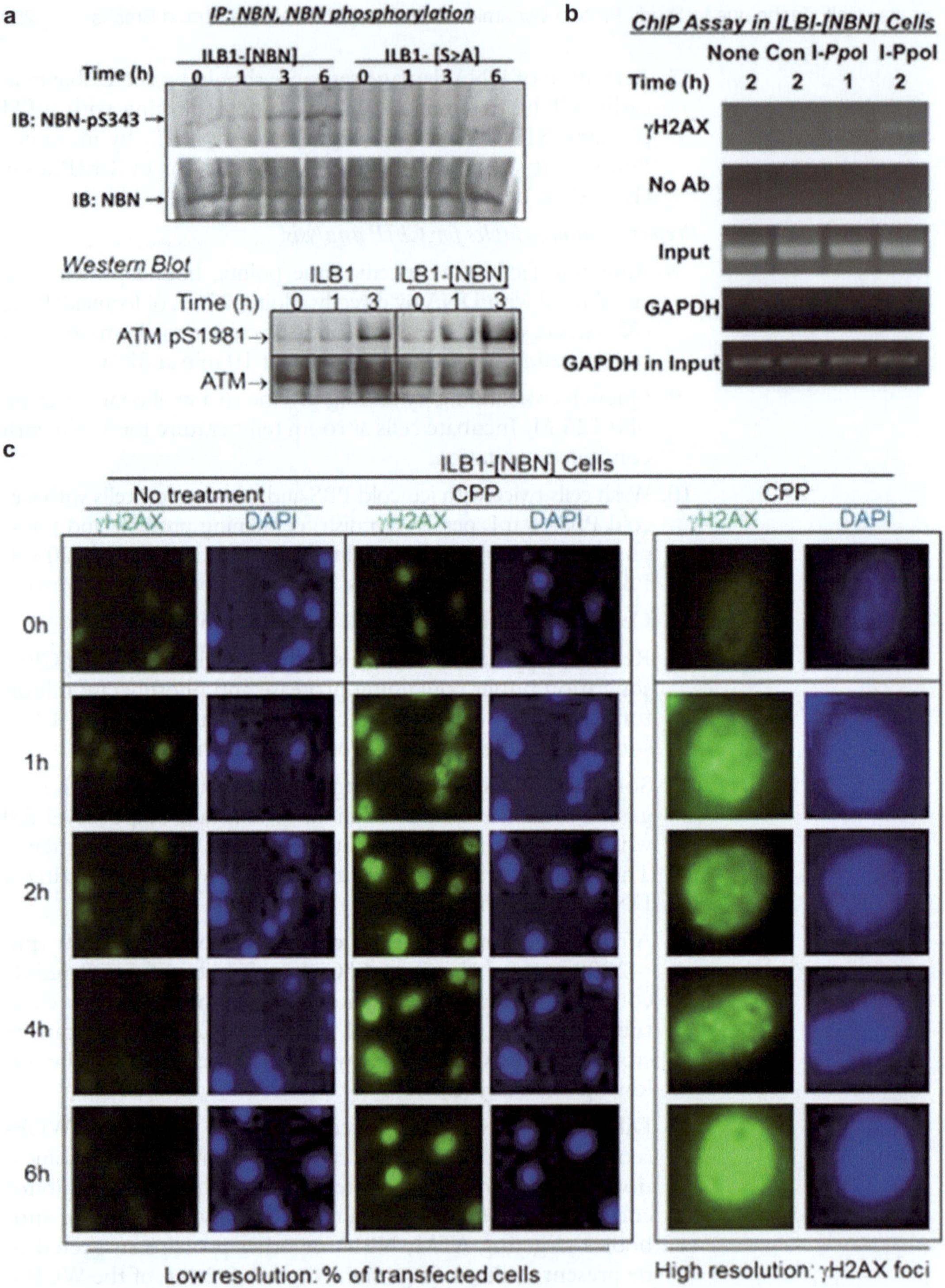

Fig. 2 I-*Ppo*I protein transduction activates ATM-dependent DNA damage responses. (**a**) Lysates from cells treated with I-*Ppo*I for the indicated times were immunoprecipitated with an anti-NBN antibody and immunoblotted with antibodies against NBN-phospho-serine343 (NBN-pS343) and total NBN. (**b**) ILB1-[NBN] cells were transduced with PBS (None), control protein or purified I-*Ppo*I protein (I-PpoI) for 1 h and 2 h. ChIP assays were performed using either no antibody, or antibody against γH2AX. The immunoprecipitated DNA was quantitated by real-time PCR using a primer set adjacent to the I-*Ppo*I cleavage site in the DAB1 gene

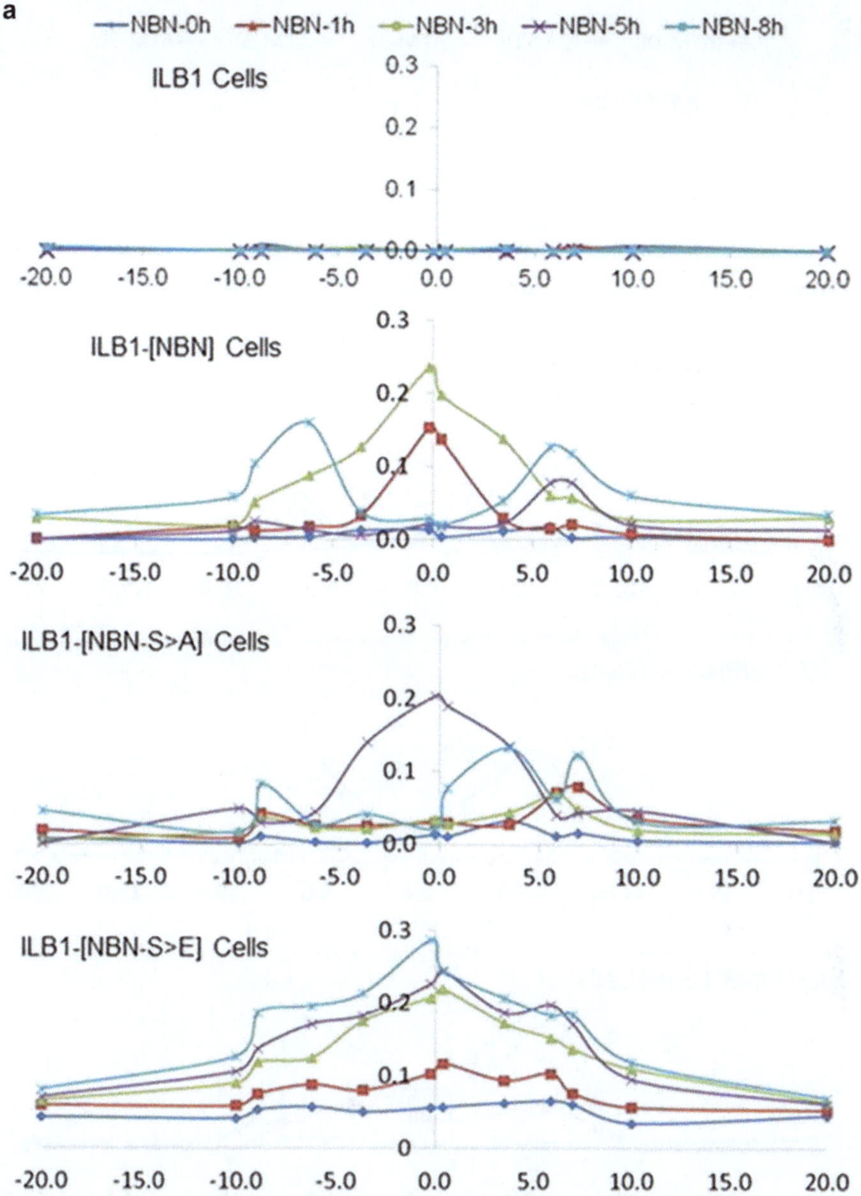

Fig. 3 Mutation of ATM phosphorylation sites in NBN alters the timing of ATM and NBN recruitment to the sites of damage. (**a**, **b**) ILB1, ILB1-[NBN], ILB1-[NBN-S>A], and ILB1-[NBN-S>E] cells were transduced with purified I-*Ppo*I protein using CPP. Cells were lysed at 0, 1, 3, 5, and 8 h after transduction. ChIP assays were performed using antibodies against NBN and ATM (**b**). Reprinted with permission from [11]

Fig. 2 (continued) on chromosome 1 or primer set targeting a housekeeping gene, GAPDH. ILB1-[NBN] cells were mock-treated (No treatment), treated with CPP alone (CPP), or transduced with either control protein (Con-treated) or purified I-*Ppo*I protein (I-PpoI-treated). (**c**) Cells were fixed at 0, 1, 2, 4, and 6 h after transduction and stained with γH2AX and 4′,6-diamidino- 2-phenylindole (DAPI). Low (×20) and high (×40) magnification images are shown for each time point. Reprinted with permission from [11]

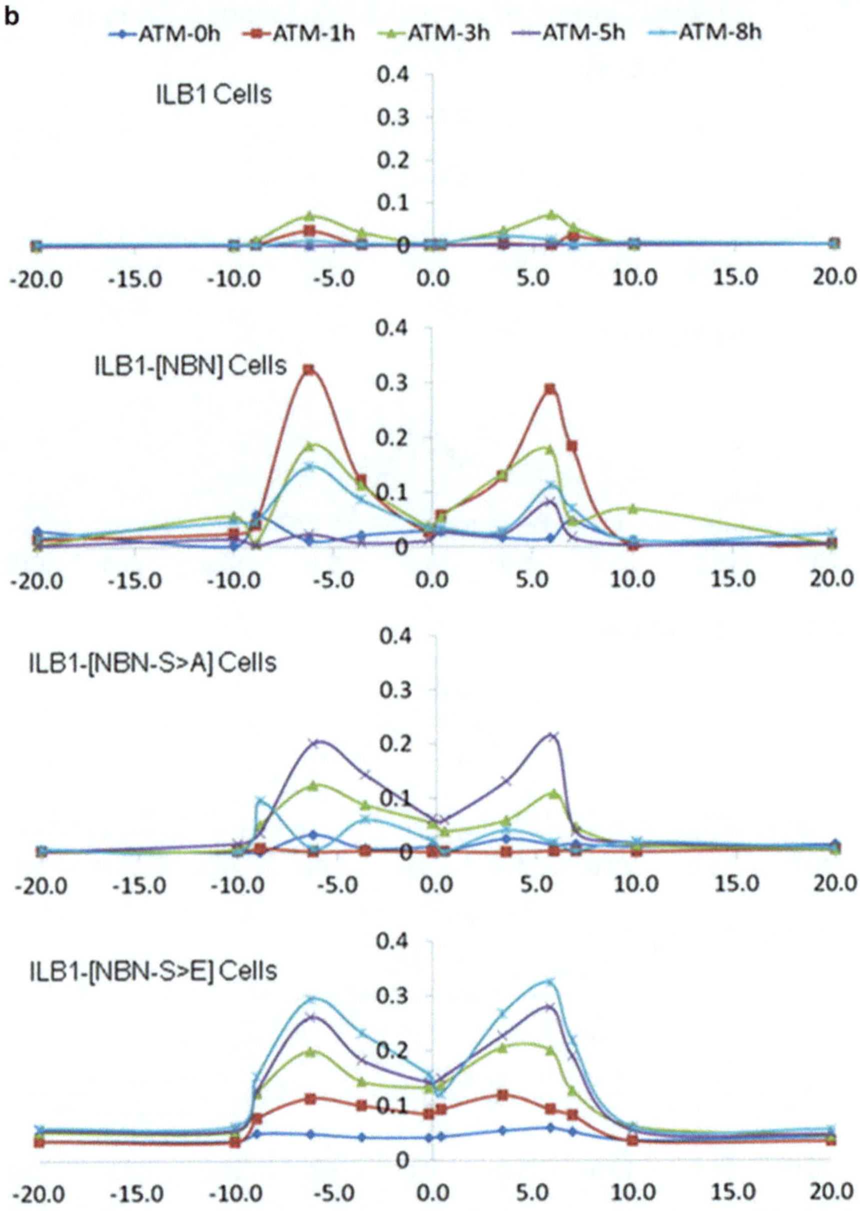

Fig. 3 (continued)

tein-G agarose beads per each immunoprecipitation reaction with a total volume of 1 mL. Wash beads three times with 1 mL of ChIP sonication buffer, add 1 bead volume of ChIP sonication buffer, 1 mg/mL BSA, and incubate overnight at 4 °C with constant agitation.

Day 3:

1. Wash blocked protein G-agarose beads twice with 1 mL of ChIP sonication buffer and then dilute with 1 bead volume of

ChIP sonication buffer containing protease and phosphatase inhibitor cocktails to get a 50% suspension of bead slurry (beads volume/total buffer volume).

2. Per each overnight-incubated immunoprecipitation reaction (1 mL total volume), add 45 μL of prepared protein G-agarose bead slurry and incubate for another 4 h at 4 °C with constant agitation.
3. After incubation, collect the beads by centrifugation at 2800 rpm, 4 °C for 5 min, then wash three times with 1 mL of ChIP sonication buffer, and then twice with 1 mL of TE buffer. Each wash needs 15 min incubation at 4 °C with constant agitation before centrifugation to remove supernatant. After the last wash with TE buffer, centrifuge and remove the last traces of TE buffer as much as possible.
4. Elute DNA/protein/antibody complex from the beads by adding 100 μL of elution buffer. Reverse crosslinking by overnight incubation of the eluted complex in 65 °C water bath. At this point, add another 100 μL of elution buffer to 2 μL of WCE, label as 1% input DNA, and incubate at 65 °C overnight. It will be used to generate standard curve for overall quantification.

Day 4:

1. After overnight incubation, shake samples vigorously on a vibrating platform for 5 × 10 s, and centrifuge at 13,000 rpm for 2 min at room temperature. Collect supernatant into a clean new tube. Be careful not to carry over any beads when transferring the eluted complexes.
2. Extract eluted DNA fragments using QIAquick PCR purification kit according to manufacturer's instruction. Resuspend DNA in 150 μL of TE buffer provided with the kit.
3. To monitor the temporal and spatial recruitment of damage response proteins around DSBs, real-time PCR analysis is carried out on 12 selected locations within 40 kb distance spanning the I-*Ppo*I cutting site in *DAB1* gene (around 20 kb each side from the cutting site).
4. For each real-time PCR reaction, use 5 μL of eluted DNA to quantitate the immunoprecipitated fragments. Pipette DNA samples into a 384-well plate and let it dry at room temperature overnight.

Day 5:

1. Prepare real-time PCR reaction solution: a 10 μL reaction containing 5 μL SYBR Green Master Mix (2×), 1 μL primer pairs, 0.4 μL DNA polymerase II, and 4.6 μL of ddH_2O.
2. Add real-time PCR reaction solution to 384-well plate containing the air-dried DNA samples at 10 μL/well.

Table 1
Oligos for real-time PCR and ChIP PCR

#	Description	Sequence (in 5′–3′ direction)	Anneal temp.	References
Oligos for real-time PCR in cutting assay (across 10 putative I-*Ppo*I recognition sites)				
1	Unaligned site	F: CAGGGGAATCCGACTGTTTA R: ATGACGAGGCATTTGGCTAC	60.0 °C	[11]
2	Chromosome X	F: GGGTGTTGATGCGATGTGAT R: TAGGGACAGTGGGAATCTCG	60.0 °C	[11]
3	Chromosome 11	F: GGTGTTGATGCCATGTGATT R: TAGGGACAGTGGGAATCTCG	60.0 °C	[11]
4	Chromosome 8	F: TTAGCACCATTCCACAAAGGC R: TAGGGACAGTGGGAATCTCG	60.0 °C	[11]
5	Chromosome 7	F: GGGTGTTGACGTGATGTGAT R: GTCTTCTTTCCCCGCTGATT	60.0 °C	[11]
6	Chromosome 3	F: GCATGTGCCTGTAGTCCCA R: TTCCACAGCTACCTTAAGAGAGAGA	60.0 °C	[11]
7	Chromosome 2	F: AGCCGACTTATAGAACTGTGTGC R: ATGACGAGGCATTTGGCTAC	60.0 °C	[11]
8	Chromosome 1	F: CCGTGGTGGGTGATGACAT R: TAGGGACAGTGGGAATCTCG	60.0 °C	[11]
9	Chromosome 1 DAB1	F: GAGGCAGGCAGGCATCTTAT R: ACCATACGTGGCAGAGTGTG	60.0 °C	[11]
10	Chromosome Y	F: CAGGGGAATCCGACTGTTTA R: ATGACGAGGCATTTGGCTAC	60.0 °C	[11]
Oligos for PCR in ChIP assay				
11	I-*Ppo*I site at chromosome 1 *DAB1* gene	TGCTGCTTTTTCTTCTTCTCC CTTCTTTCCCACCAAGTCTTC	52.8 °C	[9]
12	GAPDH	TCGGTTCTTGCCTCTTGTC CTTCCATTCTGTCTTCCACTC	55.3 °C	[9]
Oligos for real-time PCR in ChIP assay flanking DAB1 site				
13	−20 kbp	ACTGAAATGGTTAGCGTGAGG GCTAGGCCCAAAAATTTACAA	60.0 °C	[11]
14	−10 kbp	CCAGTTCAGCAATATTTCCCTAAG AGCCCCTCAGTTGAATTCCT	60.0 °C	[11]

(continued)

Table 1
(continued)

Oligos for real-time PCR in cutting assay (across 10 putative I-*Ppo*I recognition sites)				
#	**Description**	**Sequence (in 5′–3′ direction)**	**Anneal temp.**	**References**
15	−8.9 kbp	GCAAGGGCTCATGAATGATA CAAGGTAGCAAGGTAGTGTAGCA	60.0 °C	[9]
16	−6.2 kbp	TGCCTAAATGCCTCTTTCTACTG AAGCTGGAAATTGGCCTAGA	60.0 °C	[9]
17	−3.6 kbp	TTCCACACTCATTGGAGGTAA TTGGCACACACAAACAGAATAA	60.0 °C	[9]
18	−0.2 kbp	TTCACAGCACTCTCCATTCC TCTTTCCCACCAAGTCTTCA	60.0 °C	[9]
19	0.4 kbp	TTCCCATTATCTGAAGAGCGT TGGATGGCTCTGATAGTTACAAA	60.0 °C	[9]
20	3.5 kbp	GGCTTGTAACCCACAACACA GATGCTGCTCATACCCAATG	60.0 °C	[9]
21	5.9 kbp	GATGGAGGACAATTATGATGTGA TGTGACTATCCTTGCCCAAA	60.0 °C	[9]
22	7.0 kbp	GCATTCTGGAGTTCCTTGCT TAATGCACCCACTCATGCTT	60.0 °C	[9]
23	10 kbp	ATTGATTAGTAAGTGGCAGAGTCG CCAGGAAAGAATGGGTCCTA	60.0 °C	[11]
24	20 kbp	CTGAGGTGAAAAGTCCAACGATAG ATGTTGCCCCCTCATCATTA	60.0 °C	[11]

3. Real-time PCR reaction parameters are: 50 °C, 2 min; 95 °C, 10 min; 95 °C, 15 s, 60 °C, 30 s,72 °C 30 s for 40 cycles; 72 °C, 10 min.
4. Primer sequences are provided in Table 1.

3.3 DNA Cleavage Assessment

Day 1:

Cells are plated in 10-cm dishes and incubated at 37 °C in a humidified atmosphere containing 5% CO_2 to about 80–90% confluence. Normally, this corresponds to around 1×10^6 to 2×10^6 cells per 10 cm plate, cell density needs to be validated and optimized for each specific cell line.

Day 2:

1. Preparation of diluted I-*Ppo*I protein: Dilute 0.5–1 μg of I-*Ppo*I into 200 μL of cold PBS and keep on ice all the time.
2. Preparation of diluted Chariot reagent or CPP: in a separate tube, dilute 50 μL of 2 μg/μL Chariot/CPP stock solution into 200 μL of sterile H_2O for one 10-cm dish.
3. Protein transfection is carried out in the same procedure as that used for ChIP assay (*see* Subheading 3.2, Day 2, **step 4**).
4. After protein transduction, wash each plate of cells twice with 10 mL of ice-cold PBS and harvest with 500 μL of ice-cold PBS containing protease inhibitor cocktails.
5. Centrifuge cell suspensions at 1000 × *g*, 4 °C for 5 min. Remove supernatant, resuspend cell pellet in 600 μL of ChIP sonication buffer containing protease inhibitor cocktails, and incubate on ice for 10 min.
6. Sonicate cell suspension using a Branson Sonifier 250 at output 8.0, 80% cycle duty for 90 s.
7. Centrifuge cell lysates at 13,000 rpm 4 °C for 15 min, collect supernatant into a fresh new tube.
8. Extract DNA from collected whole cell extract (200 μL) using QIAquick PCR purification kit according to manufacturer's instruction. Resuspend DNA with 80 μL of TE buffer from the kit.
9. For each real-time PCR reaction, pipette 5 μL of eluted DNA into wells on a 384-well plate and let it dry at room temperature overnight.

Day 3:

10. Ten primer sets are designed to target ten endogenous I-*Ppo*I cleavage sites in the human genome. A reference primer set is designed to target a sequence that does not contain an I-*Ppo*I cutting site. Primer sequences are listed in Table 1.
11. Prepare real-time PCR reaction solution: a 10 μL reaction containing 5 μL SYBR Green Master Mix (2×), 1 μL primer pairs, 0.4 μL DNA polymerase II, and 4.6 μL of ddH_2O.
12. Add 10 μL/well of real-time PCR reaction solution to a 384-well plate that contains pre-dried DNA samples.
13. Real-time PCR reaction parameters are: 50 °C, 2 min; 95 °C, 10 min; 95 °C, 15 s, 60 °C, 30 s, 72 °C 30 s for 40 cycles; 72 °C, 10 min.
14. Analyze PCR results by comparing the yield of PCR products amplified from each of the individual cleavage sites at different time points post I-*Ppo*I transfection with that from the reference sequence.

4 Notes

1. Zinc acetate has to be added in two times to reach a final concentration of 2 mM (1 mM each). Zinc is a critical cofactor for I-*Ppo*I and its addition is essential for obtaining adequate yields and activity.
2. It is better to store cell suspension in multiple 15 mL conical tubes, not combining them in large volume.
3. $MgCl_2$ has to be added to the cell suspension right before lysing the cells using French Press Cell Disruptor.
4. Incubation time after adding Ni-NTA beads should not be longer than 30 min.
5. For multiple transfections, do not use a master mix per each tube, as this may cause aggregation.

References

1. Rogakou EP, Pilch DR, Orr AH, Ivanova VS, Bonner WM (1998) DNA double-stranded breaks induce histone H2AX phosphorylation on serine 139. J Biol Chem 273(10):5858–5868
2. Bekker-Jensen S, Lukas C, Kitagawa R, Melander F, Kastan MB, Bartek J, Lukas J (2006) Spatial organization of the mammalian genome surveillance machinery in response to DNA strand breaks. J Cell Biol 173(2):195–206. doi:10.1083/jcb.200510130
3. Lukas C, Falck J, Bartkova J, Bartek J, Lukas J (2003) Distinct spatiotemporal dynamics of mammalian checkpoint regulators induced by DNA damage. Nat Cell Biol 5(3):255–260
4. Demuth I, Bradshaw PS, Lindner A, Anders M, Heinrich S, Kallenbach J, Schmelz K, Digweed M, Meyn MS, Concannon P (2008) Endogenous hSNM1B/Apollo interacts with TRF2 and stimulates ATM in response to ionizing radiation. DNA Repair (Amst) 7(8):1192–1201. doi:10.1016/j.dnarep.2008.03.020. S1568-7864(08)00130-4 [pii]
5. Rouet P, Smih F, Jasin M (1994) Introduction of double-strand breaks into the genome of mouse cells by expression of a rare-cutting endonuclease. Mol Cell Biol 14(12):8096–8106
6. Rouet P, Smih F, Jasin M (1994) Expression of a site-specific endonuclease stimulates homologous recombination in mammalian cells. Proc Natl Acad Sci USA 91(13):6064–6068
7. Rodrigue A, Lafrance M, Gauthier MC, McDonald D, Hendzel M, West SC, Jasin M, Masson JY (2006) Interplay between human DNA repair proteins at a unique double-strand break in vivo. EMBO J 25(1):222–231. doi:10.1038/sj.emboj.7600914
8. Monnat RJ Jr, Hackmann AF, Cantrell MA (1999) Generation of highly site-specific DNA double-strand breaks in human cells by the homing endonucleases I-PpoI and I-CreI. Biochem Biophys Res Commun 255(1):88–93. doi:10.1006/bbrc.1999.0152. S0006-291X(99)90152-3 [pii]
9. Berkovich E, Monnat RJ Jr, Kastan MB (2007) Roles of ATM and NBS1 in chromatin structure modulation and DNA double-strand break repair. Nat Cell Biol 9(6):683–690. doi:10.1038/ncb1599
10. Massip L, Caron P, Iacovoni JS, Trouche D, Legube G (2010) Deciphering the chromatin landscape induced around DNA double strand breaks. Cell Cycle 9(15):2963–2972. doi:10.4161/cc.9.15.12412, 10.4161/cc.9.15.12697
11. Wen J, Cerosaletti K, Schultz KJ, Wright JA, Concannon P (2013) NBN phosphorylation regulates the accumulation of MRN and ATM at sites of DNA double-strand breaks. Oncogene 32(37):4448–4456. doi:10.1038/onc.2012.443
12. Andegeko Y, Moyal L, Mittelman L, Tsarfaty I, Shiloh Y, Rotman G (2001) Nuclear retention of ATM at sites of DNA double strand breaks. J Biol Chem 276(41):38224–38230
13. Zhao S, Weng YC, Yuan SS, Lin YT, Hsu HC, Lin SC, Gerbino E, Song MH, Zdzienicka MZ, Gatti RA, Shay JW, Ziv Y, Shiloh Y, Lee EY (2000) Functional link between ataxia-telangiectasia and Nijmegen breakage syndrome gene products. Nature 405(6785):473–477
14. Gatei M, Young D, Cerosaletti KM, Desai-Mehta A, Spring K, Kozlov S, Lavin MF, Gatti RA, Concannon P, Khanna K (2000) ATM-

dependent phosphorylation of nibrin in response to radiation exposure. Nat Genet 25(1):115–119

15. Lim DS, Kim ST, Xu B, Maser RS, Lin J, Petrini JH, Kastan MB (2000) ATM phosphorylates p95/nbs1 in an S-phase checkpoint pathway. Nature 404(6778):613–617
16. Wu X, Ranganathan V, Weisman DS, Heine WF, Ciccone DN, O'Neill TB, Crick KE, Pierce KA, Lane WS, Rathbun G, Livingston DM, Weaver DT (2000) ATM phosphorylation of Nijmegen breakage syndrome protein is required in a DNA damage response. Nature 405(6785):477–482
17. Varon R, Vissinga C, Platzer M, Cerosaletti KM, Chrzanowska KH, Saar K, Beckmann G, Seemanova E, Cooper PR, Nowak NJ, Stumm M, Weemaes CMR, Gatti RA, Wilson RK, Digweed M, Rosenthal A, Sperling K, Concannon P, Reis A (1998) Nibrin, a novel DNA double-strand break repair protein, is mutated in Nijmegen breakage syndrome. Cell 93(3):467–476. doi:10.1016/S0092-8674(00)81174-5
18. Carney JP, Maser RS, Olivares H, Davis EM, Le Beau M, Yates JR 3rd, Hays L, Morgan WF, Petrini JH (1998) The hMre11/hRad50 protein complex and Nijmegen breakage syndrome: linkage of double-strand break repair to the cellular DNA damage response. Cell 93(3):477–486
19. Falck J, Coates J, Jackson SP (2005) Conserved modes of recruitment of ATM, ATR and DNA-PKcs to sites of DNA damage. Nature 434(7033):605–611
20. Cerosaletti K, Wright J, Concannon P (2006) Active role for nibrin in the kinetics of atm activation. Mol Cell Biol 26(5):1691–1699
21. Lavin MF, Kozlov S, Gatei M, Kijas AW (2015) ATM-dependent phosphorylation of all three members of the MRN complex: from sensor to adaptor. Biomolecules 5(4):2877–2902. doi:10.3390/biom5042877

Chapter 19

Studies of the DNA Damage Response by Using the Lac Operator/Repressor (LacO/LacR) Tethering System

Rossana Piccinno, Marta Cipinska, and Vassilis Roukos

Abstract

Maintaining the integrity of genetic information is essential for the survival of cells. Recent advances in cell biological and microscopy methodologies have complemented traditional genetic and biochemical approaches, and they now permit the observation of spatiotemporal aspects of damaged chromosomal loci. In one of these approaches, integrated *LacO/TetO operator* sequences can be used as binding sites to physically tether onto chromatin any protein of interest when genetically fused to the respective *repressors (LacR/TetR)*. This methodology has been the basis of several models to probe the spatial dynamics of DNA repair in the eukaryotic nucleus and to visualize genomic loci in yeast, fly, nematodes, and in mammalian cells. Further applications are the induction of localized DNA damage by immobilizing endonucleases at different genome sites in vivo, the assessment of the hierarchy of protein interactions within repair complexes, and the activation of the DNA damage response (DDR) by the physical tethering of DSB-repair factors on chromatin in the absence of damage. We outline here a protocol for the quantification of DDR activation upon the prolonged immobilization of single repair factors on chromatin or upon tethering of the endonuclease FokI. The outlined protocol requires basic cell culture and microscopy skills and allows the tethering of any protein of interest within 2–3 days.

Key words DNA damage response, ATM, Mre11, Lac operator, Lac repressor, Tethering, Chromatin, γH2AX, 53BP1, Immunofluorescence, Imaging

1 Introduction

The integrity of the genome is constantly challenged by DNA lesions that can occur as byproducts of normal cell metabolism, as a result of the presence of chemical or environmental damaging agents [1] or as a part of developmental programs occurring in lymphocytes and germ cells during the process of antibody diversification [2]. The cellular response to these lesions is collectively termed DNA damage response (DDR), and is an essential cellular process that protects the genome by recognizing, processing, and finally repairing potentially lethal double strand breaks (DSBs) that could trigger genome instability [3] or tumorigenic translocations

Sergei V. Kozlov (ed.), *ATM Kinase: Methods and Protocols*, Methods in Molecular Biology, vol. 1599,
DOI 10.1007/978-1-4939-6955-5_19, © Springer Science+Business Media LLC 2017

[4, 5]. The DDR involves hierarchically organized signaling cascades composed of DNA damage sensors, intermediate signaling factors, and downstream effectors promoting the resolution and repair of DNA damage. The response to DNA damage is initiated by the action of the MRE11-RAD50-NBS1 (MRN) [6] complex that senses DSBs and by the activation of three PI3K-like protein kinases, the ataxia telangiectasia mutated kinase (ATM), the DNA-dependent protein kinase (DNA-PK), and/or ATR (ATM and Rad3-related) [7]. The signal is then transduced via phosphorylation of the histone variant H2AX (γ-H2AX) by ATM, which creates a platform for the recruitment of the MDC1 mediator protein [8]. Importantly, the size of the γ-H2AX chromatin domain can extend up to one megabase from the site of the DSB, thus leading to the demarcation of the area on which DNA repair proteins can accumulate. As such, MDC1 acts as a hub to recruit several factors, including chromatin remodelers and the RNF8/RNF168 E3 ubiquitin ligases complexes that catalyze the ubiquitylation of histones and other DDR substrates [9]. Altogether, they modulate the recruitment of several downstream DDR factors, including 53BP1. Downstream DDR effectors include the kinases CHK1 and CHK2, which act as distributors of the signal from the damaged region to the nucleoplasm [10] and trigger cell-cycle arrest to permit repair or, in case of irreparable damage, initiate apoptosis or senescence [11].

The molecular characterization of the DDR components has been outlined over the last two decades by using mainly biochemistry, yeast, and mouse genetics and molecular methods. Importantly, these approaches were not able to fully assess the key question of how the accumulation of the DDR signaling and repair proteins is regulated in space and time. Macroscopically discernible repair foci can be formed in response to DSBs induced by γ irradiation [12], X rays [13], α-particles and heavy ions [14, 15], radiomimetic drugs [16], or laser-microirradiation [17, 18]. Using these as sources of DSBs and a battery of cell biological imaging methodologies, such as immunofluorescence, live-cell imaging of fluorescently labeled repair proteins, or fluorescence recovery after photobleaching (FRAP) and fluorescence correlation spectroscopy (FCS), recent studies have demonstrated that repair proteins have distinct recruitment kinetics and dynamics at the damaged site [3, 17, 19, 20]. Despite the fundamental findings obtained by these studies, how DNA repair takes place within the different chromatin domains and in the context of the highly compartmentalized nucleus is poorly understood. Novel methodologies utilizing endonucleases to induce DSBs [21–23] at specific genomic sites have only started shedding light on these questions [24–27].

Different endonuclease-based systems have been generated to induce DSBs at specific sites in the genome. These cell-based systems are taking advantage of the presence of endogenous *I-PpoI*

[21] or the *AsiSI* [22] restriction sites in the mammalian genome or of the engineered chromosomal integration of the yeast *ISceI* restriction site [28, 29], which is absent from the mammalian genome. The restriction sites are recognized after exogenous expression of the respective endonucleases and DSBs can be induced in a controlled manner. In order to investigate the spatial and temporal regulation of DNA repair at these sites, visualization of the different induced break sites is required. However, by using these approaches it was technically impossible to mark and track multiple distinct chromosome loci in living cells. Integration of the endonuclease *ISceI* sites flanked by the bacterial *Lac* or *Tet operator/repressor* system has overcome this limitation. The operator *LacO* or *TetO* array sequences serve as marks of the chromosome ends after in vivo cutting as they can be visualized as discrete dots owing to the binding of fluorescently tagged *LacR* and/or *TetR* repressor proteins to their respective arrays. This approach has been widely used to probe the spatial dynamics of specific chromosome loci in different organisms from yeast to mammalian cells [28–35]. This strategy is particularly useful for the analysis of the motion properties of specific intact and damaged chromosome loci [28, 29, 35–38] and has been recently used to develop a cellular system to induce and track different DSBs in different chromosomes to reveal the spatial dynamics of chromosome translocations in living cells [28]. The methodology is easily adaptable to applications that require visualization of a chromosomal loci in the mammalian genome [39–42] or the integration of any reporter of interest [43–45]. Furthermore, the high-affinity binding of the *LacR* and *TetR* to their cognate operators ($Kd^{Lac} \sim 10$ pM [46]) allows the specific targeting and tethering of proteins of interest (expressed as chimeric fusions with the *LacR/TetR* repressors) to operator sites that have been integrated in various genomic locations [47–50]. As an example, to mimic the formation of a repair foci in the absence of DNA damage, the key players of the DNA damage response expressed as *LacR* fusions were immobilized on *LacO* arrays in mouse cells. The local accumulation of the repair factors such as MRE11, ATM, NBS1, or MDC1 resulted in activation of the DDR in the absence of physical damage [49]. Together with a similar study in yeast [48], this work revealed the hierarchy in interactions of DDR proteins and suggested the presence of amplification signaling loops [48, 49]. Similarly, the forced targeting of different chromatin components on chromatin has been shown to induce changes in the higher-order chromatin structure. HP1 tethering to chromatin causes chromatin condensation [51] and is sufficient for the activation of upstream events of the DDR signaling pathway in the absence of damage [52]. In addition, physical tethering of various methyltransferases and lysine acetyltransferase at the *LacO* locus of osteosarcoma U2OS cells has been shown to deposit the respective histone modifications at the

locus [53], suggesting that targeting of chromatin modifiers on chromatin can be used as an elegant way to ectopically induce particular chromatin domains. Further applications of the tethering approach include the de novo creation of nuclear bodies [54, 55], or the testing of protein-protein interactions by a fluorescent "two-hybrid" approach, in vivo [56, 57]. A similar approach was used to target the FokI endonuclease to *LacO* repeats to induce multiple DSBs within the locus and reveal that DDR silences transcription *in cis* [45].

Here, we present a protocol for the physical tethering of DSB-repair factors, expressed as *LacR* fusions, on chromatin of mouse and human cells with prior integrations of *LacO* arrays [28, 52]. We then outline the immunofluorescence protocol to detect a possible activation of DDR, as it is marked by phosphorylation of H2AX (γH2AX) and the recruitment of downstream repair proteins, in the absence of exogenous DNA damage [49] (Figs. 1 and 2). This protocol is simple, can be completed within 2–3 days, and is readily adaptable to experiments that require the physical immobilization of any protein of interest on chromatin. As an alternative example, the tethering of the endonuclease FokI to *LacO* induces multiple DSBs within the locus as marked by γH2AX (Fig. 3). A protocol to generate cell lines carrying stable integrations of the *LacO/TetO* repeats and to induce chromosome damage and chromosome translocations by the expression of the endonuclease *ISceI* has been published elsewhere [58].

2 Materials

2.1 Cell Lines

Mouse NIH3T3duo cells [28, 58] engineered to contain a single integration of several *LacO* operator repeated sequences in the genome, or U2OS human osteosarcoma cells containing two integrations of the *LacOISceITetO* plasmid [29, 52, 58], were used to tether the different fusions. Both the cell lines were described elsewhere [28, 29, 58] and can be purchased by Kerafast (Boston, USA). A protocol for the derivation of similar cellular systems by the integration of *LacO* or *TetO* operator repeats in specific sites of the mammalian genome has been previously outlined [58].

2.2 Plasmids

The plasmids used in this study are: the control fusion of the Lac repressor tagged with the mCherry fluorescent protein (mCherryLacR) and the fusions of the DDR proteins $\text{mCherryLacRATM}_{1300\text{-}3060\text{aa}}$, mCherryLacRMRE11, mCherryLacRNBS1, and mCherryLacRMDC1. These plasmids have been previously described [49] and can be requested from the Misteli laboratory. The mCherryLacR-FokI plasmid has been also previously described [45].

2.3 Transfection and Cell Culture

1. 15 mL centrifuge tubes.
2. Centrifuge for 15 mL and 50 mL tubes.

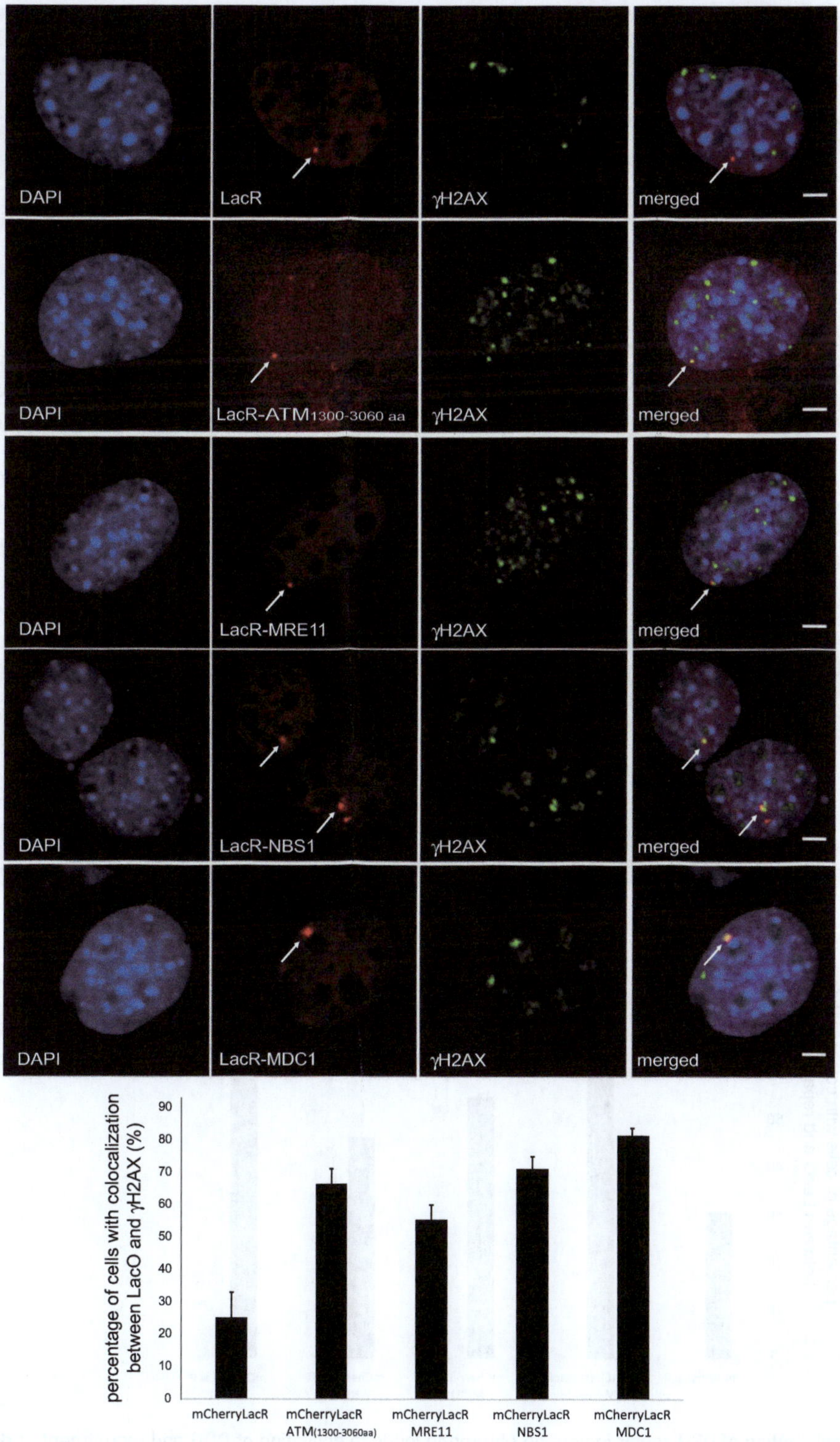

Fig. 1 Tethering of DSB-repair factors on chromatin elicits DDR activation in mouse cells. NIH3T3duo cells [28] were transfected with the indicated plasmids for 20 h. Cells were then fixed and stained by immunofluorescence using an antibody against γH2AX. *Arrows* mark the local accumulation of tethered *LacR* repressor fusions to the *Lac*O operator and the potential colocalization with γH2AX foci (*upper panel*, scale bar corresponds to 3 μm). Quantification of the percentage of cells with colocalization of the *LacO* locus and γH2AX foci. *Values* represent the mean ± SD (quantification on at least 250 cells per construct) from two independent experiments (*lower panel*)

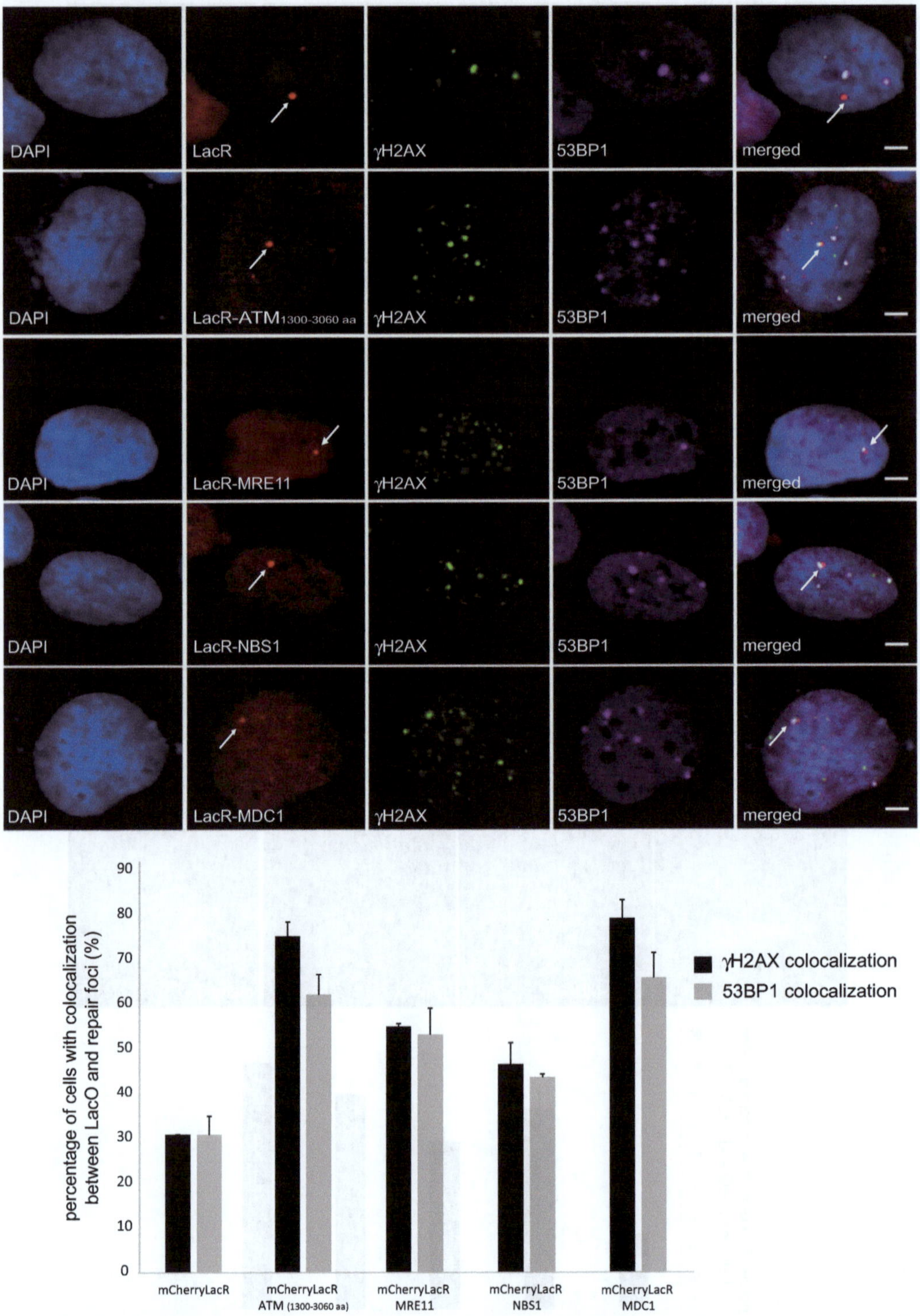

Fig. 2 Immobilization of DSB-repair factors on chromatin leads to activation of DDR and recruitment of downstream factors in human cells. Human U2OS cells with integrations of *LacO* arrays [52] were transfected with the indicated plasmids for 20 h, fixed and stained by immunofluorescence using antibodies against γH2AX and the downstream repair factor 53BP1. *Arrows* mark the local accumulation of tethered *LacR* repressor fusions to the *Lac*O operator and the potential colocalization with γH2AX and 53BP1 foci (*upper panel*, scale bar corresponds to 3 μm). Quantification of the percentage of cells with colocalization of the *LacO* locus with γH2AX/53BP1 foci is shown. *Values* represent the mean ± SD (at least 100 cells were measured per construct) from two independent experiments (*lower panel*)

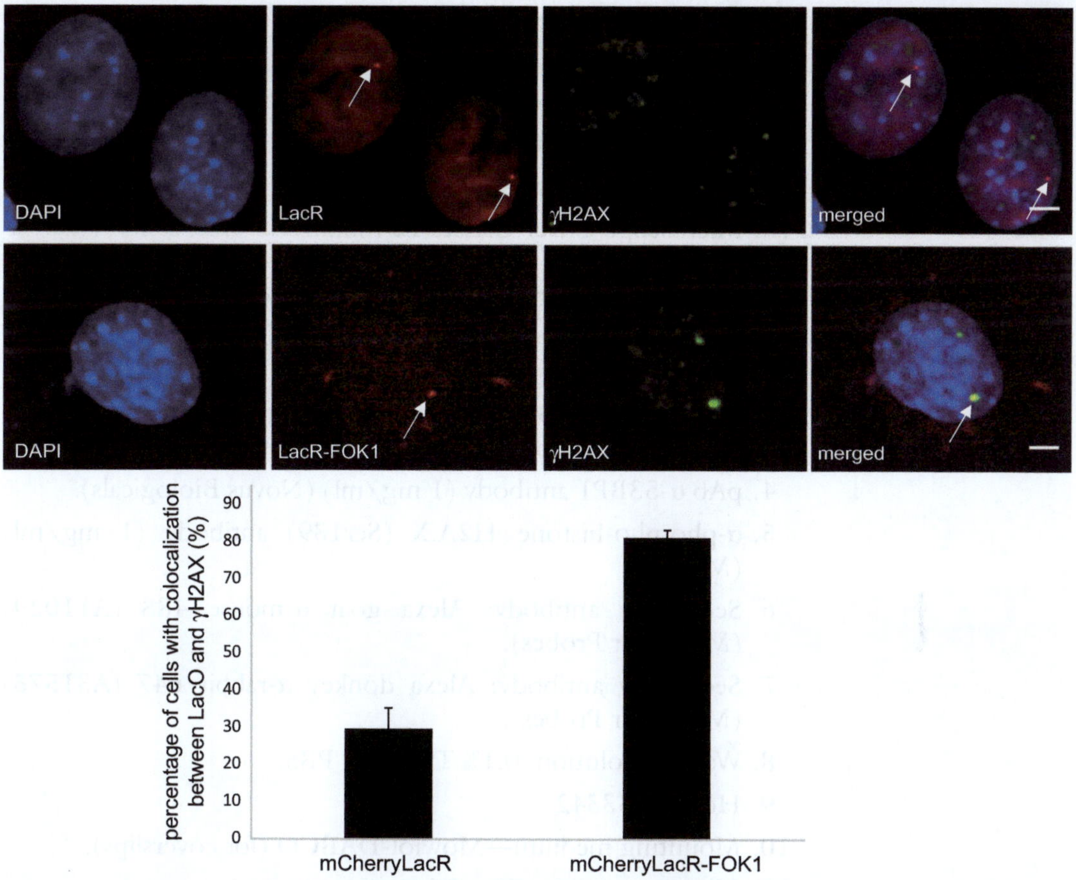

Fig. 3 Tethering of the endonuclease FokI on chromatin activates DDR. NIH3T3duo cells [28] were transfected with the indicated plasmids for 24 h. Cells were then fixed and stained by immunofluorescence using an antibody against γH2AX. *Arrows* mark the local accumulation of tethered *LacR* repressor fusions to the *Lac*O operator and the potential colocalization with γH2AX foci (*upper panel*, scale bar corresponds to 3 μm). Quantification of the percentage of cells with colocalization of the *LacO* locus with γH2AX is shown. *Values* represent the mean ± SD (quantification on at least 100 cells per construct) from two independent measurements

3. Tabletop centrifuge.
4. 5 ml pipettes.
5. 10 ml pipettes.
6. Growth media: DMEM medium (Life Technologies) supplemented with 10% fetal bovine serum (FBS), 100 U/mL penicillin, and 100 μg/mL streptomycin, L-Glutamine 2 mM.
7. Trypsin-EDTA 0.25%.
8. 1×PBS.
9. Counting slides TC20 (BioRad).
10. Automated Cell Counter TC20 (BioRad).
11. Microscope Leica DM IL LED.
12. Cell Line Nucleofector® Kit R (Lonza).

13. Cell Line Nucleofector® Kit V (Lonza).
14. Nucleofector 2b Device (Lonza).
15. CellCarrier-96 well microplate, with Lid (Perkin Elmer).
16. Tissue Culture Plate 24 well.
17. Tissue culture plates 15 cm.
18. Microscope Cover Glasses 12 mm.

2.4 Immuno-fluorescence and Imaging

1. Fixation solution: 4% Paraformaldehyde solution (PFA) in PBS.
2. Permeabilization solution, 0.3% Triton-X in PBS.
3. Blocking solution, 3% Bovine Serum Albumin (BSA) in PBS. Also used for diluting primary and secondary antibodies.
4. pAb α-53BP1 antibody (1 mg/ml) (Novus Biologicals).
5. α-phospho-histone H2A.X (Ser139) antibody (1 mg/ml) (Merck).
6. Secondary antibody: Alexa goat α-mouse 488 (A11029) (Molecular Probes).
7. Secondary antibody: Alexa donkey α-rabbit 647 (A31573) (Molecular Probes).
8. Washing solution: 0.1% Tween in PBS.
9. Hoechst 33342.
10. Mounting medium—Mowiol-DABCO (for coverslips).
11. Microscopy slides 76 × 26.
12. Orbital platform shaker (e.g., The Belly Dancer, IBI Scientific).
13. Any epifluorescence or confocal microscope (here: Opera Phenix, Perkin Elmer).

3 Methods

We provide the protocol for tethering of different single repair factors (ATM, MRE11, NBS1, MDC1) or the zinc finger endonuclease FokI, to chromatin of mouse NIH3T3 and human U2OS osteosarcoma cells. We assess the activation of the DDR by immunostaining using an antibody against the DSB marker γH2AX and the repair factor 53BP1. The cells used in this protocol have been engineered to contain integrations of the *LacO* operator at defined chromosome sites [28, 58] and their generation has been outlined elsewhere [58].

3.1 Transfection

1. Culture the cells in 10% FBS DMEM in 15 cm tissue culture plates. Pre-warm 1× sterile PBS, and 0.25% trypsin and culture media at 37 °C.

2. Aspirate the medium and wash cells with 1× PBS twice.
3. Add 0.25% trypsin and incubate for 3 min at 37 °C.
4. Detach the cells from the bottom of the dish by and neutralize trypsin by adding 5 ml of growth medium.
5. Place cells in a 15 ml tube and centrifuge at 500 × *g* for 2 min at room temperature (RT), discard the supernatant.
6. Resuspend cells in 6 ml of growth media and determine cell density by adding 20 μl on a counting slide. Insert in the automated cell counter.
7. Place 10^6 cells of human U2OS cells or 1.3×10^6 NIH3T3duo cells [28] in a 1.5 mL Eppendorf tube per transfection and centrifuge at 500 × *g* for 2 min at RT; discard the supernatant.
8. Resuspend cells in 100 μl of Cell Line Nucleofector® Solution (kit R for NIH3T3duo, kit V for U2OS cells) (*see* **Note 1**).
9. Mix 5 μg of the control plasmid mCherryLacR and 10 μg of any other plasmid listed in Subheading 2.2 with the cell suspension.
10. Incubate for 2 min.
11. Transfer cell/DNA suspension into a cuvette (avoid air bubbles).
12. Apply the Nucleofector® Program X-001 for U2OS or U-030 for NIH3T3duo. When the program is finished add 1 ml of culture medium to the cuvette.
13. Seed 2.5×10^4 cells in 96-well imaging plates or 1×10^5 in each well of a 24-well plate, each well containing a circular coverslip.
14. Culture cells for 16–24 h in a 37 °C/5% CO_2 incubator (*see* **Note 2**).

3.2 Immunofluorescence

1. In order to fix the cells, aspirate the medium and wash with 1× PBS twice (*see* **Note 3**).
2. Add 200 μl 4% PFA solution in each well of a 96-well plate or 0.5 ml of 4% PFA solution into each well of a 24-well plate to fix for 10 min at RT (*see* **Note 4**).
3. Wash cells with 1× PBS three times.
4. Permeabilize cells with 0.3% Triton-X in PBS for 5 min.
5. Wash the cells three times with 1× PBS using gentle shaking.
6. Block wells or coverslips with 3% BSA in PBS for 60 min.
7. Add the primary antibodies, anti-53BP1 and anti-phosphohistone H2A.X (Ser139), both diluted 1:2000 in blocking solution and incubate overnight at 4 °C. Create a wet chamber to avoid evaporation and use parafilm to seal the plate.
8. Next day, wash cells three times with 0.1% PBS-Tween for 5 min each under gentle shaking.

9. Add secondary antibodies, the Alexa goat α-mouse 488/Alexa donkey α-rabbit 647 both diluted 1:1000 in blocking solution and incubate for 60 min.
10. Wash cells with 0.1% Tween in PBS three times for 5 min under gentle shaking.
11. Incubate cells with Hoechst (diluted 1:2000 in 1× PBS) for 15 min at RT.
12. Wash cells with 1× PBS twice, under gentle shaking.
13. Add 6 μl of mounting medium (Mowiol and DABCO) on a microscopy slide, cover it with the glass coverslip having cells facing down, and dry for 1–2 h at RT before imaging. When transfected cells are seeded in 96-well plates, keep cells covered in 1×PBS until imaging.

3.3 Fluorescence Imaging and Measurements

To image the immunofluorescence samples an epifluorescence or confocal microscope equipped with high numerical aperture (NA) objectives is needed (*see* **Note 5**) The minimum requirements include the appropriate light source, excitation and emission filters to acquire images in four different channels including the visualization of nuclei stained with Hoechst, the tethered repair proteins tagged with mCherry and immunostaining of γH2AX and 53BP1 marked by secondary antibodies Alexa 488 and 647, respectively (Figs. 1 and 2). After acquisition, colocalization of the tethered repair factor with the foci formed by the local phosphorylation of γH2AX or the recruitment of 53BP1 is assessed and plotted (Figs. 1 and 2). Positive colocalization is defined as complete or partial overlapping of the respective foci. The images presented here have been acquired using the Phenix High-Content System (Perkin Elmer) using the 405, 488, 561, 647 nm laser lines for excitation in confocal mode using a 40× water objective lens (1.1 NA) without binning. Five planes separated by 0.5 μm were acquired in *z* and were maximum projected before analysis.

4 Notes

1. In this protocol, we use nucleofection to deliver constructs expressing the proteins of interest as fusions to *LacR* into cells. Nucleofection frequently results in high transfection efficiency, which facilitates both imaging and analysis, but can also lead to high background of damage as visualized by the presence of high γH2AX foci formation background (multiple foci outside the *LacO* area). These are not present in non-transfected cells and do not correspond to the intrinsic DNA damage of the utilized cell line. In case of high background due to nucleofection, alternative methods of transfections, such as cationic lipid-based methods or calcium phosphate transfections, may also be tested.

2. The choice for the assay endpoint after transfection is critical and should be lower than 24 h. Prolonged binding of the *LacR* fusions to the respective operators may result in interference with physiological processes such as replication and transcription [59]. This may result in increased background damage at the *LacO* site as judged by increased colocalization of γH2AX with the control plasmid mCherryLacR.
3. Avoid exposing the prepared slides or 96-well plates or the secondary antibodies to light.
4. Upon fixation, coverslips or 96-well plates can be stored in 1xPBS at 4 °C for up to 1 week.
5. Due to the reduced out-of-focus light, the use of a confocal microscope is preferred over a wide-field option, as this will allow a better assessment of colocalization and therefore increase the precision in measurements.

Acknowledgments

We thank members of the Roukos group for helpful comments and Gianluca Pegoraro for critical reading of the manuscript. This work is supported by the "DFG Major Research Instrumentation Programme" (INST 247/845-1 FUGG).

References

1. Jackson SP, Bartek J (2009) The DNA-damage response in human biology and disease. Nature 461(7267):1071–1078. doi:10.1038/nature08467
2. Alt FW, Zhang Y, Meng FL, Guo C, Schwer B (2013) Mechanisms of programmed DNA lesions and genomic instability in the immune system. Cell 152(3):417–429. doi:10.1016/j.cell.2013.01.007
3. Polo SE, Jackson SP (2011) Dynamics of DNA damage response proteins at DNA breaks: a focus on protein modifications. Genes Dev 25(5):409–433. doi:10.1101/gad.2021311
4. Roukos V, Burman B, Misteli T (2013) The cellular etiology of chromosome translocations. Curr Opin Cell Biol 25(3):357–364. doi:10.1016/j.ceb.2013.02.015
5. Roukos V, Misteli T (2014) The biogenesis of chromosome translocations. Nat Cell Biol 16(4):293–300. doi:10.1038/ncb2941
6. Stracker TH, Petrini JH (2011) The MRE11 complex: starting from the ends. Nat Rev Mol Cell Biol 12(2):90–103. doi:10.1038/nrm3047
7. Kastan MB, Lim DS (2000) The many substrates and functions of ATM. Nat Rev Mol Cell Biol 1(3):179–186. doi:10.1038/35043058
8. Coster G, Goldberg M (2010) The cellular response to DNA damage: a focus on MDC1 and its interacting proteins. Nucleus 1(2):166–178. doi:10.4161/nucl.1.2.11176
9. Reinhardt HC, Yaffe MB (2013) Phospho-Ser/Thr-binding domains: navigating the cell cycle and DNA damage response. Nat Rev Mol Cell Biol 14(9):563–580. doi:10.1038/nrm3640
10. Lukas C, Falck J, Bartkova J, Bartek J, Lukas J (2003) Distinct spatiotemporal dynamics of mammalian checkpoint regulators induced by DNA damage. Nat Cell Biol 5(3):255–260. doi:10.1038/ncb945
11. Ciccia A, Elledge SJ (2010) The DNA damage response: making it safe to play with knives. Mol Cell 40(2):179–204. doi:10.1016/j.molcel.2010.09.019
12. Carney JP, Maser RS, Olivares H, Davis EM, Le Beau M, Yates JR 3rd, Hays L, Morgan WF, Petrini JH (1998) The hMre11/hRad50 protein complex and Nijmegen breakage syndrome: linkage of double-strand break repair to the cellular DNA damage response. Cell 93(3):477–486
13. Nelms BE, Maser RS, MacKay JF, Lagally MG, Petrini JH (1998) In situ visualization of DNA

double-strand break repair in human fibroblasts. Science 280(5363):590–592
14. Jakob B, Scholz M, Taucher-Scholz G (2003) Biological imaging of heavy charged-particle tracks. Radiat Res 159(5):676–684
15. Stap J, Krawczyk PM, Van Oven CH, Barendsen GW, Essers J, Kanaar R, Aten JA (2008) Induction of linear tracks of DNA double-strand breaks by alpha-particle irradiation of cells. Nat Methods 5(3):261–266. doi:10.1038/nmeth.f.206
16. Povirk LF (1996) DNA damage and mutagenesis by radiomimetic DNA-cleaving agents: bleomycin, neocarzinostatin and other enediynes. Mutat Res 355(1–2):71–89
17. Bekker-Jensen S, Lukas C, Kitagawa R, Melander F, Kastan MB, Bartek J, Lukas J (2006) Spatial organization of the mammalian genome surveillance machinery in response to DNA strand breaks. J Cell Biol 173(2):195–206. doi:10.1083/jcb.200510130
18. Kong X, Mohanty SK, Stephens J, Heale JT, Gomez-Godinez V, Shi LZ, Kim JS, Yokomori K, Berns MW (2009) Comparative analysis of different laser systems to study cellular responses to DNA damage in mammalian cells. Nucleic Acids Res 37(9):e68. doi:10.1093/nar/gkp221
19. Essers J, Houtsmuller AB, van Veelen L, Paulusma C, Nigg AL, Pastink A, Vermeulen W, Hoeijmakers JH, Kanaar R (2002) Nuclear dynamics of RAD52 group homologous recombination proteins in response to DNA damage. EMBO J 21(8):2030–2037. doi:10.1093/emboj/21.8.2030
20. Essers J, Vermeulen W, Houtsmuller AB (2006) DNA damage repair: anytime, anywhere? Curr Opin Cell Biol 18(3):240–246. doi:10.1016/j.ceb.2006.03.004
21. Berkovich E, Monnat RJ Jr, Kastan MB (2008) Assessment of protein dynamics and DNA repair following generation of DNA double-strand breaks at defined genomic sites. Nat Protoc 3(5):915–922. doi:10.1038/nprot.2008.54
22. Iacovoni JS, Caron P, Lassadi I, Nicolas E, Massip L, Trouche D, Legube G (2010) High-resolution profiling of gammaH2AX around DNA double strand breaks in the mammalian genome. EMBO J 29(8):1446–1457. doi:10.1038/emboj.2010.38
23. Rouet P, Smih F, Jasin M (1994) Introduction of double-strand breaks into the genome of mouse cells by expression of a rare-cutting endonuclease. Mol Cell Biol 14(12):8096–8106
24. Aymard F, Bugler B, Schmidt CK, Guillou E, Caron P, Briois S, Iacovoni JS, Daburon V, Miller KM, Jackson SP, Legube G (2014) Transcriptionally active chromatin recruits homologous recombination at DNA double-strand breaks. Nat Struct Mol Biol 21(4):366–374. doi:10.1038/nsmb.2796
25. Lemaitre C, Grabarz A, Tsouroula K, Andronov L, Furst A, Pankotai T, Heyer V, Rogier M, Attwood KM, Kessler P, Dellaire G, Klaholz B, Reina-San-Martin B, Soutoglou E (2014) Nuclear position dictates DNA repair pathway choice. Genes Dev 28(22):2450–2463. doi:10.1101/gad.248369.114
26. Clouaire T, Legube G (2015) DNA double strand break repair pathway choice: a chromatin based decision? Nucleus 6(2):107–113. doi:10.1080/19491034.2015.1010946
27. van Sluis M, McStay B (2015) A localized nucleolar DNA damage response facilitates recruitment of the homology-directed repair machinery independent of cell cycle stage. Genes Dev 29(11):1151–1163. doi:10.1101/gad.260703.115
28. Roukos V, Voss TC, Schmidt CK, Lee S, Wangsa D, Misteli T (2013) Spatial dynamics of chromosome translocations in living cells. Science 341(6146):660–664. doi:10.1126/science.1237150
29. Soutoglou E, Dorn JF, Sengupta K, Jasin M, Nussenzweig A, Ried T, Danuser G, Misteli T (2007) Positional stability of single double-strand breaks in mammalian cells. Nat Cell Biol 9(6):675–682. doi:10.1038/ncb1591
30. Belmont AS, Li G, Sudlow G, Robinett C (1999) Visualization of large-scale chromatin structure and dynamics using the lac operator/lac repressor reporter system. Methods Cell Biol 58:203–222
31. Gonzalez-Serricchio AS, Sternberg PW (2006) Visualization of *C. elegans* transgenic arrays by GFP. BMC Genet 7:36. doi:10.1186/1471-2156-7-36
32. Lassadi I, Bystricky K (2011) Tracking of single and multiple genomic loci in living yeast cells. Methods Mol Biol 745:499–522. doi:10.1007/978-1-61779-129-1_29
33. Lisby M, Mortensen UH, Rothstein R (2003) Colocalization of multiple DNA double-strand breaks at a single Rad52 repair centre. Nat Cell Biol 5(6):572–577. doi:10.1038/ncb997
34. Meister P, Gehlen LR, Varela E, Kalck V, Gasser SM (2010) Visualizing yeast chromosomes and nuclear architecture. Methods Enzymol 470:535–567. doi:10.1016/S0076-6879(10)70021-5
35. Masui O, Bonnet I, Le Baccon P, Brito I, Pollex T, Murphy N, Hupe P, Barillot E, Belmont AS, Heard E (2011) Live-cell chromosome dynamics and outcome of X chromosome pairing events during ES cell differentiation. Cell 145(3):447–458. doi:10.1016/j.cell.2011.03.032
36. Dion V, Gasser SM (2013) Chromatin movement in the maintenance of genome stability.

Cell 152(6):1355–1364. doi:10.1016/j.cell.2013.02.010
37. Dion V, Kalck V, Horigome C, Towbin BD, Gasser SM (2012) Increased mobility of double-strand breaks requires Mec1, Rad9 and the homologous recombination machinery. Nat Cell Biol 14(5):502–509. doi:10.1038/ncb2465
38. Mine-Hattab J, Rothstein R (2012) Increased chromosome mobility facilitates homology search during recombination. Nat Cell Biol 14(5):510–517. doi:10.1038/ncb2472
39. Chubb JR, Boyle S, Perry P, Bickmore WA (2002) Chromatin motion is constrained by association with nuclear compartments in human cells. Curr Biol 12(6):439–445
40. Thomson I, Gilchrist S, Bickmore WA, Chubb JR (2004) The radial positioning of chromatin is not inherited through mitosis but is established de novo in early G1. Curr Biol 14(2):166–172
41. Vazquez J, Belmont AS, Sedat JW (2001) Multiple regimes of constrained chromosome motion are regulated in the interphase Drosophila nucleus. Curr Biol 11(16):1227–1239
42. Reddy KL, Zullo JM, Bertolino E, Singh H (2008) Transcriptional repression mediated by repositioning of genes to the nuclear lamina. Nature 452(7184):243–247. doi:10.1038/nature06727
43. Chuang CH, Carpenter AE, Fuchsova B, Johnson T, de Lanerolle P, Belmont AS (2006) Long-range directional movement of an interphase chromosome site. Curr Biol 16(8):825–831. doi:10.1016/j.cub.2006.03.059
44. Janicki SM, Tsukamoto T, Salghetti SE, Tansey WP, Sachidanandam R, Prasanth KV, Ried T, Shav-Tal Y, Bertrand E, Singer RH, Spector DL (2004) From silencing to gene expression: real-time analysis in single cells. Cell 116(5):683–698
45. Shanbhag NM, Rafalska-Metcalf IU, Balane-Bolivar C, Janicki SM, Greenberg RA (2010) ATM-dependent chromatin changes silence transcription in cis to DNA double-strand breaks. Cell 141(6):970–981. doi:10.1016/j.cell.2010.04.038
46. Sadler JR, Sasmor H, Betz JL (1983) A perfectly symmetric lac operator binds the lac repressor very tightly. Proc Natl Acad Sci U S A 80(22):6785–6789
47. Bergmann JH, Rodriguez MG, Martins NM, Kimura H, Kelly DA, Masumoto H, Larionov V, Jansen LE, Earnshaw WC (2011) Epigenetic engineering shows H3K4me2 is required for HJURP targeting and CENP-A assembly on a synthetic human kinetochore. EMBO J 30(2):328–340. doi:10.1038/emboj.2010.329
48. Bonilla CY, Melo JA, Toczyski DP (2008) Colocalization of sensors is sufficient to activate the DNA damage checkpoint in the absence of damage. Mol Cell 30(3):267–276. doi:10.1016/j.molcel.2008.03.023
49. Soutoglou E, Misteli T (2008) Activation of the cellular DNA damage response in the absence of DNA lesions. Science 320(5882):1507–1510. doi:10.1126/science.1159051
50. Luijsterburg MS, Acs K, Ackermann L, Wiegant WW, Bekker-Jensen S, Larsen DH, Khanna KK, van Attikum H, Mailand N, Dantuma NP (2012) A new non-catalytic role for ubiquitin ligase RNF8 in unfolding higher-order chromatin structure. EMBO J 31(11):2511–2527. doi:10.1038/emboj.2012.104
51. Verschure PJ, van der Kraan I, de Leeuw W, van der Vlag J, Carpenter AE, Belmont AS, van Driel R (2005) In vivo HP1 targeting causes large-scale chromatin condensation and enhanced histone lysine methylation. Mol Cell Biol 25(11):4552–4564. doi:10.1128/MCB.25.11.4552-4564.2005
52. Burgess RC, Burman B, Kruhlak MJ, Misteli T (2014) Activation of DNA damage response signaling by condensed chromatin. Cell Rep 9(5):1703–1717. doi:10.1016/j.celrep.2014.10.060
53. Burman B, Zhang ZZ, Pegoraro G, Lieb JD, Misteli T (2015) Histone modifications predispose genome regions to breakage and translocation. Genes Dev 29(13):1393–1402. doi:10.1101/gad.262170.115
54. Kaiser TE, Intine RV, Dundr M (2008) De novo formation of a subnuclear body. Science 322(5908):1713–1717. doi:10.1126/science.1165216
55. Shevtsov SP, Dundr M (2011) Nucleation of nuclear bodies by RNA. Nat Cell Biol 13(2):167–173. doi:10.1038/ncb2157
56. Zolghadr K, Mortusewicz O, Rothbauer U, Kleinhans R, Goehler H, Wanker EE, Cardoso MC, Leonhardt H (2008) A fluorescent two-hybrid assay for direct visualization of protein interactions in living cells. Mol Cell Proteomics 7(11):2279–2287. doi:10.1074/mcp.M700548-MCP200
57. Zolghadr K, Rothbauer U, Leonhardt H (2012) The fluorescent two-hybrid (F2H) assay for direct analysis of protein-protein interactions in living cells. Methods Mol Biol 812:275–282. doi:10.1007/978-1-61779-455-1_16
58. Roukos V, Burgess RC, Misteli T (2014) Generation of cell-based systems to visualize chromosome damage and translocations in living cells. Nat Protoc 9(10):2476–2492. doi:10.1038/nprot.2014.167
59. Jacome A, Fernandez-Capetillo O (2011) Lac operator repeats generate a traceable fragile site in mammalian cells. EMBO Rep 12(10):1032–1038. doi:10.1038/embor.2011.158

Chapter 20

Imaging of Fluorescently Tagged ATM Kinase at the Sites of DNA Double Strand Breaks

Anthony J. Davis, Shih-Ya Wang, David J. Chen, and Benjamin P.C. Chen

Abstract

The combination of live cell imaging and laser micro-irradiation is an important technique to investigate the recruitment and kinetics of DNA repair molecules to DNA damage sites. In this chapter, we describe the detailed methods to study the dynamics of fluorescently tagged ATM protein kinase at laser-induced DNA double strand break (DSB) sites. The same protocol can be applied to analyze the recruitment and kinetics of other potential or known DNA repair proteins to DSB sites.

Key words ATM kinase, Laser micro-irradiation, DNA double strand breaks, Green fluorescent protein

1 Introduction

A cell's ability to quickly respond to and repair DNA damage is paramount, as destabilization of a cell's genome can result in senescence, apoptosis, or carcinogenesis. The importance of DNA repair pathways is supported by the observation that a host of genetic disorders are found in humans with germline mutations that encode for proteins responsible for the cellular response to DNA damage. One such disorder is ataxia-telangiectasia (A-T), which is a rare, autosomal-recessive disorder, characterized by progressive cellular ataxia, neuro-degeneration, cell cycle checkpoint defects, decreased DNA double strand break (DSB) repair capacity, immune deficiencies, genomic instability, and increased risk of carcinogenesis. The gene responsible for the A-T phenotype encodes a serine/threonine protein kinase termed ataxia-telangiectasia mutated (ATM). ATM's primary function is that it plays a central role in the cellular response to DSBs via its ability to phosphorylate an extensive range of substrates. In this regard, ATM acts like a transducer to relay a strong signal to numerous downstream effectors, which include proteins responsible for modulating the cell cycle, DNA

Sergei V. Kozlov (ed.), *ATM Kinase: Methods and Protocols*, Methods in Molecular Biology, vol. 1599,
DOI 10.1007/978-1-4939-6955-5_20, © Springer Science+Business Media LLC 2017

repair, modulating signaling pathways, and apoptosis. Following DSB induction, ATM undergoes catalytic activation, is recruited to DNA damage sites, and is autophosphorylated.

A significant number of studies have focused on analyzing the key players and posttranslational modifications that modulate ATM's recruitment to DSBs and its dynamics/kinetics at DSBs. Traditionally, localization of ATM to DSBs was analyzed via immunofluorescence techniques that visualize nuclear foci formation of ATM at or near DNA damage sites. A major limitation of this technique is that to observe DSB-induced ATM foci, cells must be fixed and thus determining the real-time dynamics/kinetics of ATM at DNA damage sites is unattainable. To this our end, our group and others utilize live cell imaging of fluorescently tagged ATM to study ATM's DNA damage responses in real time in living cells. Coupling irradiation via a micro-laser with fluorescent microscopy allows for continuous monitoring of real-time recruitment, retention, and dissociation of fluorescent-tagged ATM to DSBs in vivo [1–4]. Further, we have employed this technique to study the recruitment and kinetics of a significant number of other DNA repair proteins to laser-generated DSBs. In this chapter, we will provide the reagents and experimental procedures required to analyze the real-time dynamics of fluorescently tagged ATM at the sites of DSBs and its recovery after photobleaching.

2 Materials

2.1 Plasmids

Full-length wild-type ATM cDNA was obtained from Dr. Michael B. Kastan [5]. The ATM cDNA was subcloned into a modified pcDNA3 vector that carries a FLAG and YFP tag upstream of the multiple clone site to generate a FLAG-YFP-tagged ATM (YFP-ATM) [3]. The QuikChange Site-Directed Mutagenesis kit (Stratagene) was applied to introduce an alanine point substitution at serine 1981 to generate the FLAG-YFP-tagged ATM S1981A mutant expression vector (YFP-ATM S1981A) [3] (*see* **Note 1**).

2.2 Cell Culture and Transfection

1. ATM-deficient cells such as A-T patient-derived and SV40-transformed AT5BIVA (AT5) fibroblasts [6] to express YFP-FLAG-tagged ATM constructs.
2. Minimum Essential Medium Eagle Alpha Modification supplemented with 10% fetal bovine serum.
3. CO_2-independent medium supplemented with 10% fetal bovine serum.
4. Transfection reagents such as Lipofectamine 2000 reagent (Invitrogen) or electroporation with Nucleofector (Lonza, *see* **Note 2**).

5. 35-mm glass-bottom culture dishes (MatTek).
6. Digital temperature control system with a heating insert P (PeCon GmbH) and the Tempcontrol 37-2 control unit (PeCon GmbH).

2.3 Microscopes and Imaging

1. An inverted Axiovert 200M microscope equipped with a Plan-Apochromat 63×/NA 1.40 oil immersion objective and an AxioCam HRm digital camera (Carl Zeiss MicroImaging, Inc.) for time-lapse imaging.
2. A confocal LSM 510 Meta microscope (Carl Zeiss MicroImaging, Inc.) for FRAP analysis. Computer with Carl Zeiss AxioVision 4.8.2. software.

2.4 Laser Micro-Irradiation Unit

Spectra-physics nitrogen laser (Andor technology, VSL-337ND-S): The Spectra-Physics VSL-337ND-S nitrogen laser emits 4 ns pulses in the UV at 337 nm. The pulse repetition rate may be varied from 1 to 60 Hz with a pulse energy of up to 300 μJ. The nitrogen laser is directly coupled to the epifluorescence path of the microscope through a Micropoint Laser Illumination and Ablation System (Photonic Instruments, Inc.) and focused through a Plan-Apochromat 63× oil immersion objective. The nitrogen wavelength can be changed from 337 nm to 365 nm by BPBD 365 dye solution. Power output is controlled by the Micropoint System ranged from 1% to 89%.

3 Methods

3.1 YFP-ATM Accumulation at Laser Micro-Irradiation Damage Site

3.1.1 Transient Expression of ATM

1. For transient expression, transfect the ATM constructs into AT5 cells via Lipofectamine 2000 or Nucleofector following the manufacturer's protocol (*see* **Note 3**).
2. 24 hours after the transfection, split cells and plate 1 × 10^5 cells onto a 35-mm glass-bottom culture dish and allow to attach and grow for 24 h (*see* **Notes 4** and **5**).

3.1.2 Microscope and Laser-Irradiation Setup

1. Tempcontrol—switch on at least 15 min prior to use for live cell imaging (37 °C)
2. Turn on MCU28 motor stage controller
3. Turn on laser via the key switch. Set the repetition rate to 12 or maximum position depending on experiment.
4. Switch on MicroPoint Laser Illumination and Ablation System and set number to 75 for DSB induction.
5. Turn on microscope and the EXFO:UV lamp.
6. Select Objective on 63× (oil immersion)
7. Use 70/30 beam splitter for laser damage along with GFP filter on the UV adaptor.

8. Filter set selection: select designed GFP emission filter set for laser damage with GFP excitation filter on the UV adaptor.
9. Replace normal growth medium with heated CO_2-independent medium for micro-irradiation experiment.
10. Apply one drop of immersion oil (Carl Zeiss, 0000001111807) on the objective and lower objective with "focus down" bottom before placing the sample on it.
11. Attach the 35-mm dish onto the heated dish holder to stabilize the dish and then place this onto the heating insert.
12. Focus a cell and put "Cross mark" (only on the right side-eye) in the center of the nucleus.
13. Set up imaging parameter utilizing AxioVision software. The parameters utilized for multidimensional acquisition are:
    (a) "Dye"—YFP
    (b) "Exposure"—"fixed"—"400 ms"
    (c) "Hardware setting"
        - During acquisition—"workgroup: GFP"
        - After acquisition—"workgroup: shutter closed"

    (d) For initial accumulation (up to 10 min) utilize "Time lapse setting"
        - Interval—20 s
        - Cycles—32

    (e) For 2 h kinetics, utilize hand held timer and take images at specific time points (*see* **Notes 6** and 7).

3.1.3 DSB Repair Kinetics with Laser Micro-Irradiation

1. Introduce DSBs in the nuclei of cultured cells by micro-irradiation with a pulsed nitrogen laser (365 nm, 10 Hz pulse). Set the output of the laser power at 75% of the maximum.
2. Initial accumulation (10 min) of ATM to laser-generated DSBs. For micro-irradiation and time-lapse imaging of initial accumulation of proteins to DSBs, maintain cells in CO_2-independent medium at 37 °C. Start image acquisition before laser micro-irradiation to obtain an image of the unirradiated cell. After first image acquisition, induce DSBs in cell nuclei by micro-irradiation with pulsed nitrogen laser allowing four laser pulses to hit a defined region of a single nucleus. As defined in Subheading 3.1.2, **step 13**, the defined parameters in AxioVision software will acquire images before and after irradiation via a 400-ms exposure time. Capture images every 20 s for the duration of the 10 min time course. Convert signal intensities of accumulated YFP at the micro-irradiated site into a numerical value by the use of the Carl Zeiss Axiovision software version 4.8.2 (see below).

3. Kinetics of ATM at laser-induced DSBs (2 h). Apply the same procedure as described above for 2-h micro-irradiation and time-lapse imaging except capture images at defined time points, typically at 3, 10, 20, 30, 60, 90, and 120 min post micro-irradiation. Set a fixed exposure time to capture all images before and after irradiation.

3.1.4 Calculation of Relative Fluorescent Intensity for Protein Recruitment Kinetics

1. Open AxioVision software and appropriate image set of the same cell.
2. Circle damage site (D, white cycles in Fig. 1) and a background area (B, yellow cycles in Fig. 1) at each time point (*see* **Note 8**).
3. Obtain densitometric mean (fluorescence of the spot) via the software, which is based on the fluorescence intensity and area size (*see* **Note 9**).
4. To eliminate the background fluorescence, laser-induced fluorescence intensity accumulated at the damaged spot D_t is subtracted with the background fluorescence intensity B_t at an undamaged spot in the same nuclei at each time point (Fig. 1). The absolute fluorescence intensity $I_t = D_t - B_t$.
5. Compensate nonspecific photobleaching and UV lamp output fluctuation by normalizing the absolute fluorescence intensity

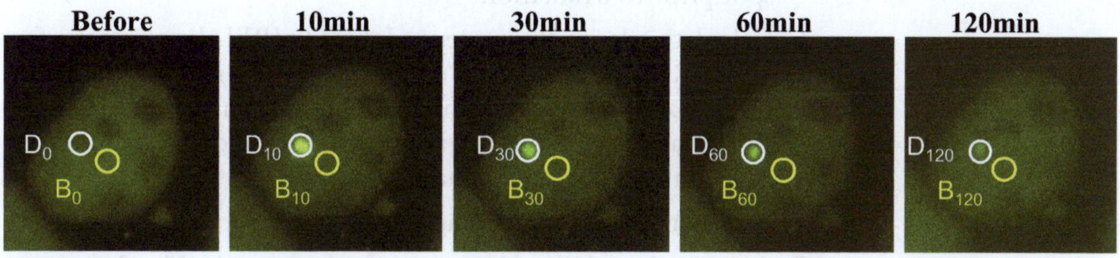

$I_0=D_0\text{-}B_0$ $I_{10}=D_{10}\text{-}B_{10}$ $I_{30}=D_{30}\text{-}B_{30}$ $I_{60}=D_{60}\text{-}B_{60}$ $I_{120}=D_{120}\text{-}B_{120}$

	Normalized intensity (NI_t)	**Relative fluorescence intensity (RFI)**
	$\mathbf{NI_t = (I_t/B_t)*D_0}$	$\mathbf{RFI_t = (NI_t-NI_0)/(NI_{max}-NI_0)}$
Before	$NI_0 = (I_0/B_0)*D_0$	$RFI_0 = (NI_0-NI_0)/(NI_{10}-NI_0) = 0$
10 min	$NI_{10} = (I_{10}/B_{10})*D_0$	$RFI_{10} = (NI_{10}-NI_0)/(NI_{10}-NI_0) = 1$
30 min	$NI_{30} = (I_{30}/B_{30})*D_0$	$RFI_{30} = (NI_{30}-NI_0)/(NI_{10}-NI_0) <1$
60 min	$NI_{60} = (I_{60}/B_{60})*D_0$	$RFI_{60} = (NI_{60}-NI_0)/(NI_{10}-NI_0) <1$
120 min	$NI_{120} = (I_{120}/B_{120})*D_0$	$RFI_{120} = (NI_{120}-NI_0)/(NI_{10}-NI_0) <1$

Fig. 1 Calculation of relative fluorescent intensity for recruitment kinetics. *Dt* fluorescence intensity accumulated at the damaged spot at a given time; *Bt* background fluorescence intensity at an undamaged spot a given time; *It* absolute fluorescence intensity of the damaged spot at a given time; *NIt* normalized fluorescence intensity of the damaged spot at a given time; *RFIt* relative fluorescence intensity of the damaged spot at a given time compared to that of maximum normalized fluorescence intensity

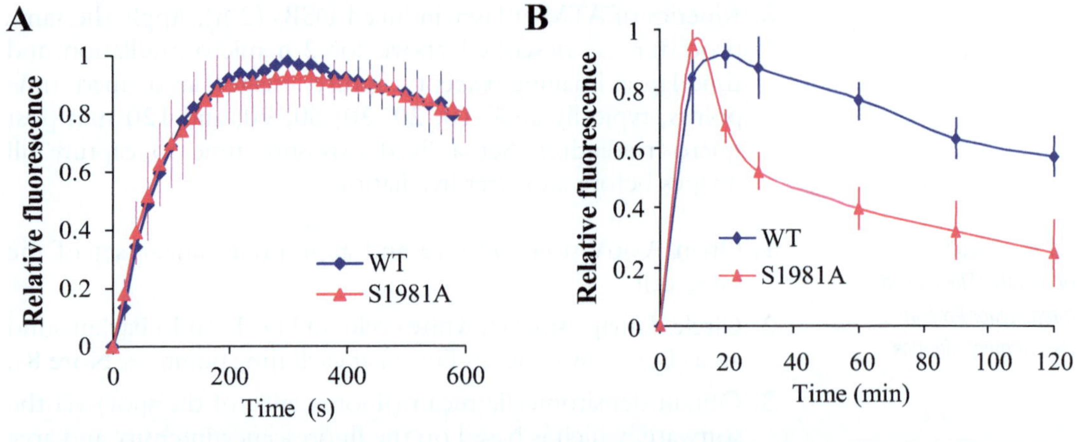

Fig. 2 Ablation of autophosphorylation at serine 1981 of ATM accelerates dissociation of ATM from DSBs. Initial accumulation kinetics (**a**) and 2-h time course (**b**) of YFP-ATM wt or YFP-ATM S1981A at laser-generated DSBs. *Error bars* represent the SD. © So et al. [3]. Originally published in JOURNAL CELL BIOLOGY. http://jcb.rupress.org/content/187/7/977

$I_{(t)}$ accumulation at the damaged spot of each time point based on background intensity prior to laser damage using the formula: Normalized fluorescence intensity $NI_t = (I_t/B_t) \times D_0$, where B_t represents the undamaged site background intensity of each time point, D_0 represents the intensity of the damaged spot prior to irradiation.

6. Calculate relative fluorescence intensity (RFI) by using the formula: $RFI_t = (NI_t - NI_0)/(NI_{max} - NI_0)$, where NI_0 means normalized fluorescence intensity of the damaged spot prior to laser irradiation and NI_{max} is the maximum normalized fluorescence intensity of the damaged spot.
7. Each data point is the average of 10 independent measurements.
8. Average the relative intensity to make a graph (Fig. 2).

3.1.5 Enhancement of Laser Micro-Irradiation with DNA Photosensitizer

1. DNA photosensitizer, such as Hoechst 33258 (Sigma-Aldrich B2883) and bromodeoxyuridine (BrdU, Sigma-Aldrich B5002), can also be used to assist in increasing signal (Fig. 3).
2. For sensitization with Hoechst, change to CO_2-independent medium and then add 1 μl of 10 mg/ml Hoechst 33258 in 2.5 ml medium and incubate the cells for minutes before initiating the experiment.
3. For BrdU sensitization, treat cells in regular medium with 10 μM BrdU for 16–24 h and then change to CO_2-independent medium, followed by laser initiation.

3.2 Fluorescence Recovery After Photobleaching (FRAP)

1. Perform FRAP measurements by photobleaching the entire ATM accumulation spot that formed after laser-induced DSBs and by measuring the subsequent recovery of fluorescence signal at the photobleached region.

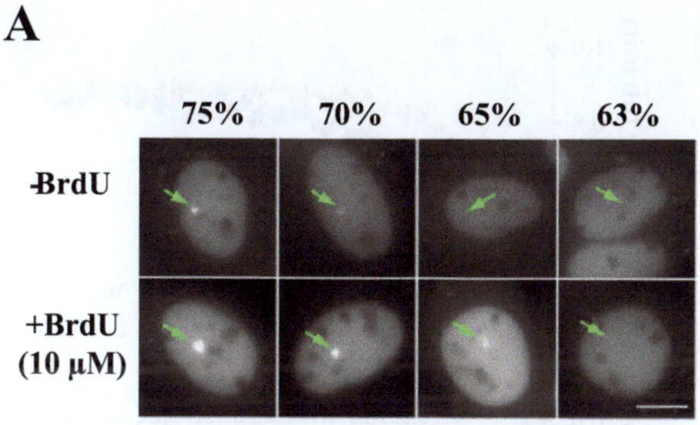

B

Laser output (%)	YFP-ATM wt recruitment to laser irradiated regions	
	-BrdU	+BrdU
63	-	+/-
65	-	+
70	+/-	++
75	+	+++

Fig. 3 Local accumulation of ATM in cells exposed to laser microirradiation in the presence or absence BrdU. (**a**) AT5BIVA cells stably expressing YFP-ATM wt were treated or untreated with BrdU (10 μM) for 16 h, and micro-irradiated with indicated laser output. Images were taken at 10 min after micro-irradiation. Bar, 10 mm. (**b**) Summary of YFP-ATM wt recruitment to irradiated region in AT5BIVA cells under different conditions. © So et al. [3]. Originally published in JOURNAL CELL BIOLOGY. http://jcb.rupress.org/content/187/7/977

2. Preform imaging and FRAP measurements on an LSM 510 Meta confocal microscope using a Plan-Apochromat 63×/NA 1.40 oil immersion objective. Acquire images by scanning with the 514-nm line of the argon laser of the LSM 510 Meta microscope using the standard image acquisition function of the LSM 510 software version 4.0SP1.
3. Set the tube current of the argon laser at 6.1 A, and the laser intensity at 0.3%. Set the pinhole at 2.51 Airy units, corresponding to an optical slice of 1.9 μm. Introduce DSBs in the nuclei of cultured cells by laser micro-irradiation with the pulsed nitrogen laser (*see* Subheading 3.1.3).
4. Perform FRAP measurements at 10–20 min after the introduction of DSBs, when accumulation of YFP-ATM reaches maximum.
5. Perform photobleaching with a pulse at 100% laser intensity (514-nm line of an argon laser, tube current set at 6.1 A). Match the dimensions of the circular FRAP area to those of the accumulation spot (~1.9 μm²).
6. Acquire image of the nucleus before photobleaching, immediately after photobleaching and at 2-s time intervals 50 cycles for 180 s.
7. In every image, measure fluorescence of the photobleached accumulation spot (Fas), from which the background fluorescence intensity was subtracted (*see* **step 4** in Subheading 3.1.4).
8. Normalize Fas intensity to the prebleach value: $F_{norm} = Fas_{(t)}/Fas_{(0)}$.
9. Set the prebleach intensity $Fas_{(0)}$ as the maximum intensity and normalize to 1. Set the intensity immediately postbleach as 0.
10. Each data point is the average of 10 independent measurements (Fig. 4).

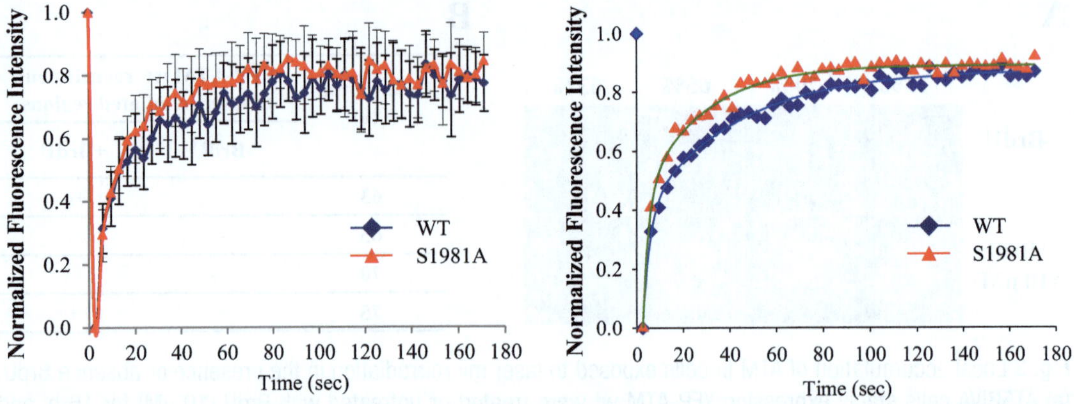

Fig. 4 Phosphorylation of ATM influences the exchange rate between DNA-bound and free forms of ATM protein at DNA double strand breaks sites. FRAP curves of wild-type AT M and 1981A in AT5BIVA cells at DNA double strand break sites. The data show the normalized data from the average of three independent experiments, fitted to an Exponential Rise to Max model. Prebleach intensity levels were normalized to 1 and post-bleach intensity levels were normalized to 0. In each experiment at least 10 cells were measured. ©Davis et al. [4]. Originally published in CELL CYCLE. http://www.tandfonline.com/doi/full/10.4161/cc.9.13.12148

4 Notes

1. Full-length ATM expression constructs are large in size and could recombine during DNA amplification. *E. coli* strains deficient in endonucleases and recombination (e.g., XL10) are recommended to improve the quality of DNA preparation. To amplify the ATM constructs, transform 50–100 ng of plasmid into XL10 cells, and then plate on LB + AMP plate for overnight incubation at 37 °C. Pick a single colony (small in size, not big ones) and grow up in 200 ml LB + AMP in a shaker incubator at 37 °C for 24 h. Next day, purify the plasmid DNA by the Qiagen midi-prep kit according to manufacturer's instructions. A typical yield for the ATM cDNA containing vectors is approximately 100 μg.
2. Apply Nucleofector protocol X-001 with solution V to electroporate YFP-FLAG-ATM constructs into AT5BIVA cells.
3. Transient expression can be performed in the human cell line HT1080.
4. Transient expression utilizing Nucleofector typically results in approximately 20% of cells expressing YFP-ATM.
5. It should be noted that stably expressed of YFP-ATM is normally applied in laser-micro-irradiation assays. To make stable cell lines, linearize the vector with a restriction enzyme and verify the complete digestion with agarose gel electrophoresis. Purify the linearized vector by phenol extraction (twice) followed by ethanol precipitation and wash with 70% ethanol and

the pellet allowed to air dry. Resuspend the pellet in TE or water and calculate the DNA concentration. Perform transfection as described. Following transfection, plate 1×10^4 cells on 100 mm (four total dishes) with normal growth medium. 24 h post transfection, add the selection drug G418 (500 μg/ml), and incubate the cells for 10 days. Pick single colonies and check expression of YFP-ATM via microscopy (YFP signal) and Western blot analysis (antibodies that recognize FLAG tag or ATM).

6. Acquire images before irradiation and then 3, 10, 20, 30, 60, 90, and 120 min post-irradiation.
7. Adjust image quality by clicking "min/max," "best fit," and by contrast light. Save all images in a .zvi format.
8. The circle of the DNA damage site (accumulation site of YFP-tagged ATM) is typically set a 1 micron. But, this is also adjusted based on the size of the spot. The background site is typically 2 micron.
9. Perform all calculations by utilizing Microsoft Excel. The densitometric means are placed in Excel for calculations.

Acknowledgments

The authors are supported by National Institutes of Health grants (CA162804, CA166677) and Cancer Prevention Research Institute of Texas (RP110465).

References

1. Uematsu N, Weterings E, Yano K, Morotomi-Yano K, Jakob B, Taucher-Scholz G, Mari PO, van Gent DC, Chen BP, Chen DJ (2007) Autophosphorylation of DNA-PKCS regulates its dynamics at DNA double-strand breaks. J Cell Biol 177:219–229
2. Yano K, Morotomi-Yano K, Wang SY, Uematsu N, Lee KJ, Asaithamby A, Weterings E, Chen DJ (2008) Ku recruits XLF to DNA double-strand breaks. EMBO Rep 9:91–96
3. So S, Davis AJ, Chen DJ (2009) Autophosphorylation at serine 1981 stabilizes ATM at DNA damage sites. J Cell Biol 187:977–990
4. Davis AJ, So S, Chen DJ (2010) Dynamics of the PI3K-like protein kinase members ATM and DNA-PKcs at DNA double strand breaks. Cell Cycle 9:2529–2536
5. Bakkenist CJ, Kastan MB (2003) DNA damage activates ATM through intermolecular autophosphorylation and dimer dissociation. Nature 421:499–506
6. Murnane JP, Fuller LF, Painter RB (1985) Establishment and characterization of a permanent pSV ori—transformed ataxia-telangiectasia cell line. Exp Cell Res 158:119–126

Chapter 21

Live Cell Imaging to Study Real-Time ATM-Mediated Recruitment of DNA Repair Complexes to Sites of Ionizing Radiation-Induced DNA Damage

Burkhard Jakob and Gisela Taucher-Scholz

Abstract

Measurements of protein recruitment and the formation of repair complexes at DNA double-strand breaks in real time provide valuable insight into the regulation of the early DNA damage response. Here, we describe the use of live cell microscopy in combination with ionizing radiation as a tool to evaluate the influence of ATM and its site-specific phosphorylation of target proteins on these processes. Recommendations are made for the preparation of the cells and the design of specialized cell chambers for the localized (and/or targeted) irradiation with charged particles at accelerator beamlines as well as the microscopic equipment and protocol to obtain high-resolution, sensitive fluorescence measurements.

Key words DNA damage, Double strand breaks (DSB), ATM, Live-imaging, Protein recruitment, GFP, Microscopy

1 Introduction

Genome integrity is crucial for cellular homeostasis and to prevent malignant transformation or premature aging, but can be threatened by DNA damage generated either endogenously or by exogenous sources. DNA double strand breaks (DSBs) are particularly dangerous lesions that can be induced by ROS (reactive oxygen species) produced in metabolism, but are usually attributed to exogenous DNA damaging agents like ionizing radiation.

After a genotoxic insult, in response to DSBs sensor and mediator proteins are rapidly recruited to the sites of damaged DNA, where they form microscopically discernible ionizing radiation-induced foci (IRIF) in a regulated manner [1]. At this very early stage, the ATM kinase is activated through multiple posttranslational modifications and interaction with the MRE11/RAD50/NBS1 (MRN) complex. Activated ATM then phosphorylates a plethora of substrates in various pathways of the DSB response including the autophosphorylation of several residues. A

Sergei V. Kozlov (ed.), *ATM Kinase: Methods and Protocols*, Methods in Molecular Biology, vol. 1599,
DOI 10.1007/978-1-4939-6955-5_21, © Springer Science+Business Media LLC 2017

prominent example of DSB-dependent phosphorylation is the histone variant H2AX, which is phosphorylated at Serine 139 forming γ-H2AX in the chromatin domain surrounding the DSB [2]. This γ-H2AX serves as a platform for MDC1 binding via its BRCT domain, which results in a further recruitment of the MRN complex and additional ATM activation [3, 4]. ATM activation thus leads to the amplification and spreading of the DNA damage response signals helping to efficiently restore the damaged chromatin [5, 6]. In this way, ATM is one of the first and most prominent factors involved in regulating DNA repair and maintaining genomic integrity.

The recruitment of proteins to DNA damage sites may be directly modified by ATM kinase activity through the regulatory phosphorylation of specific sites and/or depend on other factors that are recruited or activated in an ATM-dependent manner. For the measurement of the rapid accumulation of proteins at lesions as well as for the evaluation of subtle kinetic changes in the early DNA damage response it is not possible to rely on static images of fixed cells. The availability of chimeric fluorescent proteins has opened the possibility of studying the accumulation of repair proteins at DNA damage sites in living cells and in real time after localized irradiation [7–10] (*see* **Note 1**). The expression of plasmids encoding for a mutated repair protein under study (point mutations: phosphomimetic, nonphosphorylatable residues, truncated proteins) can yield information on the role of ATM-dependent posttranslational modifications on the recruitment of this specific protein [6, 11, 12]. Alternatively or complementary, appropriate ATM-defective cell lines or kinase inhibitors can be used to obtain this type of information. Additional repair and survival studies yield information on the efficiency and quality of the occurring DSB repair. Complementing these classical radiobiological endpoints by the kinetic information derived from real-time recruitment studies as addressed in this chapter can provide insights into the regulation of proteins involved in early DNA DSB repair.

To study real-time protein recruitment to damaged DNA, it is beneficial to apply a radiation technique generating locally confined lesions. The localized appearance of the protein accumulation facilitates the kinetic analysis as it can be easily spotted and regions of interest (ROIs) assigned to the damage sites. Due to the abundancy and ease of usage at confocal microscopes, localized DNA damage is mainly generated using UV—or in more sophisticated systems—two-photon laser irradiation. However, laser irradiation is a non-ionizing type of radiation and often requires additional sensitization of the chromatin to produce a significant fraction of DSBs. Using charged particle irradiation as offered by accelerators or by an alpha particle source provides several benefits regarding the type and control of radiation quality [13] (*see* **Note 2**). From the physical point of view charged particles are true ionizing

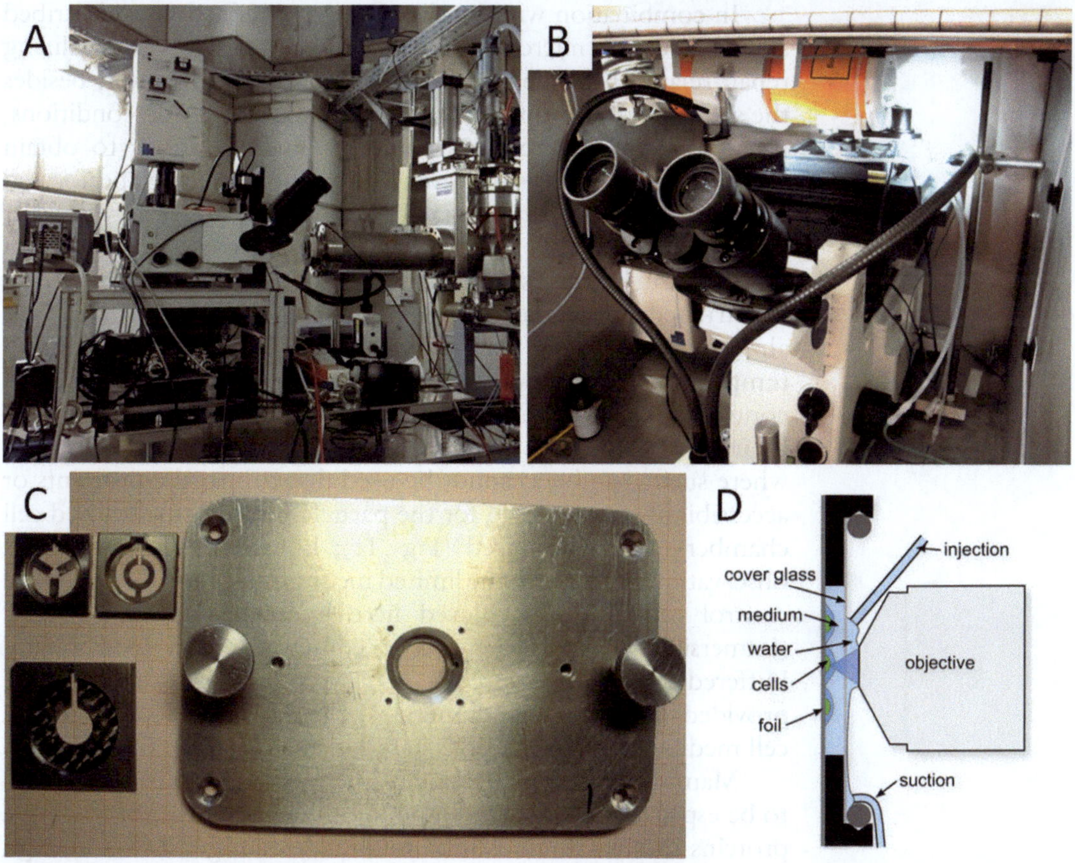

Fig. 1 Microscopy setup for real-time recruitment studies to ionizing irradiation-induced DNA damage. (**a**) At the horizontal GSI beamline, a remote controllable microscope is turned by 90° facing toward the beam exit pipe and window. Special closed chambers with spacers to provide a medium reservoir are needed for this irradiation geometry (**c**). (**b**) Same microscope (now standing) in the 35 kV x-ray cabinet equipped with an environmental chamber. In this configuration, glass-bottom petri-dishes or chambered coverglass can be used. (**c**) Collection of different types of stainless-steel cell chambers for the irradiation at the ion beamline at GSI (**d**) Schematic view of closed cell dishes during irradiation and imaging using a water immersion lens. (**a**) Reprinted from [22] Nuclear Instruments and Methods in Physics Research B, Abdollahi et al. Upgrading the GSI beamline microscope with a confocal fluorescence lifetime scanner to monitor charged particle-induced chromatin decondensation in living cells 2015, with permission from Elsevier; (**d**) Reprinted from [23] Nuclear Instruments and Methods in Physics Research B 306; Merk et al., Photobleaching setup for the biological end-station of the Darmstadt heavy-ion microprobe, 81–84, 2013, with permission from Elsevier

radiation like x- or γ-rays acting by releasing kinetic electrons from the target material. A further advantage in comparison to laser irradiation is the defined and—by the selection of ion type and energy—controllable dose deposition [13] (*see* **Note 3**). Due to the limited penetration range of the charged particles, for these experiments cells are usually grown on thin foils and mounted in special cell chambers where they are irradiated through the foil (Fig. 1a, c) (*see* **Note 4**).

In combination with localized damage induction as described above, live cell microscopy offers the advantage of visualizing repair-related biological processes in real time. However, besides the selection of appropriate constructs and transfection conditions, time-lapse imaging requires additional considerations to obtain useful data. The measurements should be preferably performed under cell culture conditions. This is relatively easy for small or integrated irradiation systems, where the microscope or part of the stage can be enclosed in a commercially available or laboratory-constructed environmental chamber as this is the case for laser, alpha particles, or x-rays (*see* Fig. 1b). This enclosure allows temperature control by warm air or stage heating elements and pH control by adjusting CO_2 levels. To minimize evaporation it may be useful to cover the cell-dish by lid, foil or mineral oil. In cases where such chambers cannot be used due to spatial constraints or accessibility reasons (e.g., for the particle beams), small closed cell chambers are mainly used (Fig. 1c); however, they restrict the observation time due to the limited medium reservoir. Temperature control can also be achieved here by stage and/or objective (immersion lenses) heaters. The commonly used bicarbonate-buffered cell media require 5% CO_2 levels. If no CO_2 supply can be provided, the normal medium may be replaced by Hepes-buffered cell media or Leibovitz's L15-based media (*see* **Note 5**).

Mammalian cells expressing chimeric fluorescent proteins seem to be especially sensitive to high intensity excitation light when the proteins of interest are bound to already damaged DNA as, e.g., induced by irradiation. This is an important aspect, as this chapter deals with the accumulation of fluorescently tagged proteins at DNA DSBs in connection with the regulation of the damage response. Therefore, the repair processes should, as far as possible, not be influenced by the observation itself. Highly sensitive detectors (we use Andor IXON EMCCD cameras) are mandatory to be able to reduce the amount of excitation light, especially if longer (more frequent) observations or 3D imaging is planned.

The settings for light intensity and integration time will depend on the fluorescence protein and its expression level (*see* **Note 6**), as well as the magnification of the microscope system and the choice of appropriate filters. Tests will be necessary for each microscopic configuration to figure out the optimal imaging conditions for sufficient detected photons allowing a robust analysis at a moderate excitation. Compromises have to be made regarding the frequency of acquired timeframes and the amount of z-slices. Beside maintaining the cells in favorable conditions and reducing a potential influence of photodamage on the studied process, this strategy also helps in minimizing photobleaching (*see* **Note 7**).

In this section, we provide a detailed description of a live-cell imaging approach to study the kinetics of the recruitment of repair-related proteins to localized DNA damage induced by charged particle irradiation in real time. This method is applicable to deduce the influence of ATM on early DNA repair events by inhibiting/silencing ATM activity or by site-directed mutagenesis of putative phosphorylation sites in the observed protein [6, 11, 12]. First, we briefly describe mammalian cell preparation and transfection with the expression plasmids for the fluorescently tagged protein. Second, the mounting of cell chambers for the irradiation at the beamline, the irradiation procedure, and time-lapse imaging using an inverted remote-controlled microscope will be addressed. Finally, we provide a description of the analysis of the real-time protein recruitment at the DNA DSBs locally produced at the sites of the traversing particles.

2 Materials

The following materials are used to examine ionizing radiation-dependent protein accumulation at DNA damage sites.

2.1 Cell Culture and Preparation

1. Mammalian cell line (*see* **Note 8**).
2. Appropriate culture medium supplemented with 10% fetal bovine serum and antibiotics (e.g., 100 U/ml Penicillin and 100 μg/ml Streptomycin).
3. Lumox® or polypropylene foil as cell substrate (*see* **Note 9**).
4. Punch of appropriate size.
5. 35-mm Petri-dishes or multi-well plates.
6. Culture chambers suitable for high-resolution fluorescence microscopy and irradiation (special cell chambers for horizontal beamlines, *see* Fig. 1c).
7. Plasmid (expression vector) coding for the protein under investigation coupled to GFP or GFP variant.
8. Transfection reagent (*see* **Note 10**).
9. Phenol red-free cell medium supplemented with 10% fetal bovine serum for time-lapse imaging (*see* **Note 11**).
10. Optional: ATM inhibitor, e.g., Ku55933.

2.2 Microscopy

1. Automated inverted microscope with high NA lens, automated stage for repositioning, and sensitive detector (*see* **Note 12**).
2. Control software.
3. Optional: Environmental chamber with temperature/CO_2 control or stage/objective heater.

2.3 Irradiation Facility

1. Access to accelerator facility, x-ray tube, or α-particle source (*see* **Notes 13** and **14**).

3 Methods

The following procedure describes the techniques to monitor the real-time recruitment kinetics in living cells after ionizing radiation using fluorescently tagged DSB repair proteins.

3.1 Cell Culture and Preparation of Culture Chambers

1. Grow cells in culture flasks under established conditions until a sufficient number of cells are available to seed on the foils as specified below.
2. Cut out lumox foils with a diameter fitting into the culture chamber using a punch (*see* **Note 15**).
3. Sterilize foils by submerging in 70% Ethanol and dry them in a sterile work bench.
4. Place the foils in 35-mm Petri-dishes or a multi-well plate with the hydrophilic side up (*see* **Note 9**).
5. Center-plate cells on foils at a density of around 30,000 cells in 500 μl of medium 1-2 days before the transfection/experiment, respectively. For x-rays or high energetic particles (HZE) with sufficient penetration depth, coverglass chambers or glass-bottom petri-dishes can be used instead of foils. At the time of the experiment, the cells should cover approximately 60–80% of the available surface (*see* **Note 16**).
6. If chemical inhibitors are to be used, these inhibitors should be added at the desired time prior to irradiation either by exchanging the cell medium to a pre-warmed medium containing the appropriate inhibitor concentration or by gently adding and mixing an aliquot of prediluted inhibitor to the cell medium until the desired concentration is reached. Add also inhibitor to the imaging medium to keep the inhibition of ATM during the time of irradiation and image acquisition.

3.2 Transfection

1. Transfect cells using Jetprime® or alternative transfection reagent according to manufacturer's instructions 1 day before the experiment (*see* **Note 17**).
2. Alternatively, especially for primary cells, electroporation (e.g., Amaxa nucleofector®) can be used. Select the program and transfection solution recommended by the manufacturer for the cell line. For AT1BR cells, we used 1×10^6 cells in solution V, program X01, and 2–3 μg of plasmid DNA for transfection (*see* **Note 18**). After electroporation in suspension, cells are diluted in pre-warmed medium and seeded on foils or into chambers. We generally use in the order of 1×10^5 cells in 500 μl of medium for seeding on 18-mm diameter foils.

3.3 Time-Lapse Microscopy and Irradiation

1. Turn on the computer and the microscope including the control devices. These devices may comprise fluorescence illumination sources, detectors, filter wheels, stage, and piezo controllers.
2. Turn on the stage/objective heater or the microscope/stage incubator and wait until temperature is stable at 37 °C.
3. Start the software controlling the microscope system and check remote control connection and capability (*see* **Note 19**). Verify that there is sufficient free hard disc capacity before starting a time-lapse experiment (*see* **Note 20**).
4. Select the desired imaging parameters like illumination wavelength, excitation and emission filters, camera (detector) setting and integration time, imaging frequency, total imaging time (optimal settings should be figured out and tested previously on control samples or regions that will not be recorded during the experiment) (*see* **Note 21**).
5. Fill chamber with 200 μl pre-warmed imaging medium, transfer foil with cells into chamber (cells pointing toward medium). Depending on the design, the chamber might be closed by a sterilized o-ring. Remove excess medium using a pipette. Gently dry backside of foil by a piece of tissue. Add a drop of appropriate immersion liquid to the lens and transfer the chamber to the microscope (*see* **Note 22**). Adjust/check setting of the correction collar of the objective to match the coverglass thickness.
6. Ensure a secured, rigid mounting of the chamber, thus avoiding a displacement during stage motion for repositioning.
7. Focus on the foil/cells using transmitted light (with oblique illumination) if possible. If you have to adjust the focal position using fluorescence illumination, try to be as quick as possible to minimize the amount of excitation light.
8. Select transfected cells showing an appropriate expression level using fluorescence excitation and store the imaging positions and focal position in the software (*see* **Notes 23** and **24**). Enable recording z-stacks if necessary and adjust spacing and number of z-slices (*see* **Note 25**).
9. Start time-lapse acquisition (*see* **Note 26**). Depending on the imaging frequency, at least 3–5 preirradiation images should be recorded from each position prior to irradiation to allow the generation of a baseline.
10. Irradiate samples by beam-request, or by opening a shutter (x-rays, α-particles) so that the ionizing radiation can hit the cells. The way of applying the radiation will depend on the irradiation setup (*see* **Note 27**).
11. At the end of the time-lapse experiment (normally up to 10–20 min for the closed chambers—it can be longer if an

open system and a stage incubator are used) remove the sample from the microscope (*see* **Note 28**), dislodge the foil, and clean and dry the chamber for reuse. Repeat from point no. 5 using a cleaned or fresh chamber.

12. At the end of the total imaging period, shut down the system, clean lens and chamber, and turn off all control devices.

3.4 Analysis of Protein Recruitment Kinetics

1. Use a suitable software program to analyze the recorded image series (e.g., Image J; http://imagej.nih.gov/ij). Assure proper timing or directly read out the time-stamps from your imaging control software (*see* **Notes 29** and **30**).
2. Images can be filtered by a denoise (median) filter to remove the shot noise.
3. In case there is an observable motion of the cell/nucleus, image registration algorithms have to be applied. We use the rigid transformation (translation/rotation) correction of the Image J plugin Stackreg [14].
4. Define three different types of regions of interest (ROIs) by surrounding the contours (Fig. 2). First, the background intensity (ROI III) is evaluated by selecting an area without fluores-

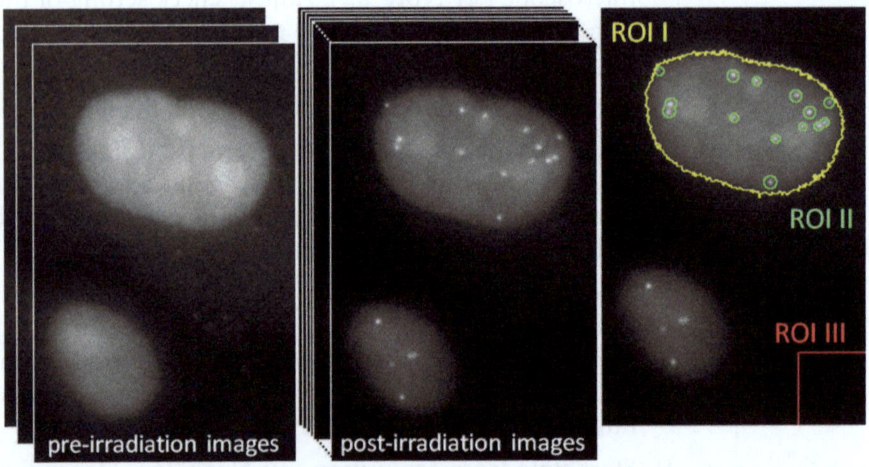

Fig. 2 Quantification of real-time protein recruitment in living cells. *Right*: Pre-irradiation images of the time-lapse sequence used for defining the baseline of the relative fluorescence intensity. *Middle*: Series of post-irradiation images showing the local accumulation of the protein of interest at the damage sites. *Right*: selection of three different regions of interest (ROI) from a post-irradiation image. ROI I comprises the total nuclear fluorescence used for the correction of photobleaching, ROI II encircles the induced sites of DNA damage (here all sites of ion traversals) and ROI III defines the background intensity (some detectors have an offset value). The mean intensity of these ROIs is measured through the whole pre-and post-irradiation series. The intensity of ROI III is subtracted from ROI I and II at all points in time. The relative fluorescence intensity at the damage sites is than calculated by dividing the corrected values of ROI II by ROI I and subtraction of the mean ratios from the pre-irradiation period. Normalization of the plateau value to one can be achieved by dividing the obtained relative fluorescence intensity by the intensity of the plateau (compare Fig. 3a right)

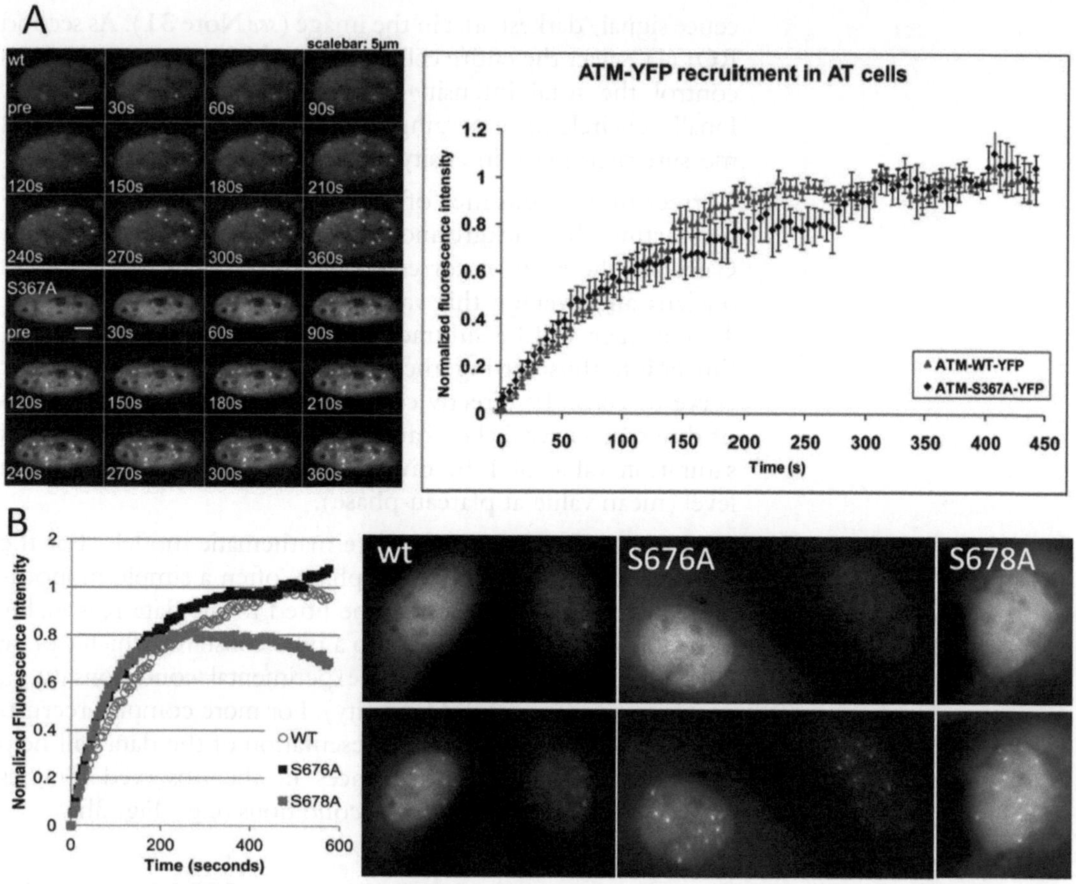

Fig. 3 Real-time quantification of protein recruitment to DNA damage sites induced by ionizing radiation. (**a**) *Left*: Selected time-lapse images of YFP-ATM (wt and S367A mutant) expressing AT cells before and after irradiation with uranium ions. *Right*: Quantification of the recruitment of YFP-ATM revealed similar accumulation kinetics for the wt and the S367A mutant. Error bars: 95% confidence interval; $n = 10$. (**b**) Recruitment of different phosphorylation site mutants of MRE11 (YFP-MRE11 (WT), YFP-MRE11S676A (S676A), or YFP-MRE11S678A (S678A)) to DNA damage induced in U2OS cells by uranium ions. *Left*: Quantification of accumulation kinetics shows that prevention of ATM-dependent phosphorylation of single phosphorylation sites of MRE11 by point-mutation does not abolish the accumulation of MRE11 at the DNA damage site. *Right*: Representative images showing the local accumulation of the different YFP-MRE11 proteins before (*upper row*) and 5 min after irradiation (*lower row*) with uranium ions. (**a**) This research was originally published in [12]: The Journal of Biological Chemistry. Kozlov et al. Autophosphorylation and ATM activation: additional sites add to the complexity. *JBC* 2011; 286(11):9107–19. © the American Society for Biochemistry and Molecular Biology. (**b**) *Left*: This research was originally published in [6]: Nucleic Acids Research. Kijas et al. ATM-dependent phosphorylation of MRE11 controls extent of resection during homology-directed repair by signaling through Exonuclease 1. *NAR* 2015; 43(17):8352–67 published by Oxford University Press

cence signal/darkest area in the image (*see* **Note 31**). As second ROI (I), select the entire cell nucleus, which serves as a basis to control the total intensity and correct for photobleaching. Finally, encircle the emerging IRIF (ROI II) in the nucleus and measure their mean intensity throughout the time series.

5. Correct mean intensities of the nucleus (I) and IRIF (II) by subtracting the background (III) values for each timepoint. Calculate the ratio of corrected intensity values of IRIF over nucleus and average this value for the preirradiation frames. This average will be subtracted from the ratio values of each timepoint, thus setting the starting line of the recruitment curve to zero. To directly compare the steepness and kinetics of different curves, they can additionally be normalized to a saturation value of 1 by dividing all values by the saturation level (mean value at plateau-phase).
6. Data can be fitted to appropriate mathematic models. For the early repair factors without lag phase, often a simple monoexponential saturation curve can be fitted to the data reasonably well. The exponential fit leads to a time constant, which can be compared under the different experimental conditions (e.g., presence/absence of ATM activity). For more complex recruitment behavior, a graphical representation of the data will help to assess and describe differences in the observed kinetics between different experimental conditions (e.g., Fig. 3b).

4 Notes

1. For the interpretation of the results it must be kept in mind that the fluorescent tag and/or the expression level might alter the protein function and its cellular environment.
2. Alternatively to using charged particles, x-rays can be used as a source of ionizing irradiation. However, generally x-rays provide a homogeneous and global dose delivery to the cells. It is difficult to achieve a localized subnuclear irradiation by x-rays, but microcollimators [15] or soft x-ray microbeams [16] have been applied.
3. Modern ion microbeam facilities allow even subnuclear targeted particle irradiation or a dose escalation by targeting multiple ions to the same spot [17].
4. For x-rays as well as high energetic particles (HZE particles) the cells can also be grown in normal glass bottom petri dishes or chambered coverslips. Due to their larger range, the irradiation with higher energy particles can be done parallel to the cell layer giving rise to a striped damage pattern that facilitates the analysis of the influence of the chromatin organization [18] or the motion of single IRIFs [19].

5. For sensitive cells it is often beneficial to get the cells adapted to the imaging medium 1 day before the experiment.
6. Generally, a low or moderate protein expression level (similar to the endogenous one) is favorable in order to avoid a perturbation of protein regulation or even the formation of aggregates. In addition, protein accumulations at DNA damage sites usually show a higher contrast to the nuclear wide protein signal at lower expression levels, thus facilitating the analysis.
7. To reduce background fluorescence and excitation light absorption, Phenol red that is commonly used as a pH indicator in commercial cell media should be omitted.
8. To obtain as large as possible fractions of cells expressing the tagged protein, cell lines allowing efficient transfection should preferably be used. Examples are the commonly used human Osteosarcoma cells U2OS or HeLa cells, a human epitheloid cervix carcinoma cell line. In the context of downstream TP53 or cell cycle regulation-related topics, cells with wild-type TP53 status should be chosen. ATM-deficient lines may well be used to address ATM-dependent recruitment.
9. To facilitate cell attachment and growth the surface of PP foils must be coated with Cell Tak (BD Biosciences) or other coatings supporting cell adhesion. Lumox® foil is gas permeable and has two different surfaces (a hydrophilic and a water-repellent one). Use the hydrophilic one for cell seeding. Testing of the surface can be done with a drop of water. For x-ray or high-energy particles commercially available glass-bottom petri dishes or chambered coverglasses can be used.
10. We use Jetprime® according to the manufacturer's instructions. Alternatively, especially for primary cells like fibroblasts, electroporation (e.g., Amaxa nucleofector®) can be used. After electroporation in suspension cells are seeded into the chambers. Optimal conditions (solution, plasmid concentration, program, and timing) have to be tested and will depend on the cell line and plasmid used.
11. When no CO_2 supply is present, the cell medium should be buffered by adding 10–20 mM Hepes buffer; alternatively C Leibovitz's L15 medium can be directly used.
12. The microscope should be designed for sensitive fluorescence detection with a high numerical aperture lens (NA), a detector with high quantum efficiency and a minimum part of optical elements in the light path to maximize photon transmission and recording. Water immersion lenses avoid optical aberrations and light loss when imaging through a layer of cell medium and are thus preferable over oil immersion lenses. The whole microscope system should include a xy-stage and motorized or piezo focusing. In addition, a software is needed

capable of controlling all auxiliary devices (stage, excitation/emission wavelength control, focusing devices), recording and revisiting of multiple positions including the selected focal position and the parameters of digital imaging acquisition.

We use an Olympus IX71 inverted microscope equipped with an Olympus 60× 1.2NA (UPLSAPO 60XW) water immersion lens with a working distance of 0.28 mm, a PolyV monochromator (Till-photonics) for excitation, appropriate high-quality excitation/dichroic/emission filter sets (Semrock Brightline®), and an Andor IXON 888 EMCCD camera with a back-illuminated sensor and a quantum efficiency of >90% (Fig. 1). If necessary, a Prior filterwheel can be installed in front of the detector for better emission separation in multicolor experiments. For the repositioning we use a Prior xy stage in combination with a piezo stage insert (P-737 PIFOC® Specimen-Focusing Z Stage, Physik Instrumente (PI) GmbH & Co. KG). An additional objective scanner (P-P-721 PIFOC® Piezo Flexure Objective Scanner, Physik Instrumente (PI) GmbH & Co. KG) with analogue voltage control allows independent z-stacking and remote focus control/correction during and after irradiation. All components are controlled by Andor IQ software in the fast lambda/z mode. For the use at the x-ray tube (Fig. 1b) the microscope is equipped with a modified stage incubation system (Tokai Hit Corp., Japan).

13. Charged particles usually need ion accelerators that in part also offer access to external groups or allow irradiation in the form of collaborations with local researchers. The institutions/homepages will inform about regulations and application procedures. Offered ion species and energies depend on the accelerator type and availability. Targeted irradiation with a precisely defined number of traversals (in one spot) can be delivered at so-called microbeam facilities [17]. Alternatively, alpha irradiation [20] or x-rays [16, 21] can be used (*see* also **Note 2**).

14. For all types of ionizing radiation safety regulations from the local authorities have to be followed.

15. We use Lumox® foils with a diameter of 18 mm for the beamline microscope. If the penetration range of the ion is sufficiently high (e.g., for 11.4 MeV/u carbon ions), we use 18 mm round thin coverglasses (Menzel #00; 0.055–0.08 mm) instead of Lumox® foil as cell substrate. The chamber design at the microbeam is as shown in Fig. 1c (lower left) where a square polypropylene foil is glued to a stainless-steel chamber by molten stearin [23]. In this case, cells are directly seeded into the sterilized chambers, which are placed in Petri-dishes.

16. Cell density and timing of transfection have to be tested prior to the experiment to find optimal transfection conditions.

Remind that transfection might lead to cell loss. Ideally, at the time of the experiment cells will have a density allowing optimal imaging conditions. They should not form heaps, but dishes should contain sufficient cells with an appropriate expression level for quick selection of imaging positions in a given area.

17. Use tagged empty vector (chromophore only) as a control to demonstrate that the observed recruitment is depending on your protein of interest.
18. Regarding transfection efficacy and cell survival, optimal conditions have to be tested for each cell line and plasmid.
19. Due to radiation protection regulations, for the experiments at an accelerator beamline the microscope has to be remote controllable. After inserting the cell dish and selecting the imaging positions and parameters, we leave the irradiation room during ion beam exposure. In that time we operate the microscope by an external terminal connected to the PC-running the control software. In addition, we can provide additional focus adjustment by changing an offset on the analogue voltage driving the objective piezo.
20. Time-lapse imaging readily generates large amounts of data, e.g., a single measurement with five positions and a z-stack of five slices and 100 time points sums up to around 5 Gb of data per channel for a 16 bit camera with 1000×1000 pixels.
21. The length of the experiment should be sufficient to cover the kinetic phase of interest as complete as possible, e.g., until accumulation reaches a plateau (*see* Fig. 3). Imaging frequency should map the kinetics sufficiently to be able to discriminate different phases (several data points per phase) and fit the data. It may be useful to adapt the sampling frequency according to this need during the course of the experiment. Sampling at high frequency when fast changes are expected and at lower frequency when a plateau is reached helps to minimize photobleaching and phototoxicity without sacrificing temporal information.
22. We use Zeiss Immersol® w instead of water for our 60× 1.2 NA water immersion lens. This immersion liquid has the same index of refraction as water, but a higher viscosity. For a horizontal configuration of the microscope at the beamline we apply the immersion liquid on the coverglass directly before mounting the sample on the microscope stage. It is important that the drop is not lost before getting into contact with the objective lens. Alternatively, immersion liquid can be continuously applied by a pump (*see* Fig. 1d). Bad image quality might arise from the lack of immersion liquid.

23. Try to minimize the exposure time of the cells during selection to avoid unnecessary phototoxicity and photobleaching. Shut off the fluorescence excitation light whenever illumination is not needed.
24. Time needed by the system for revisiting the number of selected fields and acquiring the images (stacks) has to match the desired frequency of imaging according to the velocity of the recruitment. If this is not matched, reduce the number of selected fields before starting the time-lapse measurement.
25. Generally, adherent cells with stretched out, planar nuclei facilitate the microscopy and limit the number of required z-slices. For perpendicular ion irradiation, where the ions traverse the total thickness of the nucleus generating multiple DSBs/μm along their path, even a single focal layer might be sufficient to analyze the recruitment, thus avoiding unnecessary phototoxic stress. For low angle irradiation (e.g., using high energetic particles) or for sparsely ionizing radiation (x-rays) few slices with a distance of around 0.5–1 μm are a compromise between observing more DSBs in focus at the same time avoiding bleaching and phototoxicity.
26. If enforced by radiation protection, at this step you have to leave the room and operate the microscope remotely from a second terminal (*see* **Note 19**).
27. The duration of irradiation should be fast (optimally within one period of two consequent images of one position). Time of irradiation in relation to the progressing time-lapse acquisition must be recorded. This can be done manually or in the form of a trigger, recorded as an event in the software. Unirradiated samples should be recorded under identical imaging conditions as controls.
28. Before dismounting, nonimaged areas should be visually inspected and compared to the recorded ones to rule out the influence of additional phototoxic stress in the obtained results.
29. Duration of 3D imaging of multiple positions may need longer time than anticipated. Please note that the selected imaging frequency might be overruled if the acquisition is to slow. In this case actual frame time readout from the metadata is mandatory. Alternatively, the timing has to be adjusted before starting of the experiment.
30. In case 3D stacks have been recorded, one can work with a central slice including the feature of interest or use z-projections (average or maximum projection). If only few (e.g., 3–5) largely spaced z-slices were recorded to increase the robustness of focus, a maximum projection will facilitate the analysis especially in cases where only a weak accumulation of the protein is observed, as this projection emphasizes the bright signals.

As long as only the kinetic behavior of the accumulation and not the absolute intensity increase is evaluated, there is no relevant difference in the results of both projection types. Nevertheless, for comparative measurements, only the same type of analysis should be used.

31. This is especially important as some detectors like the used EMCCD camera have a positive offset.

Acknowledgment

This work was supported by BMBF Grants 02NUK037A, 02NUK001A and DFG GRK1657.

References

1. Shiloh Y (2014) ATM: expanding roles as a chief guardian of genome stability. Exp Cell Res 329(1):154–161
2. Rogakou EP, Pilch DR, Orr AH, Ivanova VS, Bonner WM (1998) DNA double-stranded breaks induce histone H2AX phosphorylation on serine 139. J Biol Chem 273(10): 5858–5868
3. Jungmichel S, Stucki M (2010) MDC1: the art of keeping things in focus. Chromosoma 119(4):337–349
4. Jungmichel S, Clapperton JA, Lloyd J, Hari FJ, Spycher C, Pavic L, Li J, Haire LF, Bonalli M, Larsen DH, Lukas C, Lukas J, MacMillan D, Nielsen ML, Stucki M, Smerdon SJ (2012) The molecular basis of ATM-dependent dimerization of the Mdc1 DNA damage checkpoint mediator. Nucleic Acids Res 40(9):3913–3928
5. Bhatti S, Kozlov S, Farooqi AA, Naqi A, Lavin M, Khanna KK (2011) ATM protein kinase: the linchpin of cellular defenses to stress. Cell Mol Life Sci 68(18):2977–3006
6. Kijas AW, Lim YC, Bolderson E, Cerosaletti K, Gatei M, Jakob B, Tobias F, Taucher-Scholz G, Gueven N, Oakley G, Concannon P, Wolvetang E, Khanna KK, Wiesmüller L, Lavin MF (2015) ATM-dependent phosphorylation of MRE11 controls extent of resection during homology directed repair by signalling through Exonuclease 1. Nucleic Acids Res 43(17): 8352–8367
7. Lukas J, Lukas C, Bartek J (2004) Mammalian cell cycle checkpoints: signalling pathways and their organization in space and time. DNA Repair (Amst) 3(8–9):997–1007
8. Lukas C, Bartek J, Lukas J (2005) Imaging of protein movement induced by chromosomal breakage: tiny 'local' lesions pose great 'global' challenges. Chromosoma 114(3):146–154
9. Tobias F, Durante M, Taucher-Scholz G, Jakob B (2010) Spatiotemporal analysis of DNA repair using charged particle radiation. Mutat Res 704(1-3):54–60
10. Tobias F, Löb D, Lengert N, Durante M, Drossel B, Taucher-Scholz G, Jakob B (2013) Spatiotemporal dynamics of early DNA damage response proteins on complex DNA lesions. PLoS One 8(2):e57953
11. Gatei M, Jakob B, Chen P, Kijas AW, Becherel OJ, Gueven N, Birrell G, Lee JH, Paull TT, Lerenthal Y, Fazry S, Taucher-Scholz G, Kalb R, Schindler D, Waltes R, Dörk T, Lavin MF (2011) ATM protein-dependent phosphorylation of Rad50 protein regulates DNA repair and cell cycle control. J Biol Chem 286(36): 31542–31556
12. Kozlov SV, Graham ME, Jakob B, Tobias F, Kijas AW, Tanuji M, Chen P, Robinson PJ, Taucher-Scholz G, Suzuki K, So S, Chen D, Lavin MF (2011) Autophosphorylation and ATM activation: additional sites add to the complexity. J Biol Chem 286(11):9107–9119
13. Splinter J, Jakob B, Lang M, Yano K, Engelhardt J, Hell SW, Chen DJ, Durante M, Taucher-Scholz G (2010) Biological dose estimation of UVA laser microirradiation utilizing charged particle-induced protein foci. Mutagenesis 25(3):289–297
14. Thévenaz P, Ruttimann UE, Unser M (1998) A pyramid approach to subpixel registration

based on intensity. IEEE Trans Image Process 7(1):27–41

15. Pataky K, Villanueva G, Liani A, Zgheib O, Jenkins N, Halazonetis DJ, Halazonetis TD, Brugger J (2009) Microcollimator for micrometer-wide stripe irradiation of cells using 20–30 keV X rays. Radiat Res 172(2):252–259
16. Folkard M, Schettino G, Vojnovic B, Gilchrist S, Michette AG, Pfauntsch SJ, Prise KM, Michael BD (2001) A focused ultrasoft x-ray microbeam for targeting cells individually with submicrometer accuracy. Radiat Res 156(6): 796–804
17. Barberet P, Seznec H (2015) Advances in microbeam technologies and applications to radiation biology. Radiat Prot Dosimetry 166(1-4):182–187
18. Jakob B, Splinter J, Conrad S, Voss KO, Zink D, Durante M, Löbrich M, Taucher-Scholz G (2011) DNA double-strand breaks in heterochromatin elicit fast repair protein recruitment, histone H2AX phosphorylation and relocation to euchromatin. Nucleic Acids Res 39(15): 6489–6499
19. Jakob B, Splinter J, Durante M, Taucher-Scholz G (2009) Live cell microscopy analysis of radiation-induced DNA double-strand break motion. Proc Natl Acad Sci U S A 106(9): 3172–3177
20. Stap J, Krawczyk PM, Van Oven CH, Barendsen GW, Essers J, Kanaar R, Aten JA (2008) Induction of linear tracks of DNA double-strand breaks by alpha-particle irradiation of cells. Nat Methods 5(3):261–266
21. Khan R, Becker A, Taucher-Scholz G, Durante M, Fehrenbacher G, Jakob B (2014) Construction of a X-ray cabinet for live cell experiments. GSI Sci Rep 1:242
22. Abdollahi E, Taucher-Scholz G, Durante M, Jakob B (2015) Upgrading the GSI beamline microscope with a confocal fluorescence lifetime scanner to monitor charged particle induced chromatin decondensation in living cells. Nucl Instrum Methods Phys Res, Sect B 365(Pt B):626–630
23. Merk B., Voss KO., Müller I.,·Fischer BE., Jakob B., Taucher-Scholz G., Trautmann C., and Durante M. (2013) Photobleaching setup for the biological end-station of the darmstadt heavy-ion microprobe. Nucl Instrum Methods Phys Res, Sect B 306, 81–84.

Chapter 22

Analyzing Heterochromatic DNA Double Strand Break (DSB) Repair in Response to Ionizing Radiation

Karolin Klement and Aaron A. Goodarzi

Abstract

DNA damaging agents such as ionizing irradiation induce lesions in the DNA such as double strand breaks (DSBs). Depending on cell type, 10–25% of these DSBs are induced in heterochromatin. Heterochromatic DSBs are resolved with slow kinetics (compared to DSBs in euchromatin) and require ATM activity for repair. Investigating the underlying causes of the slow component of DSB repair and the role of individual response factors in this process provides insight into DSB response pathways and will further the understanding of diseases where such pathways are dysfunctional due to mutation. Here, we describe a method to detect DSB repair foci in the heterochromatin of human cells. We provide a detailed protocol for cell culture preparation, immunofluorescence microscopy, and a computer-assisted approach to analyze overlap between DSB foci and heterochromatin.

Key words Double strand break (DSB), DNA damage repair, ATM (ataxia telangiestasia mutated), Heterochromatin, γH2AX-foci, Microscopy, Ionizing radiation

1 Introduction

Our cells are exposed constantly to DNA-damaging chemicals and radiation, generating lesions such as DNA double strand breaks (DSBs). Fortunately, cells have mechanisms to repair DNA damage quickly and efficiently. Two pathways responsible for the repair of DSBs are non-homologous end joining and homologous recombination.

ATM (Ataxia telangiectasia mutated) is a serine/threonine protein kinase and a member of the PIKK (Phosphatidylinositol 3-kinase-related kinase) family of protein kinases. Following DSB induction, a complex consisting of MRE11, NBS1, and RAD50 (MRN complex) initially binds to the DSB, and recruits and activates ATM. ATM then phosphorylates several target proteins, one of which is H2AX S139, thereby generating γH2AX [1, 2]. Then, a series of additional recruitment steps and positive feedback loops to enhance the DSB signaling cascade leads to the accumulation of

Sergei V. Kozlov (ed.), *ATM Kinase: Methods and Protocols*, Methods in Molecular Biology, vol. 1599,
DOI 10.1007/978-1-4939-6955-5_22, © Springer Science+Business Media LLC 2017

mediator proteins such as MDC1 and 53BP1 and ubiquitin ligases RNF8, RNF168, and RNF20–RNF40 [3–7]. DSB signal enhancement and chromatin modifications via ubiquitylation and SUMOylation are some prerequisites for the repair of DSBs through NHEJ and HR [8].

Homologous Recombination (HR) occurs predominantly in the G2 phase of the cell cycle, where sister chromatids are available to assist the homology-directed repair. The main steps of HR involve 3′ single strand DNA (ssDNA) generation; coating of the ssDNA with the single stranded DNA binding protein RPA, followed by exchange and filament formation via RAD51; subsequent heteroduplex DNA strand invasion (to undamaged sister chromatids), Holliday junction formation, and finally branch migration and resolution [9]. A detailed review on the mechanism of HR is provided elsewhere [9].

Non-homologous end joining (NHEJ) occurs during all phases of the cell cycle and is therefore the preferred mechanism of DSB repair [9]. Importantly, it involves the binding of the DNA ends by the Ku-heterodimer (Ku70–Ku80) with subsequent recruitment of DNA-PKcs (another member of the PIKK family), resembling the DNA-PK holoenzyme. In order to promote end ligation, the DSB ends are modified by DNA end processing enzymes such as Artemis or PNKP (Polynucleotide kinase 3′phosphatase), generating 5′phosphate and 3′OH groups prior to ligation. Then, DNA ligation is performed by the DNA ligase IV, XRCC4, and XLF complex. Comprehensive reviews on the step-by-step actions of NHEJ are provided elsewhere in excellent reviews [9–11].

While DSBs induced by ionizing radiation can be repaired very efficiently within 24 h, the kinetics of repair might differ for individual DSB repair foci. Many studies have suggested the existence of a fast and a slow component for DSB repair [12–14]. Notably, slow DSB repair is linked to lesions located within heterochromatin (HC) and ATM-dependent signaling [15]. In addition, while the slow component represents NHEJ in G0 and G1 phases of the cell cycle, it is mainly linked to HR in G2 phase [16].

The underlying causes of the slow component of DSB repair are still elusive, although it is known that heterochromatin is a barrier to γH2AX expansion and signaling, which could account for the delay in timely repair [8, 17]. In addition, γH2AX-foci form preferably in euchromatin (EC), and γH2AX foci in HC appear at later time points at the periphery of densely packed HC regions [15, 18, 19]. In agreement with these findings, γH2AX-foci move throughout HC, bending around densely compacted regions and localize to the periphery of HC [20, 21].

ATM is required for the repair signaling of 10–25% of total DSBs, which are mostly located in heterochromatin and repaired with slow kinetics [15]. Additional factors known to be needed for DSB repair in heterochromatin include the MRN complex, the mediator proteins MDC1 and 53BP1, the Artemis nuclease, and the ubiquitin ligases RNF8 and RNF168 [19, 22, 23]. In order for DSB repair to function, the densely packed chromatin has to become "relaxed"—or altered to a nucleosome spacing configuration compatible with DSB repair processes. Several chromatin remodeling factors and complexes play an important role in this process. One of ATM's targets, the transcription factor KAP-1 (KRAB-associated protein 1), is phosphorylated upon DSB induction at S824 throughout the nucleus at early times following DSB induction [24]. Subsequently, phosphorylated KAP-1 (pKAP-1) accumulates selectively at the sites of (late repairing) DSBs in HC. Phosphorylation of KAP-1 at late-repairing lesion requires the action of 53BP1, which promotes hyper-accumulation of ATM activity at persisting DSB lesions over time [19]. Several key chromatin remodelers are responsible for maintaining and changing chromatin structure during the HC-DSB response. In undamaged cells, KAP-1 interacts with the CHD3 isoform 1 (CHD3.1)-containing NuRD complex via direct interactions between CHD3.1 (which is an ATP-dependent chromatin remodeler) and SUMOylated sites on KAP-1 at K804/K779/K554 [25, 26]. Upon phosphorylation of KAP-1 at S824, the interaction between KAP-1 and CHD3.1 is diminished, causing dispersal of CHD3.1 from the chromatin around the site of the DSB [27]. Furthermore, the ATP-dependent ISWI-class chromatin remodeling complex ACF1-SNF2H is then recruited in a process requiring H2B monoubiquitylation at K120 by RNF20-RNF40 [7, 28–32]. ACF1-SNF2H complex functions downstream of pKAP1-mediated CHD3.1 dispersal, and displays competitive affinities for histone-binding sites with CHD3.1 [31].

Several approaches exist to analyze DSB foci in heterochromatin, one of which is a microscopy-based approach using cells containing large areas of heterochromatin, such as mouse embryonic fibroblasts (MEFs) [15, 19]. The heterochromatin in MEFs arranges into dense clustered regions, termed chromocenters, which are visible and easy to distinguish from EC by widefield epifluorescence microscopy. The appearance of HC and EC is different in other mammalian cells, such as human diploid fibroblasts. Primary human fibroblasts are an important tool for researchers working on diseases of the DNA damage response, because many cell lines have been established from skin tissue of patients with mutations in DNA damage response genes (e.g., ATM in Ataxia telangiectasia; A-T). The heterochromatin of human cells does not form clearly visible chromocenters in the same manner as in mouse cells and is, therefore, much more diffuse and not as easy to distinguish from euchromatin by eye.

In order to analyze DSB foci in HC of human cells, we developed a high-resolution (confocal) microscopy-based method,

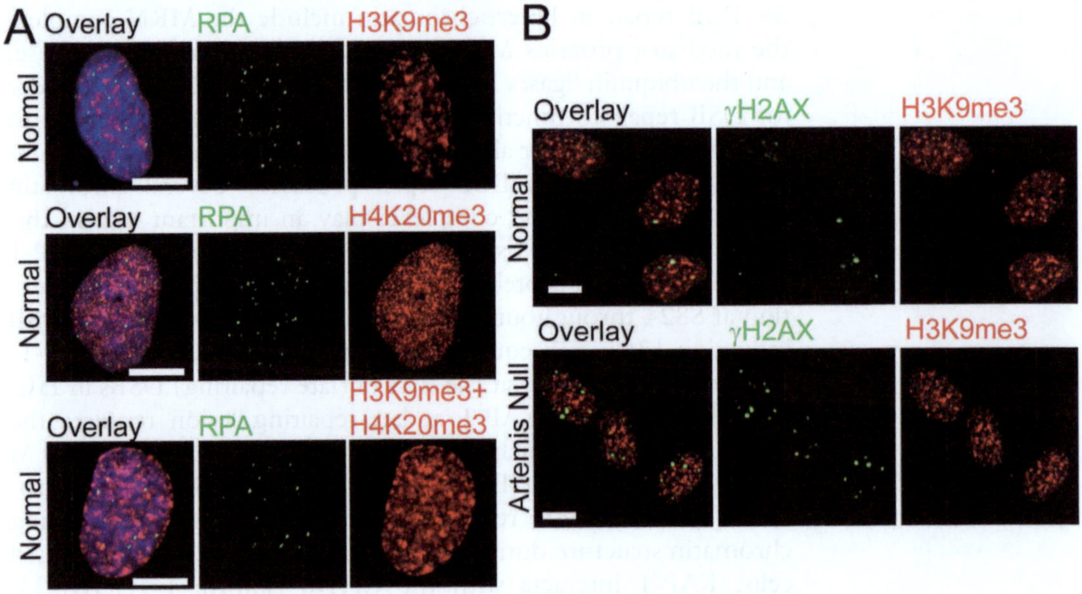

Fig. 1 Representative images of DSB repair foci and heterochromatin staining. (**a**) Primary human fibroblasts (1BR3) were fixed 8 h post 3 Gy γIR and immunostained for RPA (*green*), H3K9me3, H4K20me3, or both as indicated (*red*) and DNA (ToPro3, *blue*). Scale bar 10 μm. (**b**) Normal primary human fibroblasts (*upper panel*) and primary Artemis-null fibroblasts (*lower panel*) were fixed 24 h post 3 Gy γIR and immunostained for γH2AX (*green*) and H3K9me3 (*red*). Note the higher amount of DSB repair foci remaining in Artemis-null cells. Artemis and ATM act in the same pathway of DSB signaling [22]. Scale bar 10 μm

combined with a computer-software-assisted analysis to determine the overlap of DSB foci and markers of heterochromatin precisely. This method was used previously to analyze DSB repair foci in heterochromatin during resection in G2 phase [33]. DSB foci can be visualized by staining for DSB signaling markers, such as γH2AX and 53BP1 or RPA (the latter for resection in G2 phase) (Fig. 1a, b). Heterochromatin in human diploid cells can be visualized by staining for distinctive histone modifications characteristic for HC, such as trimethylated lysine 9 at histone 3 (H3K9me3) or trimethylated lysine 20 at histone 4 (H4K20me3) (Fig. 1a). In our hands, using both markers in the same channel will deliver coverage of the majority of heterochromatic areas within the cell (Fig. 1a).

Together, co-staining for DSB foci and HC markers with high-resolution microscopy imaging and software-assisted analysis delivers a simple and reliable method to determine the percentage of DSB foci localized to heterochromatin. Future research on factors of the DNA damage response, including the analysis of new repair factors, will benefit from this method, as it is an easy method to determine whether a factor is involved in heterochromatic DSB signaling.

Here, we describe the method of cell preparation for the analysis of DSB repair in HC in G0/G1 and G2 phases of the cell cycle (via γH2AX staining), cell preparation for the analysis of resection foci in G2 (via RPA staining), the immunofluorescence staining methods, high-resolution microscopy, and a detailed protocol on how to analyze the overlap of DSB foci in HC using computer software.

2 Materials

2.1 Tissue Culture and Preparation of Cells for Treatment with DNA-Damaging Agent

1. Normal human diploid fibroblasts (e.g., 48BR or 1BR3).
2. AT1BR cells (containing an ATM genomic deletion [specific genetic lesion not known], no ATM transcript, and null ATM expression) (*see* **Note 1**).
3. Growth media: Minimum Essential Media (MEM), supplemented with 15% (vol/vol) FBS, GlutaMAX (Gibco) (optional: penicillin and streptomycin).
4. Autoclaved phosphate-buffered saline (PBS) pH 7.5.
5. Trypsin/PBS (pH 7.5).
6. Round, sterile, 3-cm cell culture dishes.
7. Sterile, autoclaved, microscopy grade coverslips (*see* **Note 2**).
8. Aphidicolin (1 mg/mL stock) (*see* **Note 3**).
9. Source of DNA damage: Ionizing radiation (IR): γ-rays (^{137}Cs) delivered by Gammacell 1000 Elite (MDS Nordion) (*see* **Note 4**).

2.2 Indirect Immunofluorescence Procedure

1. Autoclaved 1× phosphate-buffered saline (PBS) pH 7.5.
2. Fixative: PBS containing 3% (w/v) paraformaldehyde and 2% (w/v) sucrose.
3. PBS containing 0.2% (v/v) Triton X-100.
4. PBS containing 2% (w/v) BSA (*see* **Note 5**).
5. PBS containing 0.1 μg/mL DAPI (*see* **Note 6**).

Table 1
Primary antibodies

Antibody	Company	Dilution for IF	Reference
Anti-gamma H2A.X (phospho S139) antibody [9F3]	Abcam (ab26350)	1:800	[31]
Anti-53BP1 antibody	Abcam (ab21083)	1:200–1:800	[19]
Anti-histone H3 (tri methyl K9) antibody—ChIP grade	Abcam (ab8898)	1:800	[33]
Anti-histone H4 (tri methyl K20) antibody—ChIP grade	Abcam (ab9053)	1:800	[33]
Anti-replication protein A (Ab-2) mouse mAb (RPA34-19)	Calbiochem (#NA18)	1:50	[16]

6. Anti-γH2AX antibody or anti-53BP1 antibody (to visualize DSB foci) or anti-RPA antibody (to visualize sites of resection in G2) (*see* Table 1).
7. Anti-H3K9me3 antibody and anti-H4K20me3 (to visualize heterochromatin) (*see* Table 1).
8. Appropriate fluorescently labeled secondary antibodies.
9. Clean microscope slides.
10. Antifade fluorescence mounting medium (*see* **Note 7**).

2.3 High-Resolution Microscopy

1. Microscope with the ability to distinguish reliably between the fluorescence-conjugated groups (For high resolution, a confocal microscope, e.g., from Carl Zeiss; Leica; Nikon; Olympus or similar).
2. Software to process z-stacks and to perform deconvolution and 3D rendering (*see* **Note 8**).

2.4 Data Analysis

1. Image J software [34–36] and relevant plug-ins (*see* **Note 9**).

3 Methods

3.1 Cell Culture for Preparation of Cells for Ionizing Radiation

3.1.1 Quiescent Cells for Analysis in G0

1. Plate cells into 3-cm cell culture dishes (containing sterile microscopy coverslips) at a concentration of 1 × 10^5 cells/mL for a total volume of 2 ml for each 3-cm dish.
2. Let cells grow to confluence and keep confluent for >1 week, to achieve >99% quiescence before initiating the experiment.
3. Irradiate cells with 3 Gy (*see* **Note 10**).
4. Choose appropriate time points such as 0 min, 30 min, 2 h, 8 h, and 24 h. for fixation of cells (*see* **Note 11**). Follow protocol from **step 1** of Subheading 3.2.

3.1.2 Cells in G2 Phase of the Cell Cycle and Preextraction Procedure to Visualize RPA Foci

1. Seed 1 × 10^5 cells/mL for a total volume of 2 ml for each 3-cm dish.
2. The next day, perform the experiment.
3. Shortly before irradiation of cells, add Aphidicolin (1 μg/ml).
4. Irradiate cells with 3 Gray.
5. Choose appropriate time points such as 0 min, 30 min, 2 h, 8 h for fixation of cells (*see* **Note 12**).
6. Carefully remove culture media (*see* **Note 13**).
7. Carefully add 0.2% Triton X-100/PBS for 30 sec at room temperature (RT) (*see* **Note 14**).
8. Carefully remove the preextraction solution. Do not wash with PBS.

9. Add fixative for 10 min at RT.
10. Continue protocol at **step 5** of Subheading 3.2.

3.2 Indirect Immunofluorescence to Visualize Repair Foci and Heterochromatin

1. Discard cell culture media and rinse cells three times with ~2 mL of PBS at RT (*see* **Note 15**).
2. Incubate the coverslips in fixative for 10 min at RT.
3. Rinse the cells 3× with ~2 mL of PBS at RT (*see* **Note 16**).
4. Cells are permeabilized by incubation in ~2 mL 0.2% Triton X-100/PBS for 3 min at RT (*see* **Note 17**).
5. Rinse the cells 3× with ~2 mL of PBS at RT.
6. Incubate the cells with primary antibody (dilution in 2% (w/v) BSA/PBS) for 1 h at RT. Refer to Table 1 for antibody dilutions.
7. Wash the cells 3 × 5 min with ~2 mL of PBS at RT.
8. Incubate the cells with secondary antibody (dilution in 2% (w/v) BSA/PBS) for 45 min at RT in the dark.
9. Incubate cells in ~2 mL of DAPI solution for 5–10 min at RT in the dark.
10. Wash the cells 3 × 5 min with ~2 mL of PBS at RT.
11. Label microscopy slides and mount the coverslips using antifade mounting solution. If necessary, seal the edges of the coverslip using nail polish. Let it dry and store according to manufacturer's instructions.

3.3 High-Resolution Microscopy Using a Confocal Microscope

1. Capture images of single cells (*see* **Note 18**): Use a good objective for imaging single cells, e.g., Plan-Apochromat 63×/1.40 Oil DIC M27 or similar.
2. Take z-stacks of individual cells. Since most fibroblasts are adherent to the culture dish, the dimensions in z-direction are, for the fibroblasts used in this study, ~2 μm. The z-stacks should span the whole cell. In our hands, z-stacks were taken of 2.5–3.5 μm thickness, with 0.4 μm steps (6–8 images in z-stack), in order to capture the whole cell.
3. Take images of sufficient size, e.g., dimensions minimum of 512 × 512; pixel dwell minimum of 1.6 μm; average minimum 4 (*see* **Note 19**).
4. Perform 3D rendering. The goal of this step is to project the information containing in every plane of individual pictures in the z-stack into one 2D image. 3D rendering can be performed with microscope software or ImageJ.
5. Save 2D images as JPEG or TIFF for analysis in Image J.

3.4 Data Analysis Using Image J (as Outlined in Fig. 2)

1. Open 2D images (saved as JPEG or TIFF), split channels, and convert to 8-bit gray scale. The two channels for analysis

include DSB repair foci (e.g., γH2AX-foci) and heterochromatin (i.e., H3K9me3 and H4K20me3).

2. For analysis, grayscale images have to be converted to strictly black and white pixels, which is achieved by threshold settings. Set Threshold and Apply (*see* **Note 20**) (Fig. 3).
3. Optional: Occasionally, two foci might be so close to one another that they appear as one focus. There is a function in Image J to separate these foci. In Image J: select Process—Binary—Watershed.
4. To analyze the total amount of foci: Select Analyze—Analyze Particles. Determine the size of the γH2AX-foci (pixel^2). This is dependent on the size of the foci in the respective experiment and might therefore need some adjustment. The results can be seen when setting "show outlines."
5. Determine colocalization of DSB repair foci and heterochromatin: Choose the colocalization plug-in and choose channels for analysis. Threshold setting is possible at this time too, but if already performed earlier leave setting as "50" and every "white" pixel will count as the display value 255 (positive), while "black" pixels count as 0 (negative). Two points are considered colocalized if their respective intensities are strictly higher than the threshold of their channels (which is 50 by default, and if their ratio (of intensity) is strictly higher than the ratio setting value (which is 50% by default: ratio (0–100%)). Overlapping pixels will appear in the result window.
6. To determine the amount of colocalized foci: Select Analyze—Analyze Particles (*see* **Note 21**).
7. Perform the Image J-assisted analysis on at least 50 cells of three independent experiments each. Enter data into a spreadsheet, and analyze mean values, standard deviation and other statistics of three independent repeats. Display the result in a column graph showing total foci number and number of foci colocalized with heterochromatin (Fig. 2).

4 Notes

1. Optional: Use of other cell lines with deficiency in DSB signaling, e.g., CJ179 (Artemis genomic deletion [specific genetic lesion not known], no transcript, and null expression), or F02385 (Artemis genomic deletion [specific genetic lesion not known], no transcript, and null expression).
2. Use square or round coverslips of high grade suitable for high-resolution microscopy. Thickness should be 0.13–0.17 mm.
3. If applicable; for G2 analysis only.

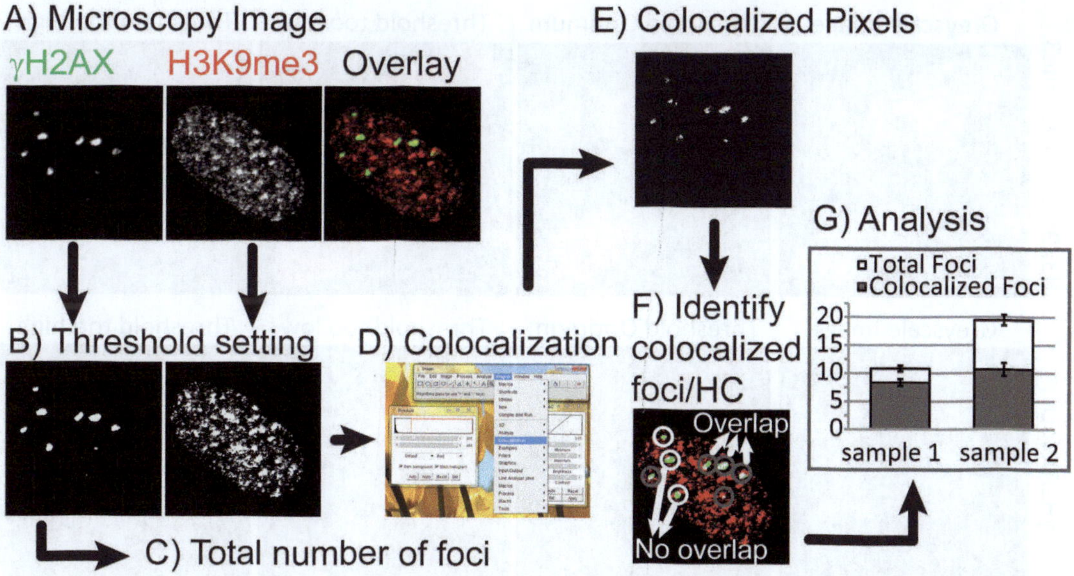

Fig. 2 Workflow of image analysis using Image J. (**a**) Individual channels of the 2D image generated from the z-stack display DSB repair foci (e.g., γH2AX) and heterochromatin (e.g., H3K9me3). (**b**) Setting the threshold using the "Set Threshold" function allows converting the grayscale image into strictly black and white pixels. (**c**) The total amount of DSB repair foci can be analyzed by the function: Select Analyze—Analyze Particles. (**d**) The amount of colocalized DSB repair foci with heterochromatin can be determined by using the colocalization plug-in. (**e**) The results window will show colocalized foci. (**f**) The amount of colocalized DSB repair foci can be analyzed by the function: Select Analyze—Analyze Particles. (**g**) Final results can be displayed using a column graph showing mean values and standard deviations of at least 50 cells of three independent experiments, each

4. The source of DNA damage can be of various origins. In this particular instance, we used gamma irradiation. However, other sources of DNA damage may be used instead, such as X-rays, alpha particles, or DNA damaging drugs such as Neocarcinostatin (NCS) or Bleomycin.
5. Prepare fresh and filter through 0.45-μm sterile filter.
6. Use 1 mg/mL 4″,6-diamidino-2-phenylindole (DAPI) stock.
7. Examples for mounting media are: Fluoromount G, Vectashield, Prolong Gold. Refer to your microscope manual about which mounting medium is the one of choice for your setting.
8. This step may be performed with the microscope software or in Image J or any other program suitable.
9. The program is free of charge and can be downloaded here: http://imagej.nih.gov/ij/; Colocalization plug-in: http://rsb.info.nih.gov/ij/plugins/ and http://rsb.info.nih.gov/ij/plugins/colocalization.html.
10. Optional: Medium exchange after irradiation.

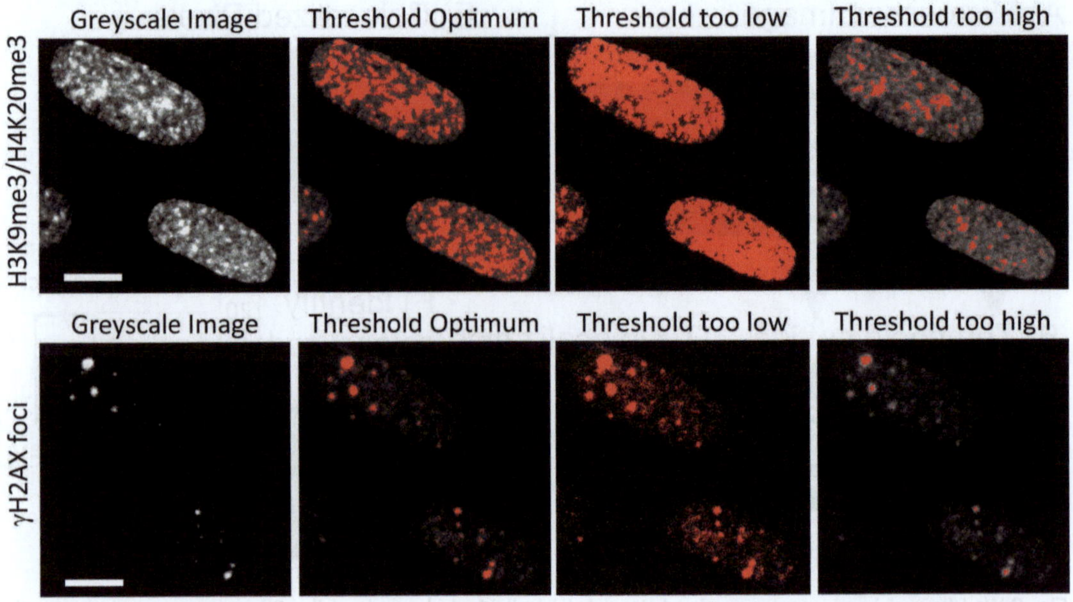

Fig. 3 Troubleshooting threshold settings. Primary human fibroblasts were irradiated with 3 Gy, fixed 24 h post γIR, immunostained for γH2AX and H3K9me3 and H4K20me3 and imaged as z-stacks using a confocal microscope. As a result of 3D rendering, a 2D image was generated showing grayscale images of the H3K9me3/H4K20me3 (*upper panel*) and γH2AX channel (*lower panel*). Optimum threshold settings will cover the stained areas only, not too much or too little. For heterochromatin, all gray areas should be selected, including the lighter areas around the very dark appearing heterochromatin centers. If the threshold is set too low, the whole nucleus will be selected, including areas of chromatin that do not stain or very weakly stain for the histone modification marks, thereby leading to false-positive results. If the threshold is set too high, not all the heterochromatin will be selected, and areas of heterochromatin will be excluded from the analysis, thereby leading to false-negative results. For DSB repair foci, the threshold should be set so that the whole focus is selected. Threshold setting too high will select for background signal, leading to false-positive overlap. If the threshold is set too low, the focus will not be selected completely, and overlap will be missing in the final analysis. Scale bar 10 μm

11. Follow the instructions for fixing the cells for each time point precisely.
12. Follow the instructions for fixing the cells for each time point precisely.
13. Be very sure not to disrupt the cells. They can be removed easily if too vigorous. Do not wash with PBS. Add the preextraction solution right away.
14. Be very exact with this time. Depending on the cell type, this time might be a bit different. Try several conditions between 20 s and 1 min.
15. Do this carefully to prevent cells from dislodging.
16. At this point, you may store the slides for a couple of days to weeks at 4 °C, sealed with parafilm. You may add sodium azide (0.05%) to prevent growth of bacteria/fungi.

17. This step is critical for the staining procedure. Ensure the cells are sufficiently permeabilized to promote equal antibody penetrance into the cells.
18. We recommend consulting with your microscopy manager or a service agent from the microscope company to ensure proper use of the confocal microscope. Capturing images of high quality and optimum resolution is necessary for successful completion of this protocol.
19. Optional: To further increase resolution and image quality, perform deconvolution (refer to the microscopes software for instructions or use another program).
20. Perform this step manually each time. While some reports might suggest a macro to set this threshold, slight variance in staining between slides and intensity settings of the laser during microscopy might have an effect on how to set the threshold. Refer to Fig. 3 to determine optimum threshold settings.
21. Always take a look at the result window with the outlines to make sure that the overlapped area resembles a single focus. Some of the results have to be revisited, as one focus might overlap with HC on two areas (avoid it being count as two) or the area is too small in comparison to the whole foci (exclude false-positives).

References

1. Rogakou EP, Pilch DR, Orr AH, Ivanova VS, Bonner WM (1998) DNA double-stranded breaks induce histone H2AX phosphorylation on serine 139. J Biol Chem 273:5858–5868
2. Burma S, Chen BP, Murphy M, Kurimasa A, Chen DJ (2001) ATM phosphorylates histone H2AX in response to DNA double-strand breaks. J Biol Chem 276:42462–42467
3. Al-Hakim A, Escribano-Diaz C, Landry MC, O'Donnell L, Panier S, Szilard RK, Durocher D (2010) The ubiquitous role of ubiquitin in the DNA damage response. DNA Repair (Amst) 9:1229–1240
4. Poulsen M, Lukas C, Lukas J, Bekker-Jensen S, Mailand N (2012) Human RNF169 is a negative regulator of the ubiquitin-dependent response to DNA double-strand breaks. J Cell Biol 197:189–199
5. Danielsen JR, Povlsen LK, Villumsen BH, Streicher W, Nilsson J, Wikstrom M, Bekker-Jensen S, Mailand N (2012) DNA damage-inducible SUMOylation of HERC2 promotes RNF8 binding via a novel SUMO-binding zinc finger. J Cell Biol 197:179–187
6. Shiloh Y, Shema E, Moyal L, Oren M (2011) RNF20-RNF40: a ubiquitin-driven link between gene expression and the DNA damage response. FEBS Lett 585:2795–2802
7. Moyal L, Lerenthal Y, Gana-Weisz M, Mass G, So S, Wang SY, Eppink B, Chung YM, Shalev G, Shema E et al (2011) Requirement of ATM-dependent monoubiquitylation of histone H2B for timely repair of DNA double-strand breaks. Mol Cell 41:529–542
8. Goodarzi AA, Jeggo PA (2012) The heterochromatic barrier to DNA double strand break repair: how to get the entry visa. Int J Mol Sci 13:11844–11860
9. Goodarzi AA, Jeggo PA (2013) The repair and signaling responses to dna double-strand breaks. Adv Genet 82:1–45
10. Lieber MR (2010) The mechanism of double-strand DNA break repair by the nonhomologous DNA end-joining pathway. Annu Rev Biochem 79:181–211
11. Neal JA, Meek K (2011) Choosing the right path: does DNA-PK help make the decision? Mutat Res 711:73–86
12. Goodarzi AA, Jeggo P, Lobrich M (2010) The influence of heterochromatin on DNA double strand break repair: getting the strong, silent type to relax. DNA Repair 9:1273–1282

13. Iliakis GE, Cicilioni O, Metzger L (1991) Measurement of DNA double-strand breaks in CHO cells at various stages of the cell cycle using pulsed field gel electrophoresis: calibration by means of 125I decay. Int J Radiat Biol 59:343–357
14. Lobrich M, Rydberg B, Cooper PK (1995) Repair of x-ray-induced DNA double-strand breaks in specific Not I restriction fragments in human fibroblasts: joining of correct and incorrect ends. Proc Natl Acad Sci U S A 92:12050–12054
15. Goodarzi AA, Noon AT, Deckbar D, Ziv Y, Shiloh Y, Lobrich M, Jeggo PA (2008) ATM signaling facilitates repair of DNA double-strand breaks associated with heterochromatin. Mol Cell 31:167–177
16. Beucher A, Birraux J, Tchouandong L, Barton O, Shibata A, Conrad S, Goodarzi AA, Krempler A, Jeggo PA, Lobrich M (2009) ATM and Artemis promote homologous recombination of radiation-induced DNA double-strand breaks in G2. EMBO J 28:3413–3427
17. Kim JA, Kruhlak M, Dotiwala F, Nussenzweig A, Haber JE (2007) Heterochromatin is refractory to gamma-H2AX modification in yeast and mammals. J Cell Biol 178:209–218
18. Cowell IG, Sunter NJ, Singh PB, Austin CA, Durkacz BW, Tilby MJ (2007) GammaH2AX foci form preferentially in euchromatin after ionising-radiation. PLoS One 2:e1057
19. Noon AT, Shibata A, Rief N, Lobrich M, Stewart GS, Jeggo PA, Goodarzi AA (2010) 53BP1-dependent robust localized KAP-1 phosphorylation is essential for heterochromatic DNA double-strand break repair. Nat Cell Biol 12:177–184
20. Jakob B, Splinter J, Conrad S, Voss KO, Zink D, Durante M, Lobrich M, Taucher-Scholz G (2011) DNA double-strand breaks in heterochromatin elicit fast repair protein recruitment, histone H2AX phosphorylation and relocation to euchromatin. Nucleic Acids Res 39:6489–6499
21. Chiolo I, Minoda A, Colmenares SU, Polyzos A, Costes SV, Karpen GH (2011) Double-strand breaks in heterochromatin move outside of a dynamic HP1a domain to complete recombinational repair. Cell 144:732–744
22. Riballo E, Kuhne M, Rief N, Doherty A, Smith GC, Recio MJ, Reis C, Dahm K, Fricke A, Krempler A et al (2004) A pathway of double-strand break rejoining dependent upon ATM, Artemis, and proteins locating to gamma-H2AX foci. Mol Cell 16:715–724
23. Woodbine L, Brunton H, Goodarzi AA, Shibata A, Jeggo PA (2011) Endogenously induced DNA double strand breaks arise in heterochromatic DNA regions and require ataxia telangiectasia mutated and Artemis for their repair. Nucleic Acids Res 39:6986–6997
24. Ziv Y, Bielopolski D, Galanty Y, Lukas C, Taya Y, Schultz DC, Lukas J, Bekker-Jensen S, Bartek J, Shiloh Y (2006) Chromatin relaxation in response to DNA double-strand breaks is modulated by a novel ATM- and KAP-1 dependent pathway. Nat Cell Biol 8:870–876
25. Schultz DC, Friedman JR, Rauscher FJ III (2001) Targeting histone deacetylase complexes via KRAB-zinc finger proteins: the PHD and bromodomains of KAP-1 form a cooperative unit that recruits a novel isoform of the Mi-2alpha subunit of NuRD. Genes Dev 15:428–443
26. Li X, Lee YK, Jeng JC, Yen Y, Schultz DC, Shih HM, Ann DK (2007) Role for KAP1 serine 824 phosphorylation and sumoylation/desumoylation switch in regulating KAP1-mediated transcriptional repression. J Biol Chem 282:36177–36189
27. Goodarzi AA, Kurka T, Jeggo PA (2011) KAP-1 phosphorylation regulates CHD3 nucleosome remodeling during the DNA double-strand break response. Nat Struct Mol Biol 18:831–839
28. Collins N, Poot RA, Kukimoto I, Garcia-Jimenez C, Dellaire G, Varga-Weisz PD (2002) An ACF1-ISWI chromatin-remodeling complex is required for DNA replication through heterochromatin. Nat Genet 32:627–632
29. Lan L, Ui A, Nakajima S, Hatakeyama K, Hoshi M, Watanabe R, Janicki SM, Ogiwara H, Kohno T, Kanno S et al (2010) The ACF1 complex is required for DNA double-strand break repair in human cells. Mol Cell 40:976–987
30. Smeenk G, Wiegant WW, Marteijn JA, Luijsterburg MS, Sroczynski N, Costelloe T, Romeijn RJ, Pastink A, Mailand N, Vermeulen W et al (2013) Poly(ADP-ribosyl)ation links the chromatin remodeler SMARCA5/SNF2H to RNF168-dependent DNA damage signaling. J Cell Sci 126:889–903
31. Klement K, Luijsterburg MS, Pinder JB, Cena CS, Del Nero V, Wintersinger CM, Dellaire G, van Attikum H, Goodarzi AA (2014) Opposing ISWI- and CHD-class chromatin remodeling activities orchestrate heterochromatic DNA repair. J Cell Biol 207:717–733
32. Nakamura K, Kato A, Kobayashi J, Yanagihara H, Sakamoto S, Oliveira DV, Shimada M, Tauchi H, Suzuki H, Tashiro S et al (2011) Regulation of homologous recombination by RNF20-dependent H2B ubiquitination. Mol Cell 41:515–528
33. Kakarougkas A, Ismail A, Klement K, Goodarzi AA, Conrad S, Freire R, Shibata A, Lobrich M,

Jeggo PA (2013) Opposing roles for 53BP1 during homologous recombination. Nucleic Acids Res 41:9719–9731

34. Schneider CA, Rasband WS, Eliceiri KW (2012) NIH Image to ImageJ: 25 years of image analysis. Nat Methods 9:671–675

35. Rasband WS ImageJ. U.S. National Institutes of Health, Bethesda, MD. http://imagej.nih.gov/ij/

36. Abramoff MD, Magalhaes PJ, Ram SJ (2004) Image processing with ImageJ. Biophotonics Int 11:36–42

Chapter 23

Phenotypic Analysis of ATM Protein Kinase in DNA Double-Strand Break Formation and Repair

Elisabeth Mian and Lisa Wiesmüller

Abstract

Ataxia telangiectasia mutated (*ATM*) encodes a serine/threonine protein kinase, which is involved in various regulatory processes in mammalian cells. Its best-known role is apical activation of the DNA damage response following generation of DNA double-strand breaks (DSBs). When DSBs appear, sensor and mediator proteins are recruited, activating transducers such as ATM, which in turn relay a widespread signal to a multitude of downstream effectors. *ATM* mutation causes Ataxia telangiectasia (AT), whereby the disease phenotype shows differing characteristics depending on the underlying *ATM* mutation. However, all phenotypes share progressive neurodegeneration and marked predisposition to malignancies at the organismal level and sensitivity to ionizing radiation and chromosome aberrations at the cellular level. Expression and localization of the ATM protein can be determined via western blotting and immunofluorescence microscopy; however, detection of subtle alterations such as resulting from amino acid exchanges rather than truncating mutations requires functional testing. Previous studies on the role of ATM in DSB repair, which connects with radiosensitivity and chromosomal stability, gave at first sight contradictory results. To systematically explore the effects of clinically relevant *ATM* mutations on DSB repair, we engaged a series of lymphoblastoid cell lines (LCLs) derived from AT patients and controls. To examine DSB repair both in a quantitative and qualitative manners, we used an EGFP-based assay comprising different substrates for distinct DSB repair mechanisms. In this way, we demonstrated that particular signaling defects caused by individual *ATM* mutations led to specific DSB repair phenotypes. To explore the impact of ATM on carcinogenic chromosomal aberrations, we monitored chromosomal breakage at a breakpoint cluster region hotspot within the *MLL* gene that has been associated with therapy-related leukemia. PCR-based *MLL*-breakage analysis of HeLa cells treated with and without pharmacological kinase inhibitors revealed ATM-dependent chromatin remodeling at the *MLL* break site giving access to DNA repair proteins but also nucleases triggering *MLL* rearrangements. This chapter summarizes these methods for functional characterization of ATM in patient LCLs and human cell lines.

Key words Ataxia telangiectasia, Breast cancer risk, Chromatin decondensation, DSB repair assay, Endonuclease G, Genomic PCR, *MLL* breakpoint cluster region, Replication stress

1 Introduction

ATM is a relatively large protein with a molecular weight of 350 kDa, comprising 3056 residues [1, 2]. The functionally most important part of the ATM protein is its carboxy-terminal

Sergei V. Kozlov (ed.), *ATM Kinase: Methods and Protocols*, Methods in Molecular Biology, vol. 1599, DOI 10.1007/978-1-4939-6955-5_23, © Springer Science+Business Media LLC 2017

catalytically active site, which contains a phosphoinositol-3 kinase (PI3K) signature despite a protein kinase activity of ATM [3, 4]. This PI3K domain constitutes ~10% of the ATM protein, while the rest contains regulatory and interaction domains determining the specificity of ATM [2]. ATM preferentially targets S/T-Q motifs, i.e., phosphorylates Serine or Threonine preceding Glu [5, 6]. Together with the catalytic subunit of DNA-dependent protein kinase (DNA-PKcs) and ATM- and RAD3-related (ATR), ATM is one of the major players in response to genotoxic stress [2]. These three PI3K-related kinases (PIKKs) can activate each other, so ATR phosphorylates ATM Serine 1981 in cells exposed to DNA damaging modalities that induce stalled replication forks [7]. ATM Serine 1981 phosphorylation has been used as a marker to detect ATM activation, it increases for example with the dose of γ-ray treatment in a linear manner up to the stoichiometric maximum [8, 9]. Three additional phosphorylation sites have been identified in activated ATM [10, 11]. Phosphorylation at these sites seems to be required for the retention of activated ATM at DSBs although not for the initial recruitment. When DSBs are formed, a portion of ATM is rapidly recruited to the DNA damage sites [12], where quiescent homodimers dissociate into active monomers [8].

The delicate balance in cellular DNA damage signaling is disrupted by the loss of functional ATM in AT patients. Clinical features include progressive neurodegeneration, telangiectasia, immunodeficiency, thymic and gonadal atrophy, and marked predisposition to malignancies [13, 14]. Cells derived from AT patients show defective cell cycle checkpoint responses to ionizing radiation (IR), profound radiosensitivity, and high levels of chromosome aberrations, indicating the importance of ATM for radioprotection and maintenance of chromosome stability [13]. Differences in the individual expression of the AT phenotype were described as exemplified best by the so-called AT-variant patients with another or a milder clinical phenotype and variable radiosensitivities associated with less deleterious missense, in-frame-deletion, or extreme C-terminal truncating *ATM* mutations [15]. Of further interest regarding differences in clinical phenotypes and biochemical functions associated with specific *ATM* mutations, more than 70% of the pathogenic mutations are ATM truncating mutations, associated with loss of ATM protein. The remaining pathogenic missense and in-frame deletion mutations are localized in the C-terminal part spanning the FAT (FRAP, ATM, TRRAP) and PI3K domains and are associated with residual ATM protein levels [16–18]. Interestingly, loss of functional ATM and pharmacological inhibition of ATM kinase activity were reported to result in different cellular phenotypes [1]. While inhibition of ATM in normal cells results in defective sister chromatid exchanges (SCE) during DNA replication, SCE works normally in AT cells lacking ATM. The authors concluded that functionally inactive ATM physically blocks the repair of replication-associated damage. From this, differences in the AT phenotype could also be explained by expression of dominant negative ATM versus loss of ATM function [19].

Adding to the complexity of AT phenotypes, Tomimatsu et al. [20] showed that many DDR activities considered ATM-dependent were markedly attenuated in AT patients rather than completely lost, because distinct ATM functions are taken over by DNA-PK and ATR. MRE11-RAD50-Nibrin (MRN) complex components exert functions upstream and downstream of ATM [2]. Consequently, mutations in the corresponding genes underlying ataxia telangiectasia-like disorders (ATLD), Nijmegen breakage syndrome (NBS)-like disorder, and NBS, respectively, cause phenotypes partially resembling AT [21–23]. While bi-allelic *ATM* mutations cause AT, mono-allelic AT mutations confer risk to breast cancer [17, 18]. Consistently, when Keimling et al. [16] analyzed DSB repair in a series of *ATM*-mutated patient LCLs DSB repair pathway usage resembled the dysfunctional pattern after inactivation of the breast cancer susceptibility gene products BRCA1, BRCA2, RAD50, or Nibrin. In particular, AT cells with in-frame deletion of amino acids 2546–2548 in the ATM FAT domain showed a shift from error-free homologous recombination (HR) to error-prone single-strand annealing (SSA) and error-prone microhomology-mediated end joining (MMEJ) like *BRCA1*-mutated cells, whereas AT cells with V2716A amino acid exchange within the Nibrin-binding segment were compromised in homology-directed repair like NBS-like or NBS patient cells [16]. Because many DSB repair factors such as BRCA1, MRE11, and Nibrin are ATM substrates [2], these data suggested that differences in AT cell functions may be the result of inactive kinase activities toward a certain fraction but not all ATM kinase substrates.

There are differing theories about how ATM is activated in response to DSBs: Bakkenist et al. [8] showed that chromatin conformational changes following DSB formation rather than direct contact with broken DNA trigger ATM signaling. Others postulated that direct interaction of ATM with broken DNA causes the activation [24], specifically the contact with single-stranded stretches at DSBs [25]. The interaction of ATM with Nibrin, which together with MRE11 and RAD50 constitutes the MRN complex, seems to be crucial for ATM recruitment to and retention at DSB sites [2]. The sensor protein MDC1 (mediator of DNA damage checkpoint protein 1) is anchored at DSBs via interaction with phosphorylated histone H2AX (γH2AX) [26] and in parallel binds ATM, facilitating ATM to phosphorylate additional H2AX which in turn binds more MDC1—an effective process for amplification and maintenance of γH2AX histone marks at the DSB [26, 27]. Furthermore ATM-dependent phosphorylation of MDC1 enhances its oligomerization at DSB sites and thus the DDR [28, 29]. Higher order chromatin serves as a barrier to the expansion of

ATM-dependent H2AX phosphorylation [30, 31]. Notably, ATM was also reported to phosphorylate the transcriptional co-repressor Krüppel-associated box (KRAB) associated protein (KAP)-1 at Serine 824 causing transient chromatin relaxation [32]. Moreover, ATM-dependent phosphorylation of the E3 ubiquitin ligase RNF20-RNF40 triggers monoubiquitylation of histone H2B. At DSBs this series of posttranslational modifications permits recruitment of NHEJ and HR repair factors [33]. However, the same ATM-dependent signaling cascade is also active in response to replication stress, e.g., after Aphidicolin-mediated DNA polymerase inhibition. Under these conditions the cascade gives access to nucleases such as Endonuclease G, which accidentally cleaves the *Mixed Lineage Leukemia* (*MLL*) breakpoint cluster region (*MLL*bcr) leading to leukemogenic *MLL* rearrangements [34]. Taken together, ATM not only responds to chromatin perturbations but also induces chromatin decondensation at DSBs as a prerequisite for most DNA metabolic processes.

When DSBs are generated, the MRN complex binds [35] and acts like a clamp, facilitating the unwinding of local DNA [36] and DNA end resection initiated by MRE11's endonuclease activity. MRE11 nicks the DNA upstream of the DSB followed by 3′→5′end resection by MRE11 and subsequently other nucleases to produce 3′ single-stranded DNA overhangs [37], which are required for RAD51 binding and initiation of HR [38]. The MRN complex activates ATM, which in turn phosphorylates MRE11, RAD50, and Nibrin [39, 40]. ATM phosphorylates MRE11 at S676 and S678 in response to DSBs [39]. ATM-dependent phosphorylation of MRE11 seems to limit the extent of resection through Exonuclease 1. Mutating the ATM phosphorylation sites on MRE11 leads to a defect in HR [39]. Phosphorylation of Nibrin regulates the S-phase checkpoint [41] and controls the accumulation of Nibrin and ATM at DSB sites [42]. RAD50 is phosphorylated by ATM at S635 in response to DSB [43], important for ATM-dependent cell cycle checkpoint control and DSB repair [39]. ATM was also seen to become activated independently of DNA damage and the MRN complex in cells challenged with hyperthermia [44], hypoxia [45], or hyperoxia through replication stress [46]. Moreover, chromatin-monitoring KAT5 (Tip60) acetyltransferase is another upstream component required for acetylation and maximal activation of ATM in cells exposed to low doses of ionizing radiation [47]. The first ATM-mediated phosphorylation substrate characterized in detail was p53 [48, 49]. p53 plays a central role in the modulation of the cellular transcriptome following DSB-induced activation by ATM [50]. While ATM-activated p53 suppresses DSB repair and leads to the expression of pro-apoptotic factors [51], ATM also activates NF-κB (nuclear factor-κB), which promotes DSB repair and the expression of several anti-apoptotic genes [52]. Another ATM-dependent process, the phosphorylation of nuclear proteasome activator PA28γ (PSME3) leads to the rapid recruitment of proteasomes to sites of DNA damage. This process is important to ensure efficient NHEJ versus HR repair of DSBs [53]. Altogether, ATM represents a master switch for the maintenance of the balance between different DNA repair and DDR pathways ultimately leading to life and death decisions in the damaged cell.

In this chapter, we describe more recently established methods for the functional characterization of the cellular ATM status. First, we provide details for the analysis of individual DSB repair activities in LCLs derived from AT patients and healthy controls with an EGFP-based assay comprising different substrates for the DSB repair mechanisms NHEJ, HR, and SSA, as well as combinations thereof [54, 55] (Figs. 1 and 2). Second, we describe the analysis of breakage at a 0.4 kb hotspot of therapy-induced rearrangements in the human *MLL*bcr. Here, we exemplarily use HeLa cells, which we treated with and without Caffeine, a pharmacological inhibitor

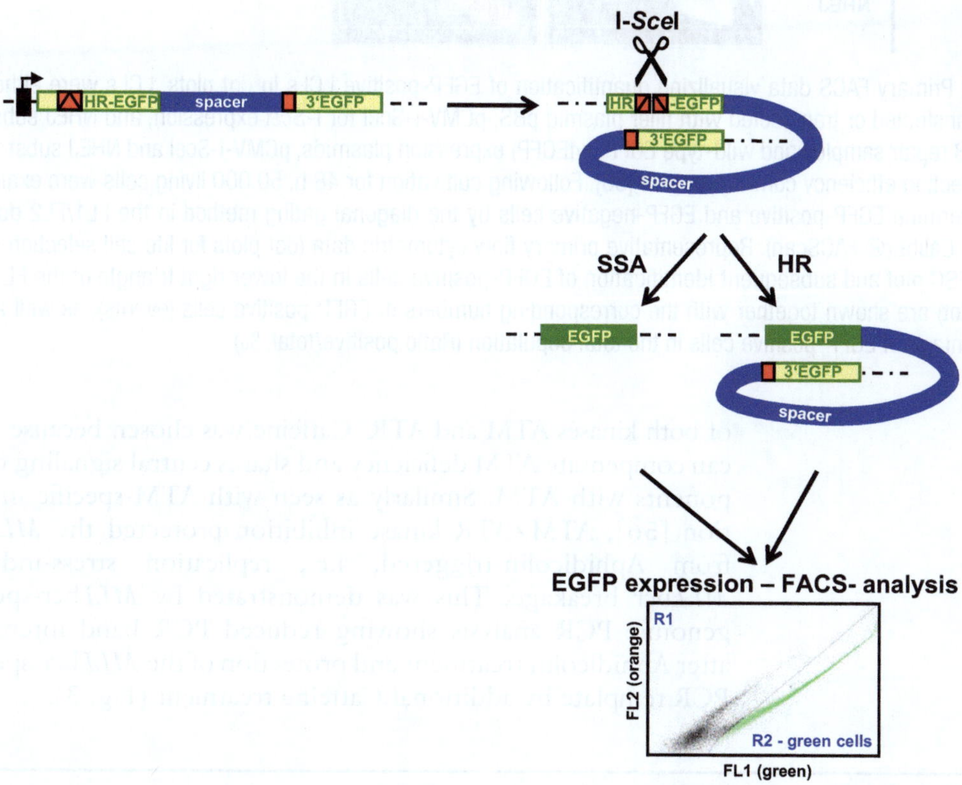

Fig. 1 The EGFP-based DSB repair assay system. DSB repair initiated after cleavage by the endonuclease I-*Sce*I (scissors) at its recognition site (*red triangle*) is representatively shown for construct HR-EGFP/3′EGFP that monitors SSA and HR [54]. Restoration and expression of wild-type EGFP indicates successful DSB repair as detected by FACS analysis. Quantification of green fluorescent cells in repair assays is normalized by the use of the individually determined transfection efficiency to calculate the DSB repair frequency. Kinked arrow, CMV promoter; deletion, red square

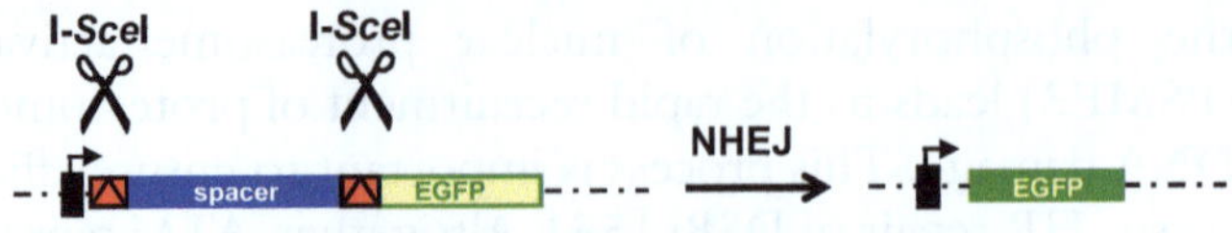

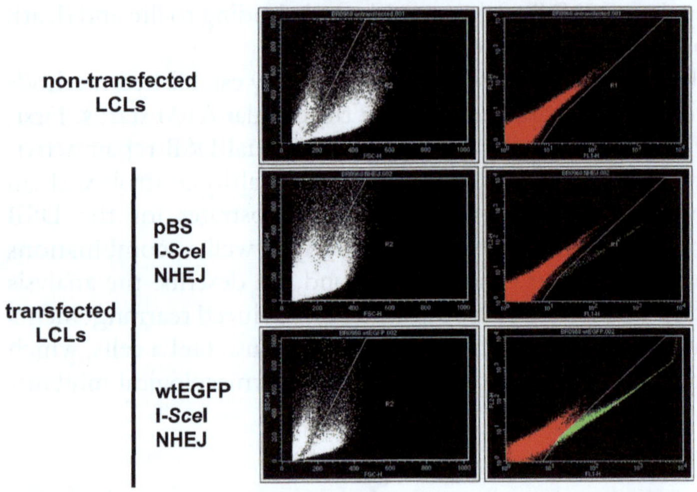

	EGFP-positive cells	
	events	Ratio positive/total (%)
non-transfected	0	0.00
pBS I-*Sce*I NHEJ	192	0.04
wtEGFP I-*Sce*I NHEJ	6632	13.26

Fig. 2 Primary FACS data visualizing quantification of EGFP-positive LCLs in dot plots. LCLs were either left nontransfected or transfected with filler plasmid pBS, pCMV-I-SceI for I-*Sce*I expression, and NHEJ substrate in DSB repair samples and wild-type EGFP (wtEGFP) expression plasmids, pCMV-I-SceI and NHEJ substrate in transfection efficiency control samples [56]. Following cultivation for 48 h, 50,000 living cells were examined to determine EGFP-positive and EGFP-negative cells by the diagonal gating method in the FL1/FL2 dot plot (FACS Calibur® FACScan). Representative primary flow cytometric data (dot plots for life cell selection in the SSC/FSC plot and subsequent identification of EGFP-positive cells in the lower right triangle of the FL1/FL2 dot plot) are shown together with the corresponding numbers of EGFP-positive cells (events), as well as the percentage of EGFP-positive cells in the total population (Ratio positive/total, %)

of both kinases ATM and ATR. Caffeine was chosen because ATR can compensate ATM deficiency and shares central signaling components with ATM. Similarly as seen with ATM-specific inhibition [56], ATM/ATR kinase inhibition protected the *MLL*bcr from Aphidicolin-triggered, i.e., replication stress-induced *MLL*bcr breakage. This was demonstrated by *MLL*bcr-specific genomic PCR analysis showing reduced PCR band intensities after Aphidicolin treatment and protection of the *MLL*bcr-specific PCR template by additional Caffeine treatment (Fig. 3).

2 Materials

2.1 Cultivation of LCLs

1. Medium without antibiotics: 500 mL RPMI Medium 1640 (1×) with L-glutamine and Phenol Red, 75 mL (15%) fetal bovine serum (FBS), 7.5 mL (1.5%) L-Glutamine.

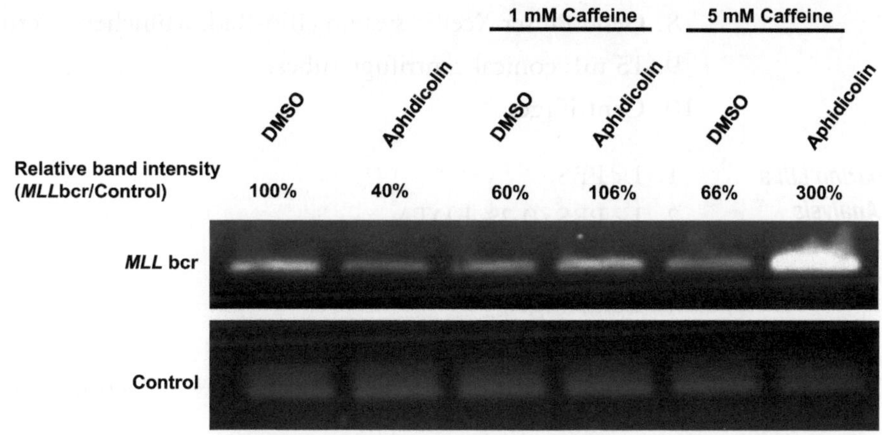

Fig. 3 PCR analysis of the effect of Aphidicholin on the integrity of the *MLL*bcr with or without inhibition of ATM and ATR by Caffeine. Genomic PCR analysis after 4 h of treatment with Aphidicolin (10 μM) showed a decrease of *MLL*bcr-specific PCR band intensity indicating breakage. Inclusion of the ATM and ATR inhibitor Caffeine (1 mM, 5 mM) protected against *MLL*bcr breakage. Band intensities were quantified and normalized to the corresponding band intensity from the *MLL*intron20 control, which was unaffected by Aphidicolin or Caffeine treatment. Mean values of relative band intensities from two experiments are shown, whereby relative band intensities for the mock-treated DMSO control were set to 100% each

2. Medium with antibiotics: 500 mL RPMI Medium 1640 (1×) with L-glutamine and Phenol Red, 75 mL (15%) FBS, 7.5 mL (1.5%) 100× L-Glutamine, Penicillin- Streptomycin.
3. 10 cm culture dishes for suspension culture cells.
4. Cell incubator.
5. Laminar flow hood.

2.2 Electroporation of LCLs

1. RPMI medium (without Phenol Red).
2. Antibiotic-free RPMI medium (*see* Subheading 2.1).
3. 6-Well cell culture plates for suspension culture cells.
4. Trypan blue solution and counting chamber.
5. DSB repair plasmid mixture (containing 10 μg each of one of the DSB repair substrates EJ5SceGFP, EJ-EGFP, 5′EGFP/HR-EGFP, HR-EGFP/3′EGFP, HR-EGFP/5′EGFP, the meganuclease expression plasmid pCMV-I-SceI and the filler plasmid pBS) at a concentration of 1 μg/μL [16, 54, 56].
6. Transfection control plasmid mixture (containing 30 μg of plasmid DNA with 10 μg each of one of the DSB repair substrates, the meganuclease expression plasmid pCMV-I-SceI and wild-type-*EGFP* expression plasmid) at a concentration of 1 μg/μL [16, 54, 56].
7. Gene Pulser®/MicroPulser™ Electroporation Cuvettes, 0.4 cm gap (Bio-Rad, München, Germany).

8. Gene Pulser Xcell™ system (Bio-Rad, München, Germany).
9. 15 mL conical centrifuge tubes.
10. Centrifuge.

2.3 Preparing LCLs for FACS Analysis

1. 1× PBS.
2. 1× PBS/0.2% EDTA.
3. 12 mL centrifuge tubes.
4. Centrifuge.
5. Ice.
6. Flow cytometer (FACSCalibur™ Becton Dickinson, Heidelberg, Germany).

2.4 Cultivation of HeLa Cells

1. HeLa medium with antibiotics (room temperature): 500 mL DMEM Medium (1×) High Glucose with Phenol Red, 50 mL (10%) FBS, 5 mL (1%) L-Glutamine with Penicillin–Streptomycin.
2. HeLa medium without antibiotics (room temperature): 500 mL DMEM Medium (1×) High Glucose with Phenol Red, 50 mL (10%) FBS, 5 mL 200 mM L-Glutamine.
3. PBS at room temperature.
4. Trypsin/EDTA at room temperature.
5. 1000 μL, 5 mL, and 10 mL pipettes.
6. Laminar flow hood.
7. 15 mL conical centrifuge tubes.
8. 10 cm tissue-culture-treated plastic dishes for adherent cells.
9. 6-well cell culture plates.
10. Cell incubator.
11. Centrifuge.

2.5 Treatment of HeLa Cells with Aphidicholin and/or Caffeine and Cell Harvest

1. Antibiotic-free DMEM medium (*see* Subheading 2.4).
2. 30 mM Aphidicolin stock solution in DMSO.
3. DMSO.
4. 50 mM Caffeine stock solution (powder dissolved in distilled water at 37 °C overnight).
5. 6-well cell culture plates for adherent cells.
6. 2.5 μL, 5 mL, and 10 mL pipettes.
7. Trypsin/EDTA solution.
8. PBS.
9. 15 and 50 mL centrifuge tubes.
10. Centrifuge.

2.6 Genomic DNA Isolation with High Pure PCR Template Preparation Kit

1. High Pure PCR Template Preparation Kit (Roche, Mannheim, Germany).

Material not supplied by the kit:

2. Water bath at 70 °C.
3. Isopropanol.
4. Pipettes.
5. 2.5 mL microcentrifuge tubes.
6. Table-top centrifuge.

2.7 Genomic PCR

1. Ice.
2. NanoDrop 2000 spectrophotometer (Thermo Scientific, Dreieich, Germany).
3. PCR machine.
4. 0.2 mL PCR Tube Strips.
5. Taq PCR Core Kit (Qiagen, Hilden, Germany).
6. Distilled water.
7. 20 ng genomic DNA.
8. Pipettes for 2–20 μL.
9. Primers for PCR product *MLL*bcr, a 351 bp fragment of the *MLL* breakpoint cluster region at the intron11-exon12 boundary: Forward: 5′-ggc tca caa cag act tgg caa t-3′ and Reverse: 5′-tat ttc ccc cac ccc act cc-3′. Primers for control PCR product *MLL*intron20 (control), a 236 bp control fragment of the genomic *MLL* intron20: Forward: 5′-gca aca cag ggc cct agt taa t-3′ and Reverse: 5′-agg caa atc agt cac ctt tta atc a-3′.

2.8 Agarose Gel Electrophoresis

1. Seakem LE Agarose (BMA, Rockland, ME, USA).
2. Microwave oven.
3. Measuring cup.
4. Scale P1200 balance.
5. 50× TAE stock buffer: 2 M Tris, glacial acetic acid, 0.1 M EDTA, pH 8.0 (adjust the pH with glacial acetic acid).
6. 1× TAE Buffer: 20 mL 50× TAE stock Buffer in 1000 mL with demineralized water (Final (1×) working concentration: 0.04 M Tris–Acetate, 0.002 M EDTA).
7. Electrophoresis Cell.
8. DNA Loading Dye 6×.
9. DNA molecular weight markers, e.g., GeneRuler™ DNA Ladder Mix (100–10,000 bp) (MBI-Fermentas, St. Leon-Rot, Germany).
10. Ethidium bromide (1 μg/mL Ethidium bromide in 1× TAE-buffer).

11. Gel imaging system, e.g., ChemiImager™ 5500 (Alpha Innotech, San Leandro, CA, USA).
12. ImageJ software (https://imagej.nih.gov/ij/National Institutes of Health, Maryland, USA).

3 Methods

3.1 Cells

LCLs were established by EBV immortalization of peripheral white blood cells [57] from AT patients and patients with breast cancer identified through clinical diagnosis combined with direct sequencing of the *ATM* gene using previously reported primers and conditions [58, 59]. Control LCLs 416MI [60], TK6 [54], HA169 [61], and HA256 [23] were described previously. Control LCLs HA169 and HA256 originated from unaffected mutation non-carriers within AT families and LCL 416MI from a healthy donor. HeLa cells were described in Gole et al. [34].

3.2 Cultivation of LCLs [16, 54, 56] (See Note 1)

1. Split the cells two to three times per week by using medium with antibiotics (*see* Subheading 2.1).
2. Optimal cell density: 2×10^5/mL to 2×10^6/mL (cells are sticky, in order to count cells you have to pipette carefully up and down to avoid cell clumps and to be able to count single cells).
3. Depending on the cell density split the cells 1:2–1:10.
4. During experiments the cells should be cultured using medium without antibiotics.

3.3 Electroporation of LCLs [16, 54, 56]

1. Pre-warm the antibiotic-free RPMI medium.
2. Prepare 6-well plates by filling the appropriate number of wells with 2 mL of antibiotic-free RPMI medium.
3. Count the cells.
4. Centrifuge the required number of cells (4×10^6 cells per sample) at $300 \times g$ for 5 min.
5. Discard the supernatant.
6. Resuspend the pellet in cold RPMI medium (without Phenol Red) to a final concentration of 1×10^6 cells/100 μL.
7. Transfer 30 μL of the plasmid mixture (*see* Subheading 2.2, **item 5**) (concentration 1 μg/μL) in a cuvette.
8. Transfer 400 μL of the cell suspension in the cuvette.
9. Insert the cuvette into the cuvette holder and start the Gene Pulser Xcell Program (Voltage: 200 V, Capacitance 1050 μF).
10. Transfer the sample from the cuvette into the prepared 6-well plate immediately after the program has finished.
11. Repeat **steps 7–10** for the remaining samples.
12. Incubate the cells for 48 h before preparing them for flow cytometry.

3.4 Preparing LCLs for FACS Analysis [16, 54, 56] (See Note 2)

1. Transfer the medium with your cells to 12 mL centrifuge tubes.
2. Wash the well with 2 mL 1× PBS and transfer it to the same centrifuge tube.
3. Centrifuge at room temperature for 5 min at 300 × *g*.
4. Remove and discard the supernatant.
5. Wash the cells in 1 mL 1× PBS/0.2% EDTA.
6. Centrifuge at room temperature for 5 min at 300 × *g*.
7. Remove and discard the supernatant.
8. Resuspend the cells in 200 μL 1× PBS/0.2% EDTA.
9. Incubate suspension on ice until used for analysis (maximally 1 h).
10. Perform flow cytometry using FACS Calibur™ FACScan (Figs. 1 and 2). First, identify live cells by the use of the forward scatter channel (FSC) and side scatter channel (SSC), whereby the intensity of the FSC corresponds to the particle's size distinguishing between cellular debris and living cells and of the SSC to the granularity of a cell. Second, identify EGFP-positive cells among live cells upon excitation with the argon laser at 488 nm by the use of the diagonal gating method in the FL1/FL2 dot plot, where FL1 detects green fluorescence and FL2 orange autofluorescence of the cells (Figs. 1 and 2).

3.5 Statistical Data Analysis [56] (See Note 3)

The fraction of green fluorescent versus non-fluorescent cells quantified in repair assay samples is normalized by the use of the individually determined transfection efficiency to calculate the DSB repair frequency, thereby excluding differences in transfection, transcription, translation, proliferation, and lethality. The statistical significance of differences between mean DSB repair frequencies is determined using nonparametric Mann-Whitney test for unpaired samples with the software GraphPad Prism 5.01 (GraphPad, San Diego, CA, USA).

3.6 Cultivation of HeLa Cells [34]

1. Prepare DMEM medium with antibiotics (*see* Subheading 2.4), PBS and trypsin, adjust to room temperature, and arrange together with new 10 cm dishes and/or 6-well plates to be used.
2. Aspirate DMEM medium with antibiotics and wash the cells with 5 mL PBS.
3. Aspirate PBS.
4. Add 2 mL of Trypsin/EDTA to cover the surface of the dish.
5. Return cells to incubator for approximately 2 min.
6. Add 10 mL of DMEM medium with antibiotics to the dish, so that the FBS in the medium quenches the trypsinization.
7. Pipette media up and down purging it around the surface of the dish to remove adherent cells from the surface.

8. Transfer the cells into a 15 mL centrifuge tube.
9. Centrifuge at 300 × *g* for 5 min and discard the supernatant.
10. Resuspend the cell pellet with 1 mL of DMEM medium with antibiotics and split the cells in a new 10 cm dish as needed. Spread the cells evenly by rocking the dish back and forth. Do not spread the cells by swirling the media in a circular motion, as this results in clumping of cells in the middle of the dish.
11. Cells are stored in an incubator at 37 °C and 5% CO_2.

3.7 Treatment of HeLa Cells with Aphidicholin and/or Caffeine [34] (See Note 4)

1. Day 1: Count the cells and seed 150,000 HeLa cells per well in a 6-well plate with 2.0 mL antibiotic-free HeLa medium. For each treatment three wells are to be seeded from a corresponding cell suspension master mixture.
2. Day 2: Change the antibiotic-free medium of the cells.
3. Day 3: Treat the cells with:
   (a) DMSO (6 mL antibiotic-free medium + 2 μL DMSO).
   (b) 10 μM Aphidicholin (6 mL antibiotic-free medium +2 μL 30 mM Aphidicholin stock solution).
   (c) DMSO + 1 mM Caffeine (6 mL antibiotic-free medium + 2 μL DMSO + 120 μL 50 mM Caffeine stock solution).
   (d) DMSO + 5 mM Caffeine (6 mL antibiotic-free medium + 2 μL DMSO + 600 μL 50 mM Caffeine stock solution).
   (e) 10 μM Aphidicholin +1 mM Caffeine (6 mL antibiotic-free medium + 2 μL Aphidicholin + 120 μL 50 mM Caffeine stock solution).
   (f) 10 μM Aphidicholin + 5 mM Caffeine (6 mL antibiotic-free medium + 2 μL Aphidicholin + 600 μL 50 mM Caffeine stock solution).

   After 4 h of treatment harvest the cells and start with the genomic DNA extraction.

3.8 HeLa Cell Harvest from 6-Well Plates for DNA Extraction [34]

1. Aspirate the cell media and wash the cells with 2 mL of PBS.
2. Aspirate the PBS.
3. Trypsinize the cells with 0.5 mL trypsin per well.
4. Transfer the suspension to 15 mL centrifuge tube.
5. Wash the wells with 2 mL of PBS and add to the suspension.
6. Centrifuge at 300 × *g* for 10 min.
7. Discard the supernatant.
8. Resuspend the cells in 200 μL of PBS.

3.9 Genomic DNA Isolation with High Pure PCR Template Preparation Kit [34, 55]

(Follow the manufacturer's instructions: Roche, Mannheim, Germany)

1. Before starting the purification, warm the Elution Buffer to 70 °C.
2. Add 200 μL of sample material in PBS (*see* Subheading 3.8), 200 μL Binding Buffer and 40 μL Proteinase K (reconstituted) solution into a nuclease-free 1.5 mL microcentrifuge tube.
3. Mix immediately and incubate at 70 °C for 10 min.
4. Add 100 μL Isopropanol and mix well.
5. Insert one High Pure Filter Tube into one Collection Tube and pipet the sample into the upper buffer reservoir of the filter tube.
6. Insert the entire High Pure Filter Tube assembly into the table-top centrifuge and centrifuge for 1 min at 8000 × *g*.
7. After centrifugation remove the Filter Tube from the Collection Tube; discard the flow-through and the Collection Tube.
8. Combine the Filter Tube with a new Collection Tube and add 500 μL Inhibitor Removal Buffer to the upper reservoir of the Filter Tube. Afterward centrifuge again for 1 min at 8000 × *g*.
9. Remove the Filter Tube from the Collection Tube, discard the flow-through and the Collection Tube.
10. Combine the Filter Tube with a new Collection Tube and add 500 μL Wash Buffer to the upper reservoir of the Filter Tube. Centrifuge for 1 min at 8000 × *g*.
11. Remove the Filter Tube from the Collection Tube; discard the flow-through and the Collection Tube.
12. Combine the Filter Tube with a new Collection Tube and add 500 μL Wash Buffer to the upper reservoir of the Filter Tube. Centrifuge for 1 min at 8000 × *g*.
13. After discarding the flow-through, centrifuge the entire High Pure assembly for other 10 s at a full speed to ensure the removal of residual Wash Buffer.
14. Discard the Collection Tube.
15. For the DNA-elution: insert the Filter Tube into a clean, sterile 1.5 mL microcentrifuge tube and add 200 μL pre-warmed Elution Buffer to the upper reservoir of the Filter Tube. Centrifuge the tube assembly for 1 min at 8000 × *g*.
16. The microcentrifuge tube now contains the eluted, purified DNA, which can be used directly or stored for later analysis. From now on, the DNA should stay at 4 °C.

3.10 Genomic PCR [34, 55]

1. Measure the DNA concentration of your samples with NanoDrop 2000 (Thermo Scientific, Dreieich, Germany). Make sure that your samples are on ice.

2. Mix the DNA with distilled water in the 0.2 mL PCR tubes as you calculated before and add 10 μL of the PCR mixture.

   PCR mixture for each sample:

   - 10.0 ng of genomic DNA in 40 μL distilled water for samples (50 μL distilled water without DNA for negative control sample).
   - 5.0 μL of 10× Taq buffer (Taq PCR core kit, Qiagen).
   - 1.5 μL of 5 mM dNTP (Taq PCR core kit, Qiagen).
   - 0.5 μL of forward primer (100 pM/μL).
   - 0.5 μL of reverse primer (100 pM/μL).
   - 0.2 μL of 5 U/μL Taq Polymerase (Taq PCR core kit, Qiagen).
   - 2.3 μL distilled water.

3. Put the PCR tubes into the PCR machine and run with the following settings:

   PCR conditions for amplification of the *MLL*bcr are: Initial denaturation for 5 min at 95 °C; amplification for 45 cycles with denaturation for 1.5 min at 95 °C, hybridization for 1 min at 68 °C, DNA synthesis for 40 s at 72 °C; final DNA synthesis for 7 min at 72 °C before cooling down to 4 °C.

   PCR conditions for amplification of the control *MLL*intron20 fragment are: Initial denaturation for 5 min at 95 °C; amplification for 45 cycles with denaturation for 1.5 min at 95 °C, hybridization for 1 min at 60 °C, DNA synthesis for 40 s at 72 °C; final DNA synthesis for 7 min at 72 °C before cooling down to 4 °C.

3.11 Agarose Gel Electrophoresis [34, 55]

1. Prepare a 1.5% agarose gel for electrophoresis: Take 1.5 g of agarose and mix it with 100 mL 1× TAE. Put it for 4 min in the microwave oven to dissolve the agarose completely. Fill the gel in the casting tray after inserting the comb into the slots and let it cool down. Look for bubbles and eliminate them.
2. When the gel is hardened, you can remove the comb, the wells are shaped.
3. Put your gel into the electrophoresis system and fill the chamber with 1× TAE.
4. Add 10 μL of GeneRuler™ DNA Ladder Mix (100–10,000 bp in the first well).
5. Mix in a 500 μL microcentrifuge tube 1 μL of 6× loading dye with 10 μL of your samples and pipette in the different wells.
6. Run gel electrophoresis at 100 V.
7. After gel electrophoresis put your gel for 10 min in an ethidium bromide bath.
8. Take a picture of your gel in the gel documentation system.

9. Quantify the bands with the software ImageJ (National Institutes of Health).
10. Normalize each band intensity for the *MLL*bcr PCR to the corresponding band of your control *MLL*intron20 PCR (Fig. 3).

4 Notes

1. For functional analysis of LCLs it is critical to determine the stability of the phenotype over a series of culture passages. Typically, LCL passage numbers should be below or equal to 25. Therefore, large batches of early passage LCLs should be prepared and frozen for phenotypic analysis by different approaches after thawing (*see* Subheading 3.2).
2. During optimization of DSB repair analyses of patient LCLs, we added several quality criteria to maximize sensitivity, specificity, and robustness of the EGFP-based testing [56], namely preparation of frozen cell stocks and plasmid batches in sufficiently large amounts to ensure the same cell and DNA quality during the whole study including all replicates. Moreover, at least two independent experiments had to be performed, whereby each sample had to be analyzed at least in triplicates plus individual transfection efficiency samples in at least duplicates. Moreover, at least one reference cell line was included in each experiment. Finally, a minimum transfection efficiency for HR measurements of ≥10% was a prerequisite for HR data evaluation (*see* Subheading 3.2).
3. To neutralize differences in the absolute values of DSB repair frequencies between experimental days, mean values for the reference samples were set to 100% per experimental day and relative DSB repair frequencies determined for individual measurements of the remaining samples (*see* Subheading 3.5).
4. Before analyzing *MLL*bcr breakage by genomic PCR, it is important to perform time course experiments for each cell type and treatment protocol to identify the time point of maximum *MLL*bcr breakage (*see* Subheading 3.7).

References

1. Choi S, Gamper AM, White JS, Bakkenist CJ (2010) Inhibition of ATM kinase activity does not phenocopy ATM protein disruption: implications for the clinical utility of ATM kinase inhibitors. Cell Cycle 9:4052–4057
2. Shiloh Y, Ziv Y (2013) The ATM protein kinase: regulating the cellular response to genotoxic stress, and more. Nat Rev Mol Cell Biol 14:197–210
3. Lovejoy CA, Cortez D (2009) Common mechanisms of PIKK regulation. DNA Repair 8:1004–1008
4. Lempiäinen H, Halazonetis TD (2009) Emerging common themes in regulation of PIKKs and PI3Ks. EMBO J 28:3067–3073

5. Kim ST, Lim DS, Canmann C, Kastan MB (1999) Substrate specificities and identification of ATM kinase family members. J Biol Chem 274:37538–37543
6. O'Neill T, Dwyer AJ, Ziv Y, Chan DW, Lees-Miller SP, Abraham RH, Lai JH, Hill D, Shiloh Y, Cantley LC, Rathbun GA (2000) Utilization of oriented peptide libraries to identify substrate motifs selected by ATM. J Biol Chem 275:22719–22727
7. Stiff T, Walker SA, Cerosaletti K, Goodarzi AA, Petermann E, Concannon P, O'Driscoll M, Jeggo PA (2006) ATR-dependent phosphorylation and activation of ATM in response to UV treatment or replication fork stalling. EMBO J 25:5775–5782
8. Bakkenist CJ, Kastan MB (2003) DNA damage activates ATM through intermolecular autophosphorylation and dimer dissociation. Nature 421:499–506
9. White JS, Choi S, Bakkenist CJ (2008) Irreversible chromosome damage accumulates rapidly in the absence of ATM kinase activity. Cell Cycle 7:1277–1284
10. Kozlov SV, Graham ME, Peng C, Chen P, Robinson PJ, Lavin MF (2006) Involvement of novel autophosphorylation sites in ATM activation. EMBO J 25:3504–3514
11. Kozlov SV, Graham ME, Jakob B, Tobias F, Kijas AW, Tanuji M, Chen P, Robinson PJ, Taucher-Scholz G, Suzuki K, So S, Chen D, Lavin MF (2011) Autophosphorylation and ATM activation: additional sites add to the complexity. J Biol Chem 286:9107–9119
12. Andegeko Y, Moyal L, Mittelman L, Tsarfaty I, Shiloh Y, Rotman G (2001) Nuclear retention of ATM at sites of DNA double strand breaks. J Biol Chem 276:38224–38230
13. McKinnon PJ (2004) ATM and ataxia telangiectasia. EMBO Rep 5:772–776
14. Lavin MF (2008) Ataxia-telangiectasia: from a rare disorder to a paradigm for cell signalling and cancer. Nat Rev Mol Cell Biol 9:759–769
15. Gilad S, Chessa L, Khosravi R, Russell P, Galanty Y, Piane M, Gatti RA, Jorgensen TJ, Shiloh Y, Bar-Shira A (1998) Genotype-phenotype relationships in ataxia-telangiectasia and variants. Am J Hum Genet 62:551–561
16. Keimling M, Volcic M, Csernok A, Wieland B, Dörk T, Wiesmüller L (2011) Functional characterization connects individual patient mutations in ataxia telangiectasia mutated (ATM) with dysfunction of specific DNA double-strand break-repair signaling pathways. FASEB J 25:3849–3860
17. Renwick A, Thompson D, Seal S, Kelly P, Chagtai T, Ahmed M, North B, Jayatilake H, Barfoot R, Spanova K, McGuffog L, Evans DG, Eccles D, Breast Cancer Susceptibility Collaboration (UK), Easton DF, Stratton MR, Rahman N (2006) ATM mutations that cause ataxia-telangiectasia are breast cancer susceptibility alleles. Nat Genet 38:873–875
18. Goldgar DE, Healey S, Dowty JG, Da Silva L, Chen X, Spurdle AB, Terry MB, Daly MJ, Buys SM, Southey MC, Andrulis I, John EM, BCFR, kConFab, Khanna KK, Hopper JL, Oefner PJ, Lakhani S, Chenevix-Trench G (2011) Rare variants in the ATM gene and risk of breast cancer. Breast Cancer Res 13:R73
19. Lavin MF, Scott S, Gueven N, Kozlov S, Peng C, Chen P (2004) Functional consequences of sequence alterations in the ATM gene. DNA Repair (Amst) 3:1197–1205
20. Tomimatsu N, Mukherjee B, Burma S (2009) Distinct roles of ATR and DNA-PKcs in triggering DNA damage responses in ATM-deficient cells. EMBO Rep 10:629–635
21. Stewart GS, Maser RS, Stankovic T, Bressan DA, Kaplan MI, Jaspers NG, Raams A, Byrd PJ, Petrini JH, Taylor AM (1999) The DNA double-strand break repair gene hMRE11 is mutated in individuals with an ataxia-telangiectasia-like disorder. Cell 99:577–587
22. Carney JP, Maser RS, Olivares H, Davis EM, Le Beau M, Yates JR 3rd, Hays L, Morgan WF, Petrini JH (1998) The hMre11/hRad50 protein complex and Nijmegen breakage syndrome: linkage of double-strand break repair to the cellular DNA damage response. Cell 93:477–486
23. Waltes R, Kalb R, Gatei M, Kijas AW, Stumm M, Sobeck A, Wieland B, Varon R, Lerenthal Y, Lavin MF, Schindler D, Dörk T (2009) Human RAD50 deficiency in a Nijmegen breakage syndrome-like disorder. Am J Hum Genet 84:605–616
24. You Z, Bailis JM, Johnson SA, Dilworth SM, Hunter T (2007) Rapid activation of ATM on DNA flanking double-strand breaks. Nat Cell Biol 9:1311–1318
25. Shiotani B, Zou L (2009) Single-stranded DNA orchestrates an ATM-to-ATR switch at DNA breaks. Mol Cell 33:547–558
26. Stucki M, Clapperton JA, Mohammad D, Yaffe MB, Smerdon SJ, Jackson SP (2005) MDC1 directly binds phosphorylated histone H2AX to regulate cellular responses to DNA double-strand breaks. Cell 123:1213–1226
27. Savic V, Yin B, Maas NL, Bredemeyer AL, Carpenter AC, Helmink BA, Yang-Iott KS, Sleckman BP, Bassing CH (2009) Formation of dynamic gamma-H2AX domains along broken DNA strands is distinctly regulated by ATM and MDC1 and dependent upon H2AX densities in chromatin. Mol Cell 34:298–310

28. Luo K, Yuan J, Lou Z (2011) Oligomerization of MDC1 protein is important for proper DNA damage response. J Biol Chem 286:28192–28199
29. Liu J, Luo S, Zhao H, Liao J, Li J, Yang C, Xu B, Stern DF, Xu X, Ye K (2012) Structural mechanism of the phosphorylation-dependent dimerization of the MDC1 forkhead-associated domain. Nucleic Acids Res 40:3898–3912
30. Kim JA, Kruhlak M, Dotiwala F, Nussenzweig A, Haber JE (2007) Heterochromatin is refractory to gamma-H2AX modification in yeast and mammals. J Cell Biol 178:209–218
31. Cowell IG, Sunter NJ, Singh PB, Austin CA, Durkacz BW, Tilby MJ (2007) GammaH2AX foci form preferentially in euchromatin after ionising-radiation. PLoS One 2:e1057
32. Ziv Y, Bielopolski D, Galanty Y, Lukas C, Taya Y, Schultz DC, Lukas J, Bekker-Jensen S, Bartek J, Shiloh Y (2006) Chromatin relaxation in response to DNA double-strand breaks is modulated by a novel ATM- and KAP-1 dependent pathway. Nat Cell Biol 8:870–876
33. Moyal L, Lerenthal Y, Gana-Weisz M, Mass G, So S, Wang SY, Eppink B, Chung YM, Shalev G, Shema E, Shkedy D, Smorodinsky NI, van Vliet N, Kuster B, Mann M, Ciechanover A, Dahm-Daphi J, Kanaar R, Hu MC, Chen DJ, Oren M, Shiloh Y (2011) Requirement of ATM-dependent monoubiquitylation of histone H2B for timely repair of DNA double-strand breaks. Mol Cell 41:529–542
34. Gole B, Baumann C, Mian E, Ireno CI, Wiesmüller L (2015) Endonuclease G initiates DNA rearrangements at the MLL breakpoint cluster upon replication stress. Oncogene 34:3391–3401
35. Robison JG, Elliott J, Dixon K, Oakley GG (2004) Replication protein A and the Mre11. Rad50.Nbs1 complex co-localize and interact at sites of stalled replication forks. J Biol Chem 279:34802–34810
36. Cannon B, Kuhnlein J, Yang SH, Cheng A, Schindler D, Stark JM, Russell R, Paull TT (2013) Visualization of local DNA unwinding by Mre11/Rad50/Nbs1 using single-molecule FRET. Proc Natl Acad Sci U S A 110:18868–18873
37. Shibata A, Moiani D, Arvai AS, Perry J, Harding SM, Genois MM, Maity R, van Rossum-Fikkert S, Kertokalio A, Romoli F, Ismail A, Ismalaj E, Petricci E, Neale MJ, Bristow RG, Masson JY, Wyman C, Jeggo PA, Tainer JA (2014) DNA double-strand break repair pathway choice is directed by distinct MRE11 nuclease activities. Mol Cell 53:7–18
38. Symington LS, Gautier J (2011) Double-strand break end resection and repair pathway choice. Annu Rev Genet 45:247–271
39. Kijas AW, Lim Y, Bolderson E, Cerosaletti K, Gatei M, Jakob B, Tobias F, Taucher-Scholz G, Gueven N, Oakley G, Concannon P, Wolvetang E, Khanna KK, Wiesmüller L, Lavin MF (2015) ATM-dependent phosphorylation of MRE11 controls extent of resection during homology directed repair by signalling through exonuclease 1. Nucleic Acids Res 43:8352–8367
40. Matsuoka S, Ballif BA, Smogorzewska A, McDonald ER 3rd, Hurov KE, Luo J, Bakalarski CE, Zhao Z, Solimini N, Lerenthal Y, Shiloh Y, Gygi SP, Elledge SJ (2007) ATM and ATR substrate analysis reveals extensive protein networks responsive to DNA damage. Science 316:1160–1166
41. Lim DS, Kim ST, Xu B, Maser RS, Lin J, Petrini JH, Kastan MB (2000) ATM phosphorylates p95/nbs1 in an S-phase checkpoint pathway. Nature 404:613–617
42. Wen J, Cerosaletti K, Schultz KJ, Wright JA, Concannon P (2013) NBN phosphorylation regulates the accumulation of MRN and ATM at sites of DNA double-strand breaks. Oncogene 32:4448–4456
43. Gatei M, Jakob B, Chen P, Kijas AW, Becherel OJ, Gueven N, Birrell G, Lee JH, Paull TT, Lerenthal Y, Fazry S, Taucher-Scholz G, Kalb R, Schindler D, Waltes R, Dörk T, Lavin MF (2011) ATM protein-dependent phosphorylation of Rad50 protein regulates DNA repair and cell cycle control. J Biol Chem 286:31542–31556
44. Hunt CR, Pandita RK, Laszlo A, Higashikubo R, Agarwal M, Kitamura T, Gupta A, Rief N, Horikoshi N, Baskaran R, Lee JH, Löbrich M, Paull TT, Roti Roti JL, Pandita TK (2007) Hyperthermia activates a subset of ataxia-telangiectasia mutated effectors independent of DNA strand breaks and heat shock protein 70 status. Cancer Res 67:3010–3017
45. Bencokova Z, Kaufmann MR, Pires IM, Lecane PS, Giaccia AJ, Hammond EM (2009) ATM activation and signaling under hypoxic conditions. Mol Cell Biol 29:526–537
46. Resseguie EA, Staversky RJ, Brookes PS, O'Reilly MA (2015) Hyperoxia activates ATM independent from mitochondrial ROS and dysfunction. Redox Biol 5:176–185
47. Bakkenist CJ, Kastan MB (2015) Chromatin perturbations during the DNA damage response in higher eukaryotes. DNA Repair (Amst) pii:S1568-7864(15)00177-9
48. Canman CE, Lim DS, Cimprich KA, Taya Y, Tamai K, Sakaguchi K, Appella E, Kastan MB,

Siliciano JD (1998) Activation of the ATM kinase by ionizing radiation and phosphorylation of p53. Science 281:1677–1679

49. Banin S, Moyal L, Shieh S, Taya Y, Anderson CW, Chessa L, Smorodinsky NI, Prives C, Reiss Y, Shiloh Y, Ziv Y (1988) Enhanced phosphorylation of p53 by ATM in response to DNA damage. Science 281:1674–1677
50. Rashi-Elkeles S, Elkon R, Shavit S, Lerenthal Y, Linhart C, Kupershtein A, Amariglio N, Rechavi G, Shamir R, Shiloh Y (2011) Transcriptional modulation induced by ionizing radiation: p53 remains a central player. Mol Oncol 5:336–348
51. Gatz SA, Wiesmüller L (2006) p53 in recombination and repair. Cell Death Differ 13: 1003–1016
52. Hadian K, Krappmann D (2011) Signals from the nucleus: activation of NF-kappaB by cytosolic ATM in the DNA damage response. Sci Signal 4:ep2
53. Levy-Barda A, Lerenthal Y, Davis AJ, Chung YM, Essers J, Shao Z, van Vliet N, Chen DJ, Hu MC, Kanaar R, Ziv Y, Shiloh Y (2011) Involvement of the nuclear proteasome activator PA28γ in the cellular response to DNA double-strand breaks. Cell Cycle 10: 4300–4310
54. Akyüz N, Boehden GS, Süsse S, Rimek A, Preuss U, Scheidtmann KH, Wiesmüller L (2002) DNA substrate dependence of p53-mediated regulation of double-strand break repair. Mol Cell Biol 22:6306–6317
55. Kraft D, Rall M, Volcic M, Metzler E, Groo A, Stahl A, Bauer L, Nasonova E, Salles D, Taucher-Scholz G, Bönig H, Fournier C, Wiesmüller L (2015) NF-κB-dependent DNA damage-signaling differentially regulates DNA double-strand break repair mechanisms in immature and mature human hematopoietic cells. Leukemia 29:1543–1554
56. Obermeier K, Sachsenweger J, Friedl TWP, Pospiech H, Winqvist R, Wiesmüller L (2016) Heterozygous *PALB2* c.1592delT mutation channels DNA double-strand break repair into error-prone pathways in breast cancer patients. Oncogene 35(29):3796–3806. doi:10.1038/onc.2015.448
57. Neitzel H (1986) A routine method for the establishment of permanent growing lymphoblastoid cell lines. Hum Genet 73:320–326
58. Sandoval N, Platzer M, Rosenthal A, Dörk T, Bendix R, Skawran B, Stuhrmann M, Wegner RD, Sperling K, Banin S, Shiloh Y, Baumer A, Bernthaler U, Sennefelder H, Brohm M, Weber BH, Schindler D (1999) Characterization of ATM gene mutations in 66 ataxia telangiectasia families. Hum Mol Genet 8:69–79
59. Dörk T, Bendix R, Bremer M, Rades D, Klöpper K, Nicke M, Skawran B, Hector A, Yamini P, Steinmann D, Weise S, Stuhrmann M, Karstens JH (2011) Spectrum of ATM gene mutations in a hospital-based series of unselected breast cancer patients. Cancer Res 61:7608–7615
60. Keimling M, Kaur J, Bagadi SA, Kreienberg R, Wiesmüller L, Ralhan R (2008) A sensitive test for the detection of specific DSB repair defects in primary cells from breast cancer specimens. Int J Cancer 123:730–736
61. Speit G, Trenz K, Schütz P, Bendix R, Dörk T (2000) Mutagen sensitivity of human lymphoblastoid cells with a BRCA1 mutation in comparison to ataxia telangiectasia heterozygote cells. Cytogenet Cell Genet 91:261–266

Chapter 24

Monitoring DNA Repair Consequences of ATM Signaling Using Simultaneous Fluorescent Readouts

Adrian Wiegmans

Abstract

The repair of deleterious DNA double strand breaks is required to maintain genome integrity. The efficacy in which this occurs relies upon the available machinery and is guided by factors that include cell cycle status, availability of donor template, and the local chromosome structure. Therefore at a single DNA breakpoint there are different outcomes that can occur. The Traffic light reporter (TLR) assay protocol is a dual fluorescent readout that has the ability to monitor simultaneous homologous recombination and non-homologous end joining activity in response to DNA damage. This provides insight to determine the upstream functionality of either pathway mediated through ATM.

Key words Traffic light reporter (TLR), Homologous recombination (HR), Non-homologous end joining (NHEJ), ATM, DNA damage foci, DNA damage response, DNA double strand breaks, High content screen, FACS analysis

1 Introduction

Fluorescent reporter systems have proven to be indispensible for the evaluation of complex biological processes, including DNA repair. The DNA damage response is required for the maintenance of genome fidelity. The two main pathways for repair of double strand breaks (DSBs) are homologous recombination (HR) and non-homologous end joining (NHEJ) [1]. It has been shown that both ATM and the DNA-dependent protein kinase utilized for NHEJ bind the region surrounding DSBs, independently of the pathway used for repair. Therefore at a single breakpoint there are different outcomes that can occur. The result of which is orchestrated by the cell cycle status, availability of donor template, which is regulated by the local chromosome structure and DNA repair protein expression and their posttranslational modifications [2]. Competition between the two canonical pathways means that rarely are the conditions optimal for a single specific mechanism, especially in the cancer setting. Here, we describe the use of a dual

Sergei V. Kozlov (ed.), *ATM Kinase: Methods and Protocols*, Methods in Molecular Biology, vol. 1599,
DOI 10.1007/978-1-4939-6955-5_24, © Springer Science+Business Media LLC 2017

fluorescent readout to monitor simultaneous HR and NHEJ activity as a readout for ATM activity [3]. Cells with integrated Traffic light reporter (TLR) cassette contain an in-frame truncated eGFP gene with out of frame full-length mCherry gene and are not fluorescent. Double strand break (DSB) is induced in situ in the eGFP gene of the TLR cassette by transfection of homing endonuclease I-SceI (infrared fluorescent protein (IFP)-tagged). Restoration of green-fluorescence signal is achieved when break is repaired by gene conversion (HR) of a blue fluorescent protein (BFP)-tagged exogenous template containing missing part of eGFP gene. Repair of the DSB by error-prone NHEJ results in a +2 shift in mCherry reading frame, resulting in red fluorescence [3]. NHEJ is analyzed from the IFP-positive cell population, while HR is analyzed from the IFP/BFP double-positive cells by FACS analysis. mCherry-positive cells representing NHEJ activity could be isolated from the IFP-positive population, and GFP-positive cells representing HR could be isolated from IFP/BFP double-positive cells (see Fig. 2).

In general the canonical function of ATM is to phosphorylate several downstream key proteins that initiate activation of the HR repair process via different DNA damage checkpoints, leading to cell cycle arrest [4]. The MRN complex (Mre11, Rad50, and Nbs1) recruits ATM to DSB sites and ATM and then phosphorylates histone H2AX and other key substrates including p53, 53-binding protein 1 (53BP1), and checkpoint kinase 2 (CHK2) to initiate DNA repair. In a homeostatic cell this would result in an avid repair of the DSBs via HR. However in the cancer setting there are a number of key points at which a deviation in ATM signaling could result in partial or complete loss of HR with corresponding activation of NHEJ or apoptosis [5].

Under different circumstances the repair pathway activated can be the result of ATM response to the machinery available for repair. The direct mutation of ATM is observed in the disease Ataxia-telangiectasia (AT), leading to chromosomal instability and increased sporadic cancer incidence due to a loss of HR and activation of NHEJ [6]. Mutation of the component of MRN, namely Mre11, causes over resection of single strand (ss) DNA resulting in a loss of recruitment for HR activity and also inappropriate ends for NHEJ activity [7]. Upstream of ATM, mutation of p53 results in constitutive activation of the DNA damage checkpoint pathway activating ATM-dependent phosphorylation of CHK2 and HR-mediated repair [8]. Meanwhile, ATM-mediated phosphorylation of 53BP1 that does not occur in S-phase of cell cycle means a lack of supporting HR machinery and activation of NHEJ [9], while mutation or loss of expression of 53BP1 results in restoration of HR [10]. Interestingly, a study into DNA repair mechanisms in cells that harbor loss of both ATM and 53BP1 demonstrated a decreased efficiency in NHEJ rather than an alternative NHEJ-mediated repair mechanism or restoration of HR [11].

The above examples show that the ability for simultaneous HR and NHEJ readout gives insight into the mechanistic functions of the DNA repair pathways via ATM.

I describe the Traffic light reporter (TLR) assay protocol to monitor simultaneously homologous recombination and non-homologous end joining activity in response to DNA damage to determine the function of ATM in these pathways.

2 Materials

2.1 Traffic Light Reporter and I-SceI Lentiviral Particles Production

1. 6-well plates.
2. 10 ml sterile tubes.
3. Transfection reagent, e.g., Lipofectamine 2000 (Thermo Fisher Scientific).
4. HEK293-FT cell line (Thermo Fisher Scientific).
5. DMEM cell culture medium: Dulbecco's Modified Eagle Medium with 10% fetal bovine serum (FBS), 2 mM L-glutamine or GlutaMax (Invitrogen), 100 U/ml penicillin, and 100 μg/ml streptomycin.
6. OPTI-MEM medium (Thermo Fischer Scientific).
7. Lentiviral Titer Test kit, Lenti-X GoStix (Clontech).
8. pCVL Traffic Light Reporter 1.1 Ef1a Puro plasmid (pCVL-TL) (available at Addgene https://www.addgene.org/).
9. pMD2G VSV-G envelope expressing plasmid (pMD2g).
10. psPAX2 packaging integration plasmid.
11. pRRL-I-SceI targeted enzyme for DNA cleavage (pRRL) plasmid: pRRL sEF1a HA.NLS. Sce(opt). T2A.IFP (available at Addgene https://www.addgene.org/).
12. pRSET-BFP-1 control transfection plasmid (pBFP).
13. 1ug/uL Polyethylenimine (PEI). To make a stock solution (may take hours to prepare): Dissolve PEI in endotoxin-free dH2O that has been heated to ~80 °C. Let it cool to room temperature. Filter sterilize (0.22 μm), aliquot and store at −20 °C; a working stock can be kept at 4 °C.

2.2 Determining the Optimal Puromycin Concentration

1. *See* Table 1 for Puromycin sensitivity for some commonly utilized cell lines. Table adapted from http://www.lifetechnologies.com/us/en/home/life-science/cellculture/transfection/selection/puromycin.html.

2.3 Generation of Traffic Light Reporter Cell Line

1. 6-well plates.
2. pCVL Traffic Light Reporter 1.1 Ef1a Puro plasmid (pCVL-TL) (available at Addgene https://www.addgene.org/).

Table 1
Suggested puromycin concentration

Cell Line	Concentration [μ/ml]	Reference
HT1080 (human fibrosarcoma)	0.2–0.4	*J Biol Chem* 282:29314–29322
	1	(2007)
	2	*J Virol* 82:3320–3328 (2008) *Cancer Res* 68: 1417–1426 (2008)
HeLa (human cervical cancer)	1–2	*Genes Cells* 12:397–406 (2007)
	2	*Mol Cell Biol* 28:4104–4115 (2008)
	2	*PNAS* 105:16490–16495 (2008)
HEK293 (human embryonic kidney)	0.25	*Nucleic Acids Res* 36:e83 (2008)
	2	*Mol Cell Biol* 28:4104–4115 (2008)
	3	*Mol Cell Biol* 27:4708–4719 (2007)
Hep G2 (human hepatocellular carcinoma)	0.3	*J Biol Chem* 283:21462–21468
	2	(2008)
		Mol Cell Biol 28:4104–4115 (2008)
		J Biol Chem 283:708–715 (2008)
Human embryonic stem (Es) cells	0.5	*Nucleic Acids Res 36:*e148 (2008)
	1	*Stem Cells* 26:850–863 (2008)
	5	*Stem Cells* 26:2245–2256 (2008)
MDF-7 (human breast cancer)	0.45	*RNA* 13:1375-1383 (2007)
	1	*Cancer Res* 68:1319–1328 (2008)
	2	*Mol Cell Biol* 28:4104–4115 (2008)
H1299 (non-small cell lung carcinoma)	1	*Nucleic Acids Res* 35:e17 (2007)
	1.5	*J Bio Chem* 283:33394–33405
	2.5	(2008)
		Proc Natl Acad Sci USA
		105:1937–1942 (2008)
MDA-MB-231 (human breast cancer)	0.5	*Mol Cell Biol* 28:997–1006 (2008)
	1.5	*Mol Cell Biol* 27:324 – 339 (2007)
	5	*Mol Cell Biol* 28:687 – 704 (2008)
A549 (lung cancer)	1.5	*Mol Cell Biol* 27:324–339 (2007)
	1.5	*J Biol Chem* 283:33394–33405
		(2008)

Table adapted from http://www.lifetechnologies.com/us/en/home/life-science/cell culture/transfection/selection/puromycin.html

3. Target cell growth media containing FBS, antibiotics, and puromycin at the optimal concentration (*see* Table 1).
4. Polybrene.
5. 100-mm cell culture dishes.
6. Cloning cylinders (PYREX Cloning Cylinders; O.D. × H: 6 × 8 mm).

2.4 Copy Number Determination

1. Genomic DNA purification kit, e.g., NucleoSpin Tissue DNA purification kit (Clontech).
2. 1× TE buffer, pH 8.0: 10 mM Tris–HCl (pH 8.0), 1 mM EDTA•Na_2.
3. TaqMan® Copy Number Assays kit (Thermo Fisher Scientific).
4. Taqman Primer and probe set targeting the reporter cassette: Traffic light copy number probe sequence (5″-/56-FAM/TTC TAA CAT/ZEN/GCG GTG ACGTGG AGG/3IABkFQ/-3″), Traffic light copy number Primer 1(5″-CCT TGA TGATGG CGG AGT TAT-3″), and Traffic light copy number Primer 2 (5″-CAT GGA CGA GCT GTA CAA GTA A-3″).

2.5 Flow Cytometry-Based Traffic Light Readout Assay

1. 6-well cell culture plates.
2. 5 ml polystyrene round-bottom FACS tubes.
3. 25-mM Biliverdine: dissolve lyophilized biliverdine (Toronto Research Chemicals, Canada) in DMSO to make a 1000× stock.

3 Methods

A flow diagram of the method is summarized in Fig. 1.

3.1 Traffic Light Reporter Lentiviral Particles Production

1. Split HEK 293FT cells 1 day before transfection in DMEM culture medium in 6-well plate at 0.75 × 10^6 cells per well (*see* **Note 1**).
2. After overnight incubation, change to 2 ml pre-warmed antibiotics-free growth media, prior to transfection bring all reagents to room temperature.
3. In a sterile tube dilute 2.5 μg total plasmid DNA in 0.25 ml OPTI-MEM (combine 1.4 μg pCVL-TL, 0.36 μg pMD2g, and 0.71 μg psPAX2 for integrating lentivirus). One microgram of pBFP plasmid is used as a transfection efficiency control. In a separate sterile tube dilute 10 μl of Lipofectamine 2000 in 0.25 ml OPTI-MEM. Incubate two tubes separately at room temperature for 5 min.
4. Add diluted DNA to the diluted Lipofectamine 2000 tube dropwise. Mix solution immediately by pipetting two to three times.

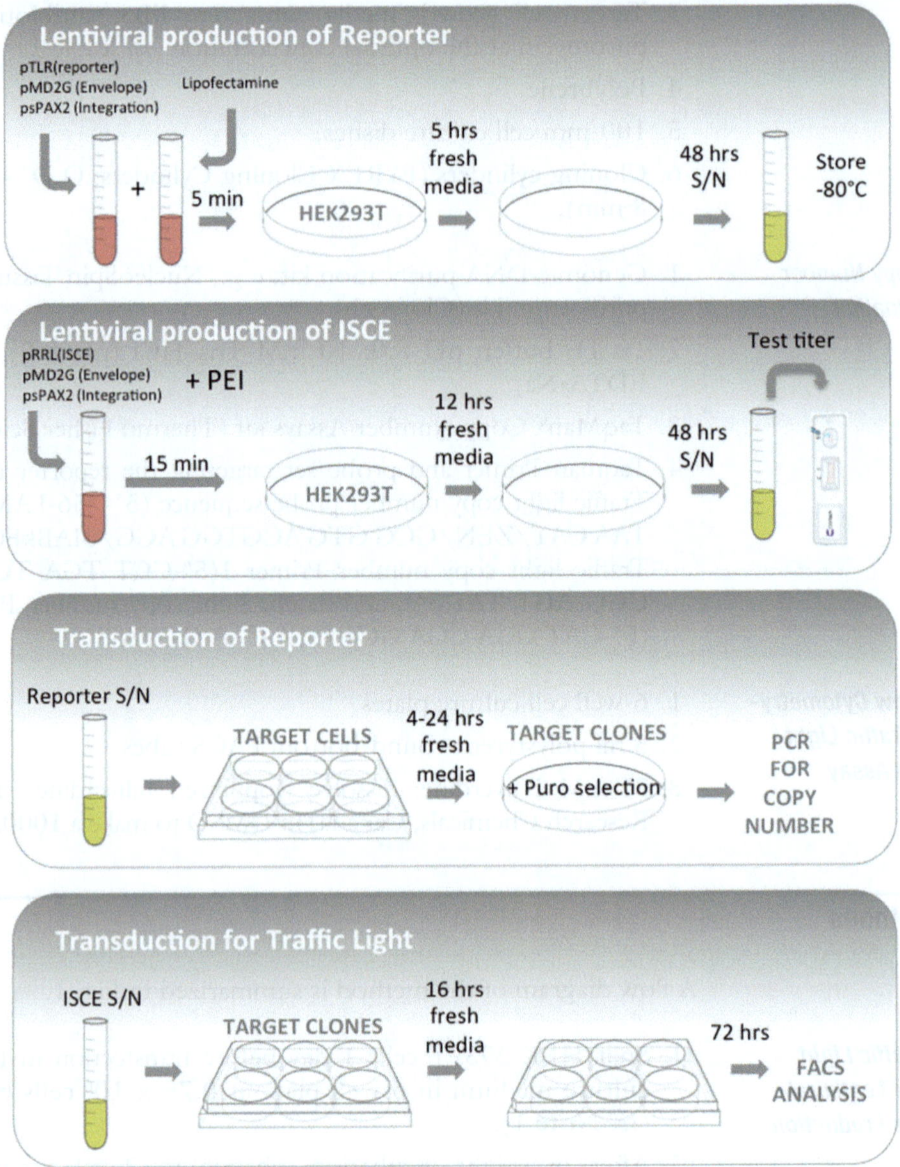

Fig. 1 Summary of traffic light reporter system

5. Incubate for 20 min at room temperature.
6. Add DNA/Lipofectamine 2000 mixture to one well.
7. Change to fresh HEK293FT cells growth media after 5 h.
8. Harvest viral supernatant at 48 h post-transfection. Centrifuge at 370 × *g* for 15 min at tabletop centrifuge to remove the cells and debris.
9. Estimate lentivirus titer using Titer Test kit Lenti-X GoStix. Take 20 μl of your lentiviral supernatant and apply to the sample well (S) of the GoStix cassette.

10. Add four drops of Chase Buffer (as supplied by manufacturer) to the sample well and wait for the bands to develop. A control band will always appear (C), and a test band (T) will start to appear within 30–120 s and reach maximum intensity at 5 min if your sample contains sufficient lentivirus. This is an estimate of greater than 5×10^5 Infectious Forming Units/ml.
11. Keep lentiviral particles stock at −80 °C (*see* **Note 3**).

3.2 I-SceI Lentivirus Particles Production

1. Split HEK293FT cells 1 day before transfection in DMEM/10% FBS medium. Use 0.8×10^6 cells for a 6-well plate, 6.0×10^6 cells for a 10-cm dish, and 12.0×10^6 cells for a 15-cm dish (*see* **Note 2**).
2. After overnight culture, change to pre-warmed growth media, prior to transfection bring all reagents to room temperature.
3. In a sterile tube dilute total plasmid DNA (μg) in OPTI-MEM (volume of media is 10% of final volume in culture vessel). Use transgene (pRRL sEF1a HA.NLS. Sce(opt). T2A.IFP):viral packaging (psPAX2):viral envelope (pMD2G) plasmid constructs at 4:2:1 DNA ratio. For a *6-well plate* mix 200 μl of medium with 3 μg of total DNA, for a *10-cm dish mix* 1 ml of medium with 7–8 μg of total DNA, and for a *15-cm dish mix* 2 ml of medium with 11–12 μg of total DNA.
4. Add Polyethylenimine (PEI) to the diluted DNA. Mix immediately by vortexing or pipeting. The volume of PEI used is based on a 3:1 ratio of PEI (μg): total DNA (μg). For 6-well plate use 9 μl of PEI, for 10-cm dish use 21 μl of PEI, for 15-cm dish use 33 μl of PEI. Incubate 15 min at RT
5. Add DNA/PEI mixture to cells.
6. Change to fresh HEK293FT cells growth media after 12 h supplemented with 15 mM HEPES.
7. Harvest transfected cells and/or viral supernatant at 48 h post-transfection and centrifuge at 2000 rpm for 15 min at the tabletop centrifuge to remove the cells and debris.
8. Keep lentiviral particles stock at −80 °C (*see* **Note 3**).

3.3 Determining the Optimal Puromycin Concentration

Day 1:

1. Plate Target Cells in Ten 6-well Plates and Grow at 37 ° C, 5% CO_2 Overnight.

Day 2:

2. The target cells should be approximately 80–90% confluent.
3. Dilute puromycin in the preferred culture media for your target cells. The final concentration of puromycin should be from 1 to 10 μg/ml in 1 μg/ml increments (*see* **Note 4**).

4. Label plates from 1 to 10 and add appropriate puromycin-containing media to cells (*see* Table 1).

Days 3 onward:

5. Examine cells each day and change to fresh puromycin-containing media every other day.
6. The minimum concentration of puromycin that results in complete cell death after 3–5 days is the concentration that should be used for selection in your experiments.

3.4 Generation of Traffic Light Reporter Cell Line

Day 1:

1. Seed Appropriate Number of Target Cells into 6-well Cluster Plate and Incubate Overnight.

Day 2:

2. Make sure the cells are at 60–80% confluency the following day.
3. Add increasing volumes of lentiviral particles to the target cells (0 μl, 0.1 μl, 0.5 μl, 1 μl, 5 μl, 25 μl into each well of 6-well plate in duplicate).
4. Incubate overnight with 4–8 μg/ml polybrene (*see* **Note 5**).

Day 3:

5. Change to fresh growth media and incubate for another 24 h. If viral toxicity is observed in your cell line, you may decrease the infection time to between 4 and 20 h. Remove the virus-containing media and replace with fresh media. Do not add puromycin until at least 24 h after infection to allow for sufficient expression of the puromycin resistance gene.

Day 4+:

6. Change to fresh puromycin-containing media and change every 2–3 days, culture for 5 days to select the positive cells (make sure no survived cells are left in wells without lentiviral transduction).

Day 9+:

7. Choose cells survived from antibiotics selection in wells with minimum amount of lentiviral particles, seed them sparsely (I use 500–1000 cells) into 100 mm tissue culture dish.
8. Keep the cells in normal culture media for ~14 days (depending on the cell proliferation rate).
9. Once separate colonies formed (50–100 cells), harvest and sub-culture individual colony by cloning cylinders (*see* **Note 6**).
10. Mark separate colonies under light microscope.

11. Remove media in the culture dish. Wash once with 10 ml of PBS.
12. Gently put one cloning cylinder onto each of the marked colony.
13. Add 50 μl of trypsin into each cloning cylinder. Incubate for ~5 min at 37 C. Check under the microscope to make sure the cells inside the cylinder are detached.
14. Add 50 μl of media to stop the reaction. Transfer the detached cells into each well of 24-well plate and continue culture.

3.5 Copy Number Determination

1. Prepare up to 1×10^7 of cells from each colony.
2. Isolate genomic DNA using NucleoSpin Tissue kit following manufacturer's instructions. Resuspend up to 1×10^7 cells in a final volume of 200 μl Buffer T1. Add 25 μl Proteinase K solution and 200 μl Buffer B3. Incubate the sample at 70 °C for 10–15 min.
3. Add 210 μl ethanol (96–100%) to the sample and vortex vigorously.
4. For each sample, place one NucleoSpin® Tissue Column into a Collection Tube. Apply the sample to the column. Centrifuge for 1 min at 11,000 × *g*. Discard the flow-through and place the column back into the Collection Tube.
5. Add 500 μl Buffer BW. Centrifuge for 1 min at 11,000 × *g*. Discard flow-through and place the column back into the Collection Tube.
6. Add 600 μl Buffer B5 to the column and centrifuge for 1 min at 11,000 × *g*. Discard flow-through and place the column back into the Collection Tube.
7. Centrifuge the column for 1 min at 11,000 × *g*. Residual ethanol is removed during this step.
8. Place the NucleoSpin® Tissue Column into a 1.5 ml microcentrifuge tube and add 100 μl pre-warmed Buffer BE (70 °C). Incubate at room temperature for 1 min. Centrifuge 1 min at 11,000 × *g*.
9. Use the TaqMan® Copy Number Assays kit as the internal control to determine integrated copy number (according to manufacturer's instructions). Prepare the PCR reaction mix. Dilute each sample to 5 ng/μl using either nuclease-free water or 1× TE buffer, pH 8.0 to make a 5× stock solution.
10. Thoroughly mix the TaqMan® Fast Advanced Master Mix. Thaw frozen samples and frozen primer/probe mix on ice. Resuspend by vortexing, and then briefly centrifuge. Calculate the total volume required for each component.
11. Add all components to a 1.5 ml microcentrifuge tube, cap the tube, and then vortex briefly to mix.

12. Centrifuge the tube briefly to spin down the contents and eliminate air bubbles. Single reaction should contain: 10 μl of TaqMan® Fast Advanced Master Mix (2×), 1 μl of Primers/probe mix (20×), 1 ul of Control primers/probe mix (20×), 2 μl of DNA Template (1 pg to 100 ng), and 6 μl of nuclease-free water in a total volume of 20 μl.
13. Transfer the appropriate volume of PCR reaction mix to each well of an optical reaction plate.
14. Cover the reaction plate, and then centrifuge the plate briefly to spin down the contents and eliminate air bubbles. Run the PCR reaction plate in a real-time PCR machine.
15. To analyze the results, open the Analysis Settings window in the real-time PCR Instrument software and set the following: Manual CT threshold—0.2, Auto baseline—On.
16. Apply the settings, and export results to copy caller software for copy number determination.

3.6 FACS-Based Traffic Light Reporter Assay Readout

Cells are transduced with the I-SceI endonuclease (IFP tagged) and repair measured after 72 h endpoint by FACS analysis comparing GFP (in IFP/BFP-positive cells) to mCherry signal (IFP-positive) ratio (Fig. 2).

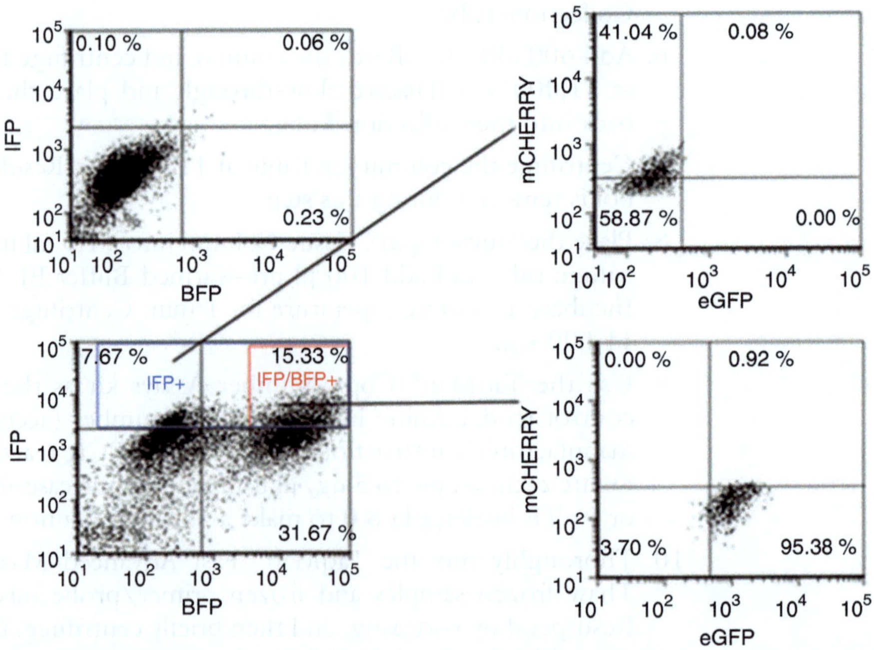

Fig. 2 Gating for analysis of FACS populations to determine HR versus NHEJ activity in DNA repair. Reprinted with permission from Wiegmans A et al. (2015) Mol Cancer Therapeutics, v14 [10], pp. 2321–2331. Copyright ©2015 American Association for Cancer Research

1. Add appropriate volume (MOI at 8–10) of I-SceI lentiviral particles to the target cells.
2. Incubate overnight with 4–8 μg/ml polybrene.
3. IFP signal from the I-SceI transgene vector can be detected 24 h post infection, if doing repair assay, wait for 72 h for the end-point results.
4. Before analyzing samples to FACS, add 25 μM of Biliverdine and incubate overnight to enhance IFP signal (*see* **Note** 7).
5. FACS samples should be run at medium flow and gated for morphology and BFP-positive cells. This positive population can then be analyzed for IFP/GFP fluorescence in a 2D or histogram plot. Gating against negative controls will provide threshold for IFP/GFP. Analyze at least 10,000 events in the double-positive population (*see* **Note 8**).

4 Notes

1. HEK293FT cells are very easy to detach, give extra care to handle them, keep the media warm all the time.
2. HEK293FT grow faster when supplied with nonessential amino acids. This means less background levels of cell death but initial plating numbers will have to be adjusted. For traffic light reporter 0.5 × 10^6 cells per well and adjust for I-Sce1 appropriately.
3. Lentiviral supernatants lose approximately 50% activity on freeze and thaw. Therefore, aliquot 2 ml volumes for storage at −80 °C.
4. For optimization of the puromycin titration the cell line may require finer increments of puromycin to determine a more precise optimal puromycin concentration.
5. Polybrene increases the efficiency of viral infection. However, polybrene is toxic to some cell lines. In these cell lines, substitute protamine sulfate for polybrene
6. If specific types of cells do not form good colonies, use limited dilution in 96-well cluster plate instead.
7. Some target cells low-level sensitivity to Biliverdine. This should be tested before beginning the FACS-based Traffic light assay readout.
8. The authors suggest positive controls such as gamma-irradiated samples for general DNA damage to determine the maximum threshold for DNA damage on the Traffic Light system and HR-deficient cell lines such as those harboring BRCA1 mutations as positive controls for NHEJ readout. Directed use of

specific ATM inhibitors such KU55933 as well as DNA-PK inhibitors (*see* review, Table 2 in [12]) on your cell line of choice will determine degree of ATM contribution versus NHEJ.

References

1. Kanaar R, Hoeijmakers JH (1997) Recombination and joining: different means to the same ends. Genes Funct 1(3):165–174
2. Kuhar R, Gwiazda KS, Humbert O, Mandt T, Pangallo J, Brault M et al (2014) Novel fluorescent genome editing reporters for monitoring DNA repair pathway utilization at endonuclease-induced breaks. Nucleic Acids Res 42(1):e4
3. Certo MT, Ryu BY, Annis JE, Garibov M, Jarjour J, Rawlings DJ et al (2011) Tracking genome engineering outcome at individual DNA breakpoints. Nat Methods 8(8): 671–676
4. Lavin MF, Shiloh Y (1997) The genetic defect in ataxia-telangiectasia. Annu Rev Immunol 15:177–202
5. Jackson SP, Bartek J (2009) The DNA-damage response in human biology and disease. Nature 461(7267):1071–1078
6. Lavin MF (2008) Ataxia-telangiectasia: from a rare disorder to a paradigm for cell signalling and cancer. Nat Rev Mol Cell Biol 9(10): 759–769
7. Kijas AW, Lim YC, Bolderson E, Cerosaletti K, Gatei M, Jakob B et al (2015) ATM-dependent phosphorylation of MRE11 controls extent of resection during homology directed repair by signalling through Exonuclease 1. Nucleic Acids Res 43(17):8352–8367
8. DiTullio RA Jr, Mochan TA, Venere M, Bartkova J, Sehested M, Bartek J et al (2002) 53BP1 functions in an ATM-dependent checkpoint pathway that is constitutively activated in human cancer. Nat Cell Biol 4(12):998–1002
9. Escribano-Diaz C, Orthwein A, Fradet-Turcotte A, Xing M, Young JT, Tkac J et al (2013) A cell cycle-dependent regulatory circuit composed of 53BP1-RIF1 and BRCA1-CtIP controls DNA repair pathway choice. Mol Cell 49(5): 872–883
10. Bothmer A, Robbiani DF, Feldhahn N, Gazumyan A, Nussenzweig A, Nussenzweig MC (2010) 53BP1 regulates DNA resection and the choice between classical and alternative end joining during class switch recombination. J Exp Med 207(4):855–865
11. Rybanska-Spaeder I, Reynolds TL, Chou J, Prakash M, Jefferson T, Huso DL et al (2013) 53BP1 is limiting for NHEJ repair in ATM-deficient model systems that are subjected to oncogenic stress or radiation. Mol Cancer Res 11(10):1223–1234
12. Jekimovs C, Bolderson E, Suraweera A, Adams M, O'Byrne KJ, Richard DJ (2014) Chemotherapeutic compounds targeting the DNA double-strand break repair pathways: the good, the bad, and the promising. Front Oncol 4:86

Chapter 25

Noncanonical ATM Activation and Signaling in Response to Transcription-Blocking DNA Damage

Jurgen A. Marteijn, Wim Vermeulen, and Maria Tresini

Abstract

Environmental genotoxins and metabolic byproducts generate DNA lesions that can cause genomic instability and disrupt tissue homeostasis. To ensure genomic integrity, cells employ mechanisms that convert signals generated by stochastic DNA damage into organized responses, including activation of repair systems, cell cycle checkpoints, and apoptotic mechanisms. DNA damage response (DDR) signaling pathways coordinate these responses and determine cellular fates in part, by transducing signals that modulate RNA metabolism. One of the master DDR coordinators, the Ataxia Telangiectasia Mutated (ATM) kinase, has a fundamental role in mediating DNA damage-induced changes in mRNA synthesis. ATM acts by modulating a variety of RNA metabolic pathways including nascent RNA splicing, a process catalyzed by the spliceosome. Interestingly, ATM and the spliceosome influence each other's activity in a reciprocal manner by a pathway that initiates when transcribing RNA polymerase II (RNAPII) encounters DNA lesions that prohibit forward translocation. In response to stalling of RNAPII assembly of late-stage spliceosomes is disrupted resulting in increased splicing factor mobility. Displacement of spliceosomes from lesion-arrested RNA polymerases facilitates formation of R-loops between the nascent RNA and DNA adjacent to the transcription bubble. R-loops signal for noncanonical ATM activation which in quiescent cells occurs in absence of detectable dsDNA breaks. In turn, activated ATM signals to regulate spliceosome dynamics and AS genome wide.

This chapter describes the use of fluorescence microscopy methods that can be used to evaluate noncanonical ATM activation by transcription-blocking DNA damage. First, we present an immunofluorescence-detection method that can be used to evaluate ATM activation by autophosphorylation, in fixed cells. Second, we present a protocol for Fluorescence Recovery After Photobleaching (FRAP) of GFP-tagged splicing factors, a highly sensitive and reproducible readout to measure in living cells, the ATM influence on the spliceosome. These approaches have been extensively used in our laboratory for a number of cell lines of various origins and are particularly informative when used in primary cells that can be synchronized in quiescence, to avoid generation of replication stress-induced dsDNA breaks and consequent ATM activation through its canonical pathway.

Key words DNA damage, Ultra violet (UV) radiation, ATM signaling, Spliceosome, U2 snRNP, U5 snRNP, Fluorescence recovery after photobleaching (FRAP), Immunofluorescence, Quiescence, Fibroblasts

Sergei V. Kozlov (ed.), *ATM Kinase: Methods and Protocols*, Methods in Molecular Biology, vol. 1599,
DOI 10.1007/978-1-4939-6955-5_25, © Springer Science+Business Media LLC 2017

1 Introduction

To counteract DNA damage and ensure genomic fidelity, cells employ intricate DNA damage response (DDR) mechanisms that convert stochastic-damage signals into organized responses, including DNA repair systems, cell cycle checkpoints, and apoptosis. DDR is coordinated by signaling networks that utilize post-translational modifications and protein-protein interactions to elicit the initial stages of the cellular response. Later DDR stages depend largely on modulation of RNA metabolism [1, 2].

A fundamental stage of RNA metabolism for the production of mature, protein-coding mRNA is the removal of introns from newly synthesized transcripts [3, 4]. This reaction is catalyzed by the spliceosome, a dynamic megaparticle composed from five structurally distinct ribonucleoprotein particles (snRNPs), each containing a unique small nuclear RNA (U1, U2, U4, U5, or U6 snRNAs), a common core of seven proteins (Sm B/B′, D3, D2, D1, E, F, and G), and a large number of snRNP-specific accessory factors. During each spliceosome cycle, exon/intron definition by U1 and U2 snRNPs stimulates binding of a preassembled U4/U6. U5 tri-particle on the nascent transcript. Following extensive conformational rearrangements and U1/U4 displacement, the two-step splicing reaction is catalyzed by the mature, catalytically active spliceosome composed of U2, U5, and U6 snRNPs.

Splicing reactions are highly complex requiring an estimated 150–200 proteins including, and in addition to snRNPS, numerous auxiliary proteins that facilitate recognition of regulatory splicing sequences and the continuous restructuring of the spliceosome [5, 6]. This complexity is farther enhanced by the fact that the majority of pre-mRNAs are alternatively spliced to produce multiple mRNA variants from a single gene [7]. To maintain splicing fidelity and in parallel accommodate splicing diversity, the spliceosome is subject to complex regulation by developmental and environmental cues. Cumulative evidence supports that splicing is also influenced by the presence of DNA damage [1, 2, 8]. The exact mechanisms by which DNA lesions regulate splicing are not fully understood but likely, most DNA-damage-induced splicing changes result from changes in the transcription elongation rates of RNAPII. Because in the majority of introns, spliceosomes assemble co-transcriptionally so that the two processes are spatiotemporally coupled [9, 10], changes in the elongation rate of RNAPII when it encounters DNA lesions, are likely to influence splicing through several mechanisms; for example by increasing the time the spliceosome has to recognize weak splice sites [11], by disrupting RNAPII interactions with splicing regulators [12–14] or by dissociation of core spliceosomes from nascent transcripts [8]. This later mechanism is a two-stage process involving a stochastic step, triggered by RNAPII stalling at DNA lesions, and a signaling-mediated step, controlled by the ATM DDR kinase [8]. Intriguingly, the interaction between spliceosome displacement and ATM signaling is reciprocal. Dissociation of spliceosomes from "stalled" RNAPII at DNA lesions activates ATM that, in turn, signals in a forward fashion to regulate spliceosome mobility, presumably in areas distal to the damage [2, 8].

ATM is a master regulator of the DDR, most extensively studied for its role in repair and downstream signaling of dsDNA breaks (DSBs) [15]. DSBs induce massive ATM activation (known as canonical ATM activation) which initiates after their detection by the MRE11–RAD50–NBS1 (MRN) complex [15, 16]. Under unperturbed conditions ATM exists as an inactive dimer. In response to DSBs, e.g., induced by ionizing radiation, canonical ATM activation requires several steps, including MRN-mediated ATM anchoring to DNA-ends, TIP60/KAT5-mediated acetylation, ATM autophosphorylation, and monomerization of the previously inactive ATM dimer [15, 17]. A large fraction of activated ATM molecules resides at DSBs, forming microscopically discernable foci, where it facilitates local chromatin changes and formation of multiprotein complexes that repair the break. In parallel, ATM signals to activate cell cycle checkpoints and modulate many aspects of the cellular response in a cell-context manner (e.g., apoptosis, DNA-damage-induced senescence).

Despite its robust activation by DSBs, the role of ATM in cellular homeostasis is not restricted to DSB-signaling, but extends to structural chromatin changes and many types of stress, which activate ATM through MRN-independent pathways [15, 17]. This noncanonical ATM activation occurs irrespective of the presence of DSBs. Transcription blocking DNA damage can activate ATM, through canonical and noncanonical pathways depending on the proliferation status of the cells. In nonreplicating cells ATM is activated via a noncanonical R-loop-dependent, but DSB-independent mechanism [8]. It is important to note that when ATM is activated by noncanonical mechanisms (in the absence of DSBs), it is uniformly distributed in the nucleoplasm as opposed to its localization in large foci when DSBs are present [8, 17].

R-loops are three-nucleic-acid structures (DNA:RNA hybrid across a region of unpaired DNA [18]) that can be formed during transcription as a consequence of negative torsional strain behind elongating RNAPII. R-loop formation is suppressed by several mechanisms including: (1) alleviation of torsional strain through the action of topoisomerases (TopoI) or (2) coupling transcription to RNA processing. This latter mechanism can both reduce complementarity and sterically hinder hybridization of nascent transcripts with DNA [2]. Coupling between transcription and splicing can be disrupted when RNAPII stalls upon encountering DNA-helix distorting lesion (e.g., UV-induced photolesions). In response, mature spliceosomes that normally form stable complexes with nascent transcripts dissociate from the complex, presumably to facilitate backtracking/removal of RNAPII from the damaged bases and subsequent repair. This results in formation of R-loops which signal for ATM activation through an undefined mechanism that in non-replicating cells is DSB-independent but R-loop dependent. In agreement, ATM can also be activated in the absence of any form of DNA damage through chemical or genetic inhibition of late-stage spliceosome assembly [8].

An intriguing aspect of ATM signaling following its activation by transcription blocking DNA damage is that it can influence alternative splicing and splicing factor mobility, while canonical ATM activation by, e.g., ionizing radiation has no influence on the spliceosome [8]. Currently, it is unclear whether this specificity in ATM function stems from differences in subnuclear localization of active ATM (in DSB-associated foci vs throughout the nucleoplasm). It is clear, however, that ATM does not initiate spliceosome mobilization, but it enhances a preexisting signal originating from stalling of RNAPII at DNA damage sites [8]. This peculiarity of ATM function when ATM is activated by transcription-blocking DNA damage can be used to evaluate noncanonical ATM activation and its influence on splicing factor mobility.

In this chapter, we will describe fluorescence microscopy methods that can be used to evaluate noncanonical ATM activation and signaling, in response to transcription-blocking DNA damage. ATM activation is evaluated by immunofluorescence-based detection of its autophosphorylated form using antibodies that specifically recognize phospho-Ser1981 ATM. In addition, we will describe in detail a live cell imaging-based approach to measure the effect of ATM on DNA damage-induced changes in spliceosome mobility. Using Fluorescence recovery after Photo bleaching (FRAP) [19], we determine protein mobility of GFP-tagged splicing factors that typically associate with late stage spliceosomes (U2 or U5 snRNP) [8]. While more traditional methodologies have been described in the literature to evaluate changes in splicing factor association with nascent RNA (and thereby, indirectly, with chromatin), such as DNA:RNA immunoprecipitation (DRIP) or fractionation procedures, we find that FRAP is a highly sensitive, rapid, and reproducible approach to evaluate changes in splicing factor behavior [8]. Furthermore, the use of living cells, where all molecules operate in their physiological environment and proper localization eliminates possibility of artifacts that could possibly be introduced by extensive experimental manipulations.

2 Materials

2.1 Serial Subcultivation of Adherent Cells and Treatments

1. Tissue culture plates and disposable pipettes.
2. Phosphate-buffered saline (PBS), pH 7.4: 136.9 mM NaCl, 2.68 mM KCl, 8.1 mM Na_2HPO_4 , 0.9 mM $CaCl_2$, 0.49 mM $MgCl_2$.
3. Growth media for primary and TERT immortalized diploid fibroblasts: Ham's F10 supplemented with 15% fetal bovine serum (FBS), 100 U/mL penicillin, and 100 mg/mL streptomycin.
4. Synchronization media: Serum-free Ham's F10 supplemented with 100 U/mL penicillin, and 100 mg/mL streptomycin.
5. Trypsin/EDTA.
6. Tissue culture incubator set at 37 °C, 5% CO_2 balanced air.
7. Z-series Coulter Counter (Beckman) for cell counting.
8. ISOTON® II Diluent (Beckman).
9. UV-C source (254 nm, TUV Lamp, Philips).
10. Pladienolide B (Santa Cruz Biotechnology). Resuspend to 5 mM with Dimethyl Sulfoxide (DMSO) and use at 1–5 μM final concentration. Store in small aliquots at −20 °C.
11. Camptothecin. Resuspend to 5 mM with Dimethyl Sulfoxide (DMSO) and use at 5–10 μM final concentration. Prepare fresh or store in small aliquots at −20 °C.
12. ATM inhibitor KU55933. Resuspend to 10 mM with Dimethyl Sulfoxide (DMSO) and use at 10 μM final concentration. Store in small aliquots at −70 °C.

2.2 Fluorescence Recovery After Photobleaching (FRAP)

1. Glass coverslips (24-mm diameter).
2. 1× PBS pH 7.4.
3. Leica TCS SP5 AOBS laser scanning confocal microscope (Leica) equipped with an environmental chamber (PeCon GmbH) and a lens warming unit (Temp control mini).

2.3 Immunofluorescence Microscopy

1. Glass coverslips (24-mm diameter).
2. 1× PBS pH 7.4.
3. Paraformaldehyde solution (PFA), 3.7%: Mix 0.37 g PFA in 10 mL 1× PBS, pH 7.5. Heat this mixture until the PFA has dissolved. Confirm pH and adjust if necessary.
4. 0.2% v/v Triton-X 100 in 1× PBS, pH 7.4.
5. 0.1% v/v Tween 20 in 1× PBS, pH 7.4.
6. Blocking solution: 1× PBS containing Bovine Serum Albumin (BSA) at 2% w/v final concentration. Mix 2 g BSA in 100 ml PBS, filter sterilize using 0.22 μm filters and store at 4 °C.

7. Primary antibody: mouse monoclonal anti-phospho-ATM (1981) from Upstate Biotechnology (antibody 05–740). Store at −20 °C in small aliquots. Dilute in blocking solution (1:1000 v/v) just before using.
8. Secondary antibodies: Goat anti-mouse Alexa Fluor555-conjugated IgG, goat anti-rabibit Alexa Fluor488-conjugated IgG (Thermo Fischer Scientific) (stored at 4 °C). Dilute 1:1000 v/v in blocking solution prior to using.
9. 4,6-Diamidino-2-phenylindole(DAPI)-containing ProLong Gold antifade reagent (Molecular Probes/Thermo Fischer Scientific).
10. Microscopic superfrost slides.
11. Zeiss Axio Imager Z2 upright laser-scanning confocal microscope.

3 Methods

3.1 Serial Cell Subcultivation of Adherent Cells and Treatments

We routinely use primary and TERT immortalized human fibroblasts but any cell type can be used (*see* **Note 1**). Noncanonical, R-loop-dependent ATM activation is triggered by transcription-blocking DNA damage induced by UVC irradiation using a 254 nm, TUV Lamp, Philips. Spliceosome inhibitors added directly in the culture media (e.g., Pladienolide B) can also be used to activate ATM through a noncanonical R-loop-dependent pathway. Alternatively, DNA damage can be induced by topoisomerase inhibitors that can both interfere with transcription, and promote R-loop formation (e.g., camptothecin and its derivatives). For FRAP experiments we use cells stably expressing GFP-tagged splicing factors that participate in the U2 (e.g., SF3a1, SF3b2) or U5 snRNP complexes (e.g., PRP8, SNRNP40). For immunofluorescence experiments we use either transfected or untransfected cells.

3.1.1 Serial Subcultivation of Primary and TERT Immortalized Human Dermal Fibroblasts

1. Aspirate growth media.
2. Wash cells twice with 1× PBS.
3. Aspirate last drops of PBS.
4. Add enough Trypsin/EDTA to cover cell monolayer (e.g., 1 ml per 100 mm plate).
5. Incubate for 2–3 min at room temperature or until cells are detached. This can be verified by microscopic examination.
6. Add 9 ml of serum-containing growth media to stop trypsinization and repeatedly pipette to ensure single-cell suspension. Verify by microscopic examination that there are no cell "clumps" present.
7. Dilute 0.5–1 ml cell suspension with Isoton II to a total volume of 10 ml (20× or 10× dilution respectively).

8. Count cells using Coulter Counter.
9. Calculate cell numbers per ml and seed cells to a density of 1×10^4 per cm^2 of growth surface area (*see* **Note 2**) in appropriate volume of growth media.
10. Place cells in a tissue culture incubator at 37 °C, 5% CO_2 balanced air.
11. Subculture when cells reach confluency.

3.1.2 Preparing Cells for Immunofluorescence Experiments and FRAP

1. Place sterilized glass coverslips into 35-mm cell culture dishes (or in 6-well plates). Seed an extra coverslip that can be used to verify the quiescent status of the cells (*see* **Note 1**).
2. Seed cells at 2–3 × 10^5 cells per dish.
3. Incubate for 24–48 h in a tissue culture incubator.
4. Wash cells twice with 1× PBS.
5. Add 2 ml of serum-free media (*see* **Note 3**).
6. Incubate cells for an additional 72 h prior to the experiment.

3.1.3 Cell Treatments

We always include in each experiment: (1) cells that do not receive the ATM inhibitor nor UV irradiation, (2) cells that are treated with the ATM inhibitor but are not irradiated, (3) cells that are not treated with the ATM inhibitor but are UV irradiated, and (4) cells that are treated with the ATM inhibitor and are UV irradiated.

1. To inhibit ATM activity add directly in the culture media 1 μl/ml from a 10 mM stock of the ATM inhibitor KU55933 (*see* **Note 4**).
2. In negative control plate, add 1 μl/ml DMSO.
3. Incubate cells for 1 h in a tissue culture incubator.
4. Transfer media into a sterile dish.
5. Wash cells twice with 1× PBS.
6. Aspirate PBS.
7. Irradiate cells with desired UVC dose. We typically irradiate cells so that they receive either 10 or 20 J/m^2 UVC (*see* **Note 5**).
8. Return media to cells and incubate for 1 h in tissue culture incubator.

3.2 Measurement of Splicing Factor Mobility by Fluorescence Recovery after Photobleaching (FRAP)

We measure changes in splicing factor mobility upon genotoxic stress and transcription inhibition, by FRAP, using cells stably expressing GFP-tagged splicing factors. For FRAP experiments we use a Leica SP5 AOBS confocal microscope equipped with three spectral PMT (Photo Multiplier Tube) detectors in combination with 458, 476, 488, 496, 514, 561, and 633 nm laser lines (Leica Microsystems). The confocal microscope is equipped with a motorized and heated stage with a coverslip holder, in combination with a PeCon small chamber incubator with heating, CO_2 and O_2 regulation (*see* **Note 6**). In addition, a lens warming unit (Temp control mini) is used to heat the lens used, to 37 °C. For our experiments we use a HCX PL APO CS 63x Magnification oil Immersion lens, with a numerical aperture of 1.40 and a working distance of 0.14 mm.

3.2.1 Switching "on" the Confocal Microscope

1. At least 15 min before starting the experiment, preheat the incubator and the heated stage at 37 °C and connect the lens heater around the 63× lens. The CO_2 level should be set at 5%.
2. Switch on the microscope, lasers, and detectors.
3. Start the LAS AF software (Leica Microsystems) which controls the confocal microscope.
4. Turn on the 488 nm laser so that its fluorescent output signal can stabilize.

3.2.2 Preparing Cells for FRAP

1. At the appropriate time-point after the treatment of the cells the 24-mm diameter coverslip with attached cells is taken out of the tissue culture plate using forceps and placed in a stainless-steel preheated (37 °C) coverslip holder.
2. Close the coverslip holder and add 1 ml of the culture medium on the top of the coverslip.
3. Take care that the coverslip holder is not leaking and that the cells do not dry out during this procedure.
4. Apply a small drop of immersion oil to the 63× lens.
5. Close the environmental chamber and wait for 5 min so that temperature (37 °C) and CO_2 (5%) are stabilized.
6. Locate cells and focus, using transmission light. Do not use Metal-Halide/Mercury lamp as this may cause bleaching of the fluorescent signal of the GFP-tagged proteins.

3.2.3 Obtaining a Representative Image of Each Cell That Will Be Analyzed by FRAP

1. Switch Argon Laser from "idle" to 24% output power.
2. Start the "FRAP Wizard" and work in the "Setup mode."
3. Switch on the "FRAP Booster."
4. Put the 488-nm laser-line output on 3% (*see* **Note 7**). All other laser-lines remain at 0% (off).
5. Select "bidirectional imaging."
6. "Activate Pinhole" and set to 2 airy units (191.1 μm).
7. Set line and frame averaging to 1.
8. Set Scan speed at 200 Hz.
9. Set imaging format at 512 × 512 lines.
10. Activate PMT1 and select emission bandwidth (begin–end) to 500–600 nm to image GFP. If needed the PMT-Trans can be activated to enable a transmission light image.

11. PMT sensitivity is set between 950 and 1100 in such a way that Mean Fluorescence intensity is above 40.
12. Using "live mode" the cells are focused using the *Z*-axis controller in fine mode. Cells with strict nuclear expression and with a clear localization of GFP-tagged splicing proteins in both the nucleoplasm and in "speckles" are selected for further analysis (*see* **Note 8**).
13. Press "capture" to obtain image.

3.2.4 Fluorescent Recovery After Photobleaching (FRAP) Procedure

1. A cell of choice is placed in exactly in the center of the imaging field using the x- and y-stage. Press CRTL + T (Target) to determine the exact center of the imaging field.
2. Set zoom at 8.0, in such a way that the selected cell is in the middle of the image and the whole nucleus is visible.
3. Rotate the cell in such a way that the oval-shaped nucleus in always exactly horizontally orientated (see below). By keeping the same orientation for all cells a similar volume of the cell is bleached (*see* Fig. 1).
4. Set scan speed at 1400 Hz.
5. Set imaging format at 512 × 32 lines representing a physical length of 30.75 μm by 1.87 μm.
6. All other settings remain the same as in Subheading 3.2.3.
7. Press "live "and position the cell in such a way that no nucleoli are present in the region of interest. Press "stop" afterward.
8. Select the "bleach setup" tab, and put the 488 nm laser power at 100% during bleaching. Using these setting the complete 512 × 32 strip will be measured and bleached (*see* Fig. 1 and **Note 9**).
9. In the "time setup" tab, we define 200 images prebleach (*see* **Note 10**), five bleach pulses (*see* **Note 11**) followed by 2000 postbleach measurements, all with a 20-ms interval. The num-

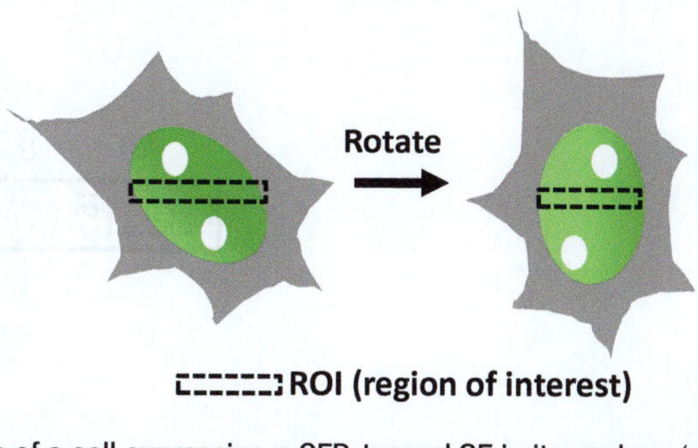

Fig. 1 Cartoon of a cell expressing a GFP-tagged SF in its nucleus (*green*). Prior to FRAP the cell is rotated and centered in the imaging field so that a 512 × 32 strip is photobleached. GFP signal intensity during FRAP is quantified in a user-defined region of interest (ROI/*dashed box*) spanning the entire length of the nucleus

ber of postbleach images can be varied depending on the mobility of the studied protein, but images should be acquired until the fluorescent signal reaches steady-state levels.

10. Press "start experiment" to start the FRAP procedure in this cell line.
11. Select a new cell and perform the whole procedure again. For each experimental condition a minimum of ten cells are analyzed. Typically, we image 15 cells per condition.

3.2.5 Quantification of FRAP and Data Analysis

1. Select "Quantification" tab in the FRAP wizard of the LAS AF software to quantify levels of fluorescence prior to, and after photobleaching, for each cell analyzed.
2. Define two regions of interests (ROI); ROI1 and ROI2. Place ROI1 outside of the cell to measure background signal. ROI2 should cover the complete width of the nucleus (Fig. 2).
3. The fluorescence nuclear signal (ROI2) should be background-corrected, by subtracting the background fluorescent signal (measured in ROI1) at the same time point (corrected signal = signal ROI2 − Signal ROI1).
4. The fluorescence signal value for each measurement should be normalized to the average fluorescence signal values of the last 50 (out of 200) prebleach measurements.
5. The background-corrected and pre-bleach-normalized values of all cells analyzed for each condition should be averaged for each FRAP point, and plotted.
6. The change in mobility (Δmobility) is calculated as the average fluorescence of treated cells—average fluorescence of untreated cells of 50 images at approximately 1 min after postbleaching. This time is chosen arbitrarily and can be varied as long as the recovery of the fluorescent signal has reached a plateau and no farther increase is observed (Fig. 3).

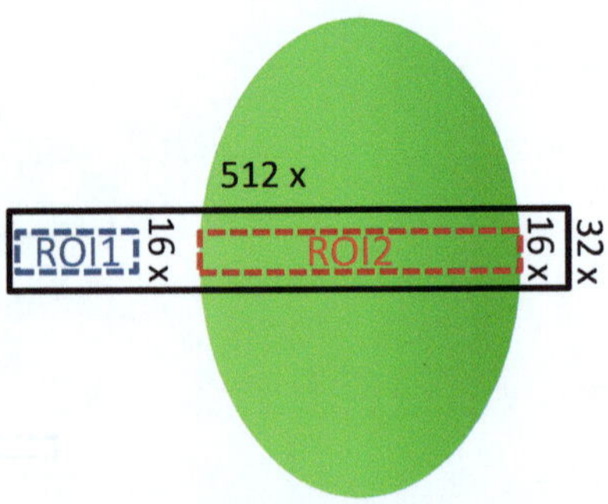

Fig. 2 Cartoon of a cell nucleus expressing a GFP-tagged SF (*green*). The bleached area (*black solid line*), background region of interest (ROI1—*blue dashed line*), and nuclear region of interest (ROI2—*red dashed line*) are indicated

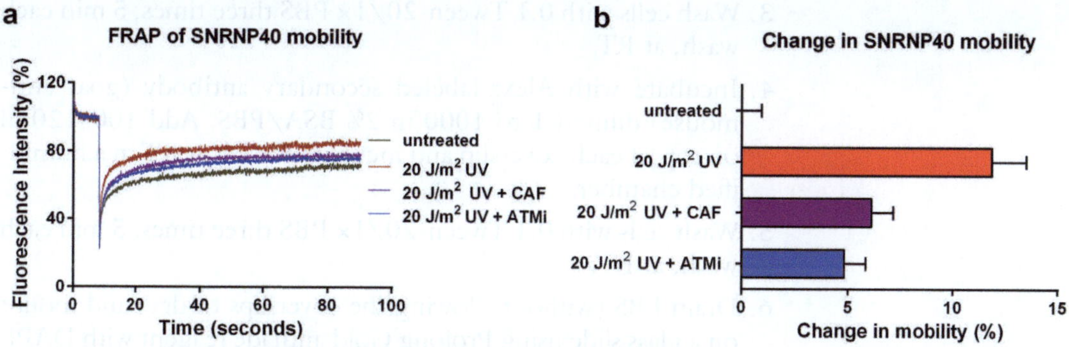

Fig. 3 ATM modulates the UV-induced spliceosome mobilization. FRAP of the U5 snRNP splicing factor SNRNP40, in quiescent human dermal fibroblasts (HDFs). (**a**) FRAP curves of SNRNP40-GFP in untreated, UV irradiated, and UV irradiated cells in the presence of a specific (ATMi KU55933, 10 μM) and nonspecific (CAF, Caffeine, 10 mM) ATM inhibitor. SNRNP40 mobility was measured 1 h post-irradiation. (**b**) Graph of SNRNP40 change in mobility calculated as the fluorescence of irradiated cells minus the fluorescence of non-irradiated cells at 1 min post-bleaching ($n = 12$, mean ± s.e.m., one-way ANOVA with Bonferroni post-test, ***P*, 0.01, ****P*, 0.001)

3.3 Assay of ATM Activation by Immunofluorescence

The following protocol describes the immunofluorescence-based detection of activated ATM in adherent mammalian cells. As a measure of ATM activity, we detect ATM autophosphorylation at Ser-1981 using an antibody that specifically recognizes this ATM modification. This approach is highly sensitive as we can detect low levels of ATM phosphorylation, but also particularly informative as it provides information about the localization of activated ATM. For example, in response to DSBs the majority of p-ATM signal localizes in large foci, together with other "markers" of DSBs. ATM kinase activated by noncanonical pathways shows more "diffused" localization.

3.3.1 Cell Fixation

Cells are seeded on coverslips, serum deprived and treated as detailed above (*see* Subheading 3.1.3).

1. Aspirate media and wash cells twice with 1× PBS.
2. Fix cells by incubating with 1 ml 3.7% PFA/PBS for 15 min at RT.
3. Wash twice with 1× PBS.
4. Permeabilize cells by incubating in 0.1% Triton X-100/PBS, for 20 min at RT.
5. Wash cells three times with 1× PBS.

3.3.2 Immunodetection

1. Block nonspecific antigens by incubating cells in 2% BSA/PBS for 60 min at RT.
2. Incubate with primary antibody (anti-phosphoATM) diluted 1 to 500 in 2% BSA/PBS. Add 100–120 μl on top of each coverslip and incubate overnight at 4 °C in a humidified chamber to avoid evaporation.

3. Wash cells with 0.1 Tween-20/1× PBS three times, 5 min each wash, at RT.
4. Incubate with Alexa-labeled secondary antibody (goat anti-mouse) diluted 1 to 1000 in 2% BSA/PBS. Add 100–120 μl on top of each coverslip and incubate for 1 h at RT in a humidified chamber.
5. Wash cells with 0.1 Tween-20/1× PBS three times, 5 min each wash, at RT.
6. Drain PBS (without allowing the coverslips to dry) and mount on a glass slide using Prolong Gold antifade reagent with DAPI.
7. Seal the edges of the coverslip with clear nail polish.
8. Visualize cells and record images using fluorescence microscope. We use a Zeiss LSM 700 Axio Imager Z2 upright laser-scanning confocal microscope but any fluorescence microscope can be used (Fig. 4).

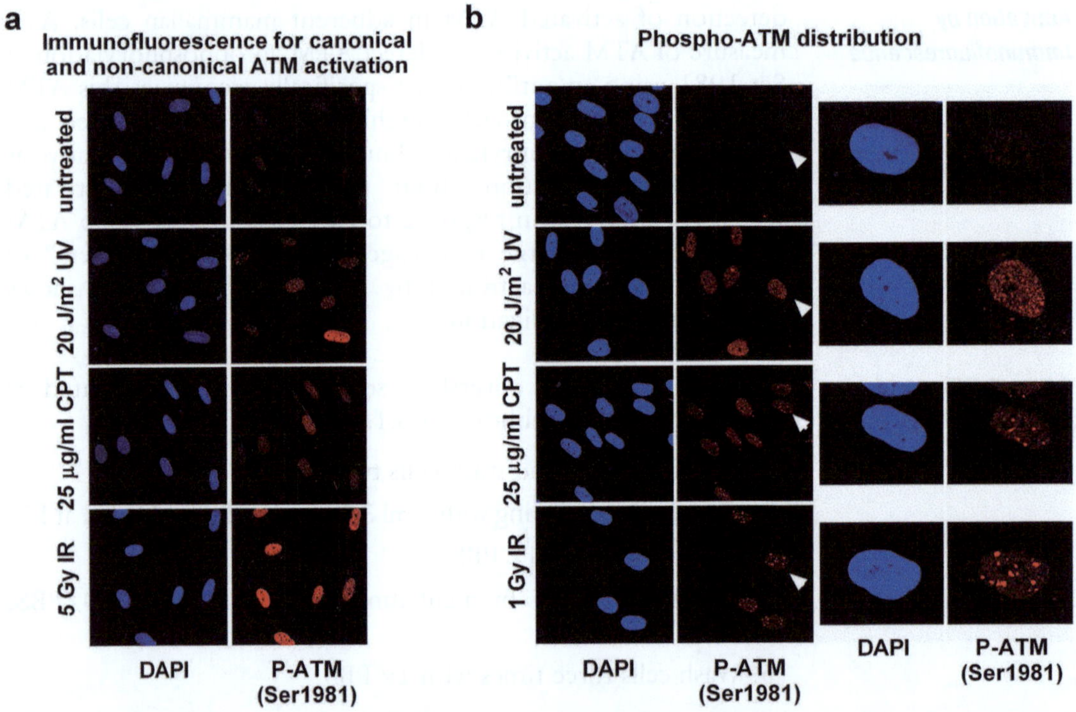

Fig. 4 Canonical and noncanonical ATM activation. (**a**) ATM autophosphorylation (Ser1981) was assayed in quiescent human dermal fibroblasts (HDFs) 1 h after the indicated treatments. In non-replicating cells UV irradiation activates ATM via a noncanonical pathway. (**b**) Differences in autophosphorylated ATM distribution in quiescent HDFs treated with various ATM activators. *Left*: multiple cells; *right*: single-cell magnification illustrating pan-nuclear localization of phosphorylated ATM after UV irradiation and focal accumulation after CPT or ionizing radiation treatments. Magnified cells are indicated by *arrows* (*left panel*). Images were obtained with a 40× and 63×, lens, respectively. Figure was adapted from Tresini et al., doi:10.1038/nature14512

4 Notes

1. The majority of our experiments are performed with quiescent cells as to control for influences of the cell cycle phase. This is particularly important for avoiding replication stress and formation of DSBs. In replicating cells (S phase) collision of replication forks with damage-arrested transcription complexes or R-loops, may result in replication fork collapse and DSBs. DSBs result in robust canonical ATM activation masking subtle changes in ATM activity by noncanonical pathways. Quiescence can be verified by immunofluorescence detection of the Ki67 antigen, a protein present in all active phases of the cell cycle but is absent in G_0 (quiescent) cells. A mouse monoclonal antibody against Ki67 is available from Abgent (ab833).
2. Seeding numbers of adherent cells are adjusted to the surface area of the tissue culture plate (i.e., the size of surface available for cells to attach). This ensures uniform and predictable growth kinetics, independent of the size of plate used. Size of surface area is provided by the manufacturer of the plates.
3. If similar experiments are performed with murine cells, complete absence of serum will result in massive cell death. Addition of 0.1–0.2% FBS in the culture media does not sustain proliferation but is sufficient to ensure cell viability.
4. Alternatively, ATM can be inhibited using Caffeine, a solution of which must be prepared fresh prior to the experiment (100 mM in PBS) and added in the cell culture medium to a 10 mM final concentration. It is important to note, however, that Caffeine is a nonspecific ATM inhibitor; it also inhibits ATR and DNA-PK, as well as other cellular kinases.
5. Quiescent human fibroblasts (both primary and TERT immortalized) can tolerate well relatively high UV doses without significant changes in viability. However, this may vary between cell lines.
6. To successfully image cellular processes in living cells, it is critical that cells are imaged in an environment that does not induce additional stress responses that may influence the experimental outcome. This is achieved by maintaining the temperature at 37 °C, keeping the pH at physiological levels (by the use of 5% CO_2 that buffers bicarbonate-based culture medium) and by minimizing evaporation to prevent changes in medium osmolality.
7. While high laser power generally results in higher signal-to-noise ratio and better images, when laser power is set too high, it will cause "monitor bleaching," i.e., bleaching of the fluorescent GFP signal while monitoring the postbleach signal. This is particularly important during FRAP experiments as during

this procedure small subcellular areas are monitored with high frequency (every 20 ms) using a high zoom factor. Monitor bleaching can be recognized by a slow decrease over-time in signal intensity during the FRAP procedure. The best method to test for monitor bleaching is to image fixed cells (e.g., using 2% PFA, 15 min, RT) under the same experimental conditions used for FRAP, but without a bleaching step. Fixation prevents diffusion of GFP-tagged proteins and the recorded signal should remain stable. "Monitor bleaching" can be avoided by decreasing the laser power, increasing the PMT settings or by imaging less frequently.

8. GFP-tagged splicing factors tend to cluster in nuclear "foci" known as splicing speckles or interchromatin granule clusters or SC35 domains. Splicing speckles can also be visualized by immunofluorescence using a monoclonal antibody against the spliceosome assembly factor SC35.

9. During imaging, presence in ROI2 of non-photobleached splicing speckles due to cell migration or chromatin movement may result in aberrant measurements. To minimize the relocalization of non-photobleached proteins into ROI2, these ROIs are selected so that they are half the width of the bleached region. ROIs are 16 lines wide, while bleached area is 32 lines wide (*see* Fig. 2). If cells or chromatin move during the FRAP analysis, ROIs are shifted upward or downward within the "32 bleached lines" in such a position that no unbleached GFP-tagged proteins are present in the ROIs during the FRAP measurement.

10. The measured GFP fluorescence intensity should be stable prior to photobleaching. GFP fluorescence normally decreases when acquiring the first few images using the above-described settings (1400 Hz, image acquisition every 20 ms). Since during data analysis of the quantified images all values post-bleaching are normalized to the prebleach values, it is crucial to acquire enough prebleach images until the GFP fluorescence stabilizes. We typically acquire 200 images prior to photobleaching and have observed that the GFP signal stabilizes after 50–60 images. The number of prebleach images should be increased if GFP fluorescence has not stabilized.

11. To accurately study changes in protein mobility, a large enough fraction of the fluorescent proteins residing in the area of interest should be photobleached. This fraction is expressed as percentage and is defined as "bleach depth." We typically bleach, using enough pulses, so that the fluorescence signal is decreased to a maximum of 30% of the prebleach levels. If this bleach depth is not achieved, the number of bleach pulses should be increased. If required, the "FRAP booster" can be switched on in the "FRAP wizard"; this increases the laser power and bleaches the GFP-tagged proteins more efficiently.

References

1. Lenzken SC, Loffreda A, Barabino SM (2013) RNA splicing: a new player in the DNA damage response. Int J Cell Biol 2013:153634
2. Tresini M, Marteijn JA, Vermeulen W (2016) Bidirectional coupling of splicing and ATM signaling in response to transcription-blocking DNA damage. RNA Biol 13(3):272–278
3. Hoskins AA, Moore MJ (2012) The spliceosome: a flexible, reversible macromolecular machine. Trends Biochem Sci 37(5):179–188
4. Wahl MC, Will CL, Luhrmann R (2009) The spliceosome: design principles of a dynamic RNP machine. Cell 136(4):701–718
5. Rino J, Carvalho T, Braga J, Desterro JM, Luhrmann R, Carmo-Fonseca M (2007) A stochastic view of spliceosome assembly and recycling in the nucleus. PLoS Comput Biol 3(10):2019–2031
6. Will CL, Luhrmann R (2011) Spliceosome structure and function. Cold Spring Harb Perspect Biol 3(7):a003707
7. Nilsen TW, Graveley BR (2010) Expansion of the eukaryotic proteome by alternative splicing. Nature 463(7280):457–463
8. Tresini M, Warmerdam DO, Kolovos P, Snijder L, Vrouwe MG, Demmers JA et al (2015) The core spliceosome as target and effector of non-canonical ATM signalling. Nature 523(7558):53–58
9. Zlotorynski E (2016) RNA metabolism co-transcriptional splicing at nucleotide resolution. Nat Rev Mol Cell Biol 17(5):264–265
10. Kornblihtt AR, Schor IE, Allo M, Dujardin G, Petrillo E, Munoz MJ (2013) Alternative splicing: a pivotal step between eukaryotic transcription and translation. Nat Rev Mol Cell Biol 14(3):153–165
11. Munoz MJ, Perez Santangelo MS, Paronetto MP, de la Mata M, Pelisch F, Boireau S et al (2009) DNA damage regulates alternative splicing through inhibition of RNA polymerase II elongation. Cell 137(4):708–720
12. Dutertre M, Sanchez G, De Cian MC, Barbier J, Dardenne E, Gratadou L et al (2010) Cotranscriptional exon skipping in the genotoxic stress response. Nat Struct Mol Biol 17(11):1358–1366
13. Paronetto MP, Minana B, Valcarcel J (2011) The Ewing sarcoma protein regulates DNA damage-induced alternative splicing. Mol Cell 43(3):353–368
14. Dutertre M, Sanchez G, Barbier J, Corcos L, Auboeuf D (2011) The emerging role of pre-messenger RNA splicing in stress responses: sending alternative messages and silent messengers. RNA Biol 8(5):740–747
15. Shiloh Y, Ziv Y (2013) The ATM protein kinase: regulating the cellular response to genotoxic stress, and more. Nat Rev Mol Cell Biol 14(4):197–210
16. Lee JH, Paull TT (2005) ATM activation by DNA double-strand breaks through the Mre11-Rad50-Nbs1 complex. Science 308(5721):551–554
17. Bakkenist CJ, Kastan MB (2015) Chromatin perturbations during the DNA damage response in higher eukaryotes. DNA Repair 36:8–12
18. Aguilera A, Garcia-Muse T (2012) R loops: from transcription byproducts to threats to genome stability. Mol Cell 46(2):115–124
19. Houtsmuller AB, Vermeulen W (2001) Macromolecular dynamics in living cell nuclei revealed by fluorescence redistribution after photobleaching. Histochem Cell Biol 115(1):13–21

Chapter 26

Study of ATM Phosphorylation by Cdk5 in Neuronal Cells

Hua She and Zixu Mao

Abstract

The phosphatidylinositol-3-kinase-like kinase ATM (ataxia-telangiectasia mutated) plays a central role in coordinating the DNA damage responses including cell cycle checkpoint control, DNA repair, and apoptosis. Mutations of ATM cause a spectrum of defects ranging from neurodegeneration to cancer predisposition. We previously showed that Cdk5 (cyclin-dependent kinase 5) is activated by DNA damage and directly phosphorylates ATM at serine 794 in postmitotic neurons. Phosphorylation at serine 794 precedes and is required for ATM autophosphorylation at serine 1981, and activates ATM kinase activity. Cdk5-ATM pathway plays a crucial role in DNA damage-induced neuronal injury. This chapter describes protocols used in analyzing ATM phosphorylation by Cdk5 in CGNs (cerebellar granule neurons) and its effects on neuronal survival.

Key words ATM, Cdk5, DNA damage, Cell cycle, Neuronal death

1 Introduction

DNA damage signals are transduced to activate a key family of phosphoinositide-3-kinase-related kinases including ATM, which phosphorylates a number of critical proteins to control cell cycle checkpoints, DNA repair, and in the case of excessive DNA damage, cell death [1, 2]. ATM has been reported to play a vital role in neurodegeneration [3, 4]. How ATM activity is regulated by DNA damage signal is not clear. Previous studies indicate that activation of ATM involves autophosphorylation of several serine residues including serine1981 [5, 6]. The widely accepted model is that autophosphorylation of ATM at serine1981 releases it from the inhibitory homodimer structure, leading to its activation. Cdk5, a small serine/threonine kinase regulated by activator p35 but not cell cycle, has been shown to play a role in neuronal stress response [7, 8]. Our previous work showed that Cdk5, activated by DNA damage, directly phosphorylates ATM at serine 794 in postmitotic neurons [9]. Phosphorylation at serine 794 precedes and is required for ATM autophosphorylation at serine 1981, and activates ATM kinase activity. Cdk5-ATM signal regulates phosphorylation and function of

Sergei V. Kozlov (ed.), *ATM Kinase: Methods and Protocols*, Methods in Molecular Biology, vol. 1599,
DOI 10.1007/978-1-4939-6955-5_26, © Springer Science+Business Media LLC 2017

ATM targets p53 and H2AX. Interruption of Cdk5-ATM pathway attenuates DNA damage-induced neuronal cell cycle reentry and expression of p53 targets PUMA and Bax, protecting neurons from DNA damage-induced death. Recently, the Cdk5-ATM pathway has also been found to serve as a critical signal to initiate intracerebral hemorrhage-induced neuronal injury and regulate cancer cell survival [10–12], which indicates that the Cdk5-ATM signal axis may regulate important functions in several cell types.

Here, we describe detailed methods for immunoprecipitating ATM and Cdk5 from rat primary CGNs, analyzing their activities, determining the levels of ATM phosphorylation by Cdk5, and measuring the effect on neuronal survival. Although we use CGNs as our model system, the general methods described here may be applicable to other systems such as cancer cells.

2 Materials

2.1 Animals

1. Postnatal day 6 or 7 (P6 or P7) Long Evans rats.

2.2 Plasticware

1. Pipettes (5 and 10 ml).
2. 60 and 100-mm cell culture dishes.
3. 6 and 24-well cell culture plates.
4. Centrifuge tubes (1.5, 5 and 50 ml).
5. Glass coverslips.

2.3 Reagents

1. AraC (Cytosine-1-β-D-arabinofuranoside) solution (1 mM, 1000×).
2. Basal Medium Eagle (BME).
3. DNase (2 mg/ml, 0.5 ml aliquot for trituration).
4. Glucose (1 M).
5. Poly-L-lysine.
6. Trypan blue.
7. Trypsin-DNase solution.
8. PBS.
9. MEM medium.
10. DMEM medium.
11. Transfection reagent (e.g., Calcium Phosphate ProFection Mammalian Transfection system, Promega).
12. Cytoplasmic and nuclear fractionation kit (e.g., EZ Nuclei Isolation Kit, Sigma-Aldrich).
13. Protease and phosphatase inhibitors tablets.
14. Protein G Plus/Protein A agarose.

15. SDS-PAGE gels.
16. [γ-^{32}P]-ATP.
17. Anti-Cdk5 antibody (Sigma, #HPA018977).
18. Anti-ATM antibody (Calbiochem, #PC116).
19. PHAS-I substrate for ATM kinase assay (Calbiochem).
20. Histone H1 for Cdk5 kinase assay (Upstate).
21. Anti-γH2AX antibody (BD Biosciences, #560445).
22. Anti-GFP antibody (BD Biosciences #ab1218).
23. Fluorescently labeled secondary antibodies.
24. Goat serum.
25. Mounting medium (e.g., VECTASHIELD Antifade Mounting Medium, Vector Laboratories).
26. Cell proliferation assay kit (e.g., Cell Proliferation Reagent WST-1, Roche).
27. Cell viability kit (e.g., The LIVE/DEAD® Viability/Cytotoxicity Assay Kit, Invitrogen).

2.4 Equipment

1. Desktop centrifuge (4 °C).
2. High-speed centrifuge (4 °C).
3. CO_2 chamber.
4. Forceps, fine.
5. Hemocytometer.
6. Hood, laminar flow.
7. Ice bucket and ice.
8. Incubator preset to 37 °C (5% CO_2).
9. Micropipettor with tips.
10. Microscope, dissection.
11. Scissors, surgical (large and small).
12. Timer for timing incubations.
13. Water bath preset to 37 °C.
14. Shaker.
15. Olympus BX51 microscope.
16. Microplate reader.
17. Electrophoresis system.
18. Membrane transfer system.

2.5 Buffers

1. NP-40 lysis buffer: 20 mM Tris–HCl, pH 8.0, 135 mM NaCl, 1 mM $MgCl_2$, 0.1 mM $CaCl_2$, 10% glycerol, 1% NP-40, 0.1 mM Na_3VO_4, 50 mM ß-glycerol phosphate, 10 mM NaF,

with protease/phosphatase inhibitors cocktail tablet (Roche) added before use.

2. HHGN dissection solution: 1× Hank's Balanced Salt Solution (HBSS), 2.5 mM HEPES-NaOH, pH 7.4, 35 mM glucose, 4 mM sodium bicarbonate. Combine above reagents and store at 4 °C for up to 2 months.
3. Cerebellar granule neurons (CGNs) culture media: Basal Medium Eagle containing 10% dialyzed FBS, 25 mM KCl, 0.1 mg/mL gentamycin, 2 mM glutamine, and 25 mM glucose.
4. ATM kinase buffer: 50 mM HEPES-NaOH, pH 7.5, 150 mM NaCl, 4 mM $MnCl_2$, 10% glycerol, 1 mM dithiothreitol, and 100 μM Na_3VO_4.
5. Cdk5 kinase buffer: 8 mM MOPS-NaOH, pH 7.0, 200 nM EDTA.
6. 2× SDS sample buffer: 2 mL Tris–HCl (1 M, pH 6.8), 4.6 mL glycerol (50%), 1.6 mL SDS(10%), 0.4 mL bromophenol blue (0.5%), 0.4 mL β-mercaptoethanol.

3 Methods

3.1 Cell Culture

Cerebellar granule neurons (CGNs) were cultured from Long Evans rats at P6 or P7 on cell culture plates coated with poly-L-lysine [13, 14]. CGNs were maintained in CGNs culture media (*see* Subheading 2.5.3). Treatments or transfection was performed 7 days after plating.

1. Prepare poly-L-lysine (1 μg/mL)-coated plates.
2. Euthanize P6/P7 Long-Evans rats using a CO_2 chamber.
3. Decapitate pups. To isolate the cerebella, one begins by grasping a pup head at the nose, inserts small surgical scissors into the foramen magnum (the aperture at the base of the skull through which the medulla and spinal cord extend), proceeds laterally and then anteriorly through the ear, and curving toward the midline. After reaching the midline from one side, reinsert scissors into the foramen magnum, and repeat the cutting procedure on the other side.
4. Using forceps, carefully lift the skull upward, and flip it back until the hindbrain is exposed. Make sure that, as the skull is lifted, the brain remains in the base of the skull.
5. If necessary, cut away any bone or other tissues that are obstructing the sides of the cerebellum. Extract the cerebellum with forceps, and place it into a dish containing at least 5 mL of HHGN dissection solution (*see* ref. 14 for clear images of cerebella isolation).

6. After isolation of the cerebella, use the dissection microscope and fine forceps to remove the meninges, as well as any large blood vessels or extraneous brain regions that are stuck to the cerebella (*see* **Note 1**).
7. In the tissue culture hood, decant the dish of cerebella into a 50 mL clean centrifuge tube. Gently swirl the tube and let cerebella to sit onto the bottom. Use a pipette to remove most of the HHGN solution carefully, and then add at least 5 mL of fresh HHGN solution. Repeat this for a total of three washes.
8. After removing the last HHGN wash, add 5 mL of trypsin-DNase solution, be sure to first mix the solution by pipetting up and down. Place the tube in a 37 °C water bath for 10 min. Gently swirl the cerebella every 5 min.
9. Wash three times with HHGN solution as in **step 7**.
10. Prepare an aliquot of DNase I for trituration by adding 250 μL of DNase I to 5 mL of BME in a separate clean tube. Mix by pipetting.
11. Remove the last HHGN wash, and add the diluted DNase solution from **step 10** to the cerebella. Dissociate by pipetting up and down with a 5 mL pipette.
12. Centrifuge cells at ~200 × *g* for 5 min at 4 °C.
13. Use a pipette to remove the supernatant. Add 5 mL of culture medium for granule neurons, and pipette up and down until a homogeneous mixture is obtained. Add 10 mL more of culture medium incrementally, pipetting up and down after dispensing every 2–3 mL to ensure that the mixture remains homogeneous (*see* **Note 2**).
14. Place the cell suspension on ice, and count a small aliquot using a hemocytometer (*see* **Note 3**).
15. Plate cells at the desired density. For morphological assays in 24-well plates with coverslips, use ten million cells per plate. For biochemical experiments in 6-well plates, use 30 million cells per plate. Scale up for larger plates.
16. Maintain cells by adding AraC solution ~24 h after plating to a final concentration of 10 μM. On day in vitro 3 (DIV 3), add 1 M glucose to a final concentration of 25 mM.

3.2 Generation of Antibody to Phospho-Ser794 ATM

Rabbit polyclonal phospho-specific antibody against ATM serine 794 was produced by twenty-first Century Biochemicals with a synthetic 16mer phosphopeptide (KLH-coupled) corresponding to residues surrounding serine 794 of human ATM (Acetyl-*CLSNATKKpSPNKIASG*-amide). The antibody was affinity purified, tested for specificity, and stored at −80 °C.

3.3 Plasmid Transfection

Primary CGNs can be transiently transfected using Calcium Phosphate ProFection Mammalian Transfection system (Promega).

1. Plating Cells for Transfection. Plate about 1×10^6 cells per 100 mm culture plate. Scale down the number of cells proportionately for 6-well or 24-well plates.
2. Preparation of DNA for Transfection. Supercoiled plasmid DNA is efficiently expressed following calcium phosphate-mediated transfection (*see* **Note 4**).
3. Thaw all system components, and warm them to room temperature. Mix each component thoroughly by swirling the container or vortexing briefly.
4. For each transfection, prepare the DNA and 2× HEPES-Buffered Saline (2× HBS, supplied with the kit) solutions in separate sterile tubes. For 100-mm plate, add DNA (10–20 μg in 20 μL) and water (418 μL) to the first tube, mix well, and then add the $CaCl_2$ solution (62 μL) and mix again. Add the specified amount of 2× HBS (500 μL) to the second tube.
5. Working in a tissue culture hood, gently vortex the tube containing the 2× HBS solution. The speed should be adjusted such that the tube can be vortexed safely with the cap off and can accommodate the addition of the prepared DNA solution. Continue to vortex while slowly adding the prepared DNA solution in Tube 1 dropwise to the 2× HBS in Tube 2. (Alternatively, using a pipette, bubble air through the 2× HBS while slowly adding the $CaCl_2$-DNA solution.) When DNA addition is complete, the solution should appear slightly opaque due to the formation of a fine calcium phosphate-DNA coprecipitate. Incubate the solution at room temperature for 30 min.
6. Remove culture medium from CGNs and save it in a clean tube. Wash CGNs with serum-free MEM medium for three times.
7. Vortex the transfection solution again just prior to adding it to the cells. Add the solution containing the $CaCl_2$-DNA precipitate dropwise and in various locations around the plates. Swirl the plates to distribute the precipitate evenly over the cells. This helps to evenly distribute the precipitate and avoid localized acidification of cells. Return the plates to a 37 °C CO_2 incubator.
8. Frequently check the formation of $CaCl_2$-DNA precipitate on cells under microscope. Stop transfection by removing $CaCl_2$-DNA precipitate and washing with serum-free MEM medium for three times. The original CGNs culture medium should be changed back 10–15 min after transfection. In general, cells may be harvested or further treated 48–72 h after the transfection.

3.4 Cdk5 shRNA Lentivirus and Infection

1. To construct the 19 nucleotide (corresponding to rat Cdk5 nt positions 639–657) [15] hairpin siRNA cassettes, two complementary DNA oligonucleotides (5′-GATCCCC-GAGGATCTTTCGACTGCTA-TTCAAGAGA-TAGCAGTCGAAAGATCCTC-TTTTTGGAAA-3′ and 5′-AGCTTTTCCAAAAA-GAGGATCTTTCGACTGCTA-TCTCTTGAA- TAGCAGTCGAAAGATCCTC-GGG-3′) are chemically synthesized.
2. The two complementary DNA oligonucleotides are annealed, and inserted into pSUPER shuttle vector immediately downstream of the H1 RNA pol III promoter.
3. The RNAi cassettes with H1 promoter are amplified by PCR and subcloned into pFUGW transfer vector [16].
4. Produce lentiviruses in HEK293T cells in 100-mm cell culture dish by cotransfection of pFUGW-CDK5i, packaging plasmid p8.91, and vesicular stomatitis virus (VSV) envelope expression plasmid (pMD.G) (5 μg each).
5. Concentrate lentiviruses by centrifugation and estimate the titer using HEK293T cells.
6. Lentiviruses are applied to CGNs at DIV 4. Serial dilutions of the viruses are tested to get the highest infectivity (at least two viruses per cell). Perform the drug treatments 72 h after infection.

3.5 Subcellular Fractionation

Cytoplasmic and nuclear fractionation can be performed using EZ Nuclei Isolation Kit (Sigma-Aldrich), which involves three cycles of thorough cell lysis and washing. Only the first cytoplasmic lysate and the last nuclear lysate are kept and used in experiments.

1. Grow CGNs in poly-L-lysine (1 μg/mL) treated dishes to desired cell density. A 100-mm diameter tissue culture dish of confluent cells contains about 0.5–3.0 × 10^7 cells per dish.
2. Wash cells. For each dish of cells, aspirate the medium and set the dish of cells on ice. Gently wash cells with 10 mL of ice-cold PBS. Carefully aspirate the wash solution.
3. Harvest and lyse cells. Add 4 mL of ice-cold Nuclei EZ lysis buffer to each dish. Harvest and lyse cells by thoroughly scraping each dish with a small bladed cell scraper. Transfer the entire cell lysate from each plate to a separate 15 ml centrifuge tube, vortex briefly, and set on ice for 5 min or until cells have been harvested from all culture dishes.
4. Collect the nuclei by centrifugation at 500 × *g* for 5 min at 4 °C. Carefully aspirate the clear supernatant from each tube and set the nuclei pellet on ice.
5. Resuspend and wash nuclei in 4 mL of ice-cold Nuclei EZ lysis buffer. Vortex nuclei pellet briefly. Add 0.5 mL cold Nuclei EZ

lysis buffer and vortex briefly at moderate to high speed to completely suspend nuclei pellet. Add the remaining 3.5 mL of Nuclei EZ lysis buffer, mix well, and set on ice for 5 min.

6. Collect washed nuclei by centrifugation as in **step 4**. Carefully aspirate the clear supernatant and set the nuclei pellet on ice.
7. Resuspend each nuclei pellet in 200 μL of ice-cold Nuclei EZ storage buffer. Vortex pellet briefly, add 200 μL cold Nuclei EZ storage buffer, and vortex as above to completely suspend the nuclei pellet. Set on ice. Triturate (pipette up and down) five to ten times with a micropipette to help break up clumps of nuclei. Carefully transfer the final nuclei suspension in storage buffer to a microcentrifuge tube for storage.

3.6 Immunoprecipitation and Immunoblotting

1. Lysates were generally prepared with NP-40 lysis buffer with added protease/phosphatase inhibitors from Roche (*see* **Note 5**).
2. Fifty to one hundred micrograms of protein was used for immunoblotting by SDS-PAGE analysis.
3. One hundred to five hundred micrograms of protein was used for immunoprecipitation by incubation with 1–2 μg of corresponding antibody at 4 °C for 4 h or overnight with gentle shaking. Cell lysates were precleaned with 20 μL protein G Plus/Protein A-Agarose for 1 h at 4 °C (*see* **Note 6**).
4. The immune complexes were collected by incubation with 20 μL protein G Plus/Protein A-Agarose for 1 h at 4 °C and washed three times with 400 μL NP-40 lysis buffer by centrifugation at 500 × *g* for 5 min at 4 °C.

3.7 Cdk5/ATM Kinase Assays

1. Active Cdk5/p25 kinase is purchased from Upstate Biotechnology. In vitro Cdk5 kinase assay is performed following the manufacturer's instructions. Incubate two micrograms of purified ATM or GST-ATM4 fragment recombinant proteins with 0.2 μg active Cdk5 kinase in 20 μL of Cdk5 kinase buffer containing 20 μM ATP, 10 μCi of [γ-^{32}**P**]-ATP at 30 °C for 15 min with gentle shaking (1000 rpm).
2. To determine kinase activity of ATM or Cdk5 in cells, wash corresponding immunoprecipitates twice with either ATM kinase buffer or Cdk5 kinase buffer, and incubate in 20 μL kinase buffer containing [γ-^{32} **P**]-ATP and 1 μg of substrate at 30 °C for 30 min with gentle shaking (1000 rpm). Use PHAS-I for ATM kinase assay and histone H1 for Cdk5 kinase assay (*see* **Note** 7).
3. Stop the reaction by adding 20 μL of 2× SDS sample buffer and boil for 5 min.
4. Run samples on SDS-PAGE.
5. Analyze phosphorylation of substrates after SDS-PAGE by autoradiography at −80 °C for appropriate time (usually signals will be visible in 3–24 h) (Fig. 1).

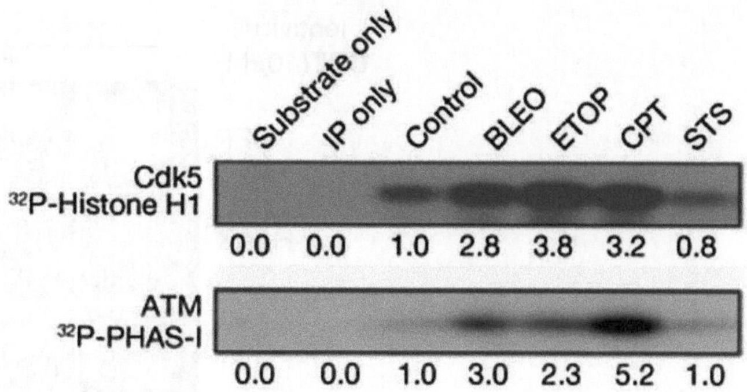

Fig. 1 Activation of Cdk5 and ATM by DNA damage. CGNs were treated with camptothecin (CPT, 10 μM), etoposide (ETOP, 10 μM), bleomycin (BLEO, 100 μg/ml), or stausporine (STS, 2 μM) for 1 h. Cdk5 and ATM kinase activity were measured using histone H1 and PHAS-I, respectively, as substrates. Adapted from [9] Nat Cell Biol, 2009, Feb; 11(2):211–218. doi:10.1038/ncb1829 with permission

3.8 Immunofluorescence

1. Culture CGNs on glass coverslips coated with poly-L-lysine (1 μg/mL).
2. Fix the cells with 3.7% neutral buffered formalin for 15 min and permeabilize with 0.3% Triton X-100 in PBS for 10 min.
3. Block coverslips in 5% normal goat serum for 30 min.
4. Stain fixed cells with anti-γH2AX or anti-GFP (1:500) antibodies overnight at 4 °C, followed by incubation with corresponding fluorescent secondary antibodies (1:500) for 1 h at room temperature.
5. Counterstain cells with Hoechst 33258, mount with mounting medium and seal with nail polish, and analyze using an Olympus BX51 fluorescence microscope. Count numbers of γ-H2AX foci in 200 or more CGNs (Fig. 2).

3.9 Survival Assays

1. Perform WST-1 assay (Roche) following the manufacturer's instructions. Culture cells in 24-well plates. Incubate cells after treatments with WST-1 substrate, a ready-to-use substrate that measures the metabolic activity of viable cells, for 1–4 h. Measure formation of colored product using a microplate reader at 440 nm (*see* **Note 8**).
2. Perform single-cell survival assays using Viability/Cytotoxicity Kit (Invitrogen) according to manufacturer's instructions. Stain CGNs with ethidium homodimer-1 (EthD-1, 1 ng/mL) without permeabilization. Count GFP-positive cells with or without EthD-1 staining using Olympus IX51 fluorescence microscope in a blind manner. Count 200 or more transfected cells for each treatment. Calculate rates of dead CGNs as percentage of cells with colocalization of EthD1 and GFP in the total number of GFP-positive cells counted.

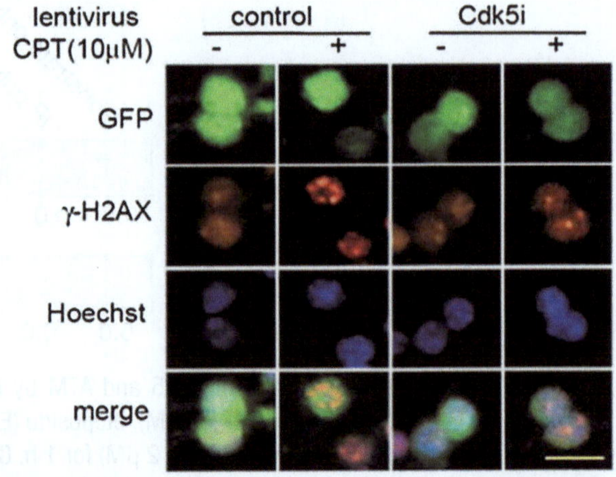

Fig. 2 Reduction of CPT-induced γ-H2AX foci formation by knocking down Cdk5. CGNs were infected with CDK5 RNAi or control lentivirus for 72 h and then exposed to CPT (10 μM) for 1 h. γ-H2AX (*red*) and GFP (*green*) were detected by immunocytochemistry, and nuclei were labeled with Hoechst (*blue*). The scale bars represent 10 μm. Adapted from [9] Nat Cell Biol, 2009, Feb; 11(2):211–218. doi:10.1038/ncb1829 with permission

4 Notes

1. Removing the meninges from cerebella should be performed at a steady pace, with a plate of ice underneath the plate of cerebella. The meninges tend to come off most easily if you start by pulling away residual choroid plexus from the ventral side, where the fourth ventricle is located.
2. During dissociation of CGNs, avoiding bubbles is critical. For approximately the first 15 rounds of pipetting, do not press the pipette tip against the very bottom of the tube; rather, allow the cerebella to fall apart into smaller chunks. Continue to pipette up and down, but begin gently pressing the pipette tip against the bottom of the tube. Press the pipette tip more firmly against the bottom of the tube, and continue the trituration until clumps of tissue are no longer visible. To perform an additional trituration, place the cell suspension to ice for 5 min, allowing any remaining undetected clumps to sink to the bottom. Transfer the top half of the cell solution to a separate Falcon tube, and triturate the bottom again using fresh DNase solution.
3. Counting the CGNs. Make a 1:10 dilution of cells by adding 100 μL of the cell suspension to 650 μL of culture medium and 250 μL of trypan blue. Count healthy, spherical cells with a halo. Do not count cells that take up the trypan blue.

4. When transfecting CGNs, the plasmid DNA to be transfected should be free of protein, RNA, and chemical contamination. Ethanol-precipitated DNA should be resuspended in a sterile solution such as TE buffer to a final concentration of 0.2–1 mg/mL. A good indicator of DNA purity is the ratio of absorbance at 260 nm (*A*260) to 280 nm (*A*280). A DNA solution with an *A*260/*A*280 ratio of 1.8 or greater is desirable.
5. All cell lysis procedures are performed on ice. Proteins and cell lysates are stored at −80 °C.
6. During immunoprecipitation, cell lysates should be precleaned with 20 μL protein G Plus/Protein A-Agarose for 1 h at 4 °C with gentle shaking.
7. In vitro kinase assays are done in a designated radioactive work specific area. The [γ-^{32}P]ATP must be handled in accordance with regulations for the use of radioactivity. Work behind a Plexiglas shield to minimize exposure to ionizing radiation. Gloves and safety spectacles should also be worn. The radioactive samples and the source of radioactivity must be shielded at all times. The [γ-^{32}P]ATP stock should be stored in a lead pot or Plexiglas box. The time spent handling radioactive samples should be minimized.
8. Reagents for cell survival assays are light sensitive. Measurement should be properly carried out shortly after all reagents are added.

Acknowledgments

This work was in part supported by grants to Z.M. (NIH AG023695, NS079858, NS095269, and ES015317.

References

1. McKinnon PJ (2004) ATM and ataxia telangiectasia. EMBO Rep 5(8):772–776. doi:10.1038/sj.embor.7400210
2. Shiloh Y, Ziv Y (2013) The ATM protein kinase: regulating the cellular response to genotoxic stress, and more. Nat Rev Mol Cell Biol 14(4):197–210
3. Li J, Chen J, Ricupero CL, Hart RP, Schwartz MS, Kusnecov A, Herrup K (2012) Nuclear accumulation of HDAC4 in ATM deficiency promotes neurodegeneration in ataxia telangiectasia. Nat Med 18(5):783–790. doi:10.1038/nm.2709
4. McKinnon PJ (2012) ATM and the molecular pathogenesis of ataxia telangiectasia. Annu Rev Pathol7:303–321.doi:10.1146/annurev-pathol-011811-132509
5. Bakkenist CJ, Kastan MB (2003) DNA damage activates ATM through intermolecular autophosphorylation and dimer dissociation. Nature 421(6922):499–506. doi:10.1038/nature01368
6. Kozlov SV, Graham ME, Peng C, Chen P, Robinson PJ, Lavin MF (2006) Involvement of novel autophosphorylation sites in ATM activation. EMBO J 25(15):3504–3514. doi:10.1038/sj.emboj.7601231
7. Su SC, Tsai LH (2011) Cyclin-dependent kinases in brain development and disease. Annu Rev Cell Dev Biol 27:465–491. doi:10.1146/annurev-cellbio-092910-154023
8. Zhu J, Li W, Mao Z (2011) Cdk5: mediator of neuronal development, death and the response to DNA damage. Mech Ageing Dev 132(8–9):389–394. doi:10.1016/j.mad.2011.04.011

9. Tian B, Yang Q, Mao Z (2009) Phosphorylation of ATM by Cdk5 mediates DNA damage signalling and regulates neuronal death. Nat Cell Biol 11(2):211–218. doi:10.1038/ncb1829
10. Wu J, Zhang X, Yan Y, Tang Z, Sun X, Huo G, Liao Z (2015) The crucial role of cyclin-dependent kinase-5-ataxia-telangiectasia mutated axis in ICH-induced neuronal injury of rat model. Mol Neurobiol 53(9):6301–6308. doi:10.1007/s12035-015-9524-4
11. Yu HP, Xie JM, Li B, Sun YH, Gao QG, Ding ZH, Wu HR, Qin ZH (2015) TIGAR regulates DNA damage and repair through pentosephosphate pathway and Cdk5-ATM pathway. Sci Rep 5:9853. doi:10.1038/srep09853
12. Pizarro JG, Folch J, Junyent F, Verdaguer E, Auladell C, Beas-Zarate C, Pallas M, Camins A (2011) Antiapoptotic effects of roscovitine on camptothecin-induced DNA damage in neuroblastoma cells. Apoptosis 16(5):536–550. doi:10.1007/s10495-011-0583-3
13. Tang X, Wang X, Gong X, Tong M, Park D, Xia Z, Mao Z (2005) Cyclin-dependent kinase 5 mediates neurotoxin-induced degradation of the transcription factor myocyte enhancer factor 2. J Neurosci 25(19):4823–4834. doi:10.1523/JNEUROSCI.1331-05.2005
14. Bilimoria PM, Bonni A (2008) Cultures of cerebellar granule neurons. CSH Protoc 2008:pdb-prot.5107. doi:10.1101/pdb.prot5107
15. Meuer K, Suppanz IE, Lingor P, Planchamp V, Goricke B, Fichtner L, Braus GH, Dietz GP, Jakobs S, Bahr M, Weishaupt JH (2007) Cyclin-dependent kinase 5 is an upstream regulator of mitochondrial fission during neuronal apoptosis. Cell Death Differ 14(4):651–661. doi:10.1038/sj.cdd.4402087
16. Lois C, Hong EJ, Pease S, Brown EJ, Baltimore D (2002) Germline transmission and tissue-specific expression of transgenes delivered by lentiviral vectors. Science 295(5556):868–872. doi:10.1126/science.1067081

Chapter 27

DNA Damage Response in Human Stem Cells and Neural Descendants

Jason M. Beckta, Bret R. Adams, and Kristoffer Valerie

Abstract

Glial cells are crucial for the normal function of neurons and are intricately involved in the pathogenesis of neurodegenerative diseases as well as neurologic malignancies. A deeper understanding of the mechanisms by which glial cells influence the development of such pathologies will undoubtedly lead to new and improved therapeutic approaches. Commercially available human embryonic stem cells (hESCs) and induced pluripotent stem cells (iPSCs), both of which can be differentiated into neural progenitors (NPs) and various neural cell lineages, have become widely used as sources for producing normal human central nervous system (CNS) cells. A better understanding of the DNA damage response (DDR) that occurs in these cells after therapeutic ionizing radiation (IR) and chemotherapy is essential for assessing the effects on healthy human brain.

Neurodegenerative features associated with conditions such as ataxia telangiectasia and Nijmegen breakage syndrome highlight the importance of DNA double strand break (DSB) repair pathways in maintaining genomic integrity in cells of the CNS. Similarly, the development of brain tumors is also intricately linked to DNA repair. The importance of ATM and the other phosphatidylinositol 3-kinase-related kinase (PIKK) family members, ATR and DNA-PKcs, is not fully defined in either CNS developmental or pathological states. While their roles are relatively well established in the DDR of proliferating cells, our recent work has demonstrated that these processes exhibit spatiotemporal evolution during cell differentiation. This chapter discusses and explores various laboratory techniques for investigating the role of ATM in hESCs and differentiated neural cells.

Key words ATM, ATR, DNA-PKcs, DNA repair

1 Introduction

The literature continues to support a critical role for stem and progenitor cells at the center of aging and disease pathogenesis as well as being potentially important therapeutic cell targets [1]. The generation of normal and pathologic human CNS cells for in vitro studies has been a long-held goal in neuroscience. Using embryonic or induced pluripotent stem cell systems, it is now a routine laboratory practice to generate neural lineages from such cells

Sergei V. Kozlov (ed.), *ATM Kinase: Methods and Protocols*, Methods in Molecular Biology, vol. 1599, DOI 10.1007/978-1-4939-6955-5_27, © Springer Science+Business Media LLC 2017

suitable for experimentation. In addition to gaining fundamental knowledge, it is anticipated that new therapeutics would emerge.

ATM, ATR, and DNA-PKcs are members of the PIKK family of kinases and represent proteins that are crucial for regulating the DDR, including cell cycle checkpoints, DSB repair, and apoptosis. ATM was originally identified as the culprit of the autosomal-recessive neurodegenerative disease, ataxia telangiectasia (A-T) [2]. The current knowledge of ATM details its involvement in a wide range of cellular processes with the maintenance of telomeres and genomic stability being some of the most thoroughly studied functions (for recent reviews, *see* [3, 4]). Clinical findings from studies on A-T patients, using cells derived from such individuals, and ATM mouse models have helped define the roles ATM play at the organismal and cellular levels (for recent reviews, *see* [5, 6]). The examination of ATM in normal human CNS cells will not only provide new information regarding its involvement in various neurological disorders but, perhaps, also what predisposes certain individuals to neoplastic disease. Due to its central role in the maintenance of genomic stability, various pharmaceutical companies have developed small-molecule inhibitors for ATM and other PIKK family members to explore new avenues in treating neurological disease and cancer.

Our laboratory initially began studying ATM to develop possible therapeutic applications for treating glioblastoma multiforme (GBM), a fatal primary brain tumor [7–10]. Additionally, we have examined the role of ATM with regard to preserving genomic integrity and the development of human neurodegenerative diseases using hESCs and progenitor cells for the purpose of testing novel pharmaceuticals in vitro and in animal models [11, 12]. There are numerous studies on ATM in cancer cells; however, the DDR signaling pathways of these aberrant cells are not representative of normal physiologic processes. Embryonic or induced pluripotent stem (iPS) cells can serve as a starting point for studying CNS development as well as producing tissue for regenerative therapy. We follow established protocols for the directed differentiation of hESCs to generate close to homogeneous populations of NPs and differentiated progenies. Additionally, we use small-molecule inhibitors to manipulate ATM and DNA-PKcs to better understand their roles in the DDR and DSB repair.

This chapter describes methodology for characterizing the DDR and DSB repair processes in human neural cells with emphasis on ATM. The focus is on using radiation-induced foci (IRIF) and DNA repair reporter assays as powerful methods to dissect the quality of DSB repair when ATM activity is manipulated. Using stem cell technology one can now obtain sufficient quantities of human neural cells suitable for such analyses.

2 Materials

2.1 hESC Cell Culture

1. hES (e.g., H9, H9-(v)1 [13, 14]) or iPS cell culture medium can be of two varieties. A budget version consists of Dulbecco's Modified Eagle's Medium ("DMEM/F12") supplemented with: 15% fetal bovine serum, 5% knockout serum replacer (KSR), nonessential amino acids, 20 mM glutamine, 1× Penicillin-Streptomycin, 0.1 mM β-mercaptoethanol, and 4 ng/mL fibroblast growth factor-2 (FGF). Store at 4 °C shielded from excessive light exposure and use within 7 days. A more expensive but less-laborious option is to use mTeSR culture medium (STEMCELL Technologies).
2. Matrigel aliquots (Corning® Matrigel® Matrix) store at −80 °C.
3. Matrigel-coated dishes for cultivation of hES/iPS cells. Thaw Matrigel aliquots on ice over several hours. Prechill all materials and reagents prior to procedure (tubes, dishes, pipette tips, serological pipettes, and media). Add an aliquot of Matrigel to an adequate volume of cold DMEM/F12 (usually 500 μl of Matrigel in 20 ml of DMEM/F12). Mix by quickly inverting the tube several times. Place enough Matrigel solution into each dish so that the bottom is completely covered (e.g., 1 ml solution for a 3-cm dish). Store at −80 °C until needed.
4. Dispase™ (STEMCELL Technologies): aliquot and store at −80 °C. Thaw to room temperature when needed.

2.2 Treatments

1. ATM kinase inhibitors KU-60019 and KU-55933 (Selleck) are dissolved in DMSO to a concentration of 3 mM (1000×) and 10 mM (1000×) and stored at −20 °C until needed.
2. DNA-PKcs inhibitor KU-57788 (Selleck) is dissolved in DMSO to a concentration of 2.5 mM (1000×), and stored at −20 °C until needed.
3. Cell Irradiator (e.g., MDS Nordion Gammacell 40 research irradiator with a Cs-137 source).

2.3 Immunofluorescence

1. Nunc™ Lab-Tek™ II Chamber Slide™.
2. 1× and 10× phosphate-buffered saline (PBS).
3. 3% paraformaldehyde (PFA). Add 3 g PFA to a glass beaker. Add 70–80 ml of distilled water and place on a hot plate with magnetic stirring. The solution needs to be heated to dissolve the PFA, but care should be taken to not allow the solution to boil. Add 1–2 drops of 1 M sodium hydroxide to improve solubility. Add 10 ml of 10× PBS. The pH should be between 7.2 and 7.4; adjust with sodium hydroxide as necessary. Add distilled water to a final volume of 100 ml. Store at 4 °C, protected from light, and use within 7 days.

4. Blocking solution: 3 ml goat serum per 100 ml of 1× casein/PBS (PBS containing 1% (w/v) casein).
5. Permeabilization buffer: 0.5 ml Triton-X 100 per 100 ml PBS.
6. Mounting medium with DAPI (e.g., VECTASHIELD Antifade Mounting Medium with DAPI, Vector Labs).
7. Coverslips.
8. Nail polish.
9. Phospho-(S1981)-ATM antibody (1:1000) (Cell Signaling Technology) (to assess ATM activation).
10. ATM antibody (1:1000) (GeneTex, Inc.) (to quantitate total ATM protein levels).
11. Oct3/4 antibody (Santa Cruz Biotechnologies) (to verify pluripotency).
12. β-Actin antibody (Santa Cruz Biotechnologies) (as a loading control for protein gels).
13. Nestin antibody (Millipore) (to assess differentiation status).
14. GFAP antibody (Millipore) (to assess differentiation status).
15. Musashi1 antibody (Millipore) (to assess differentiation status).
16. βIII tubulin antibody (Millipore) (to assess differentiation status).
17. O1 antibody (Millipore) (to assess differentiation status).
18. Sox2 antibody (Millipore) (to assess differentiation status).
19. γ-H2AX antibody (used as a surrogate marker of DSBs).
20. Alexa-488 goat anti-rabbit IgG antibody (Invitrogen).
21. Alexa-594 goat anti-mouse IgG antibody (Invitrogen)
22. Image analysis software, such as Volocity (PerkinElmer).

2.4 Differentiation of Stem Cells

1. Laminin-coated dishes: 5 μg/ml laminin mixed in cold (4 °C), sterile water. Add laminin solution to the dish and gently rock it to ensure the entire surface of the plate (or well) is covered with the laminin solution. Allow dishes to incubate at room temperature for at least 30 min, but no longer than 1 h before use. Store coated dishes at 4 °C and use within 1 month.
2. KSR Medium: DMEM/F12 supplemented with 20% KSR, 1% nonessential amino acids, 1 mM glutamine, and 1× penicillin-streptomycin.
3. N2 Medium: DMEM/F12 with 1× N2 supplement (Invitrogen), 4 ng/mL FGF-2, 2 mM glutamine, and 1× penicillin-streptomycin.
4. Neurobasal Medium: Neurobasal medium supplemented with 1× B27 (Invitrogen), 2 mM glutamine, 20 ng/ml FGF-2,

10 ng/ml Leukemia Inhibitory Factor (LIF) (Millipore), and 1× penicillin-streptomycin.

5. Differentiating Medium: Dulbecco's Minimal Essential Medium (DMEM) supplemented with 10% fetal bovine serum and 1× penicillin-streptomycin.

2.5 DNA Damage Response Analyses

2.5.1 IRIF Assay

1. Materials as described above in Subheadings 2.1–2.4 (cell culture, treatments, immunofluorescence).
2. Cells treated with DNA-damaging agents, e.g., radiation, with or without ATMi or DNA-PKi followed by a recovery time (0-24 h).

2.5.2 NHEJ Repair Assay

1. Lentivirus $WPXL_d$-2×ISceI-DsRed-neo [12, 15]. A lentivirus ($WPXL_d$-2xISceI-DsRed) harboring a repair cassette positioned upstream of the DsRed reporter gene that was recently described (*see* Supplemental Methods in [15]), with the exception that an IRES-NEO selection cassette was added [12].
2. G418 (Geneticin).
3. Adenovirus Ad-CMV-SceI expressing I-SceI endonuclease [9].
4. Trypsin.
5. 24-well plates coated with Matrigel (*see* Subheading 2.1.3).
6. Genomic DNA isolation kit (e.g., High Pure PCR Template Preparation Kit, Roche).
7. Q-PCR Relative Quantification Assay (Q-PCR kit, Applied Biosystems).
8. Sequence Detection Systems software.
9. SYBR Green PCR mix (Applied Biosystems).
10. The PCR primers for NHEJ quantification: 5′-CACGAGACTAGCCTCGAGGTTT, 5′-CTTGAAGCGC ATGAACTCCTT, β-actin, 5′-TCACCCACACTGTGCCCAT CTACGA, and 5′-CAGCGGAACCGCTCATTGCCAATGG.
11. Non-denaturing polyacrylamide gel.
12. Ethidium bromide stain.
13. Typhoon 9410 variable mode scanner (General Electric Healthcare).
14. Flow cytometer (e.g., Coulter Epics XL-MCL).

2.6 Protein Analysis

1. RIPA buffer: 10 mM Tris–Cl (pH 8.0), 1 mM EDTA, 0.5 mM EGTA, 1% Triton X-100, 0.1% sodium deoxycholate 0.1% SDS, 140 mM NaCl. Store at 4 °C. Add 1 mM PMSF and HALT™ cocktail immediately prior to use.
2. Halt™ Protease and Phosphatase Inhibitor Cocktail (100×) (Thermo Fisher Scientific). Add to the RIPA buffer immediately prior to extraction (do not store premixed aliquots).

3. 2× Laemmli sample buffer: 62.5 mM Tris–HCl (pH 6.8), 25% glycerol, 2% SDS, 0.01% Bromophenol Blue.
4. Electrophoresis equipment.
5. Criterion™ TGX™ (Tris-Glycine eXtended) precast gels for PAGE (Bio-Rad).
6. PVDF membranes.
7. Secondary antibodies of choice. We use anti-rabbit DyLight™ 800 and anti-mouse IgG Alexa 680-conjugated antibodies.
8. Image analysis software. ImageJ or Image Studio v2.0 (Li-Cor Biosciences).

2.7 Live Cell Imaging

1. A complete protocol for live cell imaging is beyond the scope of this chapter. *See* Beckta et al. [16] for comprehensive materials and methods.
2. In brief, transduce a lentivirus, $WPXL_d$-H2B-mCherry-12A-puro [17], expressing a H2B-mCherry fusion and a puromycin resistance gene, into hESCs or iPSCs and maintain cells under drug selection. Periodically check for the presence of fluorescence to ensure population quality.
3. Glass-bottom dishes are coated with Matrigel using the protocol described above (Subheading 2.1).

3 Methods

3.1 hES/iPS Cell Passaging and Growth

1. The hES/iPS cells are cultured on Matrigel™ coated dishes.
2. Add 1 ml of Dispase™ per 3-cm dish. Place on 37 °C rocker for 3 min. Colony edges can be observed detaching under 10× power.
3. Aspirate Dispase™ and wash three times with sterile PBS.
4. Add 1 ml of stem cell culture medium and scrape/gently pipette up and down to dislodge colonies and break them up into smaller clusters. Care should be taken not to create a single-cell suspension.
5. Place cells in a 15 ml conical tube containing 10 ml of culture medium.
6. Spin harvested cells at room temperature for 3 min at 200 × *g*.
7. Aspirate culture medium, gently suspend cells in 5 ml sterile PBS, and spin once again using the same procedure.
8. Resuspend cell pellet in 1-2 ml of culture medium.
9. Plate cells into dishes at the desired density.
10. Change 50–75% of the culture medium daily. Passage cells with Dispase™ at least once every 3–5 days or before the colony density becomes too high (*see* **Note 1**).

3.2 Differentiation of Stem Cells

1. Following the previously described steps (*see* Subheading 3.1), treat hES/iPS cell colonies with Dispase™ to remove the cells from the dish. In contrast to routine culturing, vigorously pipette the solution to ensure a single-cell suspension.
2. Transfer the hES/iPS cell suspension to a laminin-coated dish. For the first 24 h, culture with standard stem cell medium.
3. After 24 h, exchange stem cell medium with the 15% KSR medium, changing it daily for 1 week.
4. Next, for Neural progenitors (NP) development, grow hESCs on laminin-coated dishes in N2 medium for another 7 days.
5. Cultivate cells for an additional 7 days in B27 medium with a daily medium change.
6. Astrocytes are obtained by culturing the resultant NPs in standard DMEM/FBS cell culture medium for approximately 25 days. Media should be changed daily.
7. Perform experiments 24–48 h after passaging (*see* **Note 2**).

3.3 Immunofluorescence

1. Culture cells on chamber slides. When appropriate cell density is reached and treatment completed aspirate medium and wash once with PBS. Add enough PFA to coat the bottom of each well of the chamber slide.
2. Let it sit (without rocking) at room temperature for 15 min.
3. Wash cells three times with PBS.
4. Store fixed cells in a moisturized container at 4 °C until ready to process further.
5. Wash fixed cells one time with PBS.
6. For immunofluorescence expose cells to permeabilization solution for 10 min (without rocking). Wash cells several times with PBS.
7. Add 1 ml of blocking solution to each well, and let the slide sit at room temperature for 45 min. Aspirate blocking solution from wells.
8. Dilute the primary antibody in a volume appropriate to the number of wells that require staining (0.3–0.5 ml per chamber). Use either the supplier recommended dilution, or if not known, begin with a 1:500 dilution. Cover chambers to prevent unnecessary exposure to light and incubate (with rocking) at room temperature for 1.5 h. Alternatively, this step can be extended to overnight at 4 °C, which may permit more intense staining.
9. Aspirate primary antibody solution and wash cells four times for 15 min each with PBS at room temperature with rocking.

10. Prepare the secondary antibody in blocking solution at either the supplier recommended dilution, or if not known, begin with a 1:1000 dilution.
11. After the fourth wash of the primary antibody, aspirate the PBS and add 300–500 μl of the secondary antibody solution. Incubate for 2.5 h at room temperature. As with the primary antibody staining, cover the chambers to prevent unnecessary exposure to light.
12. Remove plastic chambers, dry, and add a drop of mounting media containing DAPI and overlay cells with a coverslip. Seal with nail polish.
13. Image and analyze cells using the appropriate instrument. In these experiments, cells are imaged using a Zeiss LSM 710 Meta imaging system and analyzed using the Volocity software (*see* **Note 3**).

A representative cell differentiation experiment and characterization by immunofluorescence of various cell markers is shown (Fig. 1).

3.4 DNA Damage Response Analyses

3.4.1 IRIF Assay

1. Culture cells from each lineage as described above (Subheading 3.3).
2. Add the ATMi (KU-55933 or KU-60019) or DNA-PKcs (KU-57788) 30 min before planned irradiation. This length of time permits the complete inhibition of ATM and DNA-PKcs kinases within the cells (*see* **Note 4**).
3. Using the appropriate controls, expose the cells to the desired dose of radiation (in our experiments, generally 2–4 Gy). Fix the cells with PFA after an appropriate time interval has passed, which will vary depending on the process being examined.
4. Identify and quantitate the average number of γ-H2AX foci per cell, a surrogate marker for DNA DSBs, using immunofluorescence methods described above (*see* Subheading 3.3). IRIF are either counted manually or by using the image quantitation algorithms within the Volocity software.

A representative repair foci experiment is shown (Fig. 2). hESCs respond very differently from the resultant astrocytes. Whereas the hESCs are refractory to inhibitors of ATM and DNA-PKcs after irradiation, repair foci formation in terminally differentiated astrocytes is completely blocked by the ATMi. In contrast, inhibiting DNA-PKcs does not block foci formation and, instead, slows down repair kinetics.

3.4.2 NHEJ Repair Assay

The NHEJ-red assay relies on a cassette containing two I-SceI recognition sequences flanking an ATG codon that, when not cut by I-SceI, acts as a decoy initiation codon, preventing translation of the DsRed reporter gene [15]. Upon cleavage with I-SceI at both recognition sequences, the decoy codon is excised within a 25-bp

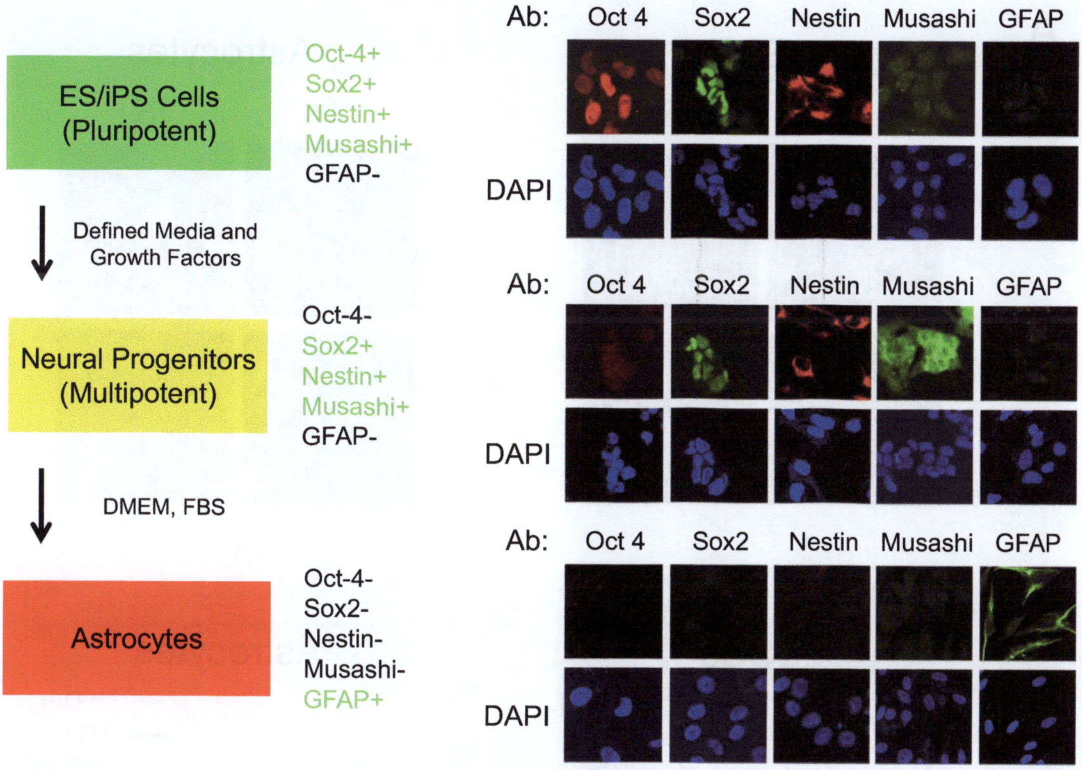

Fig. 1 Differentiation of hESCs to neural progenitors and astrocytes. Immunofluorescence demonstrates the expression of cellular markers for stem cells and their successful differentiation to neural progenitors and astrocytes

sequence stuffer fragment. If NHEJ occurs the DNA is sealed and DsRed is expressed from a downstream, previously out-of-frame ATG codon (Fig. 3). The two I-SceI sites are in opposite orientations, such that when both are cleaved with I-SceI and the stuffer fragment of the vector is excised, two partially complementary nucleotide 5′ overhangs are generated: TTAT and TATT. If these ends anneal without DNA-end resection, a two-base gap will result on both strands and would be filled in, likely by a gap-filling polymerase to generate a 125-bp product amplicon which can be quantified relative to β-actin by RT-PCR (see below). This possible repair scenario would result in a repair joint sequence reading TTATAA, which may be considered high-fidelity end joining. More extensive resection would still result in DsRed expression unless the deletion is so extensive that it removes the downstream DsRed ATG codon or removes the upstream promoter. NHEJ events are then determined by FACS of DsRed-positive cells, genomic qPCR, or by cloning and DNA sequencing.

1. Isolate hESC or iPS/NHEJ-red cells by infection with a lentivirus WPXL$_d$-2×ISceI-DsRed-IRES-NEO (hereafter referred

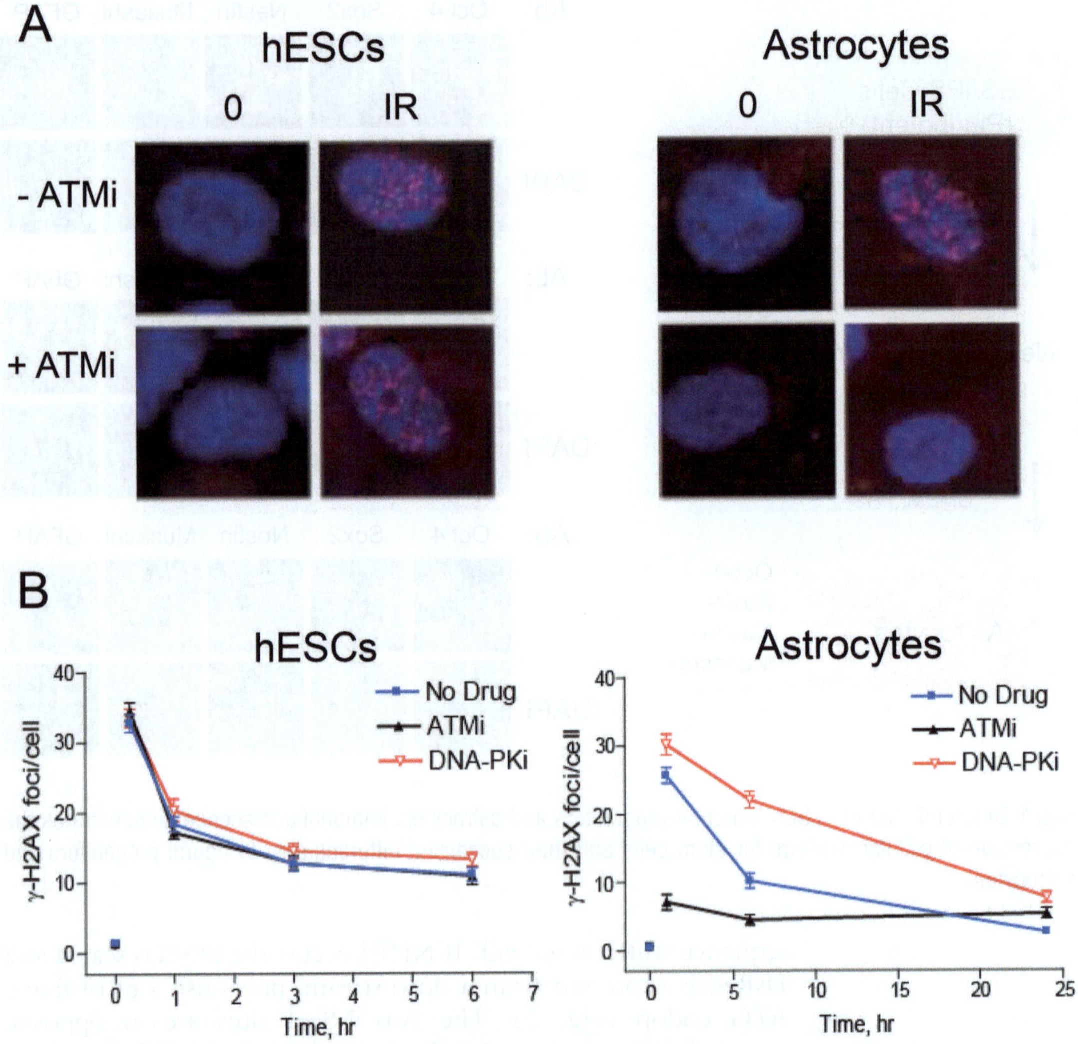

Fig. 2 Inhibiting ATM prevents repair foci formation in hESC-derived astrocytes, whereas inhibiting DNA-PKcs slows down repair kinetics. H9 hESCs and H9-derived astrocytes were irradiated (2 Gy) in the presence of 10 μM ATMi (KU-55933), 2.5 μM DNA-PKi (KU-55788), or not, followed by γ-H2AX foci formation assay at the indicated times. (**a**) Representative images of unirradiated and irradiated (IR) without (– ATMi) and with (+ ATMi) 15 min after irradiation. Nuclei were stained with DAPI. (**b**) Whereas repair kinetics is not influenced significantly in hESCs using either inhibitor, the ATMi completely inhibits foci formation and the DNA-PKi slows down repair kinetics in hESC-derived astrocytes

to as NHEJ-red cells) (*see* **Note 5**). It is important that the repair cassette is inserted in the hES/iPS cells if a comparison of NHEJ will be conducted throughout cell differentiation. Other repair reporter constructs could also be considered (*see* **Note 6**).

2. Clone cells resistant to G418 by serial dilution. Select cells for the integration of the cassette with 100 μg/ml of G418. Trypsinize cells after 1 week and plate in 24-well plates coated with Matrigel.

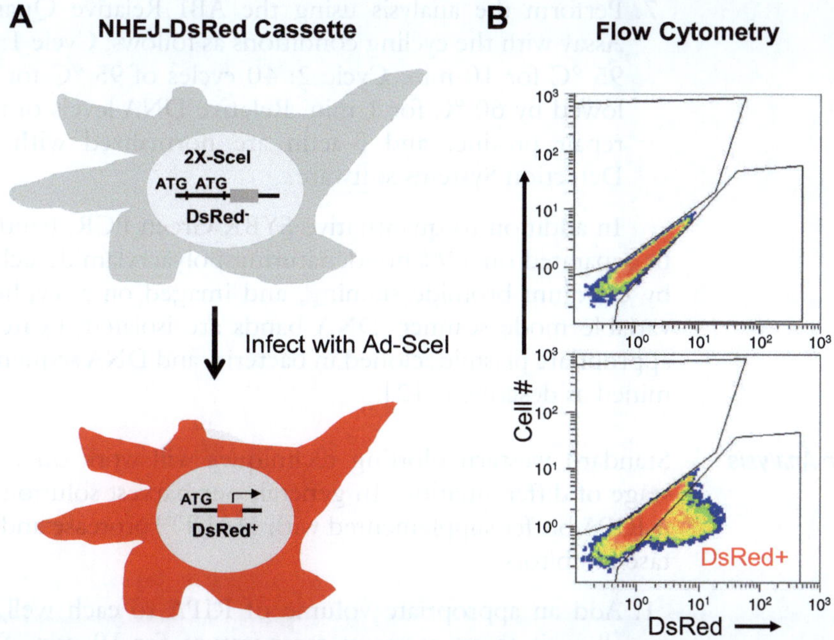

Fig. 3 hESCs utilize NHEJ. hESCs were infected with the NHEJ repair cassette lentivirus and selected with G418 (**a**). Upon infection with an adenovirus expressing I-SceI, cells expressing DsRed mark NHEJ activity as quantified by flow cytometry (**b**)

3. Screen for the integration of NHEJ-red virus by infection with an adenovirus expressing the I-SceI endonuclease from *Saccharomyces cerevisiae* (Ad-SceI) [9]. Infect the cells with adenovirus at a multiplicity of infection (MOI) of 30, while rocking the dishes at 37 °C for 4 h.
4. Analyze DsRed expression by confocal microscopy and flow cytometry (*see* **Note** 7). Pick single colonies and test for the expression of DsRed. It is recommended that several G418-resistant cell clones are screened.
5. Isolate DsRed-positive hESCs (or neural descendants) by FACS 48-72 h after infection to enrich for DsRed-positive cells that have undergone NHEJ. Analyze cells on a Coulter Epics XL-MCL flow cytometer. Briefly, trypsinize cells and resuspend in DMEM/10% FBS. Gating functions are applied to a 2-D plot of side scatter versus forward scatter, so the live cell population can be analyzed. A representative DsRed flow cytometry experiment is shown (Fig. 3).
6. Extract genomic DNA using DNA isolation kit. Perform amplification of genomic DNA on an ABI 7900HT Real-time qPCR instrument using SYBR Green and. PCR primers for the NHEJ quantification. Determine relative NHEJ levels after normalizing to β-actin.

7. Perform the analysis using the ABI Relative Quantification Assay with the cycling conditions as follows; Cycle 1: 1 cycle at 95 °C for 10 min, Cycle 2: 40 cycles of 95 °C for 15 s, followed by 60 °C for 1 min. Relative DNA levels of the NHEJ repair product and β-actin are normalized with Sequence Detection Systems software.

In addition to quantitative SYBR Green PCR, bands can also be separated on a 9% nondenaturing polyacrylamide gel, detected by ethidium bromide staining, and imaged on a Typhoon 9410 variable mode scanner. DNA bands are isolated, ligated into an appropriate plasmid, cloned in bacteria, and DNA sequences determined as described [12].

3.5 Protein Analysis

Standard western blotting techniques will work on cells at any stage of differentiation. In general, our harvest solution of choice is RIPA buffer supplemented with HALT™ protease and phosphatase inhibitors.

1. Add an appropriate volume of RIPA to each well/dish and allow it to sit at room temperature for 10 min. Scrape the wells/dishes and add the mixture to a 1.5 ml Eppendorf tube. Vigorously mix each tube with a P1000 micropipette.
2. Place the tubes on ice for an additional 10 min. Centrifuge the tubes and discard the pellet.
3. Add an appropriate volume of 2× Laemmli sample solution. At this point, tubes can be frozen for future analysis.
4. Perform gel electrophoresis and western blotting using Criterion™ TGX gels. The type of gel (5%, 10%, etc.) used varies based on the molecular weight of the protein under study.
5. After the gel has been run, transfer proteins to a PVDF membrane (*see* **Note 8**).
6. When the transfer is complete, remove the membrane from the transfer box and place in a bath of blocking solution for at least 2 h at room temperature (*see* **Note 9**).
7. When blocking is complete, incubate the membrane with primary antibodies at a 1:1000 dilution at 4 °C overnight.
8. Wash the membrane next day four times in a PBS/Tween-20 solution for 15 min each.
9. Incubate the membrane with secondary antibody at the manufacturer's specified dilution.
10. Detect and quantify protein bands using infrared-emitting dye conjugated secondary antibodies and appropriate scanner.
11. Perform densitometry using ImageJ or Image Studio v2.0.

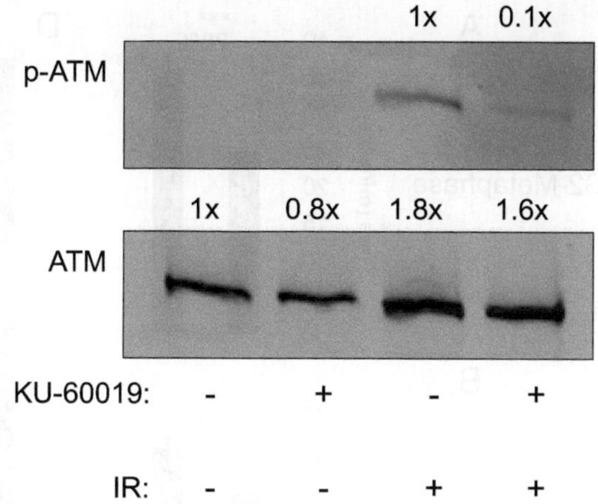

Fig. 4 ATM is present and active in hESCs and KU-60019 inhibits ATM activation in response to irradiation. Cells were treated with KU-60019 30 min prior to irradiation and harvested 30 min after irradiation with 5 Gy. Western blot analysis was performed with antibodies raised against p-(S1981)-ATM and ATM

A representative p-ATM blot is shown (Fig. 4). As seen, ATM auto-phosphorylation at Serine 1981 is inhibited by an ATMi in hESCs.

3.6 Live Cell Imaging

For further analysis of the DDR live-cell imaging can be performed. Various fluorescent protein constructs can be made or purchased and transduced/transfected into cells, permitting real-time examination of intracellular processes. As stated previously, a complete description of the materials and methods for live cell imaging is beyond the scope of this text, and the reader is encouraged to refer to Beckta et al. [16].

1. Briefly, transduce hESCs with a Histone H2B-mCherry lentivirus construct to visualize chromatin. Mitosis can thus easily be monitored.
2. Passage the cells onto glass-bottom dishes 24–48 h prior to recording. Keep the cells on an incubated stage at 37 °C and 5% CO_2 for the duration of the experiments.

A representative experiment using live cell imaging to monitor cell cycle progression through mitosis and the effect of inhibiting ATM is shown (Fig. 5). In the absence of radiation, an ATMi slows down progression through mitosis, in particular from G2-metaphase.

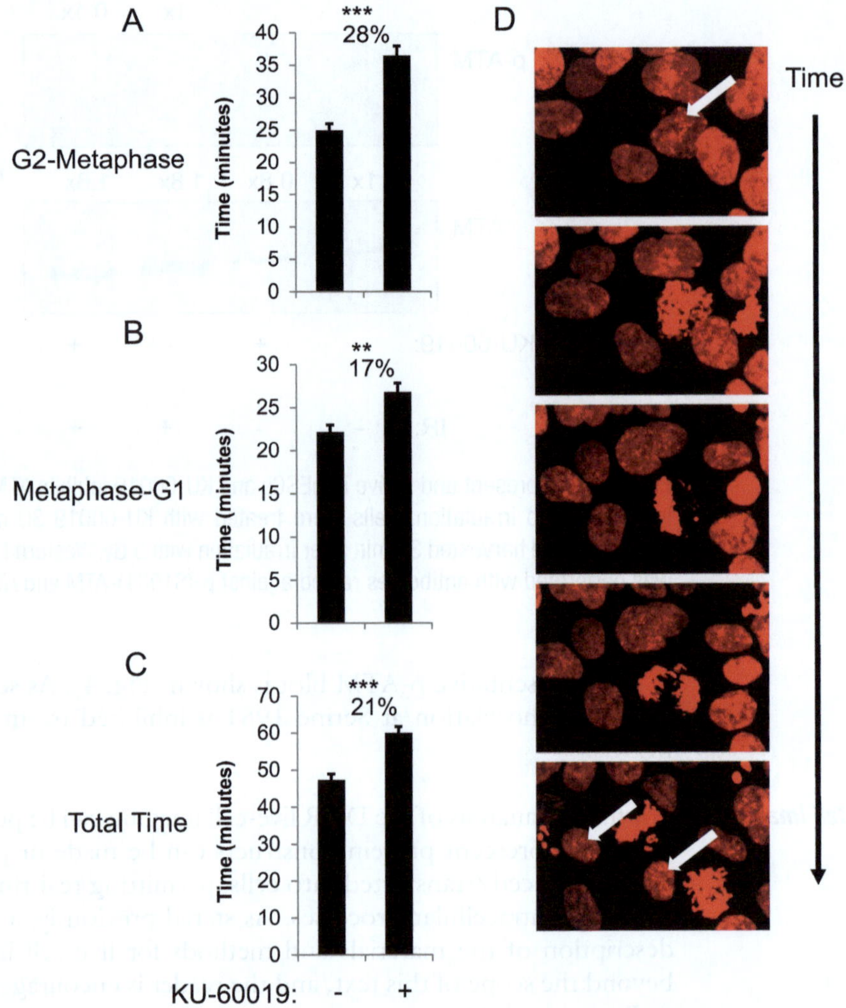

Fig. 5 Inhibition of ATM slows the progression of mitosis. hESCs were transduced with an H2B-mCherry construct and mitosis was monitored using live cell imaging as described in Methods. KU-60019 was added 30 min before recording was started. Cells were recorded for 3–4 h. Measurements were performed of G2-metaphase and metaphase to G1. The greatest difference between untreated and KU-60019-treated cells was observed in the G2-metaphase time window. (**d**) Representative images from the quantitative data presented in (**a**), (**b**), and (**c**). *Arrows* mark a cell prior to mitosis and the two daughter cells post mitosis. $^{**}p < 0.01$, $^{***}p < 0.001$

4 Notes

1. Evenly disperse cells by a quick back-and-forth movement of the dish. Do not swirl the cells as this causes inappropriate clustering in the middle of the dish. Return newly seeded dishes to the incubator. Embryonic stem cells will differentiate in any form of stress or when colonies grow too large.

Differentiation can be seen with flattening out of colonies and nucleus to cytoplasm ratio. Differentiation is typically observed beginning in the center of colonies. It is important that media is warmed to 37 °C before use.

2. It is important to understand that significant cell death occurs during directed differentiation with up to 90% of cells dying in some lineages and therefore sufficient number of cells should be plated.
3. It is important to take note of areas that have been imaged due to photo-bleaching from confocal microscopy. Areas that have been scanned already by the microscope may give falsely low signal-to-noise readings due to photo-bleaching of the secondary antibody.
4. In our experiments, we employ specific small-molecule inhibitors of ATM and DNA-PK for studying the DDR in these cells. 1000× stock solutions are kept frozen at −20 °C until ready for use. An aliquot is then thawed at room temperature immediately prior to its addition to the cell culture medium. We have found that a 1:1000 solution permits the use of a volume that can be accurately measured by standard micropipettes, while not being so large as to have the DMSO solvent exert unwanted effects on the experiment.
5. To prevent toxicity, possible differentiation, and cell death of stem cells, it is recommended to use lenti- or retrovirus delivery instead of transfection.
6. In addition to the NHEJ reporter assay we developed [12], other repair cassettes can also be used, e.g., the DR-GFP assay for determining homologous recombination repair [18], as we have reported using human glioma cells [9, 15, 19]. However, because single-copy insertion is critical for accurate results one either has to deliver the cassette to human stem cells preferably using a lentivirus at low multiplicity-of-infection to insure single-copy integration or construct knock-in cell derivatives using genetic targeting or gene editing [20].
7. It is critical to construct the repair reporter cell derivatives starting with the hES or iPS cells if relative levels of repair through cell differentiation will be determined since integration site of the reporter might influence repair.
8. We have found that longer times with lower voltages work better for large proteins, such as ATM.
9. Exposing membranes to blocking solution at a lower temperature for longer time may increase the signal-to-noise ratio and improve the result.

Acknowledgments

We thank former and present members of the Valerie lab for advice and guidance in establishing assays and experimental tools.

References

1. Mandal PK, Blanpain C, Rossi DJ (2011) DNA damage response in adult stem cells: pathways and consequences. Nat Rev Mol Cell Biol 12:198–202
2. Savitsky K, Bar-Shira A, Gilad S et al (1995) A single ataxia telangiectasia gene with a product similar to PI-3 kinase. Science 268:1749–1753
3. Bensimon A, Aebersold R, Shiloh Y (2011) Beyond ATM: the protein kinase landscape of the DNA damage response. FEBS Lett 585:1625–1639
4. Doksani Y, De Lange T (2014) The role of double-strand break repair pathways at functional and dysfunctional telomeres. Cold Spring Harb Perspect Biol 6:a016576
5. Lavin MF (2008) Ataxia-telangiectasia: from a rare disorder to a paradigm for cell signalling and cancer. Nat Rev Mol Cell Biol 9:759–769
6. Shiloh Y, Ziv Y (2013) The ATM protein kinase: regulating the cellular response to genotoxic stress, and more. Nat Rev Mol Cell Biol 14:197–210
7. Biddlestone-Thorpe L, Sajjad M, Rosenberg E et al (2013) ATM kinase inhibition preferentially sensitizes p53-mutant glioma to ionizing radiation. Clin Cancer Res 19:3189–3200
8. Golding SE, Rosenberg E, Adams BR et al (2012) Dynamic inhibition of ATM kinase provides a strategy for glioblastoma multiforme radiosensitization and growth control. Cell Cycle 11:1167–1173
9. Golding SE, Rosenberg E, Khalil A et al (2004) Double strand break repair by homologous recombination is regulated by cell cycle-independent signaling via ATM in human glioma cells. J Biol Chem 279:15402–15410
10. Golding SE, Rosenberg E, Valerie N et al (2009) Improved ATM kinase inhibitor KU-60019 radiosensitizes glioma cells, compromises insulin, AKT and ERK prosurvival signaling, and inhibits migration and invasion. Mol Cancer Ther 8:2894–2902
11. Adams BR, Golding SE, Rao RR et al (2010) Dynamic dependence on ATR and ATM for double-strand break repair in human embryonic stem cells and neural descendants. PLoS One 5:e10001
12. Adams BR, Hawkins AJ, Povirk LF et al (2010) ATM-independent, high-fidelity nonhomologous end joining predominates in human embryonic stem cells. Aging (Albany NY) 2(9):582–596
13. Thomson JA, Itskovitz-Eldor J, Shapiro SS et al (1998) Embryonic stem cell lines derived from human blastocysts. Science 282:1145–1147
14. Werbowetski-Ogilvie TE, Bosse M, Stewart M et al (2009) Characterization of human embryonic stem cells with features of neoplastic progression. Nat Biotechnol 27:91–97
15. Golding SE, Morgan RN, Adams BR et al (2009) Pro-survival AKT and ERK signaling from EGFR and mutant EGFRvIII enhances DNA double-strand break repair in human glioma cells. Cancer Biol Ther 8:730–738
16. Beckta JM, Henderson SC, Valerie K (2012) Two- and three-dimensional live cell imaging of DNA damage response proteins. J Vis Exp 4251
17. Beckta JM, Dever SM, Gnawali N et al (2015) Mutation of the BRCA1 SQ-cluster results in aberrant mitosis, reduced homologous recombination, and a compensatory increase in non-homologous end joining. Oncotarget 6:27674–27687
18. Pierce AJ, Johnson RD, Thompson LH et al (1999) XRCC3 promotes homology-directed repair of DNA damage in mammalian cells. Genes Dev 13:2633–2638
19. Golding SE, Rosenberg E, Neill S et al (2007) Extracellular signal-related kinase positively regulates ataxia telangiectasia mutated, homologous recombination repair, and the DNA damage response. Cancer Res 67:1046–1053
20. Pierce AJ, Jasin M (2005) Measuring recombination proficiency in mouse embryonic stem cells. Methods Mol Biol 291:373–384

Chapter 28

A Patient-Specific Stem Cell Model to Investigate the Neurological Phenotype Observed in Ataxia-Telangiectasia

Romal Stewart, Gautam Wali, Chris Perry, Martin F. Lavin, Francois Féron, Alan Mackay-Sim, and Ratneswary Sutharsan

Abstract

The molecular pathogenesis of ataxia-telangiectasia (A-T) is not yet fully understood, and a versatile cellular model is required for in vitro studies. The occurrence of continuous neurogenesis and easy access make the multipotent adult stem cells from the olfactory mucosa within the nasal cavity a potential cellular model. We describe an efficient method to establish neuron-like cells from olfactory mucosa biopsies derived from A-T patients for the purpose of studying the cellular and molecular aspects of this debilitating disease.

Key words Ataxia-telangiectasia, Olfactory mucosa, Stem cells, Neuronal differentiation

1 Introduction

Neurodegeneration of the cerebellum is the most debilitating aspect of ataxia telangiectasia (A-T) [1, 2]. Although considerable progress has been made in understanding ATM regulation in eukaryotic cells over the years, the cellular processes that lead to the neuropathology observed in A-T patients have yet to be elucidated. These processes can best be examined by utilizing a relevant model system that can recapitulate the cell biology of the A-T brain. To date, most studies on A-T have been carried out using non-neuronal cellular models such as fibroblasts and lymphoblastoid cells, which do not exhibit the tissue-specific functions [3, 4]. Furthermore, animal models mutated in *Atm* gene have failed to exhibit the neurodegenerative phenotype seen in A-T patients [5]. The establishment of patient-specific neural stem cells represents a novel cellular model that has the potential to address this limitation.

Sergei V. Kozlov (ed.), *ATM Kinase: Methods and Protocols*, Methods in Molecular Biology, vol. 1599,
DOI 10.1007/978-1-4939-6955-5_28, © Springer Science+Business Media LLC 2017

Located within the nasal cavity, the olfactory epithelium is the only nervous tissue that is exposed to the external environment and is therefore easily accessible for biopsy [6]. Olfactory tissue undergoes continuous neurogenesis in response to environmental insults and is sustained by a population of multipotent stem cells [7]. The olfactory mucosa has given researchers the ability to investigate anomalies in neurodevelopmental disorders such as autism and schizophrenia [8–15], as well as neurodegenerative disorders such as Parkinson's disease [14, 16], Alzheimer's disease [17, 18], hereditary spastic paraplegia [19, 20], and familial dysautonomia [21]. Olfactory mucosal cells have reportedly been used in a range of methodologies: protein studies [10], molecular studies [14, 21], and cell biological studies including neurogenesis and organelle trafficking [12, 20, 22, 23].

Using specific tissue culture methods, we generated olfactory neurosphere-derived (ONS) cells from biopsies of olfactory mucosa of A-T patients and studied A-T specific changes in this cellular model [24]. The ONS cells constitute a patient-specific neural-stem cell model that recapitulates the cellular pathology of A-T and thereby provides an excellent platform for comparative molecular studies, cell therapies, and therapeutic drug screening. The ONS model has been employed to study other neurological diseases as well [10–12, 14, 16, 19, 20, 25]. In this section, we describe methods to establish an A-T ONS cellular model to carry out various experiments in the view of understanding disease mechanisms and developing interventions.

2 Materials

All the procedures were done in a sterile environment unless specified otherwise. All waste materials were considered potentially biohazardous and disposed of according to the institute's waste disposal policy.

2.1 Tissue

1. Human olfactory mucosal tissue, removed from the nasal septum about 1 cm from the roof of the nasal cavity, the cribriform plate (described in detail in [26]).

2.2 Isolation and Culturing of Olfactory Stem Cells

1. Collagenase H, diluted at 1 mg/mL in DMEM/F12.
2. DMEM/F12 culture medium, supplemented with 10% fetal bovine serum (FBS) and 50 U/mL penicillin and 50 μg/mL streptomycin.
3. Hanks balanced salt solution (HBSS).
4. TrypLE (Thermo Fisher Scientific).
5. Dissecting microscope.

6. Micro spatula.
7. Autoclaved glass Petri dishes and sterile plastic Petri dishes.
8. Sterile disposable scalpel.
9. Sterile needles (25 G).
10. Sterile Polypropylene centrifuge tubes.
11. Tissue culture plates.
12. Tissue culture flasks.

2.3 Generation of Neurospheres and Neural Differentiation

1. Poly-L-lysine solution (1 mg/mL).
2. Sphere-inducing medium: DMEM/F12 supplemented with insulin, transferrin, selenium (ITS-X, 1%), EGF (50 ng/mL), FGF2 (25–50 ng/mL), and antibiotics.
3. Neuron-differentiating culture medium: Neurobasal medium containing 1% B-27 serum-free supplement, 2 mM glutamine, 20 μM glutamate, and antibiotics.
4. Bovine serum albumin (BSA) fraction V.
5. Phosphate-buffered saline (PBS).
6. Methanol.
7. Beta III tubulin antibody (Covance 1:400).
8. Doublecortin antibody (Cell Signalling Technology, 1:200).
9. MAP2 antibody (Sigma-Aldrich 1:100).
10. Anti-rabbit Alexa-Fluor488 secondary antibody (Invitrogen 1:400).
11. Anti-mouse Alexa-Fluor488 secondary antibody (Invitrogen 1:400).

2.4 Banking of Olfactory Stem Cells

1. Freezing medium: 90% FBS and 10% dimethyl sulfoxide (DMSO).
2. TrypLE.
3. HBSS.
4. Cryovials (labeled/2D barcoded).

3 Methods

3.1 Culture of Olfactory Stem Cells

Olfactory mucosal biopsies should be excised from human donors (*see* **Note 1**) according to the surgical procedure described by [26]. Briefly, both nasal cavities are endoscopically inspected to assess whether polyps or inflammatory lesions are present. The selected nasal cavity is then prepared with a local anesthetic, such as lidocaine with epinephrine or 10% cocaine solution. The best position for a mucosal biopsy is considered from the nasal septum about

1 cm from the roof of the nasal cavity, the cribriform plate. Using a No. 15 scalpel blade, the surgeon makes appropriate cuts on the mucosa and with the help of a pair of through-cut ethmoid forceps, >2 × 1 mm biopsy, is then collected. An Ear, Nose and Throat surgeon (otorhinolaryngologist) should carry out this procedure.

1. Place biopsies individually in sterile specimen containers filled with DMEM/F12 culture medium. Pack the specimens according to standard packaging methods and transport them to the processing laboratory in a refrigerated container.
2. Upon receipt of the biopsies in the laboratory, assign specimens unique identification numbers (IDs) and later cell line IDs as per the laboratory policy/protocol.
3. Working on each biopsy separately, remove blood if any is present from the tissue, under a dissection microscope. Wash in HBSS for three times (*see* **Note 2**).
4. Spread the mucosa out on a glass Petri dish and chop using scalpel blades to get smaller pieces of tissue (<0.05 mm^2). Collect the chopped tissues in a centrifuge tube and wash once in the DMEM/F12 medium.
5. Digest the tissue pieces with Collagenase H for 10 min at 37 °C [7]. Triturate gently approximately 20 times with a prewetted 1000 μL tip, taking care not to introduce air bubbles. Wash once in HBSS and resuspend the pellet in DMEM/F12 culture medium.
6. Seed the dissociated tissue/cell suspension in one or two wells (depending on the yield) of a 6-well tissue culture plate (*see* **Note 3**).
7. Observe the culture after 24 h. If sufficient cells/tissue have been attached to the substratum, collect the supernatant in a 10 mL tube. Pellet the tissue pieces from the supernatant and mechanically dissociate them into smaller pieces and plate them into a new well with the culture medium. Add fresh medium to the first well from which the supernatant was collected.
8. If there are very few cells/tissue attached, leave the plate undisturbed for another 24 h before again collecting the supernatant from the first well.
9. Replace the spent medium in the well every 2 days.
10. After 5–7 days, cells will spread across the well, and they will reach confluency in 10–15 days.
11. Once the culture reaches confluency, passage the cells and continue to culture/sphere/cryopreserve them according to the project needs (*see* **Note 4**).

3.2 Generation of Neurospheres

1. Prepare poly-L-lysine coated culture vessels (preferably a 6-well plate) to generate the neurospheres. Cover the surface of the culture vessels with a poly-L-lysine solution (1–5 μg/cm²) for 2 h at 37 °C. Aspirate the solution after 2 h and wash twice with HBSS.
2. Using TrypLE, harvest the primary olfactory mucosal cells grown in the DMEM/F12 culture medium for 3 days, as described in Subheading 3.1.
3. Plate these cells at approximately 16,000 cells/cm², onto the poly-L-lysine precoated wells and culture in the sphere-inducting medium.
4. Observe the culture after 24 h. If there are any tissue pieces floating in the medium, collect them in a tube and centrifuge at a lower speed (200 × *g*). Triturate those pieces with the help of a Pasteur pipette and plate into a new well.
5. Continue to feed the cells with the sphere-inducing medium every 2 days (*see* **Note 5**).
6. Neurospheres will start to form after a few days. They will be initially attached to the substratum. When they grow larger in size (>100 μm), they will detach from the substratum and float in the medium.
7. Collect floating spheres manually using a transfer pipette and use them directly for differentiation assays or to produce ONS [14] (*see* **Note 6**) (Fig. 1).

3.3 Differentiation of ONS Cells into Neuron-Like Cells

1. Collect the neurospheres generated in **step 6** of Subheading 3.2 in a centrifuge tube and pellet the spheres by centrifugation at 300 × *g*.
2. Dissociate the spheres with 1 mL of TrypLE at 37 °C for 7–8 min with intermittent, gentle mechanical trituration to obtain a single-cell suspension.
3. Seed these cells at 2000 cells/cm² on poly-L-lysine (1–5 μg/cm²) coated glass coverslips in the neuron differentiating culture medium as previously described by [24, 27].
4. Renew the medium every 3 days.
5. Monitor the cell morphology using a light microscope. Neuron-like cells will appear after 2–3 weeks.
6. Fix the cells in 100% methanol for 5 min and wash once with PBS.
7. Block the coverslips containing the fixed cells in 5% BSA fraction V in PBS for 30 min at room temperature.
8. Incubate the cells with the target primary antibodies at recommended dilutions in BSA/PBS for 1 h at room temperature

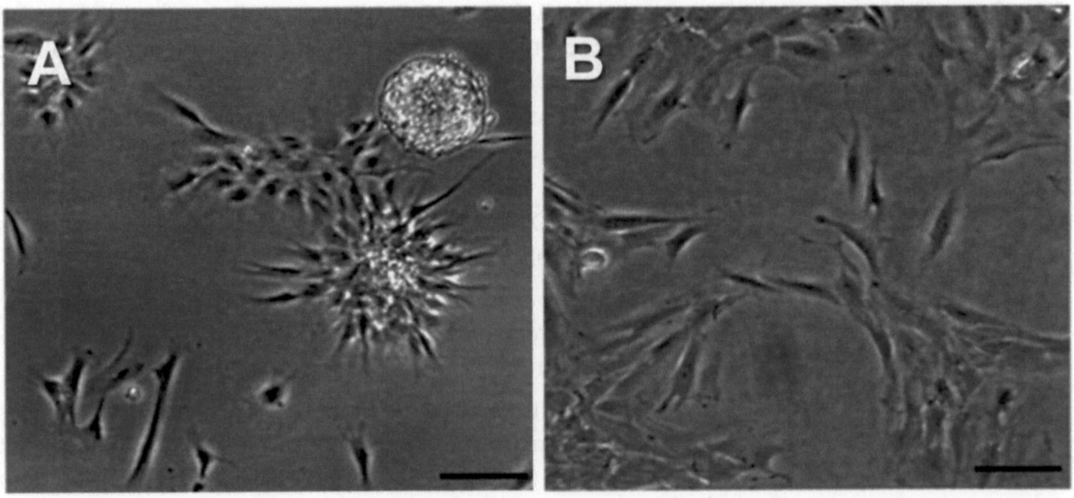

Fig. 1 Morphology of cells from an AT patient. (**a**) Cells from an olfactory mucosa biopsy of an A-T patient form neurospheres when grown in the sphere-inducing medium. Cells cluster together and form free-floating, phase-bright neurospheres (*top right*). Scale bar 100 μm. (**b**) ONS cells from an A-T patient initially grown as neurospheres, then dissociated and grown as adherent cells in a medium containing serum. Scale bar 100 μm. Reprinted from Stewart R et al., A patient-derived olfactory stem cell disease model for ataxia-telangiectasia. Hum Mol Genet, 2013, 22 (12):2495–2509 by permission of Oxford University Press

and then with a compatible secondary antibody (*see* Fig. 2 for the results of a panel of antibodies on A-T-derived ONS).

3.4 Cryopreservation and Banking of ONS

1. Collect the neurospheres generated in **step 6** of Subheading 3.2 in a centrifuge tube and pellet the spheres by centrifugation at 300 × *g*.
2. Dissociate these spheres with 1 mL of TrypLE at 37 °C for 7–8 min with intermittent, gentle mechanical trituration. Wash the cells once in a DMEM/F12 medium and seed them into new tissue culture vessels containing a DMEM/F12 culture medium at 4000–5000 cells/cm^2.
3. Test for Mycoplasma contamination using a standard detection method, such as the PCR-based or Mycoalert testing.
4. Harvest the cells at their exponential growth phase, resuspend the cells in freezing medium, and aliquot them into cryovials for freezing (*see* **Note 7**).
5. If the project requires a continuous supply of cells, it is advisable to adopt proper cell banking practices (*see* **Note 8**).

Fig. 2 (continued) (**c**) Neuronal morphology after differentiation of ONS cells from a control and a patient. The small, flat, multipolar ONS (*i, iii*) changed to larger, long, and bipolar cells indicative of immature neurons (*ii, iv*). Enlargements of outlined regions are shown on the right. Scale bars = 100 μm in **a** *i–vi* and **b** *iii–vi*; 200 μm for **b** *i* and *ii*; and 250 μm for **c** *i–vi*. Reprinted from Stewart R et al., A patient-derived olfactory stem cell disease model for ataxia-telangiectasia. Hum Mol Genet, 2013, 22 (12):2495–2509 by permission of Oxford University Press

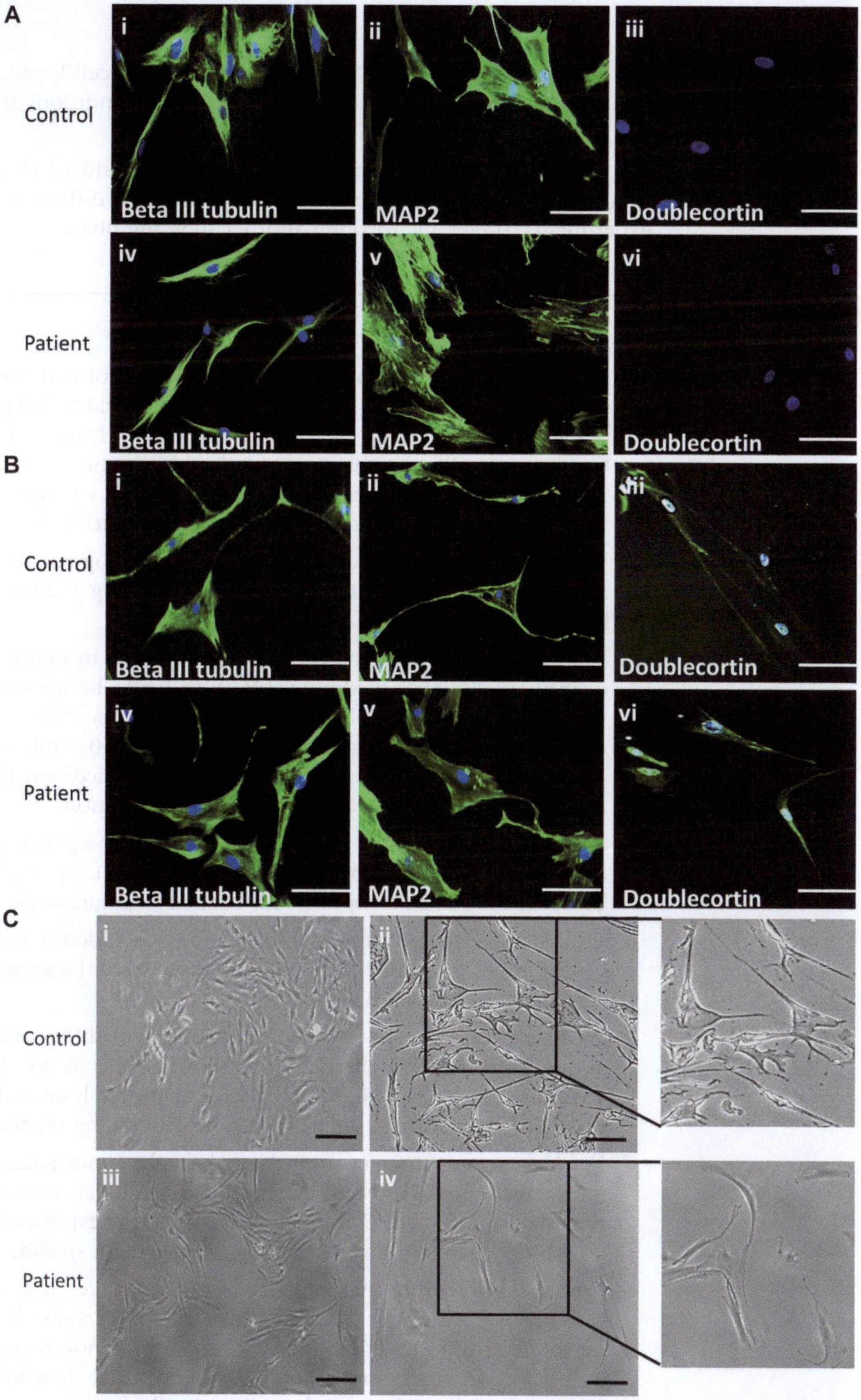

Fig. 2 Neuronal differentiation of control and A-T patient ONS cells. (**a**) ONS cells from controls and patients were immunopositive for the neuronal markers βIII-tubulin (*i, iv*), MAP2 (*ii, v*), but not doublecortin (*iii, vi*). (**b**) After neuronal differentiation, cells derived from controls and patients expressed βIII-tubulin (*i, iv*) MAP2 (*ii, v*) and doublecortin (*iii, vi*).

6. Bank the aliquots of DNA, RNA, microRNA, cell lysate, and protein samples under appropriate storage conditions if any were extracted.

The frozen ONS cells can be revived using standard thawing procedures as required. These cells can be used for differentiation experiments, preferably through another sphering phase.

4 Notes

1. Tissue samples from donors were taken with informed consent and with the approval by the Queensland Children's Health Services (RCH) and Griffith University Human Ethics Committees. Olfactory biopsies from healthy donors and heterozygous carriers were taken under local anesthesia, whereas tissue was collected from A-T children under general anesthesia.
2. Use either micro spatula or two 25 G non-pyrogenic needles attached to 1 mL disposable syringes for handling and processing of the tissues.
3. It is advisable to resuspend the dissociated tissues in a minimum quantity of culturing medium (500 μL to 1 mL) before seeding them onto the culture plate. After 1 h, top up the culture with an appropriate amount of culture medium (0.2–0.3 mL/cm^2). If cells are cultured in plates, add some culture medium/HBSS to two empty wells in order to minimize evaporation.
4. These cells can be used to generate neurospheres on a poly-L-lysine coated tissue culture substratum in a sphere-inducing medium. The cells can also be cryopreserved/banked for future work.
5. Usually, a freshly prepared culture medium is added to culture vessels at 0.2 mL/cm^2. The complete removal of the spent culture medium is not advisable.
6. Neurospheres can be freshly generated for differentiation assays from an adherent olfactory stem culture at any time: seed at around 7000–10,000 cells/cm^2 on poly-L-lysine-coated plastic tissue culture vessels with a sphere-inducing medium.
7. For longer term storage, store the vials in the vapor phase of a nitrogen cryostorage system. 2D barcoded vials or labeled normal cryo vials can be used. For cell banking and future experiments, it is essential to employ an efficient data management system.
8. Freeze more than ten vials from the early passages at 1×10^6 cells per vial as "master stocks." Then freeze the required number of "distribution stocks" for investigators. Upon receipt of a distribution stock, individual investigators will then further expand the cells and freeze the cells in batches as "user stocks" for their experimental needs. These user stocks will allow

researchers to perform multiple experiments for an extended period of time using similar starting materials. The optimal quality of ONS is maintained at least to passage 10.

These spheres or ONS are also used to extract DNA (genomics), RNA, and microRNA (transcriptomics), prepare cell lysates (metabolomics), and protein (proteomics) to analyze the samples on various technical platforms.

Acknowledgments

This work was supported by funding from the Australian Department of Health and Ageing to AMS and the National Health and Medical Research Council (NHMRC) to MFL. We thank A-T patients and their families for their support. Support and help of BrAshA-T charity (https://brashat.org.au/) are gratefully acknowledged as well as clinicians and nurses of the Australian A-T Clinic for patient recruitment and biopsies.

References

1. Boder E (1985) Ataxia-telangiectasia: an overview. Kroc Found Ser 19:1–63
2. Lavin MF (2008) Ataxia-telangiectasia: from a rare disorder to a paradigm for cell signalling and cancer. Nat Rev Mol Cell Biol 9(10):759–769. doi:10.1038/nrm2514
3. Bhatti S, Kozlov S, Farooqi AA, Naqi A, Lavin M, Khanna KK (2011) ATM protein kinase: the linchpin of cellular defenses to stress. Cell Mol Life Sci 68(18):2977–3006
4. McKinnon PJ (2012) ATM and the molecular pathogenesis of ataxia telangiectasia. Ann Rev Pathol Mech Dis 7:303–321
5. Lavin MF (2013) The appropriateness of the mouse model for ataxia-telangiectasia: neurological defects but no neurodegeneration. DNA Repair 12(8):612–619
6. Murrell W, Bushell GR, Livesey J, McGrath J, MacDonald KP, Bates PR, Mackay-Sim A (1996) Neurogenesis in adult human. Neuroreport 7(6):1189–1194
7. Murrell W, Feron F, Wetzig A, Cameron N, Splatt K, Bellette B, Bianco J, Perry C, Lee G, Mackay-Sim A (2005) Multipotent stem cells from adult olfactory mucosa. Dev Dyn 233(2):496–515. doi:10.1002/dvdy.20360
8. Abrams MT, Kaufmann WE, Rousseau F, Oostra BA, Wolozin B, Taylor CV, Lishaa N, Morel M-L, Hoogeveen A, Reiss AL (1999) FMR1 gene expression in olfactory neuroblasts from two males with fragile X syndrome. Am J Med Genet 82(1):25–30
9. Arnold SE, Han L-Y, Moberg PJ, Turetsky BI, Gur RE, Trojanowski JQ, Hahn C-G (2001) Dysregulation of olfactory receptor neuron lineage in schizophrenia. Arch Gen Psychiatry 58(9):829–835
10. English J, Fan Y, Föcking M, Lopez L, Hryniewiecka M, Wynne K, Dicker P, Matigian N, Cagney G, Mackay-Sim A (2015) Reduced protein synthesis in schizophrenia patient-derived olfactory cells. Transl Psychiatry 5(10):e663
11. Fan Y, Abrahamsen G, McGrath JJ, Mackay-Sim A (2012) Altered cell cycle dynamics in schizophrenia. Biol Psychiatry 71(2):129–135. doi:10.1016/j.biopsych.2011.10.004
12. Fan Y, Abrahamsen G, Mills R, Calderón CC, Tee JY, Leyton L, Murrell W, Cooper-White J, McGrath JJ, Mackay-Sim A (2013) Focal adhesion dynamics are altered in schizophrenia. Biol Psychiatry 74(6):418–426
13. Féron F, Perry C, Hirning M, McGrath J, Mackay-Sim A (1999) Altered adhesion, proliferation and death in neural cultures from adults with schizophrenia. Schizophr Res 40(3):211–218
14. Matigian N, Abrahamsen G, Sutharsan R, Cook AL, Vitale AM, Nouwens A, Bellette B, An J, Anderson M, Beckhouse AG, Bennebroek M, Cecil R, Chalk AM, Cochrane J, Fan Y, Feron F, McCurdy R, McGrath JJ, Murrell W, Perry C, Raju J, Ravishankar S, Silburn PA, Sutherland GT, Mahler S, Mellick GD, Wood SA, Sue CM,

Wells CA, Mackay-Sim A (2010) Disease-specific, neurosphere-derived cells as models for brain disorders. Dis Model Mech 3(11-12):785–798. doi:10.1242/dmm.005447

15. Ronnett GV, Leopold D, Cai X, Hoffbuhr KC, Moses L, Hoffman EP, Naidu S (2003) Olfactory biopsies demonstrate a defect in neuronal development in Rett's syndrome. Ann Neurol 54(2):206–218
16. Cook AL, Vitale AM, Ravishankar S, Matigian N, Sutherland GT, Shan J, Sutharsan R, Perry C, Silburn PA, Mellick GD (2011) NRF2 activation restores disease related metabolic deficiencies in olfactory neurosphere-derived cells from patients with sporadic Parkinson's disease. PLoS One 6(7):e21907
17. Arnold SE, Lee EB, Moberg PJ, Stutzbach L, Kazi H, Han LY, Lee VM, Trojanowski JQ (2010) Olfactory epithelium amyloid-β and paired helical filament-tau pathology in Alzheimer disease. Ann Neurol 67(4):462–469
18. Wolozin B, Lesch P, Lebovics R, Sunderland T (1993) Olfactory neuroblasts from Alzheimer donors: studies on APP processing and cell regulation. Biol Psychiatry 34(12):824–838
19. Abrahamsen G, Fan Y, Matigian N, Wali G, Bellette B, Sutharsan R, Raju J, Wood SA, Veivers D, Sue CM, Mackay-Sim A (2013) A patient-derived stem cell model of hereditary spastic paraplegia with SPAST mutations. Dis Model Mech 6(2):489–502. doi:10.1242/dmm.010884
20. Fan Y, Wali G, Sutharsan R, Bellette B, Crane DI, Sue CM, Mackay-Sim A (2014) Low dose tubulin-binding drugs rescue peroxisome trafficking deficit in patient-derived stem cells in hereditary spastic paraplegia. Biol Open 3(6):494–502
21. Boone N, Loriod B, Bergon A, Sbai O, Formisano-Treziny C, Gabert J, Khrestchatisky M, Nguyen C, Feron F, Axelrod FB, Ibrahim EC (2010) Olfactory stem cells, a new cellular model for studying molecular mechanisms underlying familial dysautonomia. PLoS One 5(12):e15590. doi:10.1371/journal.pone.0015590
22. Ghanbari HA, Ghanbari K, Harris PL, Jones PK, Kubat Z, Castellani RJ, Wolozin BL, Smith MA, Perry G (2004) Oxidative damage in cultured human olfactory neurons from Alzheimer's disease patients. Aging Cell 3(1):41–44
23. McCurdy RD, Féron F, Perry C, Chant DC, McLean D, Matigian N, Hayward NK, McGrath JJ, Mackay-Sim A (2006) Cell cycle alterations in biopsied olfactory neuroepithelium in schizophrenia and bipolar I disorder using cell culture and gene expression analyses. Schizophr Res 82(2):163–173
24. Stewart R, Kozlov S, Matigian N, Wali G, Gatei M, Sutharsan R, Bellette B, Wraith-Kijas A, Cochrane J, Coulthard M, Perry C, Sinclair K, Mackay-Sim A, Lavin MF (2013) A patient-derived olfactory stem cell disease model for ataxia-telangiectasia. Hum Mol Genet 22(12):2495–2509. doi:10.1093/hmg/ddt101
25. Mar JC, Matigian NA, Mackay-Sim A, Mellick GD, Sue CM, Silburn PA, McGrath JJ, Quackenbush J, Wells CA (2011) Variance of gene expression identifies altered network constraints in neurological disease. PLoS Genet 7(8):e1002207
26. Féron F, Perry C, Girard SD, Mackay-Sim A (2013) Isolation of adult stem cells from the human olfactory mucosa. Methods Mol Biol 1059:107–114
27. Girard SD, Devéze A, Nivet E, Gepner B, Roman FS, Féron F (2011) Isolating nasal olfactory stem cells from rodents or humans. J Vis Exp 54:2762. doi:10.3791/2762

Chapter 29

Lentiviral Reprogramming of A-T Patient Fibroblasts to Induced Pluripotent Stem Cells

Sam Nayler, Sergei V. Kozlov, Martin F. Lavin, and Ernst Wolvetang

Abstract

Reprogramming of cells enables generation of pluripotent stem cells and resulting progeny through directed differentiation, making this technology an invaluable tool for the study of human development and disease. Reprogramming occurs with a wide range of efficiency, a culmination of intrinsic and extrinsic factors including the tissue of origin, the passage number and culture history of the target cells. Another major factor affecting reprogramming is the methodology used and the quality of the reprogramming process itself, including for conventional viral-based approaches viral titer and subsequent viral transduction efficiency, including downstream transgene insertion and stoichiometry. Genetic background is an important parameter affecting the efficiency of the reprogramming process with reports that cells from individuals harboring specific mutations are more difficult to reprogram than control counterparts.

Ataxia-Telangiectasia (A-T) fibroblasts underwent reprogramming at reduced efficiency in contrast to their controls. To optimize reprogramming of fibroblasts from patients with A-T, we examined the response of A-T cells to various cell culture conditions after lentiviral transduction with reprogramming factors *Oc4/Sox2* (pSIN4-EF2-O2S) and *Klf4/c-Myc* (pSIN4-CMV-K2M). Parameters included media type (KSR or serum-containing DMEM), treatment with a p53 inhibitor (small-molecule cyclic pifithrin-α), and either a low or high concentration of bFGF. Post-transduction, equivalent numbers of cells from heterozygote and homozygote patients were plated and assessed at regular intervals for survival and proliferation. Our findings indicate that A-T cells responded favorably to the addition of FCS and gradual weaning away from their native media into KSR-containing stem cell media that produced suitable conditions for their reprogramming. We examined a range of properties to identify and isolate good quality iPSCs including the expression status of important stem cell transcription factors/surface proteins, methylation levels at stem cell associated regulatory loci, persistence of transgenes, karyotype status, and teratoma-forming ability.

Key words Induced pluripotent stem cells, iPSC, Reprogramming, Pluripotency, DNA damage, ATM, Ataxia-Telangiectasia

1 Introduction

Experimental models of ATM function have been restricted to systems of limited context to understanding ATM biology and its potential role in development and tissue formation. The availability of pluripotent stem cells, in particular induced pluripotent stem cells

Sergei V. Kozlov (ed.), *ATM Kinase: Methods and Protocols*, Methods in Molecular Biology, vol. 1599,
DOI 10.1007/978-1-4939-6955-5_29, © Springer Science+Business Media LLC 2017

(iPSCs) has the potential to dramatically reshape the way in which disease modelling is approached [1–3]. Animal models have failed to faithfully recapitulate the full spectrum of A-T symptoms including the neurodegenerative aspect, which remains poorly understood [4]. Existing human cellular models (such as fibroblasts and lymphoblastoid cell lines from patients) allow analysis of ATM in limited contextual space and as such little is known about the involvement of ATM or its downstream substrates in the mechanisms that exist to direct and protect development within the embryo. Perturbed regulation of DNA damage at critical temporal junctures within the developing embryo could lead to the progressive degenerative characteristics observed in this disorder. Further to this, accumulating evidence points to involvement of ATM in areas outside of its canonical role of orchestrating the DNA damage response including: meiosis [5], proteasome-mediated protein degradation [6], mitochondrial function [7], insulin resistance/glucose metabolism [7, 8], modulation of synaptic functions in neurons [9], vesicle trafficking [10], pentose phosphate pathway signaling [11], HDAC4 localization [12], and as a sensor and responder to oxidative stress [13, 14]. Emerging evidence also suggests that ATM has tissue-specific functionality and is required during development [15, 16].

hESCs (human embryonic stem cells) constitute a powerful tool for modeling of development and disease. Only one report exhaustively describes the effect of ATM knockout in hESCs by BAC-mediated transgenesis [17]. The authors show that ATM knockout results in cells that display hypersensitivity to IR (ionizing radiation) and show lack of G2/M cell cycle arrest after DNA damage, recapitulating aspects of the phenotype seen in existing cellular A-T models and demonstrating that ATM is a critical responder to DNA damage in this context. A recent technology to emerge from Japan (iPSCs) has made it possible to reprogram terminally differentiated somatic cells, such as fibroblasts, into cells that resemble hESCs, in terms of self-renewal and their ability to generate cells of all three germ layers [18].

To date, studies on A-T have failed to answer the fundamental question of why certain cell types are seemingly more affected than others. As access to pluripotent stem cells becomes mainstream, it becomes possible, to a degree, to recreate and study in vitro processes giving rise to various cell types, including neurons, using directed differentiation protocols.

We hypothesized that ATM-deficient cells may be difficult or impossible to reprogram without intervention or assistance, and that this difficulty may be at least in part due to excessive levels of reprogramming-induced or existing DNA damage, due to the poor growth characteristics of A-T cells or the inability to participate in certain pathways necessary for reprogramming. Genetic background is an important parameter affecting the efficiency of the reprogramming process with reports that cells from individuals with particular mutations are more difficult to reprogram relative to controls. These observations support the theory that modulation of pathways can either enhance or inhibit reprogramming [19]. Given findings that p53 abrogation increased the efficiency of reprogramming [19, 20], it was reasonable to assume the existence of a barrier or roadblock that might prevent cells with excessive levels of DNA damage from passing unfettered through this process. Evidence shows that DNA damage pathways are activated early during the reprogramming process [21]. Given the fact that ATM has been increasingly cited as a contributor to various metabolic pathways, and also the fact that somatic cells lacking ATM show intrinsic culture stress/elevated growth factor requirements [22], we postulated that ATM deficiency might affect reprogramming.

We show that it was indeed possible to generate iPSCs from A-T patients albeit at reduced efficiency [23]. Subsequent reports of generation of iPSCs from A-T patients and their use have followed [24–26]. This article describes methodology used in generation and characterization of bona fide iPSCs from patients with A-T by using a lentiviral transduction approach combining two separate plasmids carrying *Oct4/Sox2* (pSIN4-EF2-O2S), and *Klf4/c-myc* (pSIN4-CMV-K2M).

Although there is a knowledge gap regarding the mechanism by which a terminally differentiated cell is reprogrammed back to a pluripotent state, the general consensus is that wide-scale remodeling of the epigenetic landscape occurs [27, 28]. Modifications to chromatin assist in defining the cell's identity largely through transcriptional regulation. Although forced expression of reprogramming factors, which are master regulator stem cell transcription factors, can give rise to colonies with stem-cell-like properties, not all stem-cell-like colonies arising from these experiments are deemed fully reprogrammed iPSCs [29]. Often, cells undergo incomplete reprogramming [30], resulting in a pluripotent-like stem cell that may have aberrant proliferation and differentiation properties. A major challenge for the field remains to isolate "good-quality" or bona-fide iPSCs for cultivation and experimentation. We outline methodology used to study and validate fully reprogrammed iPSCs including the expression status of important stem cell transcription factors/surface proteins, methylation levels at stem cell associated regulatory loci, persistence of transgenes, karyotype status, and teratoma-forming ability.

2 Materials

2.1 General Equipment Required for Pluripotent Stem Cell Culture (Characterization, Maintenance, Expansion, Cryopreservation)

1. Class II biohazard safety cabinet (custom fitted to house dissecting-microscope).
2. Humidified incubator.
3. Flow cytometer.
4. Fluorescent and light microscope.
5. Thermal cycler and qPCR machine.
6. Nanodrop-1000 spectrophotometer or equivalent.

7. Irradiator or source of irradiated MEF (mouse embryonic fibroblasts) plates.
8. Cell-counting apparatus.

2.2 Reagents Required for MEF/Human Fibroblast Culture

1. DMEM (Dulbecco's Modified Eagle's medium with 10–15% FCS (fetal calf serum), MEM-Nonessential amino acids and Penicillin-Streptomycin (optional)).
2. TrypLE Select (Thermo Fisher Scientific).
3. Gelatin.
4. Tissue culture plasticware.
5. Trypan blue.

2.3 Reagents Required for HEK 293T Cell Culture, Lentiviral Production, and Reprogramming

1. DMEM cell culture medium with 10% FCS.
2. TrypLE Select (Thermo Fisher Scientific).
3. pSIN4-EF2-O2S plasmid (available from Addgene https://www.addgene.org/21162/).
4. pSIN4-CMV-K2M plasmid (available from Addgene https://www.addgene.org/21164/).
5. HEK293T cells (low passage, mycoplasma-free).
6. Geneticin.
7. Opti-MEM media.
8. Lipofectamine® LTX Reagent with PLUS™ Reagent (Thermo Fisher Scientific).
9. pVSVG packaging vector plasmid.
10. CMR delta-8.2 packaging vector plasmid.
11. 0.45 μm syringe filter.
12. 25 and 50 mL syringes.
13. cyclic-pifithrin-α (a reversible inhibitor of p53-dependent transcription and apoptosis): 10 mM stock solution in DMSO, store aliquots at −20 °C.
14. Tissue culture plastic plates and flasks.
15. Gamma-irradiated MEF feeder plates.
16. KSR complete media: (DMEM F12 supplemented with Knock-out serum replacement, MEM-NEA, Penicillin-Streptomycin).
17. Basic fibroblast growth factor (bFGF) (10 μg/mL in DMEM/F-12 with 0.1% BSA; aliquot and freeze down at −20 °C).
18. Organ-culture dishes.
19. 1 mL syringes with 22-G needles.
20. 2-mercaptoethanol.
21. Accutase cell detachment solution (STEMCELL Technologies Inc.).
22. Polybrene (optional).

2.4 Reagents Required for Pluripotent Stem Cell Culture (Maintenance, Expansion, Cryopreservation)

1. Gamma-irradiated MEF plates.
2. KSR complete media: DMEM-F12, Knock-out serum replacement, MEM-Non-Essential Amino Acids Solution, Penicillin-Streptomycin (optional).
3. bFGF (10 μg/mL in DMEM/F-12 with 0.1% BSA; aliquot and freeze down at −20 °C).
4. KSR-containing hESC culture medium: 80% DMEM-F12, 20% KSR, 2 mM L-glutamine, 1% NEAA (nonessential amino acids), 0.1 mM 2-mercaptoethanol, and 50–100 ng/mL bFGF.
5. Matrigel matrix (Corning).
6. Tissue culture plastic plates and flasks.
7. Organ-culture dishes.
8. 1 mL syringes with 22-G needles.
9. 5 mL Cryovials with hole punched in bottom with sterile needle.
10. "Mr Frosty" or equivalent freezing container (with isopropanol).
11. CryoCane™ and Cryostraws (Thermo Fischer Scientific).
12. ES HEPES medium: DMEM, 20 mM HEPES-NaOH buffer, pH 7.2–7.4, 20% FCS.
13. 1 M Sucrose solution (dissolve 3.42 g Sucrose in 10 mL of ES HEPES medium, sterilize by filtration through 0.22 μM syringe filter).
14. 10% vitrification solution: DMSO 10%, Ethylene glycol 10%, ES HEPES 80%, prepare immediately before use.
15. 20% vitrification solution: DMSO 20%, Ethylene Glycol 20%, 1 M sucrose 30%, ES HEPES 30%, prepare immediately before use.
16. Collagenase IV: Make 10 mg/mL collagenase IV in DMEM/F-12. Sterilize by filtration and freeze in aliquots.
17. Resuspension media: fresh KSR media with bFGF.
18. Freezing media: KSR with bFGF and 20% DMSO.
19. OCT4 antibody (Millipore) [1:1000] or (Santa-Cruz) [1:1000].
20. SSEA-4 antibody (Millipore) [1:100].
21. TRA-1-60 antibody (Millipore) [1:1000].
22. NANOG antibody (Millipore) [1:400].
23. TRA-1-81 antibody (Millipore) [1:1000].
24. 4% paraformaldehyde.
25. Hoechst33342 stock solution: dissolve 100 mg in 10 mL of deionized water to make a 10 mg/mL solution. Hoechst 33342

Table 1
Primers used for RT-PCR and bisulfite sequencing/methylation analysis

PRIMERS	Sequence 5′–3′
RT-PCR	
Oct 4 TG F	AGTGAGAGGCAACCTGGAGA
Sox2 TG R	GCTTAGCCTCGTCGATGAAC
Klf4 TG F	GTTCCTGCATGCCAGAGGAG
c-Myc TG R	CTGGTAGAAGTTCTCCTCCTCG
IRES F	TGGCTCTCCTCAAGCGTATT
IRES R	AGGAACTGCTTCCTTCACGA
GAPDH F	GTGGACCTGACCTGCCGTCT
GAPDH R	GGAGGAGTGGGTGTCGCTGT
Bisulfite-sequencing	
Nanog-mF3	TTAATTTATTGGGATTATAGGGGTG
Nanog-mF4	GGTTGGTTTTAAATTTTTGATTTTAG
Nanog-mR3	AAACCTAAAAACAAACCCAACAAC
Nanog-mR4	ACCAATCTCACCAAAACCATTATAA
Oct4-mF3	ATTTGTTTTTTGGGTAGTTAAAGGT
Oct4-mR3	CCAACTATCTTCATCTTAATAACATCC

is poorly soluble in water, sonication might be necessary to dissolve. Keep stock solution at 2–6 °C for up to 6 months.

26. Permeabilization solution: 0.1% Triton X100 in PBS.
27. Suitable blocking buffers for IF microscopy, i.e., 10% goat-serum in PBS or 1% BSA in PBS.
28. Alexa Fluor-conjugated secondary antibodies for IF microscopy.
29. RNA/DNA isolation kit.
30. Gentle Cell Dissociation Reagent (STEMCELL Technologies).
31. Cell scrapers.
32. Single-cell strainer.
33. Primers for RT-PCR and bisulfite sequencing (*see* Table 1).
34. Bisulfite modification kit.
35. PCR2.1 vector (e.g., TA Cloning Kit, Thermo Fischer Scientific).
36. SCID mice (CB17-SCID mice).

3 Methods

3.1 Generation of iPSCs Using Lentiviral Factors

Primary fibroblasts isolated from dermal punch biopsies from patients with Ataxia–Telangiectasia were collected through the A-T Clinic, Royal Children's Hospital & UQ Centre for Clinical Research. Sequencing confirmed two frameshift deletions in AT34

(7004delCA and 7886delTATTA), predicted to result in truncated and thus unstable ATM protein. ATh47 was included as heterozygote control. All work was carried out with informed consent from patients under the approval of the Human Research Ethics Committee (HREC/09/QRCH/103). All lines were routinely tested for the presence of Mycoplasma and discarded if found to be positive. The following protocol was used to establish fibroblasts.

1. Dissect biopsies into small pieces and incubate under coverslips in DMEM with 12% FCS until fibroblasts grow out.
2. Harvest primary human fibroblasts with TrypLE Select and expand in DMEM with 15% FCS until cryopreservation at passage 2 (cryopreservation of fibroblasts is described in Subheading 3.3.1).
3. Lentiviruses carrying *Oct4/IRES/Sox2* and *Klf4/IRES/c-myc* (pSIN4-EF2-O2S and pSIN4-CMV-K2M) were produced by transfection of HEK293T cells grown in 500 μg/mL Geneticin at 70–90% confluency with plasmids (*see* **Note 1** for tips on handling HEK293T cells and **Note 2** for handling of lentivirus).
4. Before transfection of HEK293 cells for lentiviral production, add fresh media without geneticin. Following this, allow to warm to room temperature and apply Opti-MEM media, Lipofectamine PLUS reagent at manufacturer's specifications with (per reaction) 2.5 μg Lentiviral vector (O2S or K2M), 2.5 μg pVSVG, and 5 μg of pCMR delta-8.2 packaging vector.
5. The following day observe cells under the microscope for signs of syncytia indicating successful production of viral particles.
6. Collect virus the following day by filtering supernatant with a 0.45 μm filter.
7. Transduce target fibroblasts with lentiviral constructs by application of reprogramming virus for 24 h. After transduction, allow >50,000 fibroblasts to recover for between 24 and 48 h before enzymatically passaging and transferring them to gamma-irradiated MEF feeder plates (MEF density 36,000/cm^2) in KSR media (*see* **Note 3** for optional use of Polybrene to increase transduction efficiency. **Note 4** contains general tips on starting cultures for successful reprogramming).
8. Observe plates daily for the appearance of small clusters of refractile cells signifying the formation of early putative iPSC colonies. An example is shown in Fig. 1a.
9. Record patterns of recovery and proliferation in a range of different media formulations. Quantitatively assess media formulations/treatment to optimize generation of iPSC colonies from A-T patient fibroblasts. (Fig. 2).

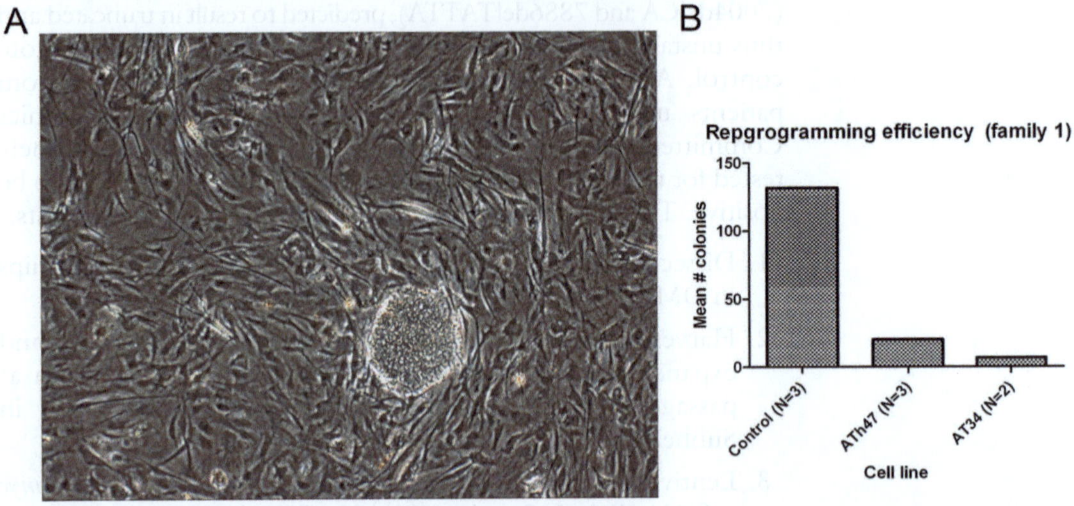

Fig. 1 (**a**) Typical early iPSC colony. (**b**) Reprogramming efficiency in control, A-T heterozygote ATh47, and A-T homozygote AT34. Reprogramming plasmids *Oct4/IRES/Sox2* and *Klf4/IRES/c-myc* (pSIN4-EF2-O2S &: pSIN4-CMV-K2M)

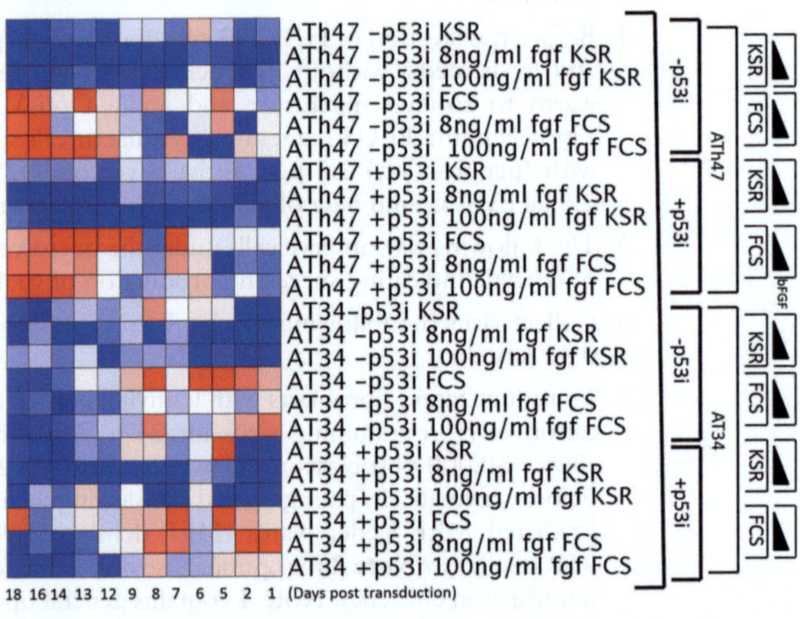

Fig. 2 Heatmap representative of data depicting proliferation after viral transduction in a variety of conditions (note heatmap timed data reads right to left from growth condition label). *Color* indicates cell density/viability ranging from *blue* (lowest) to *white* (intermediate) to *red* (highest)

10. Generate an easy to visualize heatmap to summarize and visualize this data using GenePattern [31] (Fig. 2).

Growth was noted to be generally greater in heterozygote line ATh47 when compared with homozygote line AT34, across most conditions (Fig. 2). Within the range of conditions tested for heterozygote line ATh47, growth and proliferation was markedly increased in media containing FCS when compared to KSR. In the presence of KSR, cyclic-pifithrin-α (10 μM) induced p53 inhibition resulted in increased growth. This same effect was also observed in cells grown in the presence of FCS. Similarly to in heterozygote line ATh47, homozygote line AT34 showed a marked response in growth to the presence of FCS and a similar trend for increased growth in the presence of p53 inhibitor was noted; however, this was more subtle. As a result of this analysis transduced A-T fibroblasts were weaned from DMEM with 15% FCS to 100% hESC culture medium over a period of 4 days following transduction, at 25% per day, as this was shown to greatly increase their survival and proliferation.

3.2 Cell Culture Conditions for Maintenance and Expansion of iPSCs

1. Expand hESCs and iPSCs in KSR-containing hESC culture medium at 37 °C at 5% CO2 and at high humidity.
2. Maintain cells on MEF feeder layers at desired density.
3. For experimentation, culture cells in feeder-free conditions on Matrigel or equivalent in MEF-conditioned hESC culture medium.
4. Passage cells either manually by using a 22-G needle to dissect each colony into 6–10 pieces or enzymatically (described below) before replating at a seeding ratio of between 1:2 and 1:6.
5. Replace hESC media daily and split cells at approximately 80% confluence on days 6–7.

3.3 Cryopreservation of Fibroblasts, Early iPSC Colonies, and Expanded Bulk Cultures

Cryopreservation is performed at several critical junctures. First, a stock of patient fibroblasts is cryopreserved to ensure repeated reprogramming attempts could be performed on cells of consistent passage number and genetic background. At this stage, cell pellets are also generated for the extraction of RNA/genomic DNA/protein for later use. To cryopreserve fibroblasts, the following methodology is used.

3.3.1 Cryopreservation of Fibroblasts

1. Scale up fibroblasts cultures to desired amount in T75 tissue-culture flasks in DMEM with 10–15% FCS.
2. When cells were at 75–85% confluence use PBS to wash the monolayer before the application of TrypLE for approximately 2–5 min at 37 °C.
3. Examine cells via light microscopy for retractile appearance indicating detachment from the tissue culture surface. Gently tap the flask to dislodge cells and collect by pipetting warm media onto the bottom of the flask. Add two volumes of FCS-containing DMEM to inactivate the enzyme.
4. Collect cells in a 50 mL centrifuge tube, spin at 300 × g for 3 min at room temperature, and aspirate supernatant. Use

Trypan blue to perform a viability assay and record numbers of viable cells.

5. Wash cell pellet by centrifugation at 300 × *g* for 3 min at room temperature. Aspirate supernatant.
6. Resuspend cells in 250 μL FCS-containing DMEM and add to prelabeled cryovials.
7. Add (dropwise) 250 μL of FCS-containing DMEM with 20% DMSO to cryovials from **step 6** (for a final concentration of 10% DMSO).
8. Place cells in a "Mr Frosty" or equivalent vessel containing isopropanol for a controlled decrease in temperature (−1 °C/min) at −80 °C for at least 2 h before transfer to liquid nitrogen storage.

3.3.2 Vitrification of Colony Pieces

Scarce and precious colonies arising after reprogramming are expanded clonally to organ culture dishes. These colonies are split in half for cryopreservation by vitrification and live screening for TRA-1-60 expression. A process detailing vitrification is outlined below. Additional tips can be found in **Notes 5** and **6**.

1. Prepare a 4-well vitrification plate with wells containing (A) 1 mL of ES HEPES medium, (C) 1 mL of 10% vitrification solution, and (D) 1 mL of 20% vitrification solution, well (B) left blank (*see* Subheading 2.4 for media formulations and Fig. 3).
2. Equilibrate plate in an incubator at 37 °C, 5% CO_2.
3. Insert 5 mL cryovials (with needle-punched bottom) into a CryoCane within a small liquid nitrogen canister containing liquid nitrogen.
4. Manually dissect colonies using a 22-G needle and transfer into well A of the vitrification plate using a 20 μL pipette.

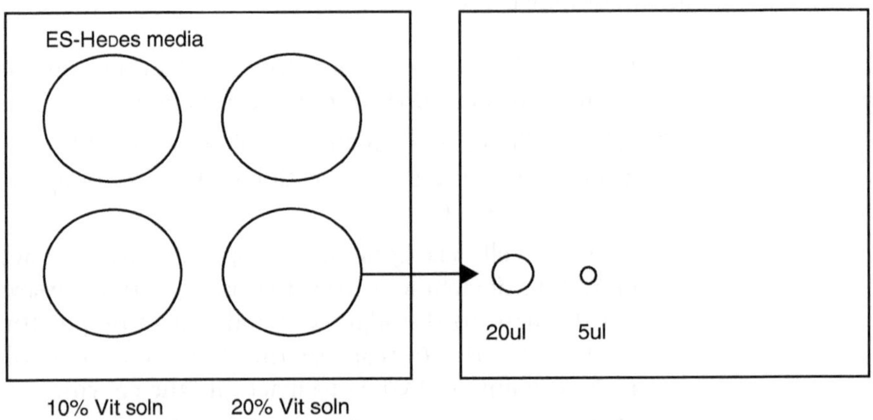

Fig. 3 Diagrammatic representation of workflow for vitrification

5. Move 8–10 pieces into well C of the vitrification plate for not more than 1 min. Use a new pipette tip to transfer 20 μL of 20% vitrification solution to the lid of the well-plate creating a high drop.
6. Take pieces from well C in minimal volume and transfer gently to well D.
7. Allow no more than 30 s before transferring pieces to the 20 μL drop.
8. Generate a second small drop containing as many pieces as possible in 5 μL on the well-plate lid. Capillary action will draw these pieces into a cryostraw if the straw is placed at a 30–45° angle touching the droplet/plate, before immediate transfer to the cryovial/cryostraw. A diagram below illustrates aspects of the procedure (Fig. 3).

3.3.3 Cryopreservation/Passaging of Bulk Culture iPSCs

Once colonies have been screened (detailed below) and expanded (usually on Matrigel or equivalent to remove the presence of contaminating MEFs) by mechanical or enzymatic passage (detailed elsewhere), bulk cultures can be established for large-scale experimentation. These should first be cryopreserved at early passage and periodically thereafter. A protocol for both passaging for expansion and for cryopreservation is detailed below (*see* **Notes 7** and **8** for additional tips on maintenance of iPSCs).

1. Grow bulk iPSC cultures for 5–7 days, approaching 80% confluence with as little differentiation as possible (Flow cytometric analysis for TRA-1-60/OCT4 can be used to establish this). Visually inspect the colonies. Pluripotent colonies will appear with uniform cellular composition (small oval-shaped cells with a prominent nuclear-to-cytoplasmic ratio) and discrete edges devoid of differentiating cell types.
2. Remove culture media and wash cells with PBS. Aspirate PBS before treatment with Collagenase IV for 5–10 min at 37 °C until edges became refractile. Using a cell-scraper, gently dislodge pieces from the monolayer and carefully transfer to a 15 mL centrifuge tube.
3. Spin tube at 300 × *g* for 1 min and aspirate supernatant.
4. Carefully, using a 5 mL serological pipette, resuspend pieces in the appropriate volume of resuspension media. If enzymatically passaging, transfer pieces to freshly prepared plates, if proceeding to cryopreservation proceed to **step 5**.
5. Transfer pieces to prelabeled cryovials.
6. Add (in a dropwise fashion) to tubes containing pieces/resuspension media an equivalent volume of freezing media.
7. Place cryovials containing cells in a "Mr Frosty" or equivalent vessel containing isopropanol for a controlled decrease in tem-

perature (−1 °C/min) (*see* **Note 9** for additional tips for recovery of cryopreserved and vitrified pieces).

3.4 Picking, Screening, and Characterization of Clones

iPSC colonies are picked at early (2 weeks post-reprogramming) and late time points (5 weeks post-reprogramming) by manual transfer using a 22-G needle and subcultured clonally on MEFs in organ culture dishes. Clones are screened for expression of TRA-1-60 and Hoechst-dye efflux, transgene persistence by RT-PCR, stem cell marker expression, and methylation status at *Oct4/Nanog* promoters. Karyotypic stability is assessed via G-band analysis (>15 metaphases per sample) by a commercial genotyping service (Sullivan Nicolaides Pathology, QLD, Australia).

3.4.1 TRA-1-60/Hoechst 33342 Dye Efflux Screening

Live immunostaining is performed to assay for expression of early pluripotency marker TRA-1-60, in combination with Hoechst 33342 efflux (i.e., active pumping out of Hoechst 33342 staining agent), as described in [32]. Select the clones that exhibit TRA-1-60 expression and exhibit a Hoechst-dim phenotype (Fig. 4a).

1. Incubate live colonies with TRA-1-60 antibody at a concentration of [1:200] in KSR-containing media for 1 h at 37 °C.
2. Wash twice live colonies with KSR-containing media and apply relevant secondary antibody for 1 h at 37 °C.
3. For the last 30 min of staining apply Hoechst 33342 at a final concentration of 0.5 μg/mL in KSR-containing media.
4. Wash cells and image in PBS before returning to KSR-containing media.

3.4.2 Immunostaining

Selected clones are expanded further and fixed to assay for immunoreactivity with a broader suite of markers including OCT4, NANOG, TRA-1-60, TRA-1-81, and SSEA-4 (Fig. 4b).

1. Wash cells in PBS and fix in 4% paraformaldehyde for 15 min at 4 °C.
2. For nuclear staining, incubate cells in permeabilization solution at room temperature for 10 min.
3. Block cells with 1% BSA/PBS or 10% goat serum.

Fig. 4 (continued) *Oct4* in samples C7 and C11 (AT34). Untransduced fibroblasts, a minus reverse transcriptase, and also no template reactions were run as controls. (**f**) CpG islands in the promoter regions of *Oct4* and *Nanog* undergo hypomethylation as part of their transition from fibroblasts to iPSCs. Fully methylated HeLa cells were used as a positive control and H9 human embryonic stem cells used as a negative control. *Open circles* depict unmethylated CG diresidues. *Full circles* represent methylated CG diresidues. (**g**) H&E stains of paraffin-embedded teratoma sections from control, ATh47, AT34, and AT30 iPSCs that show the presence of all three germ lineages, indicating pluripotency. Reproduced with permission from Nayler S et al. Induced pluripotent stem cells from ataxia-telangiectasia recapitulate the cellular phenotype. Stem Cells Transl Med. 2012, 1(7):523–35. doi: 10.5966/sctm.2012-0024

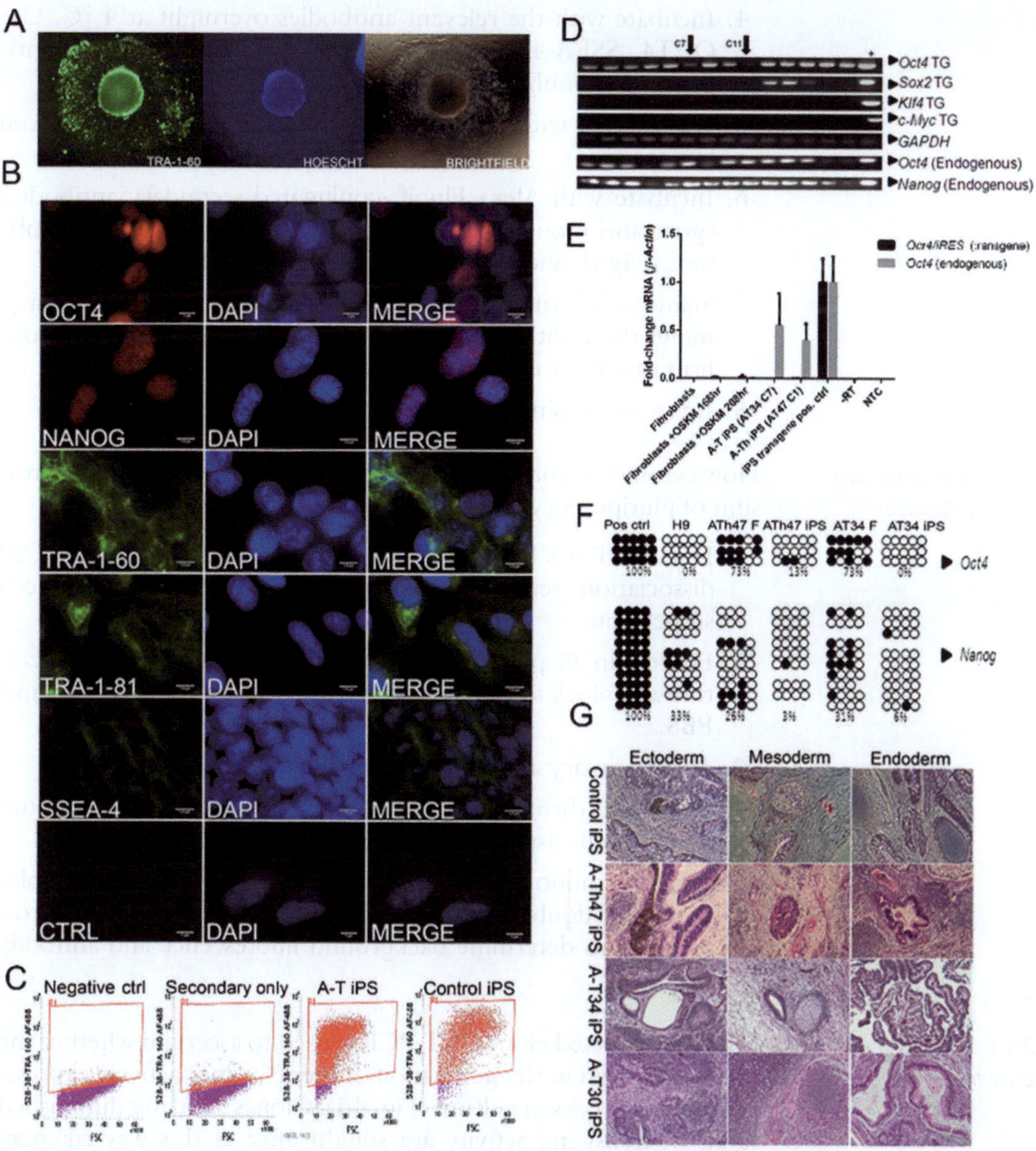

Fig. 4 (**a**) Live-staining with TRA-1-60 (*left panel*) and Hoechst dye efflux (*middle panel*). Brightfield photograph also shown (*right panel*). (**b**) iPSC colonies exhibited expression of nuclear stem cell markers OCT4 and NANOG in addition to surface markers TRA-1-60, TRA-1-81, and SSEA-4. A negative control featuring an identical preparation with omission of primary antibody was included (*bottom row*). (**c**) Flow cytometric analysis exemplifies robust and uniform expression of pluripotency marker TRA-1-60 in A-T (AT34) and control iPSCs. (**d**) RT-PCR shows by passage 10 most clones (AT34) robustly expressed Oct4/IRES transgene, with the exception of C7 (*lane 6*) and C11 (*lane 9*). Silencing of most other transgenes had occurred by this stage with the exception of two clones expressing *Sox2/IRES* (clones 12 and 13, *lanes 10* and *11*). Most clones expressed endogenous *Oct4* and *Nanog*. Amplification of *GAPDH* was used as a loading control. Positive controls (*lane 15*) were plasmid DNA and H9 human embryonic stem cell line in the case of *GAPDH*. (**e**) qPCR analysis of gene expression demonstrates transcript from the *Oct4*/IRES transgene was detected within 168 h of lentivirally transducing fibroblasts. We confirmed silencing of the *Oct4/IRES* transgene and also active expression of endogenous

4. Incubate with the relevant antibodies overnight at 4 °C. Use OCT4, SSEA-4, TRA-1-60, NANOG, and TRA-1-81 antibodies (*see* Subheading 2.4).
5. Wash cells with PBS (three times for 5 min at room temperature).
6. Incubate with Alexa Fluor®-conjugated secondary antibodies (goat anti-mouse IgG1, goat anti-mouse IgG2B, goat anti-mouse IgM, and donkey anti-rabbit IgG at 1:1000).
7. Stain nuclei with DAPI or Hoechst 33342. This preparation minus the addition of primary antibody should be used to confirm specificity of staining.
8. Image cells using fluorescent microscope.

3.4.3 Immunolabeling for Flow Cytometry

Flow cytometric analysis confirmed the robust and uniform expression of pluripotency marker TRA-1-60 (Fig. 4c).

1. Harvest iPSCs by washing once in PBS and using either Cell dissociation reagent or TrypLE to generate a single-cell suspension.
2. Fix cells in 4% paraformaldehyde for 15 min at room temperature and block for 10 min in 1% BSA/PBS or 10% goat-serum/PBS.
3. Apply primary antibodies overnight at 4 °C.
4. Strain cells through a cell strainer to remove clumps and ensure a single-cell suspension suitable for FACS.
5. Assess immunolabeling using a flow cytometer. Gate samples to exclude doublets and cellular debris. Ensure negative controls are used to determine background fluorescence and antibody specificity.

3.4.4 Transgene Extinction (RT-PCR)

From the selected clones, RT-PCR is used to ascertain whether the transgenes used in the generation of the iPSCs are actively expressing after 13 weeks in culture (Fig. 4d). Clones showing diminished levels of transgene activity are sought after as this was taken to indicate silencing of these genes and resultant acquisition of a pluripotent phenotype independent of the exogenous factors.

1. Isolate RNA using RNA isolation kit.
2. Quantify RNA concentration and 260/280 ratios by using a spectrophotometer before the synthesis of cDNA using a cDNA synthesis kit according to the manufacturer's instructions (use 500 ng of input RNA).
3. Perform RT-PCR for *Oct4* (657 bp) and *Sox2* (498 bp) transgenes according to the following conditions; 94 °C 5 min, 32 cycles (94 °C 1 min, 58 °C 30 s, 72 °C 1 min), 72 °C 7 min and *Klf4* (563 bp), *c-myc* (350 bp) transgenes as well as

GAPDH (152 bp) according to the following PCR conditions; 94 °C 5 min, 32 cycles (94 °C 1 min, 57 °C 30 s, 72 °C 1 min), 72 °C 7 min. Primer sequences are listed in Table 1.

3.4.5 Transgene Extinction (qRT-PCR)

Transgene extinction in samples C7 and C11 (AT34) was confirmed after 30 weeks in culture by qPCR amplification with primers specific for the 3′ end of *Oct4* and adjacent *IRES* sequence (Fig. 4e). Transgene transcript could be detected in virally transduced cells at 168 and 208 h after viral infection but not in untransduced fibroblasts. We failed to detect any residual *Oct4* transgene expression in AT34 (C7 and C11) after 30 cycles but could easily detect robust expression in a transgene-expressing iPSC line. No amplification of transcript was observed in "no template" or "minus reverse transcriptase" controls.

1. Isolate RNA using RNA isolation kit.
2. Quantify concentration and 260/280 ratios by using a spectrophotometer before synthesis of cDNA using a cDNA synthesis kit according to the manufacturer's instructions (use 500 ng input RNA).
3. Perform qPCR for *Oct4* transgene, endogenous *Oct4* and *β-Actin* using a thermal cycler and commercially available PCR mix, according to the following conditions: 95 °C 3 min, 30 cycles (95 °C 10 s, 60 °C 30 s). Calculate expression data using the ΔΔCt method. Primer sequences are listed in Table 1.

3.4.6 Bisulfite Sequencing

It is important to determine the methylation status of the *Oct4* and *Nanog* promoter loci, given that demethylation of CpG islands in this region is associated with a fully reprogrammed state. We use bisulfite modification of genomic DNA and PCR amplification at promoter loci to ascertain the methylation status of these promoters in our iPSCs and controls relative to their corresponding parental fibroblasts by sequencing of these loci (Fig. 4f).

1. Sort live iPSCs by flow cytometry for TRA-1-60 by first washing cells with PBS before harvesting with either Cell dissociation buffer or TrypLE for 2–5 min at 37 °C.
2. Incubate live dissociated cells with TRA-1-60 antibody (1:200, diluted in KSR media) for 45–60 min at room temperature before spinning once at 300 × *g* for 3 min and washing in KSR media.
3. Apply AlexaFluor secondary antibodies (diluted 1:500 in KSR media) for 30 min at room temperature.
4. Wash cells once in KSR media and resuspend in 1% BSA/PBS for live-sorting by Flow cytometry. Contact your cell sorter operator to discuss optimal conditions for sorting.

5. Collect positively labeled cell fraction and isolate Genomic DNA.
6. Bisulfite convert 1–2 μg of DNA using a Bisulfite conversion kit before PCR of *Oct4/Nanog* promoter regions and cloning into the PCR2.1 vector (primers listed in Table 1).
7. Screen and select clones for sequencing to confirm sites of methylation. Include synthetically hypermethylated (DNA methylated with CpG Methylase) HeLa cells and H9 hESCs as controls.

3.4.7 Teratoma Formation

A standard test for pluripotency is used to assess the potential of our A-T patients derived iPSC lines to contribute into all three germ lineages. iPSCs from both A-T homozygotes (AT34 and AT30) and A-T heterozygote parents (ATh47) formed teratomas when injected into SCID mice comprising tissue types from all three germ layers (endoderm, mesoderm, and ectoderm), indicating capacity for pluripotential tri-lineage differentiation (as shown in Fig. 4g).

1. Collect iPSCs grown on MEFs by collagenase IV treatment.
2. Resuspend approximately two million iPSCs in 50 μL of DMEM/F12 supplemented with 30% Matrigel and inject into hind-limb muscles of Methoxyflurane anesthetized 6-week-old-immune compromised SCID mice (CB17-SCID mice).
3. After 8–10 weeks, dissect and fix teratomas in 4% paraformaldehyde.
4. Embed samples in paraffin, cut sections, and stain with hematoxylin and eosin. Examine slides for the presence of representatives of the three germlayers. All mouse procedures should be conducted under local ethical guidelines and after gaining permission from the local animal ethics committee.

4 Notes

1. Care should be taken when changing media and/or observing cell morphology leading up to harvesting of lentivirus on HEK293T cells as viral infection can cause cell death and sloughing off of cells. Penicillin/Streptomycin should not be included in HEK293 culture media.
2. Appropriate care should be taken when working with lentivirus according to your institution's safety guidelines. This may include proper handling protocols for the user (Biohazard safety cabinet class 2, gloves/face mask, correct disposal of lentivirus-containing waste).

3. Including 4 μg/mL Polybrene in cell culture media of cells to be transduced with reprogramming virus during and 2 h prior to transduction can increase efficiency of this process.
4. Low-passage and actively dividing cells are important for success (both HEK293T and target cells for reprogramming). Care should be taken to not let either cell type become confluent.
5. A timer and strict adherence to the outlined time schedule is critical for success in thawing out vitrified iPSC clones.
6. The angle of orientation will affect the efficiency with which pieces are drawn up into the cryostraw.
7. If cultures exhibit higher than normal levels of differentiation, manual cutting can be used to isolate undifferentiated colonies. In addition to this microscopic examination of cultures for presence of differentiation and subsequent marking on the underside of the tissue-culture dish with a pen can allow removal using a glass pasteur pipette with aspiration.
8. An alternate cryomedia for cryopreservation of bulk cultures can also be made from 90% FCS and 10%DMSO.
9. ROCK Inhibitor Y-27632 may be used at 10 μM in cell culture media to assist recovery of cryopreserved/vitrified stem cell colonies. Make 10 mM stock of ROCK inhibitor in DMSO and freeze in aliquots. Avoid repeated freeze-thaw cycle and protect from exposure to light.

References

1. Eggan K (2013) Picking the lock on pluripotency. N Engl J Med 369(22):2150–2151
2. Sommer CA et al (2009) Induced pluripotent stem cell generation using a single lentiviral stem cell cassette. Stem Cells 27(3):543–549
3. Plath K, Lowry WE (2011) Progress in understanding reprogramming to the induced pluripotent state. Nat Rev Genet 12(4):253–265
4. Lavin MF (2013) The appropriateness of the mouse model for ataxia-telangiectasia: neurological defects but no neurodegeneration. DNA Repair (Amst) 12(8):612–619
5. Hawley RS, Friend SH (1996) Strange bedfellows in even stranger places: the role of ATM in meiotic cells, lymphocytes, tumors, and its functional links to p53. Genes Dev 10(19):2383–2388
6. Wood LM et al (2011) A novel role for ATM in regulating proteasome-mediated protein degradation through suppression of the ISG15 conjugation pathway. PLoS One 6(1):e16422
7. Ambrose M, Goldstine JV, Gatti RA (2007) Intrinsic mitochondrial dysfunction in ATM-deficient lymphoblastoid cells. Hum Mol Genet 16(18):2154–2164
8. Bar RS et al (1978) Extreme insulin resistance in ataxia telangiectasia: defect in affinity of insulin receptors. N Engl J Med 298(21):1164–1171
9. Li J et al (2009) Cytoplasmic ATM in neurons modulates synaptic function. Curr Biol 19(24):2091–2096
10. Lim DS et al (1998) ATM binds to beta-adaptin in cytoplasmic vesicles. Proc Natl Acad Sci U S A 95(17):10146–10151
11. Cosentino C, Grieco D, Costanzo V (2011) ATM activates the pentose phosphate pathway promoting anti-oxidant defence and DNA repair. EMBO J 30(3):546–555
12. Li J et al (2012) Nuclear accumulation of HDAC4 in ATM deficiency promotes neurodegeneration in ataxia telangiectasia. Nat Med 18(5):783–790
13. Guo Z, Deshpande R, Paull TT (2010) ATM activation in the presence of oxidative stress. Cell Cycle 9(24):4805–4811

14. Guo Z et al (2011) ATM activation by oxidative stress. Science 330(6003):517–521
15. Oka A, Takashima S (1998) Expression of the ataxia-telangiectasia gene (ATM) product in human cerebellar neurons during development. Neurosci Lett 252(3):195–198
16. Barlow C et al (2000) ATM is a cytoplasmic protein in mouse brain required to prevent lysosomal accumulation. Proc Natl Acad Sci U S A 97(2):871–876
17. Song H, Chung SK, Xu Y (2010) Modeling disease in human ESCs using an efficient BAC-based homologous recombination system. Cell Stem Cell 6(1):80–89
18. Takahashi K, Yamanaka S (2006) Induction of pluripotent stem cells from mouse embryonic and adult fibroblast cultures by defined factors. Cell 126(4):663–676
19. Marion RM et al (2009) A p53-mediated DNA damage response limits reprogramming to ensure iPS cell genomic integrity. Nature 460(7259):1149–1153
20. Kawamura T et al (2009) Linking the p53 tumour suppressor pathway to somatic cell reprogramming. Nature 460(7259):1140–1144
21. Polo JM et al (2012) A molecular roadmap of reprogramming somatic cells into iPS cells. Cell 151(7):1617–1632
22. Schubert R, Reichenbach J, Zielen S (2005) Growth factor deficiency in patients with ataxia telangiectasia. Clin Exp Immunol 140(3): 517–519
23. Nayler S et al (2012) Induced pluripotent stem cells from ataxia-telangiectasia recapitulate the cellular phenotype. Stem Cells Transl Med 1(7):523–535
24. Lin L et al (2015) Spontaneous ATM gene reversion in A-T iPSC to produce an isogenic cell line. Stem Cell Rep 5(6):1097–1108
25. Fukawatase Y et al (2014) Ataxia telangiectasia derived iPS cells show preserved x-ray sensitivity and decreased chromosomal instability. Sci Rep 4:5421
26. Lee P et al (2013) SMRT compounds abrogate cellular phenotypes of ataxia telangiectasia in neural derivatives of patient-specific hiPSCs. Nat Commun 4:1824
27. Maherali N et al (2007) Directly reprogrammed fibroblasts show global epigenetic remodeling and widespread tissue contribution. Cell Stem Cell 1(1):55–70
28. Nishino K et al (2010) DNA methylation dynamics in human induced pluripotent stem cells over time. PLoS Genet 7(5):e1002085
29. Zhou W et al (2013) Higher methylation in genomic DNA indicates incomplete reprogramming in induced pluripotent stem cells. Cell Reprogram 15(1):92–99
30. Kim K et al (2010) Epigenetic memory in induced pluripotent stem cells. Nature 467(7313):285–290
31. Reich M et al (2006) GenePattern 2.0. Nat Genet 38(5):500–501
32. Chan EM et al (2009) Live cell imaging distinguishes bona fide human iPS cells from partially reprogrammed cells. Nat Biotechnol 27(11):1033–1037

Chapter 30

Monitoring the ATM-Mediated DNA Damage Response in the Cerebellum Using Organotypic Cultures

Efrat Tal and Yosef Shiloh

Abstract

The *ATM* gene and its protein product, the ATM protein kinase, were identified as a result of attempts to understand the molecular basis of the genetic disorder, ataxia-telangiectasia (A-T). The cardinal symptom of A-T is neurodegeneration expressed primarily as progressive cerebellar atrophy. A major tool in the investigation of ATM functions in the cerebellum is cerebellar organotypic cultures, which allow cerebellar slices to live in culture for several weeks without losing their viability and organization. These cultures are amenable to various treatments and manipulations and provide a close look at Purkinje cells in their almost natural environment. We optimized the protocol for establishing and maintaining these cultures and provide here examples of readouts of the DNA damage response in cerebellar organotypic cultures treated with a DNA-damaging agent.

Key words Ataxia-telangiectasia (A-T), ATM kinase, DNA damage response (DDR), Cerebellum, Organotypic cultures

1 Introduction: Ataxia-Telangiectasia and ATM: Focus on the Cerebellum

Ataxia-telangiectasia (A-T; OMIN#208900) is a highly pleiotropic, autosomal-recessive human disorder [1–8] caused by loss of the ATM protein function. The responsible mutations create null alleles at the *ATM* (*A-T*, mutated) gene, which encodes the ATM protein kinase [9–11]. The prominent symptom in the typical form of A-T is progressive cerebellar ataxia that develops into general motor dysfunction, eventually confining most patients to a wheelchair toward the end of their first decade. The main underlying pathology is progressive cerebellar cortical degeneration that involves primarily Purkinje and granule neurons, but also affects basket cells. Peripheral neuropathy develops in many A-T patients during the second decade of life. Additional symptoms are oculocutaneous telangiectasias (dilated blood vessels), which usually appear in the eyes, marked immunodeficiency, and cancer predisposition, with most

Sergei V. Kozlov (ed.), *ATM Kinase: Methods and Protocols*, Methods in Molecular Biology, vol. 1599,
DOI 10.1007/978-1-4939-6955-5_30, © Springer Science+Business Media LLC 2017

malignancies being lymphoreticular. A-T patients show a striking sensitivity to the cytotoxic effect of ionizing radiation (IR), and cells derived from these patients exhibit marked chromosomal instability and sensitivity to IR and radiomimetic chemicals. This sensitivity results from a profound defect in the cellular response to DNA double strand breaks (DSBs), whose chief mobilizer is the ATM protein kinase [12–14]. Many A-T symptoms can be attributed to the abrogation of the cellular response to physiological DSBs and to DSBs induced by endogenous reactive oxygen species [13]. However, the cause of the most devastating manifestation of this disease—progressive neurodegeneration, particularly cerebellar atrophy—is still being debated [5, 15–17]. In our attempts to understand the critical role of ATM in the cerebellum, a major experimental tool is cerebellar organotypic cultures established from mice with different *Atm* genotypes [18].

Organotypic slice cultures provide a culture-based system that is much closer to a tissue compared to dissociated cultures or neuron-like cells obtained by induced differentiation, and yet are amenable to most of the manipulations that can be applied to cultured cells [19–30]. We consider cerebellar organotypic cultures a valuable tool for studying ATM's functions in the cerebellum, and therefore recently optimized the protocol for their establishment [18]. When generated from P12 mice, our cultures maintain the typical organization of the tissue for several weeks (Fig. 1), and also allow a certain degree of ex vivo development of the tissue [18]. This system increases the resolution at which we can analyze the DNA damage response (DDR) in cerebellar cells, with emphasis

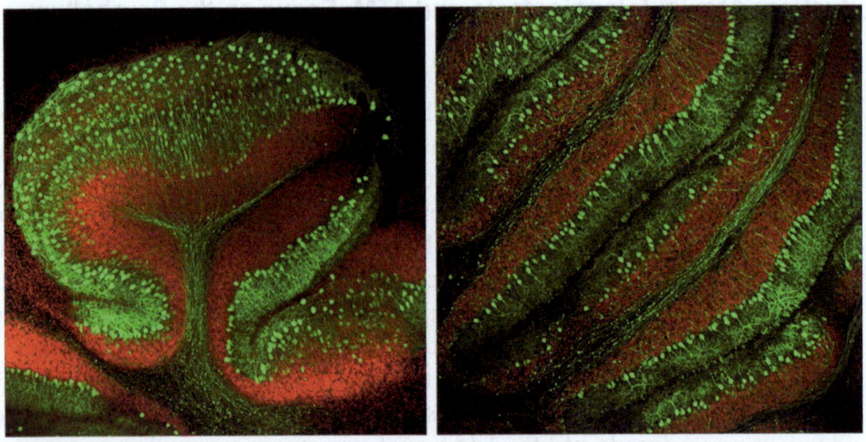

Fig. 1 Organotypic cultures established from P12 wild-type mice, 12 days after their establishment, demonstrate normal tissue organization. *Green*: Calbindin-DK28, a Purkinje cell marker; *red*: NeuN, a neuronal nucleus marker, which does not stain Purkinje nuclei; *blue*: DAPI, a fluorescent stain of DNA (from ref. [18]). Reprinted with permission from Tzur-Gilat et al. Studying the cerebellar DNA damage response in the tissue culture dish, Mechanisms of Ageing and Development, 2013, v134 (10):496–505.

on Purkinje cells, which exhibit a unique chromatin organization that is mainly euchromatic [18]. Importantly, the ATM-dependent DDR appears to be robust in Purkinje cells [18, 31, 32]. We should point out, however, that while these cultures provide Purkinje cells close to their natural environment, they carry an inherent experimental limitation: specific cell types in these cultures must be analyzed by microscopic visualization and not bulk assays. Currently, most such analyses are based on immunostaining and antibodies that specifically detect the corresponding mouse proteins.

2 Materials

All solutions must be stored at 4 °C, unless indicated otherwise.

2.1 Organotypic Culture Reagents and Equipment

All solutions must be prepared and stored under sterile conditions.

1. 50% D-Glucose solution: In a glass beaker dissolve 50 g of D-glucose in 100 ml double distilled water (*see* **Note 1**). In a sterile environment, filter the solution through a 0.22 μm filter. Divide the solution into 2 ml aliquots and store at −20 °C.
2. Dissection medium: HBSS containing 0.5% D-glucose. Add 0.5 ml of 50% D-Glucose solution to 50 ml Hank's balanced salt solution (HBSS) buffer without Mg^{2+} and Ca^{2+}.
3. BME medium: Mix 100 ml of Basal Medium Eagle (BME) with phenol red and without L-glutamine and HEPES, with 50 ml of HBSS with phenol red, 50 ml heat-inactivated horse serum (*see* **Note 2**), 2 ml of 50% D-glucose and 1 ml of 200 mM L-glutamine. Mix and filter through a 0.22 μm filter (*see* **Note 3**). Store at 4 °C.
4. Sterile cell culture inserts of 0.4 μm, 30 mm diameter (Millicell, Merck Millipore, Germany).
5. Six-well cell culture plates, flat-bottom with lid.
6. Three-cm tissue culture plates.
7. McIlwain tissue chopper (Cavey Laboratory Engineering Co. Ltd., UK).
8. Binocular microscope.
9. Light source for binocular microscope.
10. X-ray machine (suitable for cell cultures plates).
11. Etoposide. 50 mM in DMSO.
12. 70% ethanol.
13. Whatman paper.
14. Tweezers.

15. Surgical tools: mouse decapitation device, standard forceps slightly curved, surgical scissors, smooth-tip forceps, jeweler forceps, and a blunt-end spatula.

2.2 Immunostaining Reagents

All solutions must be stored at 4 °C.

1. 0.1 M Phosphate buffer: Prepare solution A by dissolving 12 g of NaH_2PO_4 monobasic anhydrous in 500 ml of double distilled water (ddw). Prepare solution B by dissolving 14.2 g of Na_2HPO_4 dibasic anhydrous in 500 ml of ddw. Prepare 0.2 M Phosphate buffer by mixing 115 ml of buffer A and 385 ml of buffer B. Before use, dilute the 0.2 M phosphate buffer 1:1 with ddw to a final concentration of 0.1 M.
2. 4% paraformaldehyde (PFA) fixative: Use 16% PFA stock solution (*see* **Note 4**). Add 10 ml of 16% PFA and 10 ml of ddw to 20 ml of 0.2 M Phosphate buffer (*see* **Note 5**).
3. PBS: Dulbecco's phosphate buffer without Mg^{2+} and Ca^{2+}.
4. Wash solution: Dissolve 1.25 ml of Triton X-100 in 500 ml of PBS (*see* **Note 6**).
5. 1 M L-lysine: Dissolve 9.13 g of L-lysine in 50 ml of PBS.
6. Blocking solution: Dissolve in a glass bottle 1 g of gelatin in 100 ml of PBS (*see* **Note** 7). Add 0.5 g of sodium azide and 1.25 ml of Titon X-100 (*see* **Note 6**). Complete volume to 500 ml with PBS. Keep 140 ml in a separate bottle for later use. Add 40 ml of 1 M L-lysine to the remaining 360 ml of solution.
7. Primary antibody dilution solution: Use the blocking solution (without the L-lysine) (*see* Subheading 2.2, **item 6**).
8. Secondary antibody dilution solution: Dissolve 0.1 g of gelatin in 50 ml of PBS (*see* **Note** 7). Add 25 μl of Triton X-100 (*see* **Note 6**).
9. Polyclonal IgG goat Anti-calbindin D28K antibody (Santa Cruz Biotechnology Inc., Santa Cruz, CA, USA). Working dilution: 1:250.
10. Monoclonal mouse anti-Neuronal Nuclei (NeuN) antibody (Merck Millipore, MA, USA). Working dilution 1:4000.
11. Monoclonal mouse IgG antibody against phosphorylated serine 139 of histone H2AX (γH2AX) (Merck Millipore, MA, USA). Working dilution 1:1000.
12. Polyclonal rabbit anti-53BP1 antibody (Novus, CO, USA). Working dilution 1:500.
13. Polyclonal rabbit anti-pS824 KAP-1 (Bethyl Laboratories, TX, USA). Working dilution 1:1000.
14. Alexa Fluor 488 donkey anti-goat IgG antibody (Molecular Probes, Netherlands).

15. Alexa Fluor 568 donkey anti-mouse IgG antibody (Molecular Probes, Netherlands).
16. Alexa Fluor 633 donkey anti-mouse IgG antibody (Molecular Probes, Netherlands).
17. Alexa Fluor 568 donkey anti-rabbit IgG antibody (Molecular Probes, Netherlands).
18. 4′,6-Diamidino-2-phenylindole (DAPI) 20 mg/ml.
19. Fluorescent mounting medium aqueous (ready to use).
20. Scalpel.
21. 48-well cell culture plate.
22. Microscope cover glasses, thickness #1.0, 0.13–0.17 mm.
23. SuperFrost Plus Slides (Fisher Scientific, PA, USA).
24. Confocal microscope.

3 Methods

3.1 Preparation of Organotypic Culture

Carry out all procedures in a sterile environment.

1. Use 6-well plates to prepare cultures from mice with different genotypes (e.g., wild-type Atm$^{+/+}$ and *Atm*$^{-/-}$). Add 1 ml of BME medium to each of six wells (*see* **Note 8**). Using sterile forceps, place one cell culture insert in each well (*see* **Note 9**). Put the plate in an incubator at 37 °C, 5% CO_2, for at least 2 h prior to dissection.
2. Prepare a surgery area: In a biological hood place the Mcllwain tissue chopper, mouse decapitation device, binocular microscope, and light source (*see* **Note 10**). Place all surgical tools in a cup filled with 70% ethanol (*see* **Note 11**). Fill 3-cm culture plates with cold dissection medium and place on ice (one plate per cerebellum).
3. Using the mouse decapitation device and the surgical tools, separate the cerebellum from a P10–P12 mouse and place it in the cold dissection medium in the 3-cm plate prepared in advance (*see* **Note 12**). Repeat with additional animals according to the experimental plan.
4. Adjust the Mcllwain tissue chopper to cut slices 400 μm thick. Place the cerebellum on the cutting platform vertical to the cutting angle and serially cut sagittal sections (*see* **Note 13**).
5. Transfer the tissue slices back to the cold dissection buffer (*see* **Note 14**). Under the binocular microscope, use two spatulas to gently separate the slices.
6. Transfer the separated tissue slices from the dissection buffer onto the cell culture inserts prepared in advance in the 6-well

plates. Up to six tissue slices can be placed on one insert (*see* **Note 15**). Repeat **steps 3–6** for the next cerebellum. Incubate the tissue cultures at 37 °C, 5% CO_2.

7. Grow the cultures for 12 days with medium change every 2–3 days. Change the medium by adding 1 ml of BME medium to each of the wells in a new 6-well plate. Using a sterile forceps, transfer the culture insert to the new plate.

3.2 Monitoring Tissue Organization Using Immunostaining

1. Wash the cultures by aspirating the medium and replacing it with 1 ml of PBS.
2. Immediately fix the tissue by replacing the PBS with 4% PFA. Apply 1 ml of PFA underneath the insert and 1 ml above the insert and leave at room temperature for 1 h (*see* **Note 16**).
3. Remove the PFA and wash the inserts by adding 1 ml of wash solution underneath the insert and 1 ml above it. Place the plate on a shaker at room temperature for 5 min. Repeat washes four times.
4. Replace the wash solution with blocking solution, 1 ml underneath the culture insert and 1 ml above it. Place the plate on a shaker and shake gently at room temperature for 1 h.
5. Separate the tissue slices from the cell culture insert: using a scalpel, cut the membrane of the cell culture insert in small circles around each individual cerebellar slice (*see* **Note 17**). Place each portion in a separate well of a 48-well plate. Add 50 μl of 0.1 M phosphate buffer to prevent drying.
6. For monitoring tissue organization, use antibodies against the Purkinje cell marker, calbindin D28k, and the neuronal marker, NeuN. Dilute the anti-calbindin antibody 1:250 and the NeuN antibody 1:4000 in the primary antibody dilution buffer. Final primary antibody volume = (100 μl) × (the number of tissue slices to be stained).
7. Remove the phosphate buffer and apply 100 μl of primary antibody mix to each well (*see* **Note 18**). Incubate with gentle shaking overnight at 4 °C.
8. From this step on, minimize the light exposure.
9. Prepare the secondary antibody solution. Dilute the donkey anti-mouse 568 antibody 1:500, and the donkey anti-goat 488 antibody in the secondary antibody dilution solution. Final secondary antibody volume = (100 μl) × (the number of tissue slices to be stained).
10. Wash the primary antibody from the samples with the wash solution: four times for 5 min with gentle shaking (*see* **Note 18**).
11. Replace the wash solution with 100 μl of secondary antibody solution. Incubate with gentle shaking for 2 h at room temperature in the dark.

12. Prepare DAPI staining: Dilute DAPI 1:20,000 in PBS (*see* **Note 19**). The volume of the DAPI solution = (100 μl) × (the number of tissue slices to be stained).
13. Wash the secondary antibody from the samples with the wash solution: four times for 5 min with gentle shaking (*see* **Note 18**).
14. Replace the wash solution with DAPI staining solution (*see* **step 12** of Subheading 3.2) by applying 100 μl of DAPI to each well. Incubate for 20 min at room temperature in the dark.
15. Place the tissues on microscope glass slides, membrane facing down and tissue facing the cover glass. Mount and cover with a cover glass. Place the slides on a flat surface and let them dry overnight in the dark.
16. Store the slides at 4 °C in a light-proof box.
17. Capture images using a confocal microscope. You should be able to identify the typical cerebellar folia and clearly identify the Purkinje cells (Fig. 1).

3.3 Observation of the DNA Damage Response in Cerebellar Organotypic Cultures

DNA damage can be induced by irradiation or chemical agents: X-irradiation and treatment with the topoisomerase II inhibitor, etoposide. DDR readouts are then monitored using immunostaining (Fig. 2).

1. Prepare fresh organotypic cultures by repeating all steps in Subheading 3.1.
2. Divide the cell culture inserts from different genotypes into separate 6-well plates according to the experimental plan. When applying X-ray irradiation prepare separate 6-well plates for the treated and untreated inserts.
3. Etoposide treatment: Prepare etoposide solution in BME medium at the desired concentration. We describe here treatment with 10 μM etoposide. Prepare 1:50 intermediate dilution in BME medium, dilute it 1:100 in BME medium to obtain a final concentration of 10 μM. Remove culture medium and replace with etoposide solution. Replace the medium in control samples with fresh BME medium. Place organotypic cultures in an incubator for 1 h. Aspirate the medium and wash with 1 ml of PBS. Immediately fix the tissue as described in **step 2** of Subheading 3.2.
4. Ionizing radiation treatment: Irradiate one plate of each genotype at the desired dose using an X-irradiator and place in the incubator for various periods of time. We refer here to the response to 5 Gy of X-rays. Place cultures in an incubator for 30 min.

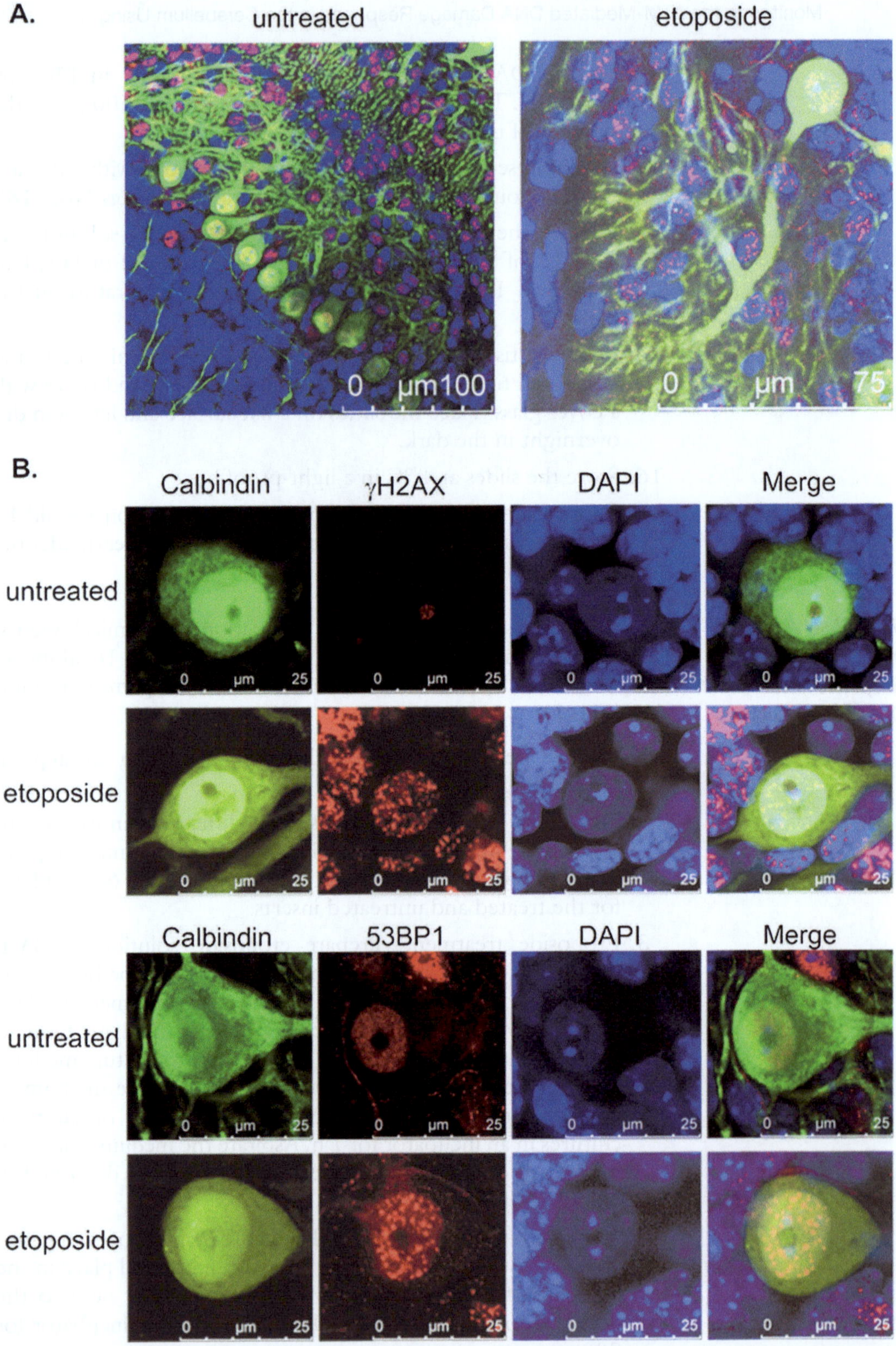

Fig. 2 Demonstration of nuclear foci formed at DSB sites 1 h after treatment of cerebellar organotypic cultures with 10 μM etoposide. (**a**) *Red*: 53BP1; *Green*: calbindin-DK28; *blue*: DAPI. (**b**) A close look at a Purkinje cell in etoposide-treated culture. The nuclear region that is strongly stained by the anti-γH2AX and anti-53BP1 antibody overlaps the nucleolus; the significance of this staining is unclear

5. Wash the cultures by aspirating the medium and replacing it with 1 ml of PBS and immediately fix the tissue, as described in **step 2** of Subheading 3.2.
6. Repeat **steps 3–5** of the protocol in Subheading 3.2.
7. For monitoring the DSB response in Purkinje cells, use primary antibodies that reveal the DSB markers, γH2AX and 53BP1 nuclear foci, and Kap-1 phosphorylation. Stain the cultures with mixtures that each contains primary antibodies diluted in primary antibody dilution buffer: (a) Anti-calbindin antibody (1:250) and anti-γH2AX antibody (1:1000). (b) Anti-calbindin antibody (1:250) and anti-53BP1 antibody (1:500) (*see* **Note 20**). (c) Anti-calbindin antibody (1:250), anti-pKAP-1 antibody (1:1000), and anti-γH2AX antibody (1:1000). The volume of each solution = (100 μl) × (the number of tissue slices to be stained).
8. Remove the buffer and add 100 μl of primary antibody mix per well—add only one primary antibody mix in to each well. Stain at least one cerebellum tissue from each genotype and treatment with each antibody mix. Incubate overnight at 4 °C with gentle shaking.
9. Prepare secondary antibody solution: (a) Donkey anti-mouse 568 antibody (1:500) and donkey anti-goat 488 antibody (1:500). (b) Donkey anti-rabbit 568 (1:500) and donkey anti-goat 488 antibody (1:500). (c) Donkey anti-rabbit 568 antibody (1:500), donkey anti-goat 488 antibody (1:500), and donkey anti-mouse 633 antibody (1:500) (*see* **Note 21**). The volume of each solution = (100 μl) × (the number of tissue slices to be stained).
10. After incubation with primary antibodies, wash the cultures with the wash solution four times for 5 min with gentle shaking.
11. Replace the wash solution with 100 μl of secondary antibody solution: cultures incubated with primary mix a, b, or c should be incubated with the matching secondary antibody mix a, b, or c. Incubate with gentle shaking for 2 h at room temperature in the dark.
12. Repeat **steps 12–17** of the protocol in Subheading 3.2.

4 Notes

1. The glucose can be easily dissolved by heating the solution to 70 °C. Let it cool down before continuing.
2. Horse serum should be stored at −20 °C. Freeze in 50 ml aliquots to minimize freezing-thawing cycles.
3. The medium is good for up to 1 month when stored at 4 °C.

4. 16% PFA can be purchased commercially. Dilute in a chemical hood.
5. Prepare fresh 4% PFA at room temperature. Although it can be stored up to 1 month in the dark at 4°C, freshly prepared solution is recommended.
6. Use a 1000 μl plastic tip with blunt end to add the Triton detergent. When ejecting the Triton into the PBS, leave the tip in the bottle to allow complete dissolution of the entire amount of the detergent. Shake gently until Triton dissolves completely.
7. Heat the PBS and gelatin solution for 1 min in a microwave oven and stir. Make sure the gelatin is completely dissolved; if not, heat it again but do not allow it to boil. Let the solution cool to room temperature.
8. Plan your experiment such that you have separate inserts for each treatment, time point, and genotype.
9. Float the membrane in the medium, with the bottom side submerged and the top side exposed to the air. Make sure no bubbles are left between the membrane and the medium.
10. Keep a sterile environment and wipe all instruments in 70% ethanol. The surgery can be carried out comfortably in a laminar flow hood.
11. Make sure all tools are accessible to avoid unnecessary movements and contamination.
12. Sacrificing the animals and separating the cerebellum should be performed by a professional investigator, in accordance with local ethics regulations.
13. To prevent the tissues from sticking to the cutting knife, soak up extra buffer around the cerebellum with a piece of Whatman paper. Be careful not to touch the tissue with the paper.
14. Using a spatula, transfer the sliced cerebellum from the cutting platform to the dissection buffer as a whole; slide the spatula underneath the cerebellum and lift the entire tissue together. To avoid damaging the tissues, separate the slices only when in the dissection buffer.
15. It is possible to produce up to 12, 400 μM slices from one cerebellum. Divide the cerebellar slices between the inserts in one 6-well plate. Place slices from one cerebellum on each insert. When placing two or more slices on an insert, make sure some space is left between them.
16. It is possible to stop at this step and proceed with the immunostaining at a later time. To do so, wash the tissues with 0.1 M phosphate buffer four times for 5 min with gentle shaking. Tissues can be stored submerged in the buffer for up to 2 weeks at 4 °C.

17. Do not separate the tissue from the membrane; cut only the membrane around the tissue. The membrane protects the tissue from being damaged and will not interfere with the staining procedure. Make sure to keep the slices wet at all times.
18. When aspirating the buffer be careful not to damage the tissue. If vacuum suction is used, use a fine tip.
19. Prepare an intermediate dilution of 1:2000 in PBS and dilute it further to 1:10 in PBS.
20. It is possible to costain with anti γH2AX and anti 53BP1 antibodies, but this may result in heavy background staining.
21. Make sure your antibodies are tagged with fluorophores with different excitation and emission wavelengths.

Acknowledgments

Work in our lab is funded by the A-T Children's Project, the Dr. Miriam and Sheldon G. Adelson Medical Research Foundation, The A-T Ease Foundation, The Israel Cancer Research Fund, The Israel Science Foundation, and the I-CORE Program of the Planning and Budgeting Committee of the Israel Ministry of Education. Y.S. is a Research Professor of the Israel Cancer Research Fund.

References

1. Perlman SL, Boder Deceased E, Sedgewick RP, Gatti RA (2012) Ataxia-telangiectasia. Handb Clin Neurol 103:307–332. doi:10.1016/B978-0-444-51892-7.00019-X
2. Lavin MF (2008) Ataxia-telangiectasia: from a rare disorder to a paradigm for cell signalling and cancer. Nat Rev Mol Cell Biol 9(10):759–769
3. Crawford TO (1998) Ataxia telangiectasia. Semin Pediatr Neurol 5(4):287–294
4. Crawford TO, Mandir AS, Lefton-Greif MA, Goodman SN, Goodman BK, Sengul H, Lederman HM (2000) Quantitative neurologic assessment of ataxia-telangiectasia. Neurology 54(7):1505–1509
5. Hoche F, Seidel K, Theis M, Vlaho S, Schubert R, Zielen S, Kieslich M (2012) Neurodegeneration in ataxia telangiectasia: what is new? What is evident? Neuropediatrics 43(3):119–129.doi:10.1055/s-0032-1313915
6. Verhagen MM, Martin JJ, van Deuren M, Ceuterick-de Groote C, Weemaes CM, Kremer BH, Taylor MA, Willemsen MA, Lammens M (2012) Neuropathology in classical and variant ataxia-telangiectasia. Neuropathology 32(3):234–244. doi:10.1111/j.1440-1789.2011.01263.x
7. Sahama I, Sinclair K, Pannek K, Lavin M, Rose S (2014) Radiological imaging in ataxia telangiectasia: a review. Cerebellum 13(4):521–530. doi:10.1007/s12311-014-0557-4
8. Taylor AM, Lam Z, Last JI, Byrd PJ (2014) Ataxia telangiectasia: more variation at clinical and cellular levels. Clin Genet. doi:10.1111/cge.12453
9. Savitsky K, Bar-Shira A, Gilad S, Rotman G, Ziv Y, Vanagaite L, Tagle DA, Smith S, Uziel T, Sfez S, Ashkenazi M, Pecker I, Frydman M, Harnik R, Patanjali SR, Simmons A, Clines GA, Sartiel A, Gatti RA, Chessa L, Sanal O, Lavin MF, Jaspers NG, Taylor AM, Arlett CF, Miki T, Weissman SM, Lovett M, Collins FS, Shiloh Y (1995) A single ataxia telangiectasia gene with a product similar to PI-3 kinase. Science 268(5218):1749–1753
10. Savitsky K, Sfez S, Tagle DA, Ziv Y, Sartiel A, Collins FS, Shiloh Y, Rotman G (1995) The complete sequence of the coding region of the ATM gene reveals similarity to cell cycle

regulators in different species. Hum Mol Genet 4(11):2025–2032
11. Ziv Y, Bar-Shira A, Pecker I, Russell P, Jorgensen TJ, Tsarfati I, Shiloh Y (1997) Recombinant ATM protein complements the cellular A-T phenotype. Oncogene 15(2):159–167. doi:10.1038/sj.onc.1201319
12. Shiloh Y, Ziv Y (2013) The ATM protein kinase: regulating the cellular response to genotoxic stress, and more. Nat Rev Mol Cell Biol 14(4):197–210
13. McKinnon PJ (2012) ATM and the molecular pathogenesis of ataxia telangiectasia. Annu Rev Pathol 7:303–321. doi:10.1146/annurev-pathol-011811-132509
14. Bhatti S, Kozlov S, Farooqi AA, Naqi A, Lavin M, Khanna KK (2011) ATM protein kinase: the linchpin of cellular defenses to stress. Cell Mol Life Sci 68(18):2977–3006. doi:10.1007/s00018-011-0683-9
15. Biton S, Barzilai A, Shiloh Y (2008) The neurological phenotype of ataxia-telangiectasia: solving a persistent puzzle. DNA Repair 7(7):1028–1038. doi:10.1016/j.dnarep.2008.03.006
16. Yang Y, Hui CW, Li J, Herrup K (2014) The interaction of the atm genotype with inflammation and oxidative stress. PLoS One 9(1):e85863. doi:10.1371/journal.pone.0085863
17. Ditch S, Paull TT (2012) The ATM protein kinase and cellular redox signaling: beyond the DNA damage response. Trends Biochem Sci 37(1):15–22. doi:10.1016/j.tibs.2011.10.002
18. Tzur-Gilat A, Ziv Y, Mittelman L, Barzilai A, Shiloh Y (2013) Studying the cerebellar DNA damage response in the tissue culture dish. Mech Ageing Dev 134(10):496–505. doi:10.1016/j.mad.2013.04.001
19. Dupont JL, Fourcaudot E, Beekenkamp H, Poulain B, Bossu JL (2006) Synaptic organization of the mouse cerebellar cortex in organotypic slice cultures. Cerebellum 5(4):243–256. doi:10.1080/14734220600905317
20. Padmanabhan J, Brown K, Shelanski ML (2007) Cell cycle inhibition and retinoblastoma protein overexpression prevent Purkinje cell death in organotypic slice cultures. Dev Neurobiol 67(6):818–826. doi:10.1002/dneu.20394
21. Julien S, Schnichels S, Teng H, Tassew N, Henke-Fahle S, Mueller BK, Monnier PP (2008) Purkinje cell survival in organotypic cultures: implication of Rho and its downstream effector ROCK. J Neurosci Res 86(3):531–536. doi:10.1002/jnr.21511
22. Kessler M, Kiliman B, Humes C, Arai AC (2008) Spontaneous activity in Purkinje cells: multi-electrode recording from organotypic cerebellar slice cultures. Brain Res 1218:54–69. doi:10.1016/j.brainres.2008.04.063
23. Lu HX, Levis H, Liu Y, Parker T (2011) Organotypic slices culture model for cerebellar ataxia: potential use to study Purkinje cell induction from neural stem cells. Brain Res Bull 84(2):169–173. doi:10.1016/j.brainresbull.2010.12.001
24. Hurtado de Mendoza T, Balana B, Slesinger PA, Verma IM (2011) Organotypic cerebellar cultures: apoptotic challenges and detection. J Vis Exp 51. doi:10.3791/2564
25. Kapfhammer JP, Gugger OS (2012) The analysis of purkinje cell dendritic morphology in organotypic slice cultures. J Vis Exp 61. doi:10.3791/3637
26. Zanjani HS, Lohof AM, McFarland R, Vogel MW, Mariani J (2012) Enhanced survival of wild-type and lurcher purkinje cells in vitro following inhibition of conventional PKCs or stress-activated MAP kinase pathways. Cerebellum. doi:10.1007/s12311-012-0427-x
27. Lebrun C, Avci HX, Wehrle R, Doulazmi M, Jaudon F, Morel MP, Rivals I, Ema M, Schmidt S, Sotelo C, Vodjdani G, Dusart I (2012) Klf9 is necessary and sufficient for Purkinje cell survival in organotypic culture. Mol Cell Neurosci. doi:10.1016/j.mcn.2012.11.010
28. Lonchamp E, Dupont JL, Beekenkamp H, Poulain B, Bossu JL (2006) The mouse cerebellar cortex in organotypic slice cultures: an in vitro model to analyze the consequences of mutations and pathologies on neuronal survival, development, and function. Crit Rev Neurobiol 18(1–2):179–186
29. Hill RA, Medved J, Patel KD, Nishiyama A (2014) Organotypic slice cultures to study oligodendrocyte dynamics and myelination. J Vis Exp 90:e51835. doi:10.3791/51835
30. Campeau JL, Wu G, Bell JR, Rasmussen J, Sim VL (2013) Early increase and late decrease of purkinje cell dendritic spine density in prion-infected organotypic mouse cerebellar cultures. PLoS One 8(12):e81776. doi:10.1371/journal.pone.0081776
31. Dar I, Biton S, Shiloh Y, Barzilai A (2006) Analysis of the ataxia telangiectasia mutated-mediated DNA damage response in murine cerebellar neurons. J Neurosci 26(29):7767–7774. doi:10.1523/JNEUROSCI.2055-06.2006
32. Dar I, Yosha G, Elfassy R, Galron R, Wang ZQ, Shiloh Y, Barzilai A (2011) Investigation of the functional link between ATM and NBS1 in the DNA damage response in the mouse cerebellum. J Biol Chem 286(17):15361–15376. doi:10.1074/jbc.M110.204172

Index

A

Accelerator facility ... 292
Acetonitrile (ACN) ... 130, 134, 168, 176, 200, 211, 217
Acetylated lysine antibody (AcLys) ... 158, 160
Acetylation ... 157–161, 320, 349
Acetyllysine-specific antibody ... 157–161
Acrylamide ... 115, 117, 119, 122, 124, 141, 149, 176
Acrylamidependant Phos-tag ... 113, 122
β-Actin antibody ... 378
Actinomycin D ... 116, 118, 120
Adenosine 5'-triphosphate (ATP) ... 45, 47, 49, 54, 145–155, 370, 373
Active caspase-3 ... 193
Adult stem cells ... 391
Affinity electrophoresis ... 113
Agarose ... 32, 37, 115–117, 119, 124, 125, 148, 152, 164, 169, 253, 284, 325–326, 330–331, 364
Alpha irradiation ... 298
Ammonium bicarbonate (AmBic) ... 130, 168, 218
Anti-53BP1 antibody ... 89, 91, 92, 307, 308, 422, 427
Anti-ATM antibody ... 129, 132, 140, 150, 158, 171, 248, 365
Antibody Cross-linking ... 200, 202–203
Anti-bromodeoxyuridine antibody (Anti-BrdU) ... 17, 20
Anti-calbindin D28K antibody ... 422
Anti-GST antibody ... 168
Anti-Neuronal Nuclei (NeuN) antibody ... 420, 422, 424
Anti-p53 antibody ... 150
Anti-phospho-Ser15-of-p53 antibody ... 150
Anti-phospho-Ser46-of-p53 antibody ... 150
Anti-phospho-Ser957 of SMC1 (NB100-205) ... 151
Anti-phospho-Thr68 of Chk2 ... 151
Anti-PML antibody ... 151
Anti-pS824 KAP-1 ... 422
Anti-γH2AX antibody ... 248, 308, 365, 426, 427
Aphidicolin ... 307, 308, 320, 322–324
Apoptosis ... 44, 61, 99, 146, 184, 187, 245, 264, 277, 278, 336, 349, 404
Apoptosis-associated (AA) DNA fragmentation ... 193
Apoptosis-associated (AA) DSBs ... 187
AraC (Cytosine-1-β-D-arabinofuranoside) ... 364
Astrocytes ... 381–384
A-T heterozygote ... 6, 8, 9, 11, 408, 416
Ataxia telangiectasia (A-T) ... 127
Ataxia telangiectasia like disorder (ATLD) ... 16, 319
Ataxia-telangiectasia and rad3-related (ATR) ... 43, 58, 127, 264, 318, 319, 322, 323, 359, 376
Ataxia telangiectasia mutated (ATM) ... 127
Ataxia Telangiectasia Mutated (*ATM*) gene ... 25
ATM acetylation ... 157, 158
ATM activation ... 26, 59, 89–90, 99, 158, 183–195, 318, 378, 387
ATM consensus motif ... 231
ATM Immunoprecipitation ... 132–134
ATM kinase ... 2, 6, 26, 45, 73, 107, 116, 118, 127–141, 153, 157–161, 197–212, 285, 287, 319, 357, 365, 366, 370
ATM kinase activation ... 128, 139
ATM kinase activity ... 97–110, 128, 288, 318, 363, 371
ATM kinase inhibitors ... 57–69, 377
ATM kinase reporter (ATMR) ... 101, 104, 109
ATM Peptides ... 130, 135–136, 206
ATM Phosphopeptides ... 130, 136–139
ATM phosphorylation ... 61, 93, 94, 127–141, 247, 252, 255, 284, 320, 357, 363–373
ATM pS1981 antibody ... 74
ATM signaling ... 319
ATM Tyr2755>A ... 147
ATM-AS (Tyr2755Ala) ... 145, 146, 148
ATM-KD ... 148
ATP ... 184
ATP analogues ... 146
ATP analogue-accepting ATM mutant (ATM-AS) ... 146
ATP-binding pocket ... 146, 147
ATP-dependent firefly (*Photinus pyralis*) luciferase ... 98
ATR kinase ... 184, 215, 322
Autophosphorylation ... 127, 128, 131, 282, 287, 349, 357, 358, 363
Auto-phosphorylation site ... 387
Autosomal recessive genetic disease ... 1
AxioVision software ... 279–281

B

Basic fibroblast growth factor (bFGF) ... 404, 405
Benzonase nuclease ... 125, 166
Bicinchoninic acid protein assay (BCA assay) ... 199, 202

Sergei V. Kozlov (ed.), *ATM Kinase: Methods and Protocols*, Methods in Molecular Biology, vol. 1599,
DOI 10.1007/978-1-4939-6955-5, © Springer Science+Business Media LLC 2017

BigDye Terminator v1.1 Cycle Sequencing Kit 32, 37
Bioinformatics 231, 242
Bioluminescence 97–110
Biphasic SDS-PAGE 132
Bisulfite sequencing 406, 415–416
Bleomycin (BLEO) 9, 10, 75, 158, 159, 165, 311, 371
Blue fluorescent protein (BFP) 336, 345
53BP1 ....264, 268, 270
BRCA1 ....2, 72, 319, 345
Breast cancer risk 319
Bromodeoxyuridine (BrdU) 282, 283

C

C-18 Zip Tips 168
Caffeine 58, 104, 322–324, 328, 359
Calbindin-DK28 ....420, 426
Camptothecin (CPT) 16, 32, 187, 351, 352, 371
Cancer 2, 10, 142, 230, 237, 336, 338
Cancer predisposition 145, 419
Cancer stem cells 74
Cdk5 kinase 365, 366, 370, 371
Cdk5 shRNA Lentivirus 369
Cdk5/p25 kinase 370
Cdk5-ATM pathway 364
Cell cycle 14, 57, 85, 93, 99, 184, 188, 245, 264, 277, 284, 304, 308–309, 335, 336, 359, 363, 387, 402
Cell cycle checkpoint 2, 26, 44, 277, 318, 320, 348, 349, 363, 376
Cell penetrating peptide (CPP) 249, 251, 254–255, 260
Cellomics™ ArrayScan VTI HCS Reader 63
Cellular assays 59
Central nervous system (CNS) 122, 375, 376
Cerebella 366, 367, 372
Cerebellar degeneration 43
Cerebellar granule neurons (CGNs) 364, 366, 368, 369, 371–373
Cerebellar organotypic cultures 420, 425–427
Cerebellum 366, 391, 429
Charged particle irradiation 291
ChIP assay 246, 247, 249, 251–260
ChIP sonication buffer 249, 253, 256, 257, 260
CHK1 kinase 58, 72, 99, 101, 264
CHK2 kinase 26, 58, 59, 72, 98, 99, 101–103, 146, 151, 264, 336
5-Chloro-2′-deoxyuridine (CldU) 15, 16
Chromatin 93, 157, 216, 249, 253, 264, 266–270, 288, 295, 304, 305, 319, 320, 349, 350, 360, 387, 403, 421
Chromatin decondensation 320
Chromatin immunoprecipitation (ChIP) 261, 307
Chromocenters 305
^{13}C isotope 210
Claspin 72
Co-immunoprecipitation 163–181
Collagenase H 392, 394
Collagenase IV 405, 411, 416
Colloidal Coomassie blue solution 133, 168
Colocalization 267–269, 272, 273, 310, 311
Colony formation efficiency (CFE) 6
Colony survival 2, 4, 7, 10
Colony survival assay (CSA) 2–6, 10, 11
Complementation 98, 99, 101–103, 105, 107, 109
Complete™ EDTA-free Protease inhibitor cocktail 166
CP-466722 58
Cross-linked antibodies 199, 201, 212
Crosslinking 140, 198–199, 203
C-terminus of ATM 157
Cyclic pifithrin-α (p53 inhibitor) 404, 409

D

de novo discovery of phosphorylation sites 216
4,6-diamidino-2-phenylindole (DAPI) 186, 188
Dimethyl labelling 230
Dimethyl pimelimidate (DMP) 198
Dispase™ 377, 380, 381
Dissection medium 421, 423
Dithiothreitol (DTT) 33, 39, 45, 47, 129, 130, 148, 152, 158, 167, 172, 217, 366
D-luciferin 101, 104, 105
DNA content 188
DNA damage 2, 6, 14, 26, 44, 59, 71–83, 85, 86, 94, 110, 129, 131, 139, 146, 165, 171, 183, 184, 187, 198, 215–225, 230, 231, 245–247, 249, 263–273, 277, 278, 285, 288, 291–295, 297–300, 311, 318, 320, 321, 336, 337, 345, 360, 363, 371, 375–389, 402, 403, 419–429
DNA damage foci 73, 80–82, 93
DNA damage repair 85, 95, 215
DNA damage response (DDR) 2, 43, 44, 59, 72, 89–90, 94, 163, 230, 231, 245–247, 253–255, 263–270, 273, 278, 288, 319, 321, 335, 348, 349, 376, 387, 389, 402, 420, 425, 429
DNA damage signaling 85, 120
DNA damage-related checkpoints 307
DNA double strand breaks (DSBs) 2, 72–73, 85, 94, 99, 127, 157, 164, 245–261, 285, 288, 290, 291, 331, 382, 420
DNA fiber assay 14
DNA fibers 14, 16–18, 20–23
DNA repair 2, 26, 44, 85, 86, 92–93, 95, 99, 184, 231, 245, 247, 264, 265, 277–278, 321, 335–346, 348, 363
DNA repair reporter assays 376
DNA Sanger sequencing 26

DNA:RNA hybrid 349
DNA-damage induced senescence 349
DNA-damage response (DDR) 58
DNA-dependent protein kinase catalytic subunit (DNA-PKc) 43, 165, 215, 304, 318, 376, 377, 382, 384
DNase-free RNase A 187
Double strand break (DSB) 13, 16, 58, 59
Doublecortin antibody 393
Doxorubicin 75
DSB repair assay 321
DsRed 379, 382, 385

E

EBV immortalization 326
Electroblotting 117, 119–120, 123, 124
Electroporation 85–95, 278, 293, 297, 323–324, 326
Electrotransfer 125
Endonuclease G 320
Endonucleases 85–95, 264, 265, 284
Enhanced chemiluminescence (ECL) 153, 159, 160, 166, 248
Etoposide (ETOP) 75, 371, 421, 425, 426
Euchromatin (EC) 304, 305
Exploratory analysis 231, 232

F

FACS analysis 321, 324, 327, 344
False discovery rate (FDR) 242
FATC domain 157
FHA2 phospho-peptide binding domain 101
FLAG-tagged wild-type ATM 148
FLAG-YFP-tagged ATM (YFP-ATM) 278–285
FLAG-YFP-tagged ATM S1981A mutant (YFP-ATM S1981A) 278, 282
Flow cytometry (FC) 2, 6, 8–9, 184, 186, 189–190, 322, 324, 326, 327, 339, 379, 385, 403, 411–415
Fluorescence recovery after photobleaching (FRAP) 264, 279, 282–284, 318, 350–357, 359, 360
Fluorescence resonance energy transfer (FRET) 44, 99, 100
Fluorescently-Tagged ATM 285
FokI endonuclease 266, 269, 270

G

G418 (Geneticin) 101, 103, 104, 107, 285, 379, 384, 385
Gene ontology enrichment 0, 231
Genomic PCR 322, 323, 325, 329–331
GFAP antibody 378
Glioma 58, 389
Glo-sensor c-AMP reagent 101
Glutathione- Sepharose 152, 167, 173
Glutathione-S-Transferase (GST) fusion protein 148, 152–153, 163–165, 173
Glutathione-S-Transferase (GST) fusion protein pull-down assay 163–181
β-glycerophosphate 129, 151, 166, 218
Green fluorescent protein (GFP) 279, 344, 345, 354, 355, 359, 360, 371, 372
GST-ATM Fusion Protein Pull-Down Assay 166–168, 171–175
GST Pull-Down 163–181
GST-p53 fusion proteins 148

H

γ-H2AX antibody 248, 365
γH2AX-foci 59
HEK293FT cells 340, 341, 345
HeLa cell line 198
Heterochromatic DNA Double Strand Break (DSB) Repair 303–305, 307–310, 312
Heterochromatin (HC) 304, 306, 308–310, 312, 313
Heterozygote 9, 407–409
High content screening (HCS) 57
High energetic particles (HZE) 292, 295
Higher-energy C-trap dissociation (HCD) cell 222
High-throughput phosphoproteomics 230
High-throughput screening 59, 99
Histone H2AX 88, 92, 95, 184, 336, 422
Histone H2AX (γH2AX) 58, 59, 72, 99, 184, 306, 312, 319, 365
Hoechst 33258 .... 282, 371
Hoechst 33342 dye efflux 412
Hoechst33342 .... 63, 405
Homing endonuclease I-SceI 246, 336
Homologous recombination (HR) 85, 303, 304, 319–321, 331, 335–337, 344
Homozygote 6, 9, 408, 409
HTRF® Assay 43–45, 47–49, 51, 53, 55
Human embryonic stem cells (hESCs) 376, 377, 380–385, 387, 388, 402, 405, 409, 412–413, 416
Human Stem Cells 389
Hydrogen peroxide (H_2O_2) 187, 230
Hydrophilic interaction chromatography (HILIC) 216

I

IFP/BFP double-positive 336
Image J software 308
Image-based high content screening (HSC) 71–83
Imaging 20, 33, 40, 57–63, 65–68, 97–110, 159, 165, 167, 270–272, 277–285, 287–300, 326, 350, 354, 355, 360, 380, 382, 387, 388
Imaging cytometry 184
Imaging-assisted cytometry 184
Immobilized metal ion affinity chromatography (IMAC) 216

Immunoaffinity enrichment 198, 200
Immunodeficiency 1, 2, 43, 58, 145, 318, 419
Immunofluorescence, ... 14, 72, 185, 264, 266–272, 278, 350, 352, 353, 357–360, 371, 377–379, 381–383
Immunofluorescence microscopy 85–95, 351–352
Immuno-MRM assay 198
Immunoprecipitation 129, 131–133, 140, 158–160, 164, 165, 171, 176, 249, 257, 350, 370, 373
Indirect immunofluorescence 307–308
Induced pluripotent stem cells (iPSC) 380, 402, 412–413, 417
Induction of DNA damage 75
Infrared fluorescent protein (IFP) 336, 337, 341, 344, 345
In-gel digestion of proteins 134
Intra-S-Phase Checkpoint 13, 14, 16–18, 20–22
Iodoacetamide (IAA) 140, 168, 199
5-Iodo-2′-deoxyuridine (IDU) 14–16
Ion microbeam facility(ies) 295, 298
Ionizing radiation 165, 169, 174, 202, 230, 245, 246, 287–289, 291–293, 295, 297–300, 308–309, 318, 320, 349, 350, 358, 373, 402, 420, 425
Ionizing radiation-induced foci (IRIF) 73, 287, 295, 376, 379, 382
I-PpoI 45, 48
I-PpoI homing endonuclease 246
I-PpoI protein purification 246, 249–251
I-*Ppo*I transduction system 247
I-*Ppo*I-mediated DSBs 249
iPSC colony 407–413
Irradiation 4, 16, 62, 73, 104, 106, 131, 139, 174, 264, 278, 280–282, 285, 288, 289, 291–295, 298, 308, 311, 353, 358, 382, 384, 387, 425
I-SceI endonuclease 344, 379, 385
I-*Sce*I homing endonuclease 246
I-SceI Lentivirus 341
Isopropyl β-D-1-thiogalactopyranoside (IPTG) 149, 167, 247, 250

K

Karyotype 403
KAT5 (Tip60) acetyltransferase 320
Kinase 2, 14, 26, 72, 97–99, 101–108, 115
Kinase inhibitors 49, 288
Kinase reporter 99
Krüppel-associated box (KRAB)-associated protein 1 (KAP1) 26, 33, 39, 58, 99, 305
KSR medium 378, 381
KU-55933 ATM kinase inhibitor 49, 55, 58, 103–105, 109, 292, 346, 351, 353, 357, 377, 382, 384
KU-57788 DNA-PKcs inhibitor 377
KU59403 58
KU-60019 ATM kinase inhibitor 61, 62, 104, 109, 377, 382, 387, 388

L

Lac operator 263–273
Lac Operator/Repressor (LacO/LacR) Tethering System 273
Lac repressor 266
LacO array sequences 265
LacO arrays 265, 266, 268
LacR repressor 267–269
Laminin 378, 381
Large genomic rearrangements 26
Large phosphoprotein 113–125
Laser micro-irradiation 279–283
Laser Scanning Cytometer (LSC) 183–195
Laser-generated DSBs 278, 280, 282
Leibovitz's L15 medium 297
Lentivirus titer 340
Liquid chromatography-tandem mass spectrometry (LC-MS/MS) 129, 130, 135–139, 216
Live cell bioluminescence imaging system 101
Live cell imaging 264, 291, 387–388
Live cell microscopy 289
Lymphoblastoid cells (LCLs) 2, 4, 14–16, 391
Lys-C endoproteinase 130
Lysine 3016 of ATM 157
Lysyl endopeptidase (LysC) 218

M

Mammalian target of rapamycin (mTOR) 43, 127, 146, 164
MAP2 antibody 393
Mass spectrometry 46, 127–141, 168–169, 176–178, 197–212
Mass spectrometry-based proteomics 215–225, 229–242
Matrigel™ 380
MaxQuant software 223
mCherryLacRATM1300-3060aa plasmid 266
mCherryLacR-FokI plasmid 266
Mediator of DNA damage checkpoint protein 1 (MDC1) 319
Methylation 403, 416
Methylation status 412, 415
Micro tip-based strong cation exchange chromatography (Micro-SCX) 221
Microcystin-LR 129
Microhomology-mediated end joining (MMEJ) 319
Missense mutation 26, 252
Mixed Lineage Leukemia (*MLL*) breakpoint cluster region (*MLL*bcr) 320
*MLL*bcr-specific PCR 322, 323
Mn^{2+}–Phos-tag SDS-PAGE 113–125
molecular imaging 98
Mouse embryonic fibroblasts (MEFs) 305, 404, 411, 412, 416

Mouse xenograft model 109
Mre11 2, 7, 10, 14, 16, 72, 184, 247, 264, 265, 270, 287, 319, 320, 336, 349
MRE11/RAD50/NBN (MRN) c 247
MS acquisition 222
MS^2 acquisition 222
Mucosal biopsy 393
Multiple ligation of probe amplification (MLPA) 26, 32, 37–38
Multiple reaction monitoring (MRM) 197
Multiplexed immuno-MRM assay 198
Musashi1 antibody 378

N

N6-1-methylbutyl ATP [N6-1-MeBu-ATP] 150
N6-2-methylbutyl ATP [N6-2-MeBu-ATP] 151
N6-benzyl ATP [N6-Bn-ATP] 150
N6-phenylethyl ATP [N6-PhEt-ATP] 150, 154
N6-substituted ATP analogues 147
NanoDrop spectrophotometer 221
Na_3VO_4 151, 166
NaF 158, 166
NANOG antibody 405
Nasal septum 392, 393
Neocarzinostatin 131
Nestin antibody 378
NETN wash buffer 167
Neural descendants 385, 389
Neural differentiation 393
Neural progenitors (NPs) 381, 383
Neurobasal medium 378, 393
Neurodegeneration 1, 2, 58, 145, 318, 363, 376, 391, 402, 420
Neuronal cells 373
Neuronal death 364
Neuronal differentiation 396–397
Neuron-like cells 395–396, 420
Neurosphere 392, 393, 395, 396, 398
Neutral-pH gel system 114, 115
NHEJ Repair Assay 379, 382–386
NIH3T3duo cells 266, 267, 269, 271
Nijmegen breakage syndrome (NBS) 7, 10, 319
Nijmegen breakage syndrome (NBS)-like disorder 319
^{15}N isotope 210
Non-Canonical ATM Activation 347–360
Non-homologous end joining 304
Non-homologous end joining (NHEJ) 2, 85, 303, 304, 320–322, 335–337, 344, 345, 379, 382, 384, 385, 389
Nonparametric rank product analysis 232, 242
Nonsense mutations 26
Normal human diploid fibroblast (NHDF) 86, 92, 95, 307
Normalization 136, 139, 232, 237, 242
NP-40 lysis buffer 365, 370
Nuclear factor-κB 320
Nuclear foci 73, 77–79, 278, 426, 427
Nucleases 320
Nucleofection 272

O

O1 antibody 378
OCT4 antibody 405
Okadaic acid 129, 166
Olfactory mucosa 392, 393, 395, 396
Olfactory neurosphere-derived (ONS) cells 392
OPTI-MEM medium 337, 341, 404, 407
Orbitrap mass analyzer 222
Organotypic Culture 420–425, 429

P

p53 26, 33, 39, 44, 45, 48, 52, 88, 146, 150, 152, 320, 336, 364, 403, 404, 409
Paraformaldehyde (PFA) 74, 151, 270, 307, 351, 377, 405, 412, 414, 416, 422
α-particle source 292, 294
Patient-specific neural stem cells 391–399
pEF vector 100, 103
Peptide cleanup 177–178
Peptide Immunoaffinity Enrichment 197–212
Peripheral blood lymphocytes (PBL) 2–3, 6, 41
PHAS-I substrate 365
Phosphatase inhibitor 40, 129, 131, 132, 166, 169, 248, 253, 257, 364, 366, 370, 379, 386
Phosphate-affinity sodium dodecyl sulfate–polyacrylamide gel electrophoresis 113
Phosphatidylinositol (PI) 127
Phosphatidylinositol (PI) 3-kinase-related kinases (PIKKs) 318
Phosphatidylinositol 3-kinase-related kinase (PIKK) 215
Phosphatidylinositol-3 kinases (PI3K) 25, 43, 58, 147, 264, 318
Phosphatidylinositol-3-kinase-related protein kinases (PIKKs) 229
Phosphoinositide 3-kinase (PI3K)-related protein kinase (PIKK) family 43, 146
Phosphoinositol-3 kinase (PI3K) signature 318
Phosphopeptide 367
Phosphopeptide analysis 128
Phosphopeptide enrichment 129, 130, 139
Phosphopeptides 135–137, 198, 216, 230, 231, 237, 242
Phosphopeptides elution 140
Phosphoprotein 113–115
Phosphoproteome 113, 141
Phosphoproteomic 113, 129
Phosphorylation 6, 9, 44, 51, 56, 58, 114, 115, 121, 146, 183–195, 215–225, 231, 232, 247, 264, 266, 272, 284, 288, 291, 305, 318, 320, 321, 336

Phosphorylation sites 14, 133
Phospho-Ser794 ATM antibody 367
Phospho-specific antibodies 153, 188
PhosSTOP™ Phosphatase inhibitor cocktail (Roche) 166
Phos-tag 114
Photo bleaching 77, 278, 281–284, 291, 294, 299, 351, 355–356, 360, 389
Phototoxicity 299, 300
PIKK regulatory domain (PRD) of ATM 157
Pluripotency 378, 412–414, 416
Polybrene 339, 342, 345, 405, 407, 417
Polyethylenimine (PEI) 337, 341
Poly-L-lysine 364, 366, 369, 371, 393, 395, 398
Polyvinylidene difluoride (PVDF) membranes 117
Post-irradiation 357
Postmitotic neurons 363
Posttranslational modifications (PTMs) 57, 72, 145, 163, 164, 278, 287, 320, 335, 348
Principle Component Analysis (PCA) 238
Progressive neurodegeneration 318, 420
Propidium iodide (PI) 186, 187
Protease inhibitor 32, 33, 74, 129, 148, 152, 167, 172–174, 199, 218, 219, 248, 253, 260
Protein assay kit 33, 39, 129, 149
Protein extraction 32–33, 39–41
Protein interactions 266, 348
Protein kinase 2, 25, 43, 44, 46–49, 52, 54, 55, 72, 97, 145, 146, 229, 264, 277, 317–331, 419
Protein phosphorylation 57, 145, 215
Protein recruitment 246, 288, 291
Protein recruitment kinetics 281–282, 294–295
Protein transduction 246, 247, 251–260
Protein-protein interaction 99, 145, 163–165, 171, 231, 266, 348
Purkinje cell 420, 421, 427
Puromycin 37, 41, 337, 341–342, 345, 380
PVDF membrane 119, 125, 159, 160, 380, 386

Q

Quantitation 382
Quantitative LC-MS/MS 130, 137
Quantitative mass spectrometry 229–242
Quantitative PCR 247
Quantitative phosphoproteomics 141
Quantitative real-time PCR 246, 249
Quiescence 308, 359
Quiescent cells 308, 359

R

R/Bioconductor software 232
Radiation-induced foci (IRIF) 73, 376
Radiomimetic 75, 165, 166, 245, 264, 420
Radioresistant DNA synthesis (RDS) 13, 14, 16–22
Radiosensitivities 1–11, 58, 318
Rat primary CGNs 364
Reactive oxygen species (ROS) 72, 82, 420
Real-time dynamics of ATM 278
Real-time PCR 249, 254–255, 257–260, 344
Recombinant p53 .... 46, 148–149
Red (PI) fluorescence 191
Region of interest (ROI) 104, 106, 108, 164, 355
Relative fluorescence intensity (RFI) 281, 282
Repair foci 304, 306, 309–312, 382, 384
Replication 13–15, 19, 44, 72, 318, 319
Replication stress 164, 320, 322, 359
Replication stress-induced dsDNA breaks 347
Reporter 71, 98, 100–103, 384, 389
Reprogramming 401–417
Resection foci 306–307
Restriction endonuclease(s) 89, 90, 100, 246
Ribonucleoprotein particles (snRNPs) 86, 348
RIPA buffer 218, 219, 379, 386
R-loops 349, 350, 352, 359
RNA processing 349
ROCK Inhibitor Y-27632 417

S

S phase 359
S/T(Q) containing phosphopeptides
S/T-Q motif 146, 318
Sarcosyl 167, 172–173
SCID mice 109, 406, 416
SCX 216
Selected reaction monitoring, (SRM) 197
Serine 794 ATM 363, 367
SILAC 219
Single-strand annealing (SSA) 319, 321
Sister chromatid exchanges (SCE) repair 318
Skyline software 206
Slow component for DSB 304
Sodium cyanoborohydride (NaBH3CN) 130, 135, 136
Sodium fluoride (NaF) 33, 218, 365
Sodium orthovanadate (Na3VO4) 32, 33, 148, 365, 366
Sonication 125, 152, 173, 253, 406
Sox2 antibody 378
S-phase 93, 185, 320, 336
Splice sites 348
Spliceosome 348–350, 352, 357
Splicing alteration 37
Splicing speckles 360
Split firefly-based bioluminescence reporter 100, 108
Split-luciferase 99
SSEA-4 antibody 405
Stable isotope dimethylation labeling 141
Stable Isotope Labeling with Amino Acids in Cell Culture (SILAC) 217–218
Stable-isotope labelled formaldehyde 13CD2O 136

Stable-isotope labelled formaldehyde CD2O ... 130
StageTip chromatography ... 136, 221
Statistical ... 229–242, 327
Stem cell ... 375, 378–381, 389, 396–397, 399
strong anion exchange (SAX) chromatography ... 216
strong cation exchange (SCX) chromatography ... 216
Structural maintenance of chromosomes 1 ... 2, 3, 6, 8, 99, 151
2-Sulfanylethanol ... 116, 122
Survival fraction ... 4, 6, 10
SYBR Green ... 248, 257, 260, 379, 386

T

TaqMan® Copy Number Assays kit ... 339
Targeted proteomics ... 197
Temozolomide ... 75
Teratoma ... 412–413, 416
Teratoma formation ... 403, 416
TERT immortalized human fibroblasts ... 351–353
Tethering ... 265–267, 269, 270
TetO array sequences ... 265
Thymidine analogs ... 14, 15, 23
Time-lapse imaging ... 279–281, 289, 291, 292, 299
Time-Lapse Microscopy ... 293–294
TiO_2 ... 216, 218
Tip60 acetyltransferase ... 320
Tip60-dependent acetylation of ATM ... 157
Titanium dioxide (TiO2) chromatography ... 136
TopBP1 ... 72
Topotecan (Tpt) ... 185
TRA-1-60 antibody ... 405, 412, 415
TRA-1-81 antibody ... 405, 412–414
Traffic light reporter (TLR) ... 336–345
Traffic light reporter cell line ... 337, 342
Transcription-blocking DNA damage ... 350, 352, 360
Transformation/transcription domain-associated protein (TRRAP) ... 43, 146, 164, 318
Transgene ... 345, 402, 403, 412–415
Triethylammonium bicarbonate (TEAB) ... 130, 134
Trimethylated lysine 20 at histone 4 (H4K20me3) ... 306, 308, 310
Trimethylated lysine 9 at histone 3 (H3K9me3) ... 306, 308, 310, 312
Tris(2-carboxyethyl) phosphine ... 199
Tris(2-carboxyethyl) phosphine hydrochloride v(TCEP) ... 130
Tris–AcOH buffer system ... 116, 120
TrypLE ... 392, 393, 395, 396, 404, 407, 409, 414, 415
Trypsin ... 16, 18, 61, 86, 88, 89, 93, 107, 130, 134, 198, 199, 220, 248, 269, 270, 307, 324, 327, 328, 343, 351, 352, 364, 367
Trypsin (Sequencing Grade) ... 169
Tryptic peptide ... 134, 139, 177

U

U2 snRNPs ... 348, 350, 352, 357
U2OS human osteosarcoma cells ... 266
U5 snRNP ... 350, 352, 357
UHPLC system ... 222
Ultra Violet (UV) radiation ... 279, 281, 288, 350, 351, 353, 357
Universal immunoprecipitation buffer (UIP) ... 129, 131, 166, 169, 174
UVC irradiation ... 352

V

Vitrification solution ... 405, 410, 411
Volocity software ... 382

W

Western blotting ... 39–40, 102, 104, 109, 115, 132, 139, 163, 167, 168, 170–171, 248, 253, 386
Whole blood ... 4, 26, 34
Wild-type ATM ... 148, 278, 423
Wortmannin ... 58

X

Xenogen IVIS live cell imaging system ... 101, 104
X-ray tube ... 292

Z

Zinc(II) chloride ZnCl2 ... 116, 119
Zinc(II) nitrate solution Zn(NO3)2 ... 122
Zn(II) -Phos-tag SDS-PAGE ... 113, 118–119, 121
Zn2+–Phos-tag complex ... 114

FSC
www.fsc.org
MIX
Papier aus verantwortungsvollen Quellen
Paper from responsible sources
FSC® C105338

If you have any concerns about our products,
you can contact us on
ProductSafety@springernature.com

In case Publisher is established outside the EU,
the EU authorized representative is:
Springer Nature Customer Service Center GmbH
Europaplatz 3, 69115 Heidelberg, Germany

Printed by Libri Plureos GmbH
in Hamburg, Germany